Lecture Notes in Mathematics　　2119

Editors-in-Chief:
J.-M. Morel, Cachan
B. Teissier, Paris

Advisory Board:
Camillo De Lellis (Zürich)
Mario di Bernardo (Bristol)
Alessio Figalli (Austin)
Davar Khoshnevisan (Salt Lake City)
Ioannis Kontoyiannis (Athens)
Gabor Lugosi (Barcelona)
Mark Podolskij (Aarhus)
Sylvia Serfaty (Paris and NY)
Catharina Stroppel (Bonn)
Anna Wienhard (Heidelberg)

More information about this series at
http://www.springer.com/series/304

Antoine Ducros • Charles Favre • Johannes Nicaise
Editors

Berkovich Spaces
and Applications

 Springer

Editors

Antoine Ducros
Pierre and Marie Curie University
Paris
France

Charles Favre
Ecole Polytechnique
Palaiseau Cedex
France

Johannes Nicaise
University of Leuven
Heverlee
Belgium

ISBN 978-3-319-11028-8   ISBN 978-3-319-11029-5 (eBook)
DOI 10.1007/978-3-319-11029-5
Springer Cham Heidelberg New York Dordrecht London

Lecture Notes in Mathematics ISSN print edition: 0075-8434
                             ISSN electronic edition: 1617-9692

Library of Congress Control Number: 2014955370

Mathematics Subject Classification (2010): 14F20, 14G22, 37F10, 11S85, 37E25, 37P50, 11S82, 12J25,
                             31C45, 32H50, 37P40, 13A18, 32J05, 20E08, 20E42,
                             51E24, 14L15, 14G22

Contents

Introduction to the Volume: Berkovich Spaces and Applications

Antoine Ducros, Charles Favre, and Johannes Nicaise

1 Foreword

Most of the materials the reader will find in this book were presented during two scientific meetings which took place in January 2008 and June 2010 in Santiago de Chile and Paris, respectively. The Chilean workshop was organized by C. Favre, J. Kiwi and J. Rivera-Letelier, and focused on the interactions between dynamical systems and non-Archimedean geometry. The meeting in Paris was set up as a summer school to introduce young researchers from different backgrounds to the foundations of Berkovich's theory and its connections to number theory, model theory, Bruhat-Tits theory and dynamics. It was organized by A. Ducros, C. Favre, N. Fournaiseau, J. Nicaise and F. Paugam; the lectures were filmed, and the videos are available on the web site of the École normale supérieure, in the section *Savoirs en multimédia*.[1]

---

[1]URL : http://archives.diffusion.ens.fr/index.php?res=cycles&idcycle=490.

A. Ducros
Université Paris 6 Institut de Mathématiques, F-75251 Paris Cedex 05, France
e-mail: ducros@math.jussieu.fr

C. Favre
CNRS–CMLS École Polytechnique, F-91128 Palaiseau Cedex, France
e-mail: charles.favre@polytechnique.edu

J. Nicaise
KULeuven, Department of Mathematics, Celestijnenlaan 200B, 3001 Heverlee, Belgium
e-mail: johannes.nicaise@wis.kuleuven.be

Apart from some more specialized talks, the programs of these two meetings consisted of the following mini-courses:

– Santiago de Chile (January 2008):

- (∗) Jean-Pierre Otal: *Compactification of spaces of representations (after Culler, Morgan and Shalen)*;
- Nicolas Ressayre: *Geometric invariant theory: constructing the moduli space of rational maps (after Silverman)*;
- Antoine Ducros: *Introduction to Berkovich spaces*.

– Paris (June 2010):

- (∗) Michael Temkin: *Introduction to Berkovich analytic spaces*;
- (∗) Antoine Ducros: *Étale cohomology of schemes and analytic spaces*;
- Vladimir Berkovich: $\mathbf{F}_1$-*geometry*;
- François Loeser: *Model theory and analytic geometry*;
- (∗) Bertrand Rémy and Amaury Thuillier: *Bruhat-Tits buildings and analytic geometry*;
- (∗) Mattias Jonsson: *Dynamics on Berkovich spaces*;
- Jérôme Poineau: *Berkovich spaces over* $\mathbf{Z}$.

The present volume contains expanded notes of the courses marked with (∗), together with the contributions of A. Ducros on the cohomological finiteness of proper morphisms and C. Favre on sequential compactness of Berkovich spaces. It is meant as an introduction to Berkovich's theory and a selection of its applications to dynamics, geometry and number theory.

It is a great pleasure to thank all lecturers for their efforts to explain technically involved theories to a broad audience, and especially to those who made an extra effort to prepare their course notes for publication in these proceedings.

The two aforementioned conferences received financial support from many institutions including the Facultad de Matemáticas de la Pontificia Universidad Católica de Chile; the Institut de Mathématiques de Jussieu; the CNRS; the Universities Paris 6 Pierre et Marie Curie and Paris 7 Denis Diderot. They also benefited from the invaluable support of the projects ANR Berko in France, the Research Network on Low Dimensional Dynamics in Chile, and the joint French-Chilean ECOS/CONICYT project.

2   Introduction

2.1   *Analytic Spaces Over a Non-Archimedean Field*

Non-Archimedean geometry is the field of mathematics that aims to extend the methods and results from complex analytic geometry to non-Archimedean valued fields, with the field $\mathbf{Q}_p$ of p-adic numbers as a prime example, and to apply these techniques to the study of various problems in geometry and number theory. There are different approaches to this aim, giving rise to different theories of non-Archimedean geometry. In this volume, the primary focus will lie on the theory developed by Berkovich at the end of the 1980s, whose foundations were established in [Ber90].

The most obvious way to develop a theory of analytic geometry over a non-Archimedean field K is to define analytic functions on open subsets of K^n as functions with values in K that can locally be written as converging power series. Once this class of functions is defined, one can use charts and atlases (or the language of locally ringed spaces) to define K-analytic varieties; see for instance [Gro] or Chap. 2 of [Igu]. More generally, one can include in such spaces points with coordinates in finite extensions of K; this leads to the notion of a *wobbly analytic space* in [Tat, §9].

These naive objects already have interesting applications in number theory, particularly through the use of p-adic integrals (see for instance [Igu]). However, the metric topology on a non-Archimedean field is fundamentally different from the metric topology on $\mathbf{C}$: the ultrametric property of the absolute value implies that the topology is totally disconnected. As a consequence, a purely local definition of an analytic function (like the one above) can never lead to a satisfactory global theory. For instance, every locally constant function on K is analytic, and there are many such functions that are not globally constant on K. This violates the principle of analytic continuation, which is one of the cornerstones of the complex analytic theory. Another problematic consequence of our naïve definition is that there are only very few isomorphism classes of K-analytic manifolds: if K is a non-Archimedean local field (i.e. a p-adic field or a finite extension of the field of Laurent series $\mathbb{F}_p((t))$), then Serre showed in [Ser] that every compact K-analytic manifold is isomorphic to a disjoint union of r open balls, with r a unique element in $\{0, \ldots, p-1\}$. Thus we cannot hope to recover a proper algebraic K-variety from its analytification, so that the GAGA principle fails in this setup. Even if we include non-K-rational points and consider wobbly spaces, similar problems arise: see for instance Corollary 2 in [Tat, §9].

To overcome these problems, J. Tate developed in the 1960s his theory of *rigid analytic spaces*. His main motivation was to provide a rigorous framework for his observations on the uniformization of p-adic elliptic curves with multiplicative reduction. The foundations of this theory were published in [Tat]. The underlying idea is to endow the wobbly spaces with some additional structure in order to *rigidify* them and restore the connection between local and global aspects. This is achieved by defining a Grothendieck topology on the wobbly space that only allows a particular type of covering. The theory of rigid analytic geometry, together with the necessary background from valuation theory and non-Archimedean analysis, is described in detail in [BGR]. Another introduction, which also discusses some interesting applications, can be found in [FvdP].

In the early 1970s, Raynaud showed in [Ray] how a certain class of rigid analytic spaces over K can be realized as generic fibers of formal schemes over the valuation ring R of K, and how different formal models of a fixed rigid analytic space can be related via so-called admissible blow-ups (formal blow-ups whose centers are contained in the special fiber of the formal scheme). This result makes it possible to deduce many fundamental properties of rigid analytic spaces from the corresponding properties of formal schemes, for which an extensive algebro-geometric machinery is available. Raynaud's theory was systematically developed by Bosch, Lütkebohmert and Raynaud in the 1990s in the series of papers [BL93a, BL93b, BLR95a, BLR95b]. The foundations of the theory were recently laid out and expanded by Abbes in [Abb].

In 1990 the book [Ber90] of V. Berkovich appeared, where still another approach to non-Archimedean geometry was presented, based on spaces of absolute values with the topology of pointwise convergence. The distinguishing feature of these new analytic spaces is their particularly tame topology, which is both locally compact and locally arcwise connected. Moreover, the homotopy type of such an analytic space usually encodes very deep arithmetic-geometric information. With hindsight, one can say that it is precisely these properties that contributed to the success of this theory and its applicability in various (unexpected) fields including automorphic forms, compactification of buildings, singularity theory, complex dynamics and complex analysis.

In the introduction of [Ber90], Berkovich compares the construction of his spaces to the embedding of $\mathbf{Q}$ in $\mathbf{R}$. Pushing this analogy a bit further, we can compare rigid geometry to Berkovich's approach as follows. The metric topology on $\mathbf{Q}$ is totally disconnected, so that the algebra of continuous complex-valued functions on $\mathbf{Q}$ is rather pathological. One way to fix this problem is to define a Grothendieck topology on $\mathbf{Q}$ with the same open sets as the metric topology, but where the class of coverings of open subsets U of $\mathbf{Q}$ is restricted to coverings that can be refined to a locally finite covering by open intervals with rational endpoints (here locally finite means that every open interval with rational endpoints contained in U intersects only finitely many members of the covering). The archetypal example of a non-admissible open covering is the decomposition

$$\mathbf{Q} =]-\infty; \sqrt{2}[\, \cup\,]\sqrt{2}; +\infty[.$$

It is not hard to see that this Grothendieck topology is connected. Furthermore the continuous complex-valued functions with respect to this Grothendieck topology (i.e., the naive continuous complex-valued functions such that the pull-back of *any* open covering of $\mathbf{C}$ is an admissible covering of $\mathbf{Q}$) are precisely the functions that extend to a continuous function on $\mathbf{R}$. The rigid approach corresponds to endowing $\mathbf{Q}$ with this Grothendieck topology, while Berkovich's approach corresponds to adding points to $\mathbf{Q}$ in order to obtain the topological space $\mathbf{R}$ itself, which has much better properties than $\mathbf{Q}$.

We should mention that there exist still other approaches to non-Archimedean geometry, each having its particular assets and applications, such as Huber's adic spaces [Hub] and the Riemann-Zariski spaces of Fujiwara and Kato [Kat]. Unfortunately, these theories fall outside the scope of this volume, but we strongly recommend the interested reader to consult the literature on these fascinating topics.

2.2 Literature

Beside Berkovich's original monographs [Ber90] and [Ber93], there are now several texts and surveys treating this subject at various levels of generality and difficulty. Let us mention [Ber98] and [Ber07] by Berkovich himself (the second one contains a very interesting account of his first steps in non-Archimedean geometry); the lecture notes [Con] and the survey [Nic], which both cover several approaches to non-Archimedean geometry (Tate, Raynaud, Berkovich); and the survey [Duc] about Berkovich's theory and its applications.

The text of M. Temkin in the present volume explains, in a systematic way and with some original insights, the foundations of the theory of Berkovich spaces and the relation with other theories of non-Archimedean geometry. It is more extensive and detailed than the surveys listed above.

2.3 Applications of Berkovich Theory

Since the origins of the theory, Berkovich spaces have been used in often surprising ways in several areas of geometry and number theory. Let us give a concise survey of some of these applications, with a view towards the chapters in this volume.

An important aspect of Berkovich geometry is the existence of a full-fledged theory of étale cohomology, which was published by Berkovich in [Ber93]. The first striking application of Berkovich theory to algebraic geometry appeared in 1994 when Berkovich proved a conjecture of Deligne concerning ℓ-adic vanishing cycles. Deligne's conjecture was deduced from a comparison theorem for étale vanishing cycles of formal schemes, which were defined by Berkovich using the theory of étale cohomology of analytic spaces. This theory has also become an important tool in the Langlands Program, where it is used to produce Galois representations over local fields in a geometric way.

The theory of étale cohomology of analytic spaces is carefully explained in the first text of A. Ducros, with a particular emphasis on the parallels (and some differences) between the étale cohomology of schemes and analytic spaces. No prior knowledge of the theory of étale cohomology of schemes is assumed.

Another remarkable feature of Berkovich geometry is that it works equally well over fields with the trivial absolute value. This makes it possible to use analytic tools in the study of algebraic schemes over arbitrary fields, by endowing the field with its trivial absolute value and applying the GAGA functor. As a toy example, one can obtain the étale cohomology spaces with compact support of an algebraic variety as derived functors of the functor of global sections with compact support, by first analytifying the algebraic variety (this is not true if one remains in the category of schemes). The second contribution of A. Ducros is a note that shows how to use Berkovich spaces over a field with the trivial absolute value to prove finiteness results in coherent cohomology for proper morphisms of noetherian schemes. Ducros' strategy avoids the reduction to projective morphisms that is used in the classical approach.

A third class of applications of Berkovich spaces to algebraic geometry arise as follows. In many setups, one studies a geometric structure by constructing certain combinatorial invariants that reflect the geometry of the situation; the most obvious example is the theory of toric varieties and toroidal embeddings. It sometimes happens that these combinatorial invariants admit a natural interpretation in terms of valuations, so that they can be embedded in a suitable Berkovich space. One can then use Berkovich geometry to analyze the given geometric structure. An intriguing example is Thuillier's generalization of Stepanov's theorem in [Thu07]. Thuillier proves that the homotopy type of the dual complex of a log resolution of a pair of varieties (X, Y) over a perfect field k is independent of the log resolution. His strategy consists in showing that the dual complex Δ embeds into a certain analytic space over k with the trivial absolute value, attached intrinsically to (X, Y), and that Δ is a strong deformation retract of this Berkovich space.

Applications of a similar flavor appear in the theory of algebraic groups over local fields (Bruhat-Tits theory). Such applications were already explored by Berkovich in his foundational book [Ber90] in the split case. The survey by B. Rémy, A. Thuillier and A. Werner in this volume presents the main results obtained by the authors in a project where they extended Berkovich's first insights to an arbitrary algebraic reductive group G defined over a non-Archimedean local field. More precisely, they describe families of compactifications of the Bruhat-Tits building of G by observing that the building can be naturally embedded into the Berkovich analytification G^{an} of G.

A final field of applications of Berkovich spaces that we discuss is the theory of dynamical systems. Here one of the main ideas is to use Berkovich geometry to translate complex dynamical problems into the study of a dynamical system acting on a more tractable structure (for instance, an $\mathbb{R}$-tree) that lives inside a suitable Berkovich space.

This is explained in detail in the chapter by M. Jonsson, which discusses a series of (largely) unexpected applications of Berkovich theory to complex dynamical systems in low dimensions. Important technical tools in this work are the Laplace operator on $\mathbb{R}$-trees and the so-called subharmonic functions; their definitions are reviewed and put into a geometric perspective. This yields a gentle introduction to potential theory over non-Archimedean analytic spaces, which is treated more thoroughly in the book [BR] in the case of the Berkovich projective line, and by A. Thuillier in his Ph.D. thesis [Thu05] in the case of curves. We also refer to the series of recent papers [BFJ12a, BFJ12b, BPS, CL, CLD, Gub] for a glimpse of the recent struggles to develop such a potential theory over Berkovich analytic spaces of arbitrary dimension.

The text of J.-P. Otal explains how one can compactify the character variety of a group G (i.e., its space of representations into $SL(2, \mathbb{C})$ modulo conjugacy) in a dynamically meaningful way. Points in the boundary correspond to representations of G into $\mathrm{SL}(2, k)$ where k is a (non-Archimedean) valued extension of $\mathbb{C}$. This remarkable compactification was constructed by Morgan and Shalen in the 1980s, and the survey by Otal offers a fresh view on their work with applications to hyperbolic geometry in the vein of Thurston's ideas. Although Berkovich spaces are not explicitly mentioned in his text, a key role in this theory is played by $\mathbb{R}$-trees, which arise as Bruhat-Tits buildings of the group $\mathrm{SL}(2)$ and can be identified with subsets of the Berkovich projective line. Beside the obvious link with the article of Rémy, Thuillier and Werner, a whole section of Otal's paper is devoted to the notion of Riemann-Zariski space. This notion also plays an important role in the text by Temkin when studying the reduction of germs of analytic spaces, and in the short contribution by C. Favre on the sequential compactness of Berkovich spaces.

2.4 An Animated Introduction

We propose in this volume an introduction to Berkovich's theory that emphasizes its applications in various fields. A first part contains surveys of foundational nature (by M. Temkin, A. Ducros, C. Favre). A second part focuses on applications to geometry (A. Ducros, and B. Rémy, A. Thuillier and A. Werner). The third and final part explores the relationship between non-Archimedean geometry and dynamics (J.-P. Otal, M. Jonsson).

This is by no means a full account of the rapidly growing theory of Berkovich spaces, but we have tried to illustrate some of the main types of applications (as explained in the previous section). We hope that this book provides the reader with enough material on basic concepts and constructions related to Berkovich spaces to move on to more advanced research articles on the subject. We also hope that the applications presented here will inspire the reader to discover new settings where these beautiful and intricate objects might arise.

3 Plan of This Volume

We now provide a more detailed description of each of the articles in this book.

3.1 Part I: Berkovich Analytic Spaces

3.1.1 M. Temkin

Introduction to Berkovich Analytic Spaces
  This text gives an overview of Berkovich's theory. The most technically involved proofs are omitted and replaced by precise references to the literature; many other

proofs are sketched, or formulated as exercises to the reader with some hints to a solution.

The text starts with the basic definitions of the theory (analytic spaces, morphisms, etc.), with some novelties: for instance, an alternative definition of analytic spaces is suggested (based upon the notion of a set-theoretic Grothendieck topology, instead of the language of atlases) and the notion of an H-strict space is introduced (it is a space defined using Zariski-closed subsets of polydiscs whose radii belong to a subgroup H of $\mathbf{R}^*_+$). Next, the author summarizes some basic facts about analytic spaces and their topology, and studies some important classes of morphisms (proper, smooth, flat, boundaryless, etc.).

Subsequently the author discusses the relations between Berkovich geometry and other theories. He explains the definitions and the basic properties of two natural functors (analytification of algebraic varieties, generic fibers of formal schemes). He compares Berkovich's viewpoint to other approaches to non-Archimedean geometry (Tate, Raynaud, Huber), and he shows how one can study analytic germs through Riemann-Zariski spaces (this method being due to the author).

The final chapter is devoted to analytic curves and explains, among other things, how the Semi-Stable Reduction Theorem helps to understand their homotopy types. The fundamental example of the affine line (which is investigated in full detail at the beginning of the paper) plays a crucial role here.

3.1.2 A. Ducros

Étale Cohomology of Schemes and Analytic Spaces

Étale cohomology was introduced in the scheme-theoretic context by Grothendieck in the 1950s and 1960s in order to provide a purely algebraic cohomology theory, satisfying the same fundamental properties as the singular cohomology of complex varieties, which was needed for proving the Weil conjectures. For deep arithmetic reasons related to the Langlands program, it turned out to be worthwhile to develop such a theory in the p-adic analytic context. This was done by Berkovich in the early 1990s.

After having given some general motivation, this article starts with the notion of a Grothendieck topology and its associated cohomology theory. Then it turns to the basic ideas and properties of étale cohomology for schemes and Berkovich spaces (these theories are closely related to each other), and to its fundamental results including various comparison theorems, Poincaré duality and so forth. Almost no proofs are given since most of them are highly technical and can be found in the references to the specialized literature provided in the text. Instead, the author has chosen to insist on examples, and shows how étale cohomology can be quite close to the classical topological intuition, while at the same time dealing in a completely natural manner with deep field-arithmetic phenomena (such as Galois theory). This often allows to think of arithmetic problems in a purely geometrical way.

3.1.3 C. Favre

Countability Properties of Berkovich Spaces

It is proven that compact Berkovich spaces are also sequentially compact when defined over the field of formal Laurent series in one variable. The proof is based on the extensive use of the Riemann-Zariski space of a variety, an object that plays a prominent role in the text of J.-P. Otal.

3.2 Part II: Applications to Geometry

3.2.1 A. Ducros

Cohomological Finiteness of Proper Morphisms in Algebraic Geometry: A Purely Transcendental Proof, Without Projective Tools

Originally, the coherent cohomological finiteness of proper morphisms between noetherian schemes was proven by reducing to the projective case, thanks Chow's lemma, and then using explicit computations, based upon the Čcch complex associated with the standard affine covering of $\mathbb{P}^n$. In this short note, we give a new proof of this fact when the schemes are of finite type over a field k. The strategy consists in endowing the latter with the trivial valuation, then proving by hand a coherent GAGA-theorem in that setting (in a very concrete and explicit way), and eventually using the analytic coherent cohomological finiteness of proper morphisms, which was proven by Kiehl using non Archimedean functional analysis.

3.2.2 B. Remy, A. Thuillier, and A. Werner

Bruhat-Tits Buildings and Analytic Geometry

Let G be a reductive algebraic group defined over a non-Archimedean local field k. During the 1960s and 1970s, F. Bruhat and J. Tits developed a new way to encode the properties of the group of rational points $G(k)$. They established a combinatorial description that can be stated in geometric terms using the *Euclidean building* of G over k. This Euclidean building is both a complete metric space and a simplicial complex and can be seen in many ways as a (singular) analogue of the Riemannian symmetric space of a semi-simple real Lie group. When Berkovich developed his theory of non-Archimedean geometry in the 1980s, he already mentioned the possibility of combining his theory with the theory of Bruhat-Tits. The crucial point is that the Bruhat-Tits building of G over k can be naturally embedded, as a topological space, into the Berkovich analytification G^{an} of G.

In this survey, the authors present the results of a joint project in which they develop and extend Berkovich's ideas. They show in particular how natural compactifications of the Bruhat-Tits building can be obtained through Berkovich geometry by procedures that are quite similar to Satake's theory for symmetric spaces. The first part of the text reviews the necessary material from Bruhat-Tits theory, including basic definitions and properties of Euclidean buildings.

3.3 Part III: Valuation Spaces and Dynamics

3.3.1 M. Jonsson

Dynamics on Berkovich Spaces in Low Dimensions
    The main goal of this survey is to review some dynamical properties of discrete dynamical systems acting on Berkovich analytic spaces of dimensions 1 and 2 and to describe some applications of this study to dynamical questions of algebraic nature.

- In the first part (Sects. 2–5), the author studies the dynamics of a rational map $R \in K(T)$ defined over an algebraically closed non-Archimedean valued field K. Since the work of Benedetto and Rivera-Letelier at the turn of the millennium, it is known that the natural setup for developing this theory is to look at the action of R on the Berkovich projective line $\mathbf{P}^1_{\text{Berk}}$. The author focuses here on one special result of the theory: the fact that preimages of (most) points are asymptotically equidistributed with respect to a canonical R-invariant probability measure. This measure is supported in $\mathbf{P}^1_{\text{Berk}}$ on the locus where the dynamics of R is unstable (the Julia set of R).

    Section 2 in the initial chapter is devoted to tree structures.[2] It also contains a construction of a potential theory on these objects, a tool that plays a fundamental role in the proof of the equidistribution result. Several equivalent definitions of $\mathbf{P}^1_{\text{Berk}}$ are then given in Sect. 3, and its topology is carefully analyzed. The action of a rational map on the Berkovich projective line is described in Sect. 4. Basic results in local non-Archimedean dynamics, the construction of the dynamical partition of $\mathbf{P}^1_{\text{Berk}}$ into the Fatou and the Julia set of R, the construction of the canonical measure, and the proof of the equidistribution theorem are finally given in Sect. 5.

- The second part (Sects. 6–10) contains a review of joint works of the author and C. Favre on the growth of two dynamical invariants associated with a polynomial map $f : \mathbf{A}^2 \to \mathbf{A}^2$. The first invariant $c(f)$ arises when f fixes a point, say the origin, and describes the local rate of contraction of f near this fixed point. The main result (Theorem B in Sect. 8) is that $c(f^n) \asymp c^n_\infty$ for some *quadratic integer* c_∞. The second invariant $\deg(f)$ is by definition the maximum of the degree of its components in some affine coordinates. Theorem C of Sect. 10 states that either $\deg(f^n) \asymp d^n_\infty$ for some quadratic integer d_∞ just as in the

---

[2]The intersection with Sect. 3 in the text of J.-P. Otal is nonempty but the main focus is quite different.

local case, or $\deg(f^n) \asymp nd_\infty^n$ with $d_\infty \in \mathbf{N}^*$ and f preserves a pencil of affine lines (up to conjugacy).

Both theorems are proved by looking at the dynamics of f on suitable subspaces of the Berkovich affine plane $\mathbf{A}^2_{\mathrm{Berk}}$ over a trivially valued field. For Theorem B, one considers the space $\mathcal{V}_0$ of semi-valuations centered at 0 and suitably normalized. This space is a tree, and its structure is described in Sect. 7. The key to the proof of Theorem B is the construction of a fixed point in $\mathcal{V}_0$. In the case of Theorem C, one looks at the set of semi-valuations in $\mathbf{A}^2_{\mathrm{Berk}}$ centered at ∞ and also suitably normalized. Its internal structure is tightly related to the geometry of the affine plane and is therefore more intricate than its local counterpart. It is described in detail in Sect. 9. The proof of Theorem C is given in Sect. 10.

Although this text does not contain any new results, it does present substantial simplifications of proofs as they were originally published. Interesting discussions in the text nicely complement the existing literature. A geometric interpretation of local degrees for rational maps in one variable is given in Sect. 4.6. The geometry of the Berkovich affine plane over a trivially valued field is described in an elementary way in Sect. 6. Arguments are given to extend results to the case of non-algebraically closed fields.

3.3.2 J.-P. Otal

Compactification of Spaces of Representations (After Culler, Morgan and Shalen)

These notes give an account of the work of Culler, Morgan and Shalen from the late 1980s concerning the compactification of the space of representations of a finitely generated group G to $\mathrm{SL}(2, \mathbf{C})$ modulo conjugacy, or, in other words, to the *character variety* $X(G)$ of this group. When G is the fundamental group of a 3-manifold M, then the geometry of $X(G)$ is intimately related to the existence of special geometric structures on M. For instance, the existence of a discrete fixed point free and faithful representation $G \to \mathrm{SL}(2, \mathbf{C})$ is equivalent to the existence of a hyperbolic metric on M.

The main goal of this survey is to explain how one can compactify $X(G)$ in a dynamically relevant way. Since $X(G)$ is an affine space over $\mathbf{C}$, the most naive idea would be to take its closure in a suitable projective space. This, however, does not lead to an interpretation of the boundary points that is suitable for applications. Culler, Morgan and Shalen designed a way to add points at infinity in $X(G)$ that correspond to representations of G into $\mathrm{SL}(2, F)$ for suitable non-Archimedean extensions F of $\mathbf{C}$. Any such representation then leads to an isometric action of the group G on the Berkovich projective line over F which turns out to be an $\mathbf{R}$-tree.

The presentation closely follows the original work of Morgan and Shalen and identifies the boundary of the compactification as (the image of) a suitable Riemann-Zariski space. The connection with Berkovich analytic spaces is not explicitly stated, but we refer to Sect. 5.7 of the paper of M. Temkin in this volume, or to the contribution by C. Favre for a description of the link between Berkovich and Riemann-Zariski spaces.

As an application of these technics, J.-P. Otal gives a complete proof of a theorem due to Thurston that played a fundamental role in Thurston's approach to the geometrization of 3-manifolds. It states that the space of characters coming from discrete and faithful representations is relatively compact in $X(G)$ whenever G is the fundamental group of a compact boundary-incompressible and acylindrical 3-manifold.

This survey is essentially self-contained and recalls all facts from valuation theory and isometric actions on **R**-trees that are necessary for the proof of Thurston's theorem.

References

[Abb]   A. Abbes, *Éléments de géométrie rigide, Volume I.* Construction et étude géométrique des espaces rigides. Progress in Mathematics, vol. 286 (Birkhäuser/Springer Basel AG, Basel, 2010), xvi+477 pp.

[BR]   M. Baker, R. Rumely, *Potential Theory and Dynamics on the Berkovich Projective Line.* Mathematical Surveys and Monographs, vol. 159 (American Mathematical Society, Providence, RI, 2010), xxxiv+428 pp.

[Ber90]   V.G. Berkovich, *Spectral Theory and Analytic Geometry Over Non-Archimedean Fields.* Mathematical Surveys and Monographs, vol. 33 (American Mathematical Society, Providence, RI, 1990), x+169 pp.

[Ber93]   V.G. Berkovich, Étale cohomology for non-Archimedean analytic spaces. Publ. Math. IHES **78**, 5–161 (1993)

[Ber98]   V.G. Berkovich, *p*-adic analytic apaces, in *Proceedings of the International Congress of Mathematicians*, Berlin, August 1998, Doc. Math. J. DMV, Extra Volume ICM II (1998), pp. 141–151

[Ber07]   V.G. Berkovich, Non-Archimedean analytic geometry: first steps, in *p-adic Geometry: Lectures from the 2007 Arizona Winter School*, University Lecture Series, vol. 45 (American Mathematical Society, Providence, RI, 2008), pp. 1–7

[BGR]   S. Bosch, U. Güntzer, R. Remmert, *Non-Archimedean Analysis. A Systematic Approach to Rigid Analytic Geometry.* Grundlehren der Mathematischen Wissenschaften, vol. 261 (Springer, Berlin, 1984), xii+436 pp.

[BL93a]   S. Bosch, W. Lütkebohmert, Formal and rigid geometry. I. Rigid spaces. Math. Ann. **295**(2), 291–317 (1993)

[BL93b]   S. Bosch, W. Lütkebohmert, Formal and rigid geometry. II. Flattening techniques. Math. Ann. **296**(3), 403–429 (1993)

[BLR95a] S. Bosch, W. Lütkebohmert, M. Raynaud, Formal and rigid geometry. III. The relative maximum principle. Math. Ann. **302**(1), 1–29 (1995)

[BLR95b] S. Bosch, W. Lütkebohmert, M. Raynaud, Formal and rigid geometry. IV. The reduced fibre theorem. Invent. Math. **119**(2), 361–398 (1995)

[BFJ12a] S. Boucksom, C. Favre, M. Jonsson, Solution to a non-Archimedean Monge-Ampère equation. Preprint: arXiv:1201.0188. To appear in J. Am. Math. Soc

[BFJ12b] S. Boucksom, C. Favre, M. Jonsson, Singular semipositive metrics in non-Archimedean geometry. Preprint: arXiv:1201.0187. To appear in J. Algebraic Geom

[BPS]   J.I. Burgos Gil, P. Philippon, M. Sombra, Arithmetic geometry of toric varieties. Metrics, measures and heights. Astérisque **360**, vi+222 pages (2014)

[CL]   A. Chambert-Loir, Mesure et équidistribution sur les espaces de Berkovich. J. Reine Angew. Math. **595**, 215–235 (2006)

[CLD]   A. Chambert-Loir, A. Ducros, Formes différentielles réelles et courants sur les espaces de Berkovich. Preprint: `arXiv:1204.6277`

[Con]   B. Conrad, Several approaches to non-Archimedean geometry. *p*-adic geometry, in *p-adic Geometry: Lectures from the 2007 Arizona Winter School*. Univ. Lecture Ser., vol. 45 (American Mathematical Society, Providence, RI, 2008), pp. 9–63

[Duc]   A. Ducros, Espaces analytiques p-adiques au sens de Berkovich. Séminaire Bourbaki. Vol. 2005/2006. Astérisque No. 311 (2007), Exp. No. 958, viii, pp. 137–176

[FvdP]   J. Fresnel, M. van der Put, *Rigid Analytic Geometry and Its Applications*. Progress in Mathematics, vol. 218 (Birkhäuser Boston, Boston, MA, 2004), xii+296 pp.

[Gro]   A. Grothendieck, Techniques de construction en géométrie analytique, II: Généralités sur les espaces annelés et les espaces analytiques. Séminaire Henri Cartan, vol. 13 No. 1 (1960–1961), exp No. 9, 1–14

[Gub]   W. Gubler, Local heights of subvarieties over non-Archimedean fields. J. Reine Angew. Math. **498**, 61–113 (1998)

[Hub]   R. Huber, *Étale Cohomology of Rigid Analytic Varieties and Adic Spaces*. Aspects of Mathematics, vol. E30 (Friedr. Vieweg & Sohn, Braunschweig, 1996), x+450 pp.

[Igu]   J. Igusa, *An Introduction to the Theory of Local Zeta Functions*. Studies in Advanced Mathematics, vol. 14, American Mathematical Society, Providence, RI; (International Press, Cambridge, MA, 2000), xii+232 pp.

[Kat]   F. Kato, Topological rings in rigid geometry, in *Motivic Integration and Its Interactions with Model Theory and Non-Archimedean Geometry, I*, ed. by R. Cluckers, J. Nicaise, J. Sebag, London Mathematical Society Lecture Notes Series, vol. 383 (Cambridge University Press, Cambridge, 2011), xii+262 pp.

[Nic]   J. Nicaise, Formal and rigid geometry: an intuitive introduction and some applications. Enseign. Math. (2) **54**(3–4), 213–249 (2008)

[Ray]   M. Raynaud, Géométrie analytique rigide d'après Tate, Kiehl,.... Mémoires de la S.M.F. **39–40**, 319–327 (1974)

[Ser]   J.-P. Serre, Classification des variétés analytiques *p*-adiques compactes. Topology **3**, 409–412 (1965)

[Tat]   J. Tate, Rigid analytic spaces. Invent. Math. **12**, 257–289 (1971)

[Thu05]   A. Thuillier, *Théorie du potentiel sur les courbes en géométrie analytique non archimédienne. Applications à la théorie d'Arakelov*. Thesis, Université de Rennes 1, 2005. Available at `tel.archives-ouvertes.fr/docs/00/04/87/50/PDF/tel-00010990.pdf`

[Thu07]   A. Thuillier, Géométrie toroïdale et géométrie analytique non archimédienne. Application au type d'homotopie de certains schémas formels. Manuscr. Math. **123**(4), 381–451 (2007)

Part I
Berkovich Analytic Spaces

Introduction to Berkovich Analytic Spaces

Michael Temkin

Contents

1   Introduction

1.1   Berkovich Spaces and Some History

1.1.1   Naive Non-Archimedean Analytic Spaces

Since non-archimedean complete real-valued fields (e.g. $\mathbf{Q}_p$) were discovered in the beginning of the last century, it was very natural to try to develop a theory of analytic spaces over such a field k analogously to the theory of real or complex analytic spaces. At least, one can naturally define analytic functions on an open subset V of the naive affine space $A_k^n = k^n$ as the convergent power series on V. This allowed to introduce some naive k-analytic spaces, but the theory was not rich enough. Actually, a global theory of such varieties does not make too much sense because the topology of k is totally disconnected. In particular, locally analytic (and even locally constant) functions do not have to be globally analytic.

M. Temkin (✉)
Einstein Institute of Mathematics, The Hebrew University of Jerusalem, Giv'at Ram, Jerusalem, 91904, Israel
e-mail: temkin@math.huji.ac.il

© Springer International Publishing Switzerland 2015
A. Ducros et al. (eds.), *Berkovich Spaces and Applications*, Lecture Notes in Mathematics 2119, DOI 10.1007/978-3-319-11029-5_1

3

1.1.2 Rigid Geometry

In his study of elliptic curves with multiplicative bad reduction over a non-archimedean field k, Tate discovered about 1960 that these curves admit a natural uniformization by G_m. The latter was given as an abstract isomorphism of groups $k^\times/q^{\mathbf{Z}} \xrightarrow{\sim} E(k)$, and even such expert as Grothendieck doubted at first that this was not an accidental brute force isomorphism. Tate suspected, however, that his isomorphism can be interpreted as an analytic one, and he had to develop a good global theory of non-archimedean analytic spaces to make this rigorous. This research resulted in Tate's definition of rigid geometry, whose starting idea was to simply forbid all bad open coverings (responsible for disconnectedness) and to shrink the set of analytic functions accordingly. As a result, one obtains a good theory of sheaves of analytic functions, but the underlying topological spaces have to be replaced with certain topologized (or Grothendieck) categories, also called G-topological spaces.

1.1.3 Berkovich Spaces

More recently, some other approaches to non-archimedean geometry were discovered: Raynaud's theory of formal models, Berkovich's analytic geometry and Huber's adic geometry. They all allow to define (nearly) the same categories of k-analytic spaces, but provide analogs of rigid spaces with additional structures invisible in the classical Tate's theory. Also, they extend the category of rigid spaces in different directions. Here we only discuss Berkovich's theory, which was developed in [Ber1] and [Ber2]. In this theory, classical rigid spaces are saturated with many new points (analogs of non-closed points of algebraic varieties), and the obtained spaces are honest topological spaces. In addition, the underlying topological spaces are rather nice (locally pathwise connected, for example).

Now, let us list some interesting features of Berkovich's theory that distinguish it from all other approaches. First, one can work with all positive real numbers almost as well as with the values of $|k^\times|$. In particular, one can study rings of power series with radii of convergence linearly independent of $|k^\times|$. The latter fact allows to include the case of a trivially valued k into the theory, and the theory of such k-analytic fields has already been applied to classical problems of algebraic geometry. Another interesting feature is that in a similar fashion one can develop (an equivalent form of) the usual theory of complex analytic spaces. Moreover, one can define Berkovich spaces that include both archimedean and non-archimedean worlds, for example, the affine line over $(\mathbf{Z}, |\ |_\infty)$.

1.2 Structure of the Notes

We do not aim to prove all results discussed in these notes (and this is impossible in a six lectures long course). Our goal is to make the reader familiar with basic definitions, constructions, techniques and results of non-archimedean analytic geometry. Therefore, we prefer to formulate difficult results as Facts, and in some cases we discuss main ideas of their proofs in Remarks. Easy corollaries from these results (that may themselves be important pieces of the theory) are then suggested to the reader as exercises. Many exercises are provided with hints, but it may be worth to first try to solve them independently (especially, those not marked with an asterisk).

1.2.1 Overview

The course is divided into five sections as follows. First, we study in Sect. 2 semi-norms, norms and valuations, basic operations with these objects, Banach rings and their spectra. Then we describe the structure of $\mathcal{M}(\mathbf{Z})$, and after that we switch completely to the non-archimedean world. We finish the section with describing affine line over an algebraically closed non-archimedean field. In Sect. 3 we introduce k-affinoid algebras and spaces and study their basic properties. In Sect. 4, this local theory is used to introduce and study global k-analytic spaces. Relations of k-analytic spaces with other categories are studied in Sect. 5. This includes analytification of algebraic varieties and GAGA, generic fibers of formal k°-schemes and Raynaud's theory, and some discussion of rigid and adic geometries. In addition, in Sect. 5.7 we study local structure of analytic spaces by use of Riemann-Zariski (or birational) spaces over the residue field $\tilde{k}$. Finally, in Sect. 6 we study k-analytic curves in details. In particular, we describe their local and global structure and explain how this is related to the stable reduction theorem for formal k°-curves.

1.2.2 References and Other Sources

The main references that helped me to prepare the course are [Ber1, Ber2] and [Ber3]. The first two are a book and a large article in which the non-archimedean analytic spaces were introduced. The third one is a lecture note of an analogous introductory course given by Berkovich in Trieste in 2009. I recommend the third source as an alternative (and shorter) expositional introductory text. It is worth to note that the first three sections of [Ber3] and of these notes are parallel, but the exposition is (often but not always) rather different. Also, this text contains much more exercises and remarks, and this seriously increases its length. Finally, the reader may wish to consult lecture notes [Con] on non-archimedean geometry (including the rigid geometry) by Brian Conrad.

1.2.3 Novelties

Some things in these notes are new. We develop a general theory of H-strict k-analytic spaces (or k_H-analytic) spaces that are built using radii of convergence from a group $|k^\times| \subseteq H \subseteq \mathbf{R}_+^\times$, with the extreme cases being strict and general k-analytic spaces. The main reference for H-strict theory is [CT]. We give Berkovich's definition of k-analytic spaces that uses atlases but also show how one can define these spaces without atlases. This relies on the new Fact 3.3.3.3, which allows to characterize affinoid spaces and their morphisms as certain Banach ringed spaces and their morphisms. Another new result is Fact 3.1.4.2, which asserts that for a non-trivially valued k any k-homomorphism of k-affinoid algebras is bounded. Each new fact is followed by an exercise with a detailed hint on proving it.

1.2.4 Conventions

Throughout these notes *ring* always means a commutative ring with unity. For any field k by k^s and k^a we denote its separable and algebraic closures, respectively. We will use underline to denote finite tuples of real numbers or of coordinates. For example, a polynomial ring $k[T_1 \ldots T_n]$ will often be denoted as $k[\underline{T}]$, where $\underline{T}$ is the tuple $(T_1 \ldots T_n)$ of coordinates. Also we will use the notation $\underline{T}^i = T_1^{i_1} \ldots T_n^{i_n}$ for $i \in \mathbf{N}^n$. For example, a power series $f(\underline{T}) \in k[[\underline{T}]]$ can be uniquely written as $\sum_{i \in \mathbf{N}^n} a_i \underline{T}^i$ with $a_i \in k$.

2 Norms, Valuations and Banach Rings

2.1 Seminorms

2.1.1 Seminormed Groups

Definition/Exercise 2.1.1.1 (i) A *seminorm* on an abelian group A is a function $| \ |: A \to \mathbf{R}_+$ which is *sub-additive*, i.e. $|a+b| \le |a|+|b|$, and satisfies $|0| = 0$ and $|-a| = |a|$. A seminorm is a *norm* if its kernel is trivial. If the seminorm is fixed then we call A a *seminormed group*. A seminorm is *non-archimedean* if it satisfies the strong triangle inequality $|a + b| \le \max(|a|, |b|)$.

(ii) The morphisms in the category of seminormed abelian groups are *bounded homomorphism*, i.e. homomorphisms $\phi: A \to B$ such that $\|\phi(a)\| \le C|a|$ for some fixed constant $C = C(\phi)$. In particular, $(A, | \ |)$ and $(A, \| \ \|)$ are isomorphic if and only if the seminorms $| \ |$ and $\| \ \|$ are *equivalent*, i.e. there exists a constant $C > 0$ such that $|a| \le C\|a\|$ and $\|a\| \le C|a|$ for any $a \in A$.

(iii) Any quotient A/H possesses a *residue seminorm* $\| \ \|$ given by $\|a + H\| = \inf_{h \in H} |a + h|$. A homomorphism of seminormed groups $\phi: A \to B$ is *admissible* if the residue seminorm on $\phi(A)$ is equivalent to the seminorm induced from B.

(iv) We provide a seminormed ring A with the *semimetric* $d(a, b) = |a - b|$. The induced *seminorm topology* is the weakest topology for which the balls $B_{a,r} = \{x \in A | \ |x - a| < r\}$ are open. This topology distinguishes points (i.e. is T_0) if and only if the seminorm is a norm. Two seminorms are equivalent if and only if their induced topologies coincide. Any bounded homomorphism is continuous with respect to the seminorm topologies (see also Exercise 2.2.1.3).

(v) The *separated completion* $\hat{A}$ of a seminormed group A is the set of equivalence classes of Cauchy sequences in A. Use continuity to extend the group structure to $\hat{A}$ and show that $\hat{A}$ is a normed group, the natural map $A \to \hat{A}$ is an admissible homomorphism (called the *separated completion homomorphism*) and its kernel is $\mathrm{Ker}(|\ |)$. In particular, $A/\mathrm{Ker}(|\ |)$ is a normed group with respect to the residue seminorm.

Remark 2.1.1.2 Usually, we will simply say "completion" in the sequel. Sometimes we will say "separated completion" in order to stress that the completion homomorphism may have a kernel.

2.1.2   Seminormed Rings and Modules

Definition/Exercise 2.1.2.1   (i) A *seminorm* (resp. *norm*) on a ring A is a seminorm (resp. norm) on the additive group of A which is *submultiplicative*, i.e. $|ab| \leq |a||b|$. If $|\ |$ is multiplicative, i.e. $|ab| = |a||b|$ and $|1| = 1$, then it is called a *real semivaluation* (resp. *real valuation*). If such a structure is fixed then the ring is called *seminormed, normed, real-valued* or *real-semivalued*, accordingly.

(ii) If A is seminormed then a seminorm on an A-module M is an additive seminorm $\|\ \|$ such that $\|am\| \leq C|a|\|m\|$ for a fixed $C = C(M)$ and any $a \in A$ and $m \in M$.

(iii) Formulate and prove the analogs of all results/definitions from Sect. 2.1.1.1, including separated completions and admissible homomorphisms.

Remark 2.1.2.2   (i) General non-archimedean semivaluations on rings are defined similarly but with values in $\{0\} \coprod \Gamma$, where Γ is a totally ordered multiplicative abelian group. Note that due to the strong triangle inequality, the definition makes sense even though there is no addition on $\{0\} \coprod \Gamma$.

(ii) When studying general semivaluations one usually does not distinguish between the *equivalent* ones, i.e. semivaluations that admit an ordered isomorphism $i:\mathrm{Im}(|\ |) \overset{\sim}{\to} \mathrm{Im}(\|\ \|)$ such that $i \circ |\ | = \|\ \|$. This is the only reasonable possibility in the case when the group of values Γ is not fixed. On the other hand, it is very important that we do distinguish equivalent but not equal real semivaluations.

(iii) The valuation terminology is not unified in the literature. For example, in adic geometry of R. Huber, any semivaluation is called a valuation.

In the following exercises we provide some definitions, examples and constructions related to seminormed rings.

Definition/Example/Exercise 2.1.2.3 Let $(A, |\ |)$ be a normed ring (the constructions make sense for seminormed rings but we will not need that).

(i) The *spectral seminorm* $\rho = \rho_A$ is the maximal *power-multiplicative* (i.e. $\rho(f^n) = \rho(f)^n$)) seminorm dominated by $|\ |$. Show that ρ exists and is defined by $\rho(f) = \lim_{n \to \infty} |f^n|^{1/n}$.

(ii) For a tuple of positive numbers $\underline{r} = (r_1 \ldots r_n)$ provide $A[T_1 \ldots T_n]$ with the norm

$$\| \sum_{i \in \mathbf{N}^n} a_i \underline{T}^i \|_{\underline{r}}^{\mathrm{ar}} = \sum_{i \in \mathbf{N}^n} |a_i| \underline{r}^i$$

(where "ar" stands for archimedean) and let

$$A\{\underline{r}^{-1}\underline{T}\}^{\mathrm{ar}} = A\{r_1^{-1}T_1 \ldots r_n^{-1}T_n\}^{\mathrm{ar}}$$

denote its completion. This ring can be viewed as the ring of convergent power series over A with polyradius of convergence $\underline{r}$. Work this out: show that $A\{\underline{r}^{-1}\underline{T}\}^{\mathrm{ar}}$ is a subring of $\hat{A}[[\underline{T}]]$ defined by a natural convergence condition.

(iii) If $(M, |\ |_M)$ and $(N, |\ |_N)$ are normed A-modules (resp. rings) then we provide $M \otimes_A N$ with the *tensor product seminorm* $\|x\| = \inf(\sum_{i=1}^n |m_i|_M |n_i|_N)$ where the infimum is taken over all representations of x of the form $x = \sum_{i=1}^n m_i \otimes n_i$. The separated completion of this seminormed module is denoted $M \hat{\otimes}_A^{\mathrm{ar}} N$ and called the (archimedean) *completed tensor product* of modules (resp. rings). We will later see that the tensor product seminorm is often not a norm.

(iv) The *trivial semi-norm* $|\ |_0$ on a ring A sends $A \setminus \{0\}$ to 1. It is power-multiplicative (resp. a valuation) if and only if A is reduced (resp. integral).

(v) For any natural $n > 1$ define the n-adic norm on $\mathbf{Q}$ by $|x|_n = n^d$, where $d \in \mathbf{Z}$ is the minimal number with $xn^d \in \mathbf{Z}_{(n)}$ (the localization of $\mathbf{Z}$ by all primes coprime with n). This norm is a valuation only when n is prime. The equivalence class of $|\ |_n$ depends only on the set $p_1 \ldots p_m$ of prime divisors of n. The completion $\mathbf{Q}_n$ with respect to $|\ |_n$ is called the ring of n-adic numbers. Show that $\mathbf{Q}_n = \oplus_{i=1}^m \mathbf{Q}_{p_i}$. In particular, the completion operation does not preserve the property of being an integral domain. Show that $\mathbf{Q}_p$ is a field. It is called the *field of p-adic numbers*.

(vi) Define t-adic valuations on $k[t]$ analogously to the p-adic valuation (they are trivial on k and are uniquely determined by $r = |t| \in (0, 1)$). Show that $k[[t]]$ is the completion.

(vii) Ostrowski's theorem provides a complete list of real semivaluations on $\mathbf{Z}$: the trivial valuation $|\ |_0$, the p-adic valuations $|\ |_{p,r} = (|\ |_p)^r$ for any $r \in (0, \infty)$, the archimedean valuations $|\ |_{\infty,r} = (|\ |_\infty)^r$ for $r \in (0, 1]$ (where $|x|_\infty$ is the usual absolute value of x), and the semivaluations $|\ |_{p,\infty}$ that take $p\mathbf{Z}$ to 0 and everything else to 1.

Remark 2.1.2.4 There is a certain analogy, that will be used later, between ideals on rings and bounded semi-norms on semi-normed rings. Exercises (iv) and (v) above indicate that multiplicative (resp. power-multiplicative) semi-norms correspond to prime (resp. reduced) ideals. In the style of the same analogy, passing from a semi-normed ring $(A, | \ |)$ to $(A/\mathrm{Ker}(\rho_A), \rho_A)$ can be viewed as an analog of reducing the ring (i.e. factoring a ring by its radical). The elements in the kernel of ρ_A are called *quasi-nilpotent* elements.

2.2 Banach Rings and Their Spectra

2.2.1 Banach Rings, Algebras and Modules

Definition 2.2.1.1 (i) A *Banach ring* is a complete normed ring $\mathcal{A}$ (i.e. the completion homomorphism $\mathcal{A} \to \hat{\mathcal{A}}$ is an isomorphism). A *Banach $\mathcal{A}$-algebra* is a Banach algebra $\mathcal{B}$ with a bounded homomorphism $\mathcal{A} \to \mathcal{B}$.
(ii) A *Banach $\mathcal{A}$-module* is a complete normed $\mathcal{A}$-module.

Instead of the polynomial rings and tensor products of modules, when working with Banach rings and modules we will use the convergent power series rings and completed tensor products.

Fact 2.2.1.2 The valuation of any complete real-valued field k uniquely extends to any algebraic extension of k.

Example/Exercise 2.2.1.3 Let k be any complete real-valued field and let $f : \mathcal{A} \to \mathcal{B}$ be a homomorphism of Banach k-modules (also called Banach k-spaces).

(i) Assume that k is not trivially valued (e.g. **R**, **C**, $\mathbf{Q}_p$ or $\mathbf{C}((t))$). Then f is bounded if and only if it is continuous.
(ii) Assume that k is trivially valued. Show that as an abstract ring, $k\{r^{-1}T\}^{\mathrm{ar}}$ is isomorphic to $k[T]$ when $r \geq 1$ and is isomorphic to $k[[T]]$ when $r < 1$. In particular, there exist continuous but not bounded homomorphisms of Banach k-algebras.

2.2.2 The Spectrum

The analogy from Remark 2.1.2.4 suggests the following definition.

Definition 2.2.2.1 (i) *Spectrum* of a Banach ring $\mathcal{A}$ is the set $\mathcal{M}(\mathcal{A})$ of all bounded real semivaluations $| \ |_x$ on $\mathcal{A}$ (i.e. $| \ |_x \leq C | \ |$ for some C) provided with the weakest topology making continuous the maps $|f| : \mathcal{M}(\mathcal{A}) \to \mathbf{R}_+$ for all $f \in \mathcal{A}$. The latter maps take $| \ |_x$ to $|f|_x$ and usually we will use the notation $x \in \mathcal{M}(\mathcal{A})$ and $|f(x)|$ instead of $| \ |_x$ and $|f|_x$.

(ii) For any point $x \in \mathcal{M}(\mathcal{A})$ the kernel of $| \ |_x$ is a prime ideal and hence $\mathcal{A}/\mathrm{Ker}(| \ |_x)$ is an integral valued ring. The completed fraction field of this ring is called the *completed residue field of* x and we denote it as $\mathcal{H}(x)$. The bounded character corresponding to x will be denoted $\chi_x : \mathcal{A} \to \mathcal{H}(x)$.

The following exercise shows that the definition of $\mathcal{M}$ is analogous to the definition of Spec.

Exercise 2.2.2.2 The points of $\mathcal{M}(\mathcal{A})$ are the isomorphism classes of bounded homomorphisms $\chi : \mathcal{A} \to k$ whose image is a complete real-valued field generated by the image of χ (i.e. $\mathrm{Im}(\chi)$ is not contained in a proper complete subfield of k).

Here are basic facts about the spectrum. As one might expect, the general line of the proof is to construct enough points by use of Zorn's lemma.

Fact 2.2.2.3 (i) Let $\mathcal{A}$ be a Banach ring. The spectrum $X = \mathcal{M}(\mathcal{A})$ is compact, and it is empty if and only if $\mathcal{A} = 0$.

(ii) The maximum modulus principle: $\rho(f) = \max_{x \in X} |f(x)|$.

Exercise 2.2.2.4 (i) Extend $\mathcal{M}$ to a functor to topological spaces, that is, for any bounded homomorphism of Banach rings $\phi : \mathcal{A} \to \mathcal{B}$ construct a natural continuous map $\mathcal{M}(\phi) : \mathcal{M}(\mathcal{B}) \to \mathcal{M}(\mathcal{A})$.

(ii) An element $f \in \mathcal{A}$ is invertible if and only if $\inf_{x \in X} |f(x)| > 0$.

(iii) If $\mathcal{A}$ is a Banach k-ring for a complete real-valued field k then

$$\mathcal{M}(\mathcal{A} \hat{\otimes}_k^{\mathrm{ar}} k^a)/\mathrm{Gal}(k^s/k) \overset{\sim}{\to} \mathcal{M}(\mathcal{A})$$

where k^a and k^s are provided with the extended valuation.

(iv)* Let $\mathcal{A}$ be finite over $\mathbf{Z}$. Show that up to equivalence there exists a unique structure of a Banach $(\mathbf{Z}, | \ |_\infty)$-algebra on $\mathcal{A}$ and describe $\mathcal{M}(\mathcal{A})$ similarly to the description of $\mathcal{M}(\mathbf{Z})$ in Exercise 2.1.2.3(vii). Analyze similarly the spectra of finite Banach $(k[T], | \ |_0)$-algebras, where k is a field and $| \ |_0$ is the trivial valuation.

Hint: take the scheme $\mathrm{Spec}(\mathcal{A})$. Keep all its closed points, and for any generic point x of a curve component of $\mathrm{Spec}(\mathcal{A})$ replace x with all valuations on $k(x)$.

2.2.3 Relative Affine Spectrum

The following definition is not standard, but it seems to be convenient.

Definition 2.2.3.1 Let $\mathcal{A}$ be a Banach ring and let C be an $\mathcal{A}$-algebra (without any norm). We define the analytic spectrum of C as the set $\mathcal{M}\mathrm{Spec}(C)$ of all real semivaluations on C that are bounded on $\mathcal{A}$ (i.e. the restriction of $| \ |_x$ to $\mathcal{A}$ is bounded). Naturally, $\mathcal{M}\mathrm{Spec}(C)$ is provided with the weakest topology making continuous each map $x \mapsto |f(x)|$ with $f \in \mathcal{A}$.

Remark 2.2.3.2 There is a natural projection $\mathcal{M}\mathrm{Spec}(C) \to \mathcal{M}(\mathcal{A})$, which is typically a non-compact map. We stress that $\mathcal{M}\mathrm{Spec}(C)$ depends on the structure of C as an $\mathcal{A}$-algebra. In some sense, it is an analog of the relative **Spec** construction in algebraic geometry.

2.2.4 The Affine Space

Definition 2.2.4.1 The *n-dimensional affine space* over a Banach ring $\mathcal{A}$ is the topological space $\mathbf{A}^n_{\mathcal{A}} := \mathcal{M}\mathrm{Spec}(\mathcal{A}[T_1 \ldots T_n])$.

Exercise 2.2.4.2 The n-dimensional affine space $\mathbf{A}^n_{\mathcal{A}}$ is the union of closed $\mathcal{A}$-*polydiscs* $\mathcal{M}(\mathcal{A}\{\underline{r}^{-1}\underline{T}\}^{\mathrm{ar}})$ of *polyradius* $\underline{r} = (r_1 \ldots r_n)$. In particular, it is locally compact.

Remark 2.2.4.3 For a complete real-valued field k one can provide $\mathbf{A}^n_k$ with the sheaf of analytic functions which are local limits of rational functions from $k(\underline{T})$. If $U \subset \mathbf{A}^n_k$ is open and V is a Zariski closed subset of U given by vanishing of analytic functions $f_1 \ldots f_n$, then factoring by the ideal generated by f_i one obtains a sheaf of analytic functions on V. Gluing such local models V with the sheaves of analytic functions one can construct a theory of k-analytic spaces without boundary. The advantage of this approach is that it works equally well over $\mathbf{C}$ and over $\mathbf{Q}_p$. The main disadvantage of this approach is that it does not treat well enough the cases of a trivially valued k and of analytic $\mathbf{Q}_p$-spaces with boundary. More details about the outlined approach can be found in [Ber1, §1.5] and [Ber3, §1.3]. We will use another approach to construct non-archimedean analytic spaces.

2.3 Non-Archimedean Setting

2.3.1 Strong Triangle Inequality

Definition 2.3.1.1 (i) A non-archimedean seminorm (resp. norm, semivaluation, etc.) is a seminorm $| \ |$ that satisfies the *strong triangle inequality* $|a + b| \leq \max(|a|, |b|)$.
(ii) A *non-archimedean field* is a complete real-valued field k whose valuation is non-archimedean.

Example 2.3.1.2 (i) Kürschák's proved that any complete real-valued field containing $\mathbf{C}$ coincides with it (Gel'fand-Mazur proved a stronger claim later). Now, applying Ostrowski's classification we obtain that any complete real-valued field, excluding $\mathbf{R}$ and $\mathbf{C}$, is non-archimedean.
(ii) For any ring A its trivial seminorm is non-archimedean.

Exercise 2.3.1.3 Show that for an archimedean field k one has an homeomorphism $\mathbf{A}^1_k \xrightarrow{\sim} k^a/\mathrm{Gal}(k^s/k)$ where the image is provided with the valuation topology, i.e. $\mathbf{A}^1_k$ coincides with the naive affine line.

In the sequel we will work only with non-archimedean seminorms, semivaluations, etc., so the word "non-archimedean" will usually be omitted. Let $\mathcal{A}$ be a non-archimedean Banach ring. The basic definitions should now be adjusted as follows.

Definition/Example/Exercise 2.3.1.4 (i) Check that $\mathcal{M}(\mathcal{A})$ is the set of all non-archimedean bounded semivaluations on $\mathcal{A}$.

(ii) The spectral seminorm $\rho_\mathcal{A}$ is non-archimedean.

(iii) Non-archimedean definitions of $\mathcal{A}\{\underline{r}^{-1}\underline{T}\}$, $M \hat{\otimes}_\mathcal{A} N$ and their norms copy their archimedean analogs with maxima used instead of sums. For example,

$$\left\| \sum_{i \in \mathbf{N}^n} a_i \underline{T}^i \right\|_{\underline{r}} = \max_{i \in \mathbf{N}^n} |a_i| \underline{r}^i$$

Check that $\| \ \|_{\underline{r}}$ is a valuation.

(iv) Calculus student's dream: a sequence a_n in $\mathcal{A}$ is a Cauchy sequence if and only if $\lim_{n \to \infty} |a_n - a_{n+1}| = 0$. In particular, a series $\sum_{n=0}^{\infty} a_n$ converges in $\mathcal{A}$ if and only if $\lim_{n \to \infty} |a_n| = 0$.

At this point we fix a non-archimedean ground field k and start to develop non-archimedean analytic geometry over k. When developing this theory we will compare it from time to time to the classical theory of algebraic varieties over a field. Analogously to the latter theory, we will first introduce k-Banach algebras of topologically finite type and their spectra, called k-affinoid algebras and spaces. Then we will construct general k-analytic spaces by pasting k-affinoid ones. Despite this general similarity, many details in our theory are subtler. We will try to indicate critical moments where the theories differ.

2.3.2 Reduction Ring

Definition 2.3.2.1 It follows from Exercise 2.3.1.4 that any non-archimedean ring $\mathcal{A}$ contains an open subring $\mathcal{A}° = \{a \in \mathcal{A}| \ \rho(a) \le 1\}$ with an ideal $\mathcal{A}°° := \{a \in \mathcal{A}| \ \rho(a) < 1\}$. The ring $\tilde{\mathcal{A}} = \mathcal{A}°/\mathcal{A}°°$ is called the *reduction ring* of $\mathcal{A}$.

Example/Exercise 2.3.2.2 If k is a non-archimedean field then $k°$ is its valuation ring with maximal ideal $k°°$ and residue field $\tilde{k}$.

2.3.3 Description of Points of $\mathbf{A}^1_k$

Definition 2.3.3.1 Let l be a non-archimedean k-field. Recall that $e_{l/k}$ is the cardinality of $|l^\times|/|k^\times|$ (may be infinite) and $f_{l/k} = [\tilde{l} : \tilde{k}]$. Transcendental analogs of these cardinals are $E_{l/k} = \mathrm{rank}_\mathbf{Q}(|l^\times|/|k^\times| \otimes_\mathbf{Z} \mathbf{Q})$ and $F_{l/k} = \mathrm{tr.deg.}(\tilde{l}/\tilde{k})$.

Exercise 2.3.3.2 Assume that l is algebraic over the completion of its subfield l_0 which is of transcendence degree n over k. Prove Abhyankar's inequality: $E_{l/k} + F_{l/k} \leq n$.

We will use this result to classify points on $\mathbf{A}^1_k = \mathcal{M}\mathrm{Spec}(k[T])$ (and a similar argument classifies points on any k-analytic curve).

Definition/Exercise 2.3.3.3 (0) A point $x \in \mathbf{A}^1_k$ is *Zariski closed* if $| \; |_x$ has a non-trivial kernel. Show that this happens if and only if $\mathcal{H}(x)$ is finite over k. Show that otherwise $\mathcal{H}(x)$ is a completion of $k(T)$ and to give a point which is not Zariski closed is the same as to give a real valuation on $k(T)$ that extends that of k. In particular, $E_{\mathcal{H}(x)/k} + F_{\mathcal{H}(x)/k} \leq 1$.

  (1) x is of type 1 if $\mathcal{H}(x) \subseteq \widehat{k^a}$.
  (2) x is of type 2 if $F_{\mathcal{H}(x)/k} = 1$.
  (3) x is of type 3 if $E_{\mathcal{H}(x)/k} = 1$.
  (4) x is of type 4 if $E_{\mathcal{H}(x)/k} = F_{\mathcal{H}(x)/k} = 0$ and x is not of type 1.
  (5)* Show that in case (1) $\mathcal{H}(x)$ may contain an infinite algebraic extension of k (e.g. if $k = \mathbf{Q}_p$ then it may coincide with $\mathbf{C}_p = \hat{\mathbf{Q}}^a_p$). In particular, the map $\mathbf{A}^1_{\widehat{k^a}} \to \mathbf{A}^1_k$ usually has infinite (pro-finite) fibers.

    Hint: fix elements $x_i \in k^s$ and take $T = \sum_{i=1}^{\infty} a_i x_i$ where $a_i \in k$ converge to zero fast enough; then use Krasner's lemma to show that $k(x_i) \subset \widehat{k(T)}$.

Remark 2.3.3.4 More generally, $\mathcal{H}(x)$ may contain an infinite algebraic extension of k for type 4 points, but not for type 2 or 3 points.

Assume now that k is algebraically closed and let us describe the points of $\mathbf{A}^1_k$ in more details.

Exercise 2.3.3.5 (i) By *closed disc* $E(a, r) \subset \mathbf{A}^1_k$ of radius r and with center at a we mean the set of points of $\mathbf{A}^1_k$ that satisfy $|(T - a)(x)| \leq r$. Show that $E(a, r) = \mathcal{M}(k\{r^{-1}(T - a)\})$

 (ii) Show that a valuation on $k[T]$ is determined by its values on the elements $T - a$ with $a \in k$. The number $r = \inf_{a \in k} |T - a|$ is called the *radius* of x (with respect to the fixed coordinate T).

 (iii) Assume that the infimum r is achieved, say $r = |T - a|$. Show that

    (a) if $r = 0$ then x is Zariski closed and of type 1 and $|f(x)| = |f(a)|$ for any $f \in k[T]$.

    (b) if $r > 0$ then x is the maximal point of the disc $E(a, r)$ (i.e. $|f(x)| \geq |f(y)|$ for any $y \in E(a, r)$ and $f = \sum_{i=0}^{n} f_i T^i \in k[T]$) and $|f(x)| = \max_i |f_i| r^i$. If r is *rational* in the sense that $r^n \in |k^\times|$ for some integral $n > 0$ then x is of type 2, and otherwise x is of type 3.

 (iv) Assume that the infimum is not achieved, say $a_j \in k$ are such that the sequence $r_j = |T - a_j|$ decreases and tends to r. Then x is of type 4, it is the only point in the intersection of the discs $E(a_j, r_j)$, and $|f(x)| = \inf_j |f(x_j)|$ where x_j is the maximal point of $E(a_j, r_j)$. In particular, for an algebraically closed ground field k, type 4 points exist in $\mathbf{A}_k^1$ if and only if k is not *spherically complete*, i.e. there exist nested sequences of discs over k without common k-points.

Actually, $\mathbf{A}_k^1$ is a sort of an infinite tree whose leaves are type 1 and 4 points.

Exercise 2.3.3.6 (i) Use the previous exercise to prove that $\mathbf{A}_k^1$ is pathwise connected and simply connected. Moreover, show that for any pair of points $x, y \in \mathbf{A}_k^1$ there exists a unique path $[x, y]$ that connects them.

Hint: $[x, y] = [x, z] \cup [z, y]$ where z is the maximal point of the minimal disc containing both x and y and the open path (x, z) (resp. (z, y)) consists of the maximal points of discs that contain x but not y (resp. y but not x).

(ii) Show that $\mathbf{A}_k^1 \setminus \{x\}$ is connected whenever x is of type 1 or 4, consists of two components when x is of type 3, and consists of infinitely many components naturally parameterized by $\mathbf{P}_{\tilde{k}}^1$ when x is of type 2. Thus, $\mathbf{A}_k^1$ is an infinite tree with infinite ramification at type 2 points. If k is trivially valued then there is just one type 2 point and no type 4 points, so the tree looks like a star whose rays connect the type 2 point (the trivial valuation) with the Zariski closed points.

Almost all non-discretely valued fields are not spherically complete.

Exercise 2.3.3.7 (i) Let k_0 be trivially valued and let k be the t-adic field $k_0((t))$. Show that $\mathbf{C}_p$ and $\widehat{k^a}$ are not spherically complete.

Hint: for example, choose centers of the discs at $\sum_{i=0}^n t^{l_i}$, where l_i is a decreasing sequence of rational numbers that tends to a positive number r.

(ii) Prove by Zorn's lemma that spherically complete, algebraically closed, and non-trivially valued non-archimedean fields exist.

(iii) Here is the only known explicit construction of such fields. Let k_0 be an algebraically closed trivially valued field and let $\Gamma \subseteq \mathbf{R}_+^\times$ be a divisible subgroup. Let $k = k_0((t^\Gamma))$ be the set of all series $\sum_{\gamma \in \Gamma} a_\gamma t^\gamma$ where $a_\gamma \in k_0$ and any increasing family of γ's with non-zero a_γ is finite (finite sums form the group ring $k_0[t^\Gamma]$). Show that one can naturally define multiplication that makes k to a spherically complete and algebraically closed non-archimedean field with group of values Γ and residue field k_0.

Remark 2.3.3.8 The construction from (iii) is very nice, but I do not know about any application of the fields $k_0((t^\Gamma))$ to non-archimedean geometry. However, existence of a spherically complete closure plays important role in non-archimedean geometry. For example, few approaches to the stable reduction theorem first prove the result over a spherically complete field, thus avoiding some troubles caused by type 4 points, and then establish the general case by a descent argument. It seems that the first such proof is due to van der Put. A similar strategy is also used in the recent work [HL] by Hrushovski-Loeser, that we will recall in Sect. 4.3.3.

3 Affinoid Algebras and Spaces

3.1 Affinoid Algebras

3.1.1 The Definition

Definition 3.1.1.1 (i) A k-*affinoid algebra* $\mathcal{A}$ is a Banach k-algebra that admits an admissible surjective homomorphism from a Banach algebra of the form $k\{\underline{r}^{-1}\underline{T}\}$. We say that $\mathcal{A}$ is *strictly k-affinoid* if one can choose $r_i \in |k^\times|$. More generally, we say that $\mathcal{A}$ is H-*strict* for a group $|k^\times| \subseteq H \subseteq \mathbf{R}_+^\times$ if one can choose such a homomorphism with $r_i \in H$.

(ii) The category of (resp. H-strict, resp. strictly) k-affinoid algebras with bounded morphisms is denoted k-$Af\,Al$ (resp. k_H-$Af\,Al$, resp. st-k-$Af\,Al$). It will also be convenient to say k_H-*affinoid algebra* instead of H-strict k-affinoid algebra.

Exercise 3.1.1.2 Check that H-strictness depends only on the group $\sqrt{H}$ consisting of all elements $h^{1/n}$ with $h \in H$ and integral $n \geq 1$.

Remark 3.1.1.3 The group $\sqrt{H}$ is not dense in $\mathbf{R}_+^\times$ if and only if $H = 1$, and 1-strict spaces are precisely the strictly analytic spaces over a trivially valued field. The case of $H = 1$ is degenerate and often demonstrates a very special behavior. We will ignore it in all cases when it requires a separate argument.

Example/Exercise 3.1.1.4 Let $\underline{r} = (r_1 \ldots r_n)$ be a tuple of positive real numbers linearly independent over $|k^\times|$. Show that the k-affinoid ring

$$K_{\underline{r}} := k\{\underline{r}^{-1}\underline{T}, \underline{r}\,\underline{T}^{-1}\} = k\{\underline{r}^{-1}\underline{T}, \underline{r}\,\underline{S}\}/(T_1 S_1 - 1 \ldots T_n S_n - 1)$$

is a field and $K_{\underline{r}} \xrightarrow{\sim} K_{r_1} \hat{\otimes}_k K_{r_2} \hat{\otimes}_k \ldots \hat{\otimes}_k K_{r_n}$.

3.1.2 Basic Properties

Here is a summary of basic properties of k-affinoid algebras. Excellence was proved very recently by Ducros in [Duc2] (and the strictly affinoid case is due to Kiehl).

Fact 3.1.2.1 (i) Any affinoid algebra $\mathcal{A}$ is noetherian, excellent and all its ideal are closed.

(ii) If $f \in \mathcal{A}$ is not nilpotent then there exists $C > 0$ such that $\|f^n\| \leq C\rho(f)^n$ for all $n \geq 1$. In particular, f is not quasi-nilpotent (i.e. $\rho(f) > 0$), and so ρ is a norm if and only if $\mathcal{A}$ is reduced.

(iii) If $\mathcal{A}$ is reduced then the Banach norm on $\mathcal{A}$ is equivalent to the spectral norm.

(iv) $\mathcal{A}$ is H-strict if and only if $\rho(\mathcal{A}) \subseteq \{0\} \cup \sqrt{H}$.

In particular, $(\mathcal{A}/\mathrm{Ker}(\rho_\mathcal{A}), \rho_\mathcal{A})$ is equivalent to the quotient of $\mathcal{A}$ by its radical (provided with the residue semi-norm). In view of Remark 2.1.2.4, this can be interpreted as equivalence of the "topological reduction" of $\mathcal{A}$ and the usual reduction of $\mathcal{A}$. The following example shows that even naively looking k-Banach algebras do not have to satisfy the same nice conditions.

Example/Exercise 3.1.2.2 Let k be complete non-perfect field with a non-trivial valuation (e.g. $k = \mathbf{F}_p((t))$). Take any element x lying in the completion of the perfect closure of k and non-algebraic over k, e.g. $x = t^{1+1/p} + t^{2+2/p^2} + \ldots$, and let K be the closure of $k(x)$ in $\widehat{k^a}$.

Show that the element $1 \otimes x - x \otimes 1$ is a quasi-nilpotent element of $K \hat{\otimes}_k K$ which is not nilpotent.

Hint: observe that $K = \mathcal{H}(z)$ for a non Zariski closed point $z \in \mathbf{A}_k^1$ of type 1.

For strictly affinoid algebras one can say more.

Fact 3.1.2.3 Let $\mathcal{A}$ be a strictly k-affinoid algebra (where the trivially valued case is allowed).

(i) Noether normalization: there exists a finite admissible injective homomorphism $k\{T_1 \ldots T_n\} \to \mathcal{A}$.
(ii) Hilbert Nullstellensatz: $\mathcal{A} \neq 0$ has a point in a finite extension of k.
(iii) The rings $k\{T_1 \ldots T_n\}$ are of dimension n.

Remark 3.1.2.4 (i) A very systematic and detailed theory of strictly affinoid algebras is developed in chapters 5 and 6 of [BGR] (they are called "affinoid algebras" in loc.cit.). In particular, loc.cit. contains the proof of all claims of Facts 3.1.2.1 and 3.1.2.3 excluding the excellence result. Summarizing in a couple of words, first one develops a Weierstrass theory (preparation and division theorems) for strictly affinoid algebras. As a corollary one deduces analogs of two famous theorems about affine algebras: Noether normalization and Hilbert Nullstellensatz. All these results are used to establish Fact 3.1.2.1 in the strict case.

(ii) Berkovich introduced non-strict algebras in [Ber1] and suggested the following descent trick to deal with them. Obviously, for any k-affinoid algebra $\mathcal{A}$ its base change $\mathcal{A}\hat{\otimes}_k K_r$ is strictly K_r-affinoid for an appropriate K_r. This allows to show that many good properties known to hold for strictly K_r-affinoid algebras also hold for general k-affinoid algebras. In particular, this approach provides a simple reduction of all claims of Fact 3.1.2.1, excluding excellence, to the known strictly affinoid case.

(iii) It was not studied in the literature whether one can develop the whole theory for all affinoid algebras. My expectations are as follows. Weierstrass theory can be developed for all affinoid algebras. Hilbert Nullstellensatz holds in a corrected form that any affinoid $\mathcal{A}$ has a point in a finite extension of some K_r, see [Duc1, Th. 2.7]. I expect that the following weak form of Noether normalization is the best one can get (see Example 6.1.2.1(ii)). There exists injective homomorphisms $f : k\{r_1^{-1}T_1 \ldots r_n^{-1}T_n\} \to \mathcal{A}'$ and $g : \mathcal{A}' \to \mathcal{A}$

such that f is finite admissible and g has dense image (then $\mathcal{M}(\mathcal{A})$ is a Weierstrass domain in a finite surjective covering $\mathcal{M}(\mathcal{A}')$ of a polydisc, as we will later see).

Fact 3.1.2.1 has the following corollary, which is very important when developing the theory of affinoid spaces.

Exercise 3.1.2.5 Assume that $\phi:\mathcal{A} \to \mathcal{B}$ is a bounded homomorphism of k-affinoid algebras, $f_1 \ldots f_n \in \mathcal{B}$ are elements and $r_1 \ldots r_n > 0$ are real numbers. Then ϕ extends to a bounded homomorphism $\psi:\mathcal{A}\{r_1^{-1}T_1 \ldots r_n^{-1}T_n\} \to \mathcal{B}$ with $\psi(T_i) = f_i$ if and only if $\rho_{\mathcal{B}}(f_i) \le r_i$.

Definition 3.1.2.6 Let $\mathcal{A}$ be a k-affinoid algebra. Any Banach $\mathcal{A}$-algebra that admits an admissible surjective homomorphism from $\mathcal{A}\{r_1^{-1}T_1 \ldots r_n^{-1}T_n\}$ is called $\mathcal{A}$-*affinoid*. Obviously, it is also a k-affinoid algebra.

3.1.3 Finite $\mathcal{A}$-Modules

It turns out that the theory of finite Banach $\mathcal{A}$-modules is essentially equivalent to the theory of finite $\mathcal{A}$-modules.

Definition 3.1.3.1 A Banach $\mathcal{A}$-module M is *finite* if it admits an admissible surjective homomorphism from a free module $\mathcal{A}^n$ provided with the norm

$$||(a_1 \ldots a_n)|| = \max_{1 \le i \le n} |a_i|.$$

Fact 3.1.3.2 (i) The categories of finite Banach $\mathcal{A}$ modules and finite $\mathcal{A}$-modules are equivalent via the forgetful functor. In particular, any $\mathcal{A}$-linear map between finite Banach $\mathcal{A}$-modules is admissible.
(ii) Completed tensor product with a finite Banach $\mathcal{A}$-module M coincides with the usual tensor product. Namely, $M \otimes_{\mathcal{A}} N \xrightarrow{\sim} M \hat{\otimes}_{\mathcal{A}} N$ for any Banach $\mathcal{A}$-module N.

Exercise 3.1.3.3 Formulate and prove an analog of Fact 3.1.3.2 for the category of finite $\mathcal{A}$-algebras. In addition, prove that any finite Banach $\mathcal{A}$-algebra is $\mathcal{A}$-affinoid.

3.1.4 Complements

Fact 3.1.4.1 (i) Fibred coproducts exist in the category k_H-$Af\,Al$ and coincide with completed tensor products.
(ii) For any non-archimedean k-field K, the correspondence $\mathcal{A} \mapsto \mathcal{A}\hat{\otimes}_k K$ provides a ground field extension functor k_H-$Af\,Al \to K_{H|K^\times|}$-$Af\,Al$ compatible with completed tensor products.

Fact 3.1.4.2 The ground field k is not trivially valued if and only if any homomorphism between k-affinoid algebras is bounded.

The latter fact was known only for strictly affinoid algebras, so we suggest a proof below.

Exercise 3.1.4.3   (i) Show that any automorphism of the k-field K_r from Example 3.1.1.4 is bounded if and only if k is not trivially valued.

Hint: you have to use arithmetical properties of K_r because $\widehat{K_r^a}$ obviously has a lot of non-bounded automorphisms.

(ii)* Prove Fact 3.1.4.2 in general.

Hint: use the notion of Shilov boundary from Sect. 3.4.1 to show that for any k-affinoid algebra $\mathcal{A}$ with an element f the spectral seminorm $\rho(f)$ can be described as follows. It is the minimal number r such that for any $a \in k^a$ with $|a| > r$ the element $f + a \in \mathcal{A} \hat{\otimes}_k k^a$ possesses a root of any natural degree prime to $\mathrm{char}(\tilde{k})$.

Although Fact 3.1.4.2 is convenient for some applications (especially in rigid geometry), it seems to be rather accidental. We will not use it; anyway, it does not hold for trivially valued ground fields. Here is one more example when trivially valued fields require an additional care. It also shows that the class of finite and admissible homomorphisms of affinoid algebras is the right analog of the class of finite homomorphisms of affine algebras.

Exercise 3.1.4.4   (i) Show that a finite homomorphism is admissible whenever k is not trivially valued, but not in general. Also, give an example of a finite bounded homomorphism which becomes non-finite after a ground field extension.

Hint: $k[T] \to k[T]$ with different norms does the job.

(ii)* Show that a homomorphism of k-affinoid algebras $\phi : \mathcal{A} \to \mathcal{B}$ is finite admissible if and only if its ground field extension $\phi \hat{\otimes}_k K$ is finite and admissible.

Hint: the difficult case is to show the descent. First find K_r such that the algebras become strict over K_r and lift to the completion L of $\mathrm{Frac}(K \otimes_k K_r)$. The descent from $\phi \hat{\otimes}_k L$ to $\phi \hat{\otimes}_k K_r$ is easy because everything is strictly affinoid, and the descent from $\phi \hat{\otimes}_k K_r$ to ϕ can be done by hands, since K_r has a nice explicit description.

3.2   Affinoid Domains

3.2.1   Affinoid Spectra

In the sequel we will develop a theory of k_H-analytic spaces built from spectra of k_H-affinoid spaces (see [CT]). The two extreme choices of H will correspond to the general k-analytic spaces and strictly k-analytic spaces from [Ber2] and [Ber3].

So, from now on an intermediate group $|k^\times| \subseteq H \subseteq \mathbf{R}_+^\times$ is fixed, and if not said otherwise all k-analytic and k-affinoid spaces are assumed to be H-strict.

Remark 3.2.1.1 When the valuation on k is not trivial, it is important to develop the theory of strictly analytic spaces because it has connections to other approaches to non-archimedean geometry: formal geometry over k°, rigid geometry and adic geometry. We prefer to develop the general H-strict theory because it includes both the theory of general k-analytic spaces and the theory of strictly k-analytic spaces, and hence we do not have to distinguish these two cases in some formulations. In addition, this seems to be a slightly more "accurate" approach that keeps track of the used parameters.

The category of k_H-*affinoid spectra* is the category opposite to the category of k_H-affinoid algebras. Its objects are topological spaces of the form $\mathcal{M}(\mathcal{A})$ with a k_H-affinoid algebra $\mathcal{A}$ and morphisms are maps of the form $\mathcal{M}(f)$ for bounded homomorphisms $f : \mathcal{A} \to \mathcal{B}$. We stress that $\mathcal{A}$ is a part of the data forming the affinoid spectrum. The k_H-affinoid spectra are objects of global nature that will later be enriched to more geometrical affinoid spaces. This will be done in the next two sections; we will localize the current construction by introducing an appropriate Grothendieck topology and structure sheaf.

3.2.2 Generalized Normed Localization

Topology on affine schemes is defined by localization. For a k_H-affinoid algebra $\mathcal{A}$ and an element $f \subset \mathcal{A}$, the localization $\mathcal{A}_f$ is not affinoid for an obvious reason— we did not worry to extend the norm. The formula $\mathcal{A}_f = \mathcal{A}[T]/(Tf - 1)$ leads to an idea to consider the k-affinoid algebras $\mathcal{A}_{r^{-1}f} = \mathcal{A}\{rT\}/(Tf - 1)$. It turns out that the latter normed localization is not general enough but its natural extension described below does the job. In the sequel, let $X = \mathcal{M}(\mathcal{A})$ be a k_H-affinoid spectrum.

Definition/Exercise 3.2.2.1 (i) Assume that elements $g, f_1 \ldots f_n \in \mathcal{A}$ do not have common zeros and $r_1 \ldots r_n \in \sqrt{H}$ are positive numbers. Show that

$$\mathcal{A}_V = \mathcal{A}\left\{\underline{r}^{-1}\frac{f}{g}\right\} := \mathcal{A}\{r_1^{-1}T_1 \ldots r_n^{-1}T_n\}/(gT_1 - f_1 \ldots gT_n - f_n)$$

is the universal $\mathcal{A}$-affinoid algebra such that $\rho_{\mathcal{A}_V}(f_i) \le r_i \rho_{\mathcal{A}_V}(g)$ for $1 \le i \le n$. Deduce that the map $\phi_V : \mathcal{M}(\mathcal{A}_V) \to X$ is a bijection onto

$$V = X\left\{\underline{r}^{-1}\frac{f}{g}\right\} := \{x \in X \mid |f_i(x)| \le r_i|g(x)|, \; 1 \le i \le n\}$$

Show that ϕ_V satisfies the following universal property: any morphism of k-affinoid spectra $\mathcal{M}(\mathcal{B}) \to X$ with image in V factors through $\mathcal{M}(\mathcal{A}_V)$.

The compact subset V is called an H-strict *rational domain* in X and by the universal property one can identify it with the k_H-affinoid spectrum $\mathcal{M}(\mathcal{A}_V)$.

(ii) For any choice of $g_1 \ldots g_m, f_1 \ldots f_n \in \mathcal{A}$ and $s_1 \ldots s_m, r_1 \ldots r_n \in \sqrt{H}$ introduce analogous domains

$$V = X\{\underline{r}^{-1}\underline{f}, \underline{s}\,\underline{g}^{-1}\} = \{x \in X \mid |f_i(x)| \leq r_i, |g_j(x)| \geq s_j\}$$

with algebras

$$\mathcal{A}_V = \mathcal{A}\{\underline{r}^{-1}\underline{f}, \underline{s}\,\underline{g}^{-1}\} = \mathcal{A}\{\underline{r}^{-1}\underline{T}, \underline{s}\,\underline{R}\}/(T_i - f_i, g_j R_j - 1)$$

and prove the analogs of all properties from (i). These H-strict domains are called *Laurent* (resp. *Weierstrass* in the case when $m = 0$).

(iii) Show that any Laurent domain is rational and Laurent domains form a fundamental family of neighborhoods of a point whenever $H \neq 1$. Show that the latter is false for $H = 1$.

(iv) Show that $V \subseteq X$ is an H-strict rational, Laurent or Weierstrass domain if and only if it is a domain of the same type and the k-affinoid algebra $\mathcal{A}_V$ is H-strict.

Hint: use Fact 3.1.2.1.

(v) For any map of k-affinoid spectra the preimage of a rational, Laurent or Weierstrass domain is a domain of the same type and given by the same inequalities. In particular, all three classes of domains are closed under finite intersections.

(vi) Show that the classes of rational and Weierstrass domains are transitive (e.g. if Y is rational in X and Z is rational in Y then Z is rational in X), but Laurent domains are not transitive. Actually, this transitivity property is the main reason to extend the class of Laurent domains.

Although, we had to consider a more general type of localizations than in the theory of affine algebra, the main difference with the theory of varieties is that affinoid domains are compact and hence have to be closed. This fact will have serious consequences when we will develop the theory of coherent sheaves.

Example/Exercise 3.2.2.2 (i) Let $X = \mathcal{M}(k\{\underline{r}^{-1}\underline{T}\})$ be a polydisc with center at 0 and of polyradius $\underline{r}$, let $s_i \leq r_i$ be positive numbers, and let $a_i \in k$ be elements with $|a_i| \leq r_i$. Then the polydisc $\mathcal{M}(k\{s_1^{-1}(T_1 - a_1) \ldots s_n^{-1}(T_n - a_n)\})$ with center at $\underline{a}$ and of polyradius $\underline{s}$ is a Weierstrass domain in X.

(ii) For $s \leq r$ the annulus $A(0; s, r) = \mathcal{M}(k\{r^{-1}T, sT^{-1}\})$ is a Laurent but not Weierstrass domain in the disc $E(0, r) = \mathcal{M}(k\{r^{-1}T\})$.

(iii) Any finite union of discs in $E(0, r)$ is a Weierstrass domain (and is a disjoint union of finitely many discs). In particular, even when $\mathcal{A}$ is an integral domain, its generalized localization does not have to be integral.

3.2.3   General Affinoid Domains

It is difficult to describe a general open affine subscheme explicitly but one can easily characterize it by a universal property. Here is an affinoid analog, which was already checked for rational domains in 3.2.2.1.

Definition 3.2.3.1 A closed subset $V \subseteq X$ is called a k_H-*affinoid domain* if there exists a morphism of k_H-affinoid spectra $\phi{:}\mathcal{M}(\mathcal{A}_V) \to X$ whose image coincides with V and such that any morphism of k_H-affinoid spectra $\mathcal{M}(\mathcal{B}) \to X$ with image in V factors through $\mathcal{M}(\mathcal{A}_V)$.

Note that $\mathcal{A}_V$ is unique up to a canonical isomorphism. The following fact allows us to identify V with the k_H-affinoid spectrum $\mathcal{M}(\mathcal{A}_V)$.

Fact 3.2.3.2 The non-empty fibers of ϕ are isomorphisms, i.e. ϕ is a bijection onto V and for any $y \in \mathcal{M}(\mathcal{A}_V)$ we have that $\mathcal{H}(\phi(y)) \overset{\sim}{\to} \mathcal{H}(y)$.

Exercise 3.2.3.3   (i) Prove Fact 3.2.3.2 for a point y with $[\mathcal{H}(y) : k] < \infty$.
 (ii)* Prove Fact 3.2.3.2 in general.
      Hint: first prove that $\mathcal{A}_V \hat{\otimes}_A \mathcal{A}_V \overset{\sim}{\to} \mathcal{A}_V$ (in particular, the separated completion homomorphism $\mathcal{A}_V \otimes_A \mathcal{A}_V \to \mathcal{A}_V \hat{\otimes}_A \mathcal{A}_V$ usually has a huge kernel); also, use without proof a non-trivial result of Gruson that the completion homomorphism $\mathcal{B} \otimes_k \mathcal{B} \to \mathcal{B} \hat{\otimes}_k \mathcal{B}$ is injective for any k-Banach algebra $\mathcal{B}$.
 (iii)* Show that V is Weierstrass if and only if the image of the homomorphism $\mathcal{A} \to \mathcal{A}_V$ is dense.
      Hint: first you should establish the following very useful fact about affinoid generators. Assume that $\phi{:}k\{r_1^{-1}T_1 \ldots r_n^{-1}T_n\} \to \mathcal{A}$ is an admissible surjection, $\| \ \|$ is the residue norm on $\mathcal{A}$ and $f_i = \phi(T_i)$. Then there exists $\varepsilon > 0$ such that for any choice of $f_i' \ldots f_n'$ with $\| f_i - f_i' \| < \varepsilon$ there exists an admissible surjection $\phi'{:}k\{r_1^{-1}T_1 \ldots r_n^{-1}T_n\} \to \mathcal{A}$ with $\phi'(T_i) = f_i'$.

The following result shows that our definition of generalized localization was general enough. It allows to use rational domains for all local computations on affinoid spectra.

Fact 3.2.3.4 (Gerritzen-Grauert theorem) Any k_H-affinoid domain $V \subseteq X$ is a finite union of H-strict rational domains $V_i \subseteq X$.

Remark 3.2.3.5   (i) This result (in the strict case) was not available in the first version of rigid geometry due to Tate. For this reason, Tate simply worked with rational domains and did not consider the general affinoid domains. In rigid geometry, the theorem was proved by Gerritzen-Grauert, and in Berkovich geometry the non-strict case was first deduced by Ducros. The two known rigid-theoretic proofs of this result are rather long and difficult. Originally, this theorem was needed to develop the very basics of analytic geometry, including Fact 3.2.3.2. Later it was shown in [Tem3] that Fact 3.2.3.2 can be proved independently (via the hint from Exercise 3.2.3.3(ii)), and then Gerritzen-Grauert theorem can be deduced rather shortly. In addition, one obtains a similar description of all monomorphisms in the category of affinoid spaces.

(ii) To be slightly more precise, the above form of the theorem is missing in the literature. It is proved in [Tem3] that one can choose the V_i's to be rational domains. The fact that the V_i's may be also chosen to be H-strict requires a simple additional argument. For $H \neq 1$ this is done similarly to the proof of [CT, 7.3], and for $H = 1$ one can deduce this from an analogous result for schemes (because for a trivially valued k, the category of k_1-affinoid algebras is equivalent to the category of finitely generated k-algebras.)

3.3 G-Topology and the Structure Sheaf

From now on we assume that $H \neq 1$.

3.3.1   G-Topology of Compact Domains

In order to define k-affinoid spaces we should provide each spectrum $X = \mathcal{M}(\mathcal{A})$ with a certain structure sheaf $\mathcal{O}_X$. Naturally, we would like $\mathcal{O}_X$ to be a sheaf of k-affinoid or k-Banach algebras but then we should study the sections over closed subsets (e.g. affinoid domains). A naive attempt to consider the topology generated by affinoid domains does not work out.

Exercise 3.3.1.1 Observe that the unit interval $[0, 1]$ is neither connected nor compact in the topology generated by closed intervals $[a, b]$. Show, similarly, that the unit closed disc $\mathcal{M}(k\{T\})$ is neither connected nor compact in the topology generated by affinoid domains.

A brilliant idea of Tate (with a strong influence of Grothendieck) is to generalize the notion of topology by allowing only certain open coverings. The resulting notion of a G-topology τ is simply a Grothendieck topology on a set τ^{op} of subsets of X such that τ^{op} is closed under finite intersections and any covering of this topology is also a set-theoretical covering. Sets of τ^{op} are also called τ-open and the coverings of this Grothendieck topology are called *admissible coverings*. (Note that X is in τ^{op} because it is the intersection indexed by the empty set.)

Definition 3.3.1.2 (i) A *compact k_H-analytic domain* Y in a k_H-affinoid spectrum X is a finite union of k_H-affinoid domains. (It is called a special domain in [Ber1].)

(ii) The *H-strict compact G-topology* τ_H^c on a k_H-affinoid spectrum X is defined as follows: $(\tau_H^c)^{op}$ is the set of all compact k_H-analytic domains and admissible coverings are the finite ones.

Remark 3.3.1.3 By Gerritzen-Grauert theorem one can replace affinoid domains with rational domains in this definition obtaining the original Tate's definition of G-topology.

3.3.2 The Structure Sheaf

Tate proved that (in the strict case) this G-topology is the right tool to define coherent sheaves of modules. In particular, $\mathcal{O}_{X_H}(V) = \mathcal{A}_V$ extends to a τ_H^c-sheaf of Banach algebras.

Fact 3.3.2.1 (Tate's acyclity theorem) For any finite affinoid covering $X = \cup_i V_i$ and finite Banach $\mathcal{A}$-module M the Čech complex

$$0 \to M \to \prod_i M_i \to \prod_{i,j} M_{ij} \to \dots$$

is exact and admissible, where $M_i = M \otimes_{\mathcal{A}} \mathcal{A}_{V_i}$, $M_{ij} = M \otimes_{\mathcal{A}} \mathcal{A}_{V_{ij}}$, etc.

Admissibility in this result is very important. In particular, it allows to define norms on the structure sheaf introduced below.

Exercise 3.3.2.2 (i) For any compact k_H-analytic domain V with a finite affinoid covering $V = \cup_i V_i$ set $\mathcal{O}_{X_H}(V) = \mathrm{Ker}(\prod_i \mathcal{A}_{V_i} \to \prod_{i,j} \mathcal{A}_{V_i \cap V_j})$ and provide it with the restriction norm. Show that $\mathcal{O}_{X_H}(V)$ depends only on V, in particular, $\mathcal{O}_{X_H}(V) = \mathcal{A}_V$ when V is affinoid. In addition, show that $\mathcal{O}_{X_H}$ is a sheaf of k-Banach algebras, in particular, the restriction morphisms are bounded.

(ii) Deduce that any polydisc $X = \mathcal{M}(k\{r_1^{-1}T_1 \dots r_n^{-1}T_n\})$ with $r_i \in \sqrt{H}$ is τ_H^c-connected, i.e. X is not a disjoint union of two non-empty compact k_H-analytic domains.

Hint: $\mathcal{O}_{X_H}(X)$ is integral.

Fact 3.3.2.3 A compact k_H-analytic domain $V \subseteq X = \mathcal{M}(\mathcal{A})$ is an affinoid domain if and only if the Banach algebra $\mathcal{A}_V = \mathcal{O}_{X_H}(V)$ is k_H-affinoid and the natural map of sets $\phi_V : V \to \mathcal{M}(\mathcal{A}_V)$ is bijective. In particular, any k_H-affinoid domain V is a k-affinoid domain.

Exercise 3.3.2.4 (i) Define the map ϕ_V in Fact 3.3.2.3.

Hint: take a finite affinoid covering $V = \cup_{i=1}^n V_i$ and show that the affinoid morphisms $V_i \to \mathcal{M}(\mathcal{A}_V)$ are compatible on intersections.

(ii)* Prove Fact 3.3.2.3.

Hint: use Gerritzen-Grauert and Tate's acyclity theorems; the main stage is to show that if $U = X\{\underline{r}^{-1}\underline{f}/g\}$ is a rational domain contained in V then $U \xrightarrow{\sim} \mathcal{M}(\mathcal{A}_V\{\underline{r}^{-1}\underline{f}/g\})$.

3.3.3 k_H-Affinoid Spaces

Definition 3.3.3.1   (i) A k_H-*affinoid space* X is a k_H-affinoid spectrum $\mathcal{M}(\mathcal{A})$
provided with the G-topology τ_H^c and the τ_H^c-sheaf of k-Banach rings $\mathcal{O}_{X_H}$,
which is called the *structure sheaf*.

 (ii) A *morphism* of k_H-affinoid spaces is a continuous and τ_H^c-continuous map
$f{:}Y \rightarrow X$ provided with a *bounded* homomorphism of sheaves $f^\#{:}\mathcal{O}_{X_H} \rightarrow$
$f_*(\mathcal{O}_{Y_H})$ in the sense that for any pair of compact k_H-analytic domains
$X' \subseteq X$ and $Y' \subseteq f^{-1}(X)$ the homomorphism $f^\#{:}\mathcal{O}_{X_H}(X') \rightarrow \mathcal{O}_{Y_H}(Y')$
is bounded.

(iii) A τ_H^c-sheaf of finite $\mathcal{O}_{X_H}$-Banach modules is *coherent* if it is of the form
$\mathcal{M}(V) = M \otimes_{\mathcal{A}} \mathcal{O}_{X_H}(V)$ for a finite Banach $\mathcal{A}$-module M.

Exercise 3.3.3.2 Show that the continuity assumption in (ii) follows from τ_H^c-
continuity and hence can be removed.

Note that k_H-affinoid spaces are "self-contained" geometric objects analogous
to affine schemes. It will later be an easy task to globalize this definition.

Fact 3.3.3.3 The categories of k_H-affinoid spectra and k_H-affinoid spaces are
naturally equivalent.

Since this fact seems to be new, we give a detailed exercise on its proof.

Exercise 3.3.3.4   (i) Reduce the previous fact to the following claim: if
$(f, f^\#){:}(Y, \mathcal{O}_{Y_H}) \rightarrow (X, \mathcal{O}_{X_H})$ is a morphism of k_H-affinoid spaces and
$\phi{:}\mathcal{O}_{X_H}(X) \rightarrow \mathcal{O}_{Y_H}(Y)$ is the induced bounded homomorphism of k_H-
affinoid algebras, then $f = \mathcal{M}(\phi)$ and $f^\#{:}\mathcal{O}_{X_H} \rightarrow f_*(\mathcal{O}_{Y_H})$ is the bounded
homomorphism of the structure sheaves induced by ϕ. In other words, $(f, f^\#)$
is, in its turn, induced by ϕ.

 (ii) Choose a point $y \in Y$ with $x = f(y)$. For any H-strict rational domain $X' = X\{r^{-1}\frac{g}{h}\}$ containing x choose an H-strict rational domain $Y' \subseteq f^{-1}(X')$
containing y. Set $\mathcal{A} = \mathcal{O}_{X_H}(X)$, $\mathcal{B} = \mathcal{O}_{Y_H}(Y)$ and $\mathcal{B}' = \mathcal{O}_{Y_H}(Y')$. Since
the homomorphism $\mathcal{A} \rightarrow \mathcal{B}'$ factors through $\mathcal{O}_{X_H}(X') = \mathcal{A}\{r^{-1}\frac{g}{h}\}$, it follows
that the homomorphism $\mathcal{B} \rightarrow \mathcal{B}'$ factors through $\mathcal{B}\hat{\otimes}_{\mathcal{A}}\mathcal{A}\{r^{-1}\frac{g}{h}\} \xrightarrow{\sim} \mathcal{B}\{r^{-1}\frac{g}{h}\}$,
and hence $Y' \subseteq Y\{r^{-1}\frac{g}{h}\}$. Therefore, any inequality $|g(x)| \leq r|h(x)|$ with
$g, h \in \mathcal{O}_{X_H}(X)$ implies that $|\phi(g)(y)| \leq r|\phi(h)(y)|$. Deduce that $\mathcal{M}(\phi)$
takes y to x and hence coincides with f.

(iii) Finish the argument by showing that $f^\#$ is also induced by ϕ.

Hint: check this for sections on rational domains and then apply Tate's
theorem.

In the sequel we will not distinguish between k_H-affinoid spectra and k_H-affinoid
spaces, that is, we will automatically enrich any k_H-affinoid spectrum with the
structure of a k_H-affinoid space. Also, we will refine the structure of k_H-affinoid
spaces a little bit more in Sect. 4.1.

3.3.4 Coherent Sheaves

We finish Sect. 3.3 with a discussion on coherent sheaves. By Tate's theorem any finite Banach $\mathcal{A}$-module gives rise to a sheaf of finite Banach $\mathcal{O}_{X_H}$-modules. The opposite result (in a slightly different formulation) was proved by Kiehl.

Fact 3.3.4.1 (i) (Kiehl's theorem) Any G-locally coherent sheaf is coherent. Namely, if for a finite Banach $\mathcal{O}_{X_H}$-module $\mathcal{M}$ there exists a finite affinoid covering $X = \cup_i V_i$ such that the restrictions $\mathcal{M}|_{V_i}$ are coherent then $\mathcal{M}$ is coherent.

(ii) Tate's and Kiehl's theorems easily imply that the categories of coherent $\mathcal{O}_{X_H}$-modules and finite Banach $\mathcal{A}$-modules are naturally equivalent.

Remark 3.3.4.2 The theory of k_H-analytic (and even k_H-affinoid) spaces does not admit a reasonable class of infinite type modules analogous to the class of quasi-coherent modules on schemes. Moreover, this theory does not even have a notion of affinoid morphisms. There exist morphisms $f : Y \to X$ with finite affinoid coverings $X = \cup_i X_i$ such that X and $f^{-1}(X_i)$ are affinoid but Y is not affinoid, see Example 4.2.1.4(ii).

3.4 The Reduction Map, Boundary and Interior

3.4.1 Reduction

In general, reduction relates (strictly) k-affinoid algebras and spaces to geometry over the residue field $\tilde{k}$.

Fact 3.4.1.1 Assume that the valuation is non-trivial. For a bounded homomorphism $\phi : \mathcal{A} \to \mathcal{B}$ of strictly k-affinoid algebras the following conditions are equivalent: ϕ is finite, $\tilde{\phi} : \tilde{\mathcal{A}} \to \tilde{\mathcal{B}}$ is finite, $\phi^\circ : \mathcal{A}^\circ \to \mathcal{B}^\circ$ is integral.

Exercise 3.4.1.2 Deduce that for any strictly k-affinoid $\mathcal{A}$ the reduction $\tilde{\mathcal{A}}$ is finitely generated over $\tilde{k}$.

Remark 3.4.1.3 (i) This result shows that the reduction functor controls strictly k-affinoid algebras very well. A similar result holds for general k_H-affinoid algebras if one replaces $\tilde{\mathcal{A}}$ with the H-graded reduction

$$\tilde{\mathcal{A}}_H = \oplus_{h \in H} \{x \in \mathcal{A} | \, \rho(x) \leq h\}/\{x \in \mathcal{A} | \, \rho(x) < h\}$$

(ii) The question whether ϕ° is finite is more subtle. Already for a finite field extension l/k it often happens that l°/k° is not finite. On the other hand if K is algebraically closed or discretely valued and the algebras are reduced then ϕ° is integral if and only if it is finite. We refer the reader to [BGR, Ch 6, §§3–4] for more details.

Now, let us study the geometric side of the reduction.

Definition 3.4.1.4 (i) *Reduction* of a strictly k-affinoid space $X = \mathcal{M}(\mathcal{A})$ is the reduced $\tilde{k}$-variety $\tilde{X} = \mathrm{Spec}(\tilde{\mathcal{A}})$.

(ii) *Reduction map* $\pi_X : X \to \tilde{X}$ sends a point x with the character $\chi_x : \mathcal{A} \to \mathcal{H}(x)$ to the point $\tilde{x} \in \tilde{X}$ induced by the character $\tilde{\chi}_x : \tilde{\mathcal{A}} \to \widetilde{\mathcal{H}(x)}$. (Note that $k(\tilde{x})$ can be much smaller than $\widetilde{\mathcal{H}(x)}$; as we know from Exercise 2.3.3.3(5) the latter field does not even have to be finitely generated over $\tilde{k}$.)

Exercise 3.4.1.5 The map π_X is *anti-continuous* in the sense that the preimage of an open set is closed and vice versa.

Fact 3.4.1.6 (i) The reduction map is surjective.

(ii) The preimage of a generic point of $\tilde{X}$ is a single point, and the union of all such points is the *Shilov boundary* $\Gamma(X)$ of X. Namely, any function $|f(x)|$ with $f \in \mathcal{A}$ accepts its maximum on $\Gamma(X)$ and $\Gamma(X)$ is the minimal closed set satisfying this property.

Remark 3.4.1.7 The same result holds for H-graded reduction if one defines $\tilde{X}_H = \mathrm{Spec}_H(\tilde{\mathcal{A}}_H)$ as the set of all homogeneous prime ideals in the H-graded ring $\tilde{\mathcal{A}}_H$.

Example 3.4.1.8 (i) The spectral seminorm on $\mathcal{A}$ is multiplicative if and only if $\tilde{\mathcal{A}}$ is integral. In this case, the spectral seminorm itself defines a point which is both the preimage of the generic point of $\tilde{X}$ and the Shilov boundary of X. For example, the Shilov boundary of a polydisc $E(0, \underline{r}) = \mathcal{M}(k\{\underline{r}^{-1}\underline{T}\})$ is a single (maximal) point.

(ii) The Shilov boundary of a closed annulus $X = A(0; s, r) = E(0, r) \setminus D(0, s) = \mathcal{M}(k\{r^{-1}T, sT^{-1}\})$ with $s < r$ consists of two points, the maximal points of the closed discs $E(0, r)$ and $E(0, s)$. Assume that $s, r \in |k^\times|$, then the reduction $\tilde{X}$ is the cross $\mathrm{Spec}(\tilde{k}[R, S]/(RS))$, where R and S are the reductions of appropriate rescalings of T and T^{-1}. The points collide when s tends to r, namely, $\Gamma(A(0; r, r))$ consists of a single point and the reduction is $\tilde{k}[R, S]/(RS - 1) = \tilde{k}[\tilde{T}, \tilde{T}^{-1}]$ in this case.

3.4.2 Relative Boundary and Interior

Definition 3.4.2.1 Let $\phi : Y \to X$ be a morphism of k-affinoid spaces and let $X = \mathcal{M}(\mathcal{A})$ and $Y = \mathcal{M}(\mathcal{B})$.

The *relative interior* $\mathrm{Int}(Y/X) \subseteq Y$ consists of points $y \in Y$ such that there exists an admissible surjective $\mathcal{A}$-homomorphism $\psi : \mathcal{A}\{r_1^{-1}T_1 \ldots r_n^{-1}T_n\} \to \mathcal{B}$ with $|\psi(T_i)(y)| < r_i$ for $1 \leq i \leq n$.

The *relative boundary* $\partial(Y/X)$ is the complement to the relative interior in Y.

The *absolute interior* $\mathrm{Int}(Y)$ and *boundary* $\partial(Y)$ are defined with respect to the morphism $Y \to \mathcal{M}(k)$.

A morphism is said to be boundaryless if its relative boundary is empty.

Remark 3.4.2.2 (i) This definition has a very geometric interpretation as follows. The homomorphism ψ induces a closed immersion of Y into a relative closed polydisc of polyradius $\underline{r}$ over X. The inequalities in the definition mean that the image of y lies in the open relative polydisc of the same polyradius.

(ii) Realization of boundary as a set is specific to Berkovich analytic geometry. For example, we will later see that boundaryless morphisms of general spaces form a very important class (e.g. any proper morphism is boundaryless). This class was introduced in the classical rigid geometry in a rather formal way, but it was not so clear how to describe obstructions for being boundaryless (note that analytic boundary consists of non-rigid points). To some extent this obstruction could be felt by working with formal models, but a concrete notion of boundary was missing.

The notions of relative boundary and interior turn out to be very important in analytic geometry. Most of the facts about them are proved by use of the reduction theory. Here is a list of their basic properties.

Fact 3.4.2.3 (i) The relative interior is open in Y and the relative boundary is closed.

(ii) The relative interiors are compatible with compositions in the sense that $\text{Int}(Z/X) = \text{Int}(Z/Y) \cap \psi^{-1}(\text{Int}(Y/X))$ for a pair of morphism $Z \xrightarrow{\psi} Y \xrightarrow{\phi} X$.

(iii) Relative boundary is G-local on the base in the sense that for a finite affinoid covering $X = \cup_i X_i$ and $Y_i = \phi^{-1}(X_i)$ one has that $\partial(Y/X) = \cup_i \partial(Y_i/X_i)$.

(iv) ϕ is *boundaryless*, i.e. has an empty boundary, if and only if it is *finite*, i.e. $\mathcal{A} \to \mathcal{B}$ is finite admissible.

(v) If Y is an affinoid domain in X then $\text{Int}(Y/X)$ is the topological interior of Y in X.

(vi) Assume that X and Y are strictly k-affinoid, $y \in Y$ and $x = \phi(y)$. Then $y \in \text{Int}(Y/X)$ if and only if the image of $\tilde{\mathcal{B}}$ in $k(\tilde{y})$ (usually denoted $\tilde{\chi}_y(\tilde{\mathcal{B}})$) is finite over the image of $\tilde{\mathcal{A}}$ in $k(\tilde{x})$.

Remark 3.4.2.4 The last condition is very convenient for explicit computations. Its analog holds for H-strict X and Y and H-graded reduction.

Now, we illustrate the introduced notions with some examples. For simplicity, all analytic spaces in the examples are assumed to be strict.

Example/Exercise 3.4.2.5 (i) Show that $\text{Int}(Y)$ is the preimage under the reduction map π_Y of the set of closed points of $\tilde{Y}$.

(ii) Let X be an *affinoid curve*, i.e. a k-affinoid space of dimension one (in the sense of Sect. 3.5 below) and without isolated Zariski closed points. Show that the boundary of X coincides with its Shilov boundary.

Hint: use Noether's normalization to prove that $\tilde{X}$ is a curve.

Next, let us consider a higher dimensional example.

Example/Exercise 3.4.2.6 Let ϕ be the projection of the unit polydisc $Y = \mathcal{M}(k\{T, S\})$ onto the unit disc $X = \mathcal{M}(k\{S\})$. We will compare $\partial(Y)$ and $\partial(Y/X)$ fiberwise over X. Let $x \in X$ be a point and let $Y_x = \psi^{-1}(x)$ denote the fiber over x, so that $Y_x = \mathcal{M}(\mathcal{H}(x)\{T\}) = E_{\mathcal{H}(x)}(0, 1)$ is a closed unit disc over $\mathcal{H}(x)$.

  (i) The fiber disc Y_x contains exactly one point in its boundary $\partial(Y_x)$, the maximal point of the disc.
 (ii) If x is not the maximal point of X then the sets $Z = \partial(Y/X) \cap Y_x$ and $\partial(Y) \cap Y_x$ coincide by Fact 3.4.2.3(ii). A non-maximal point of Y_x is contained in Z if and only if it is contained in a subdisc $E_{\mathcal{H}(x)}(a, r)$ of Y_x with $r < 1$ and $\inf |a - k^a| = 1$. In particular, if x is of type 1,3 or 4 then there are no such points and so $Z = \partial(Y_x)$, but if x is of type 2 then Z contains infinitely many open unit subdiscs.
(iii) If x is the maximal point of X then $\partial(Y/X) \cap Y_x$ is much smaller than $\partial(Y) \cap Y_x$. For example, let η and ε be the generic points of the quadrics $\tilde{T}^2 - \tilde{S} = 0$ and $\tilde{T}\tilde{S} = 1$ in $\tilde{Y} = \mathrm{Spec}(\tilde{k}[\tilde{T}, \tilde{S}])$. Show that $\pi_Y^{-1}(\eta) \subset \mathrm{Int}(Y/X)$ but $\pi_Y^{-1}(\varepsilon) \subset \partial(Y/X)$. Find a geometric explanation for this fact.

   Hint: how are these fibers embedded into a larger polydisc $\mathcal{M}(k\{r^{-1}T, S\})$ with $r > 1$?

3.5   The Dimension Theory

In the strict case one can just copy the dimension theory of affine varieties. However, we will see that in general one has to be slightly more careful.

Exercise 3.5.0.1   (i) Prove that the dimension of a strictly k-affinoid algebra $\mathcal{A}$ is preserved under any ground field extension.
     Hint: use Noether's normalization.
 (ii) Show that general k-affinoid algebras do not share this property.
     Hint: show that $K_r \hat{\otimes}_k K_r \xrightarrow{\sim} K_r\{r^{-1}T\} \xrightarrow{\sim} K_r\{T\}$.

Since the dimension stabilizes after a sufficiently large ground field extension, it is natural to define the dimension of k-affinoid spaces as follows.

Definition 3.5.0.2   Dimension $\dim(X)$ of a k-affinoid space $X = \mathcal{M}(\mathcal{A})$ is the dimension of an algebra $\mathcal{A}_K = \mathcal{A} \hat{\otimes}_k K$, where K is a non-archimedean k-field such that $\mathcal{A}_K$ is strictly K-affinoid.

Remark 3.5.0.3   Assume for concreteness that $r = (r_1)$. One can view $X = \mathcal{M}(K_r)$ as a k-curve consisting of its generic point. The only difference with the theory of algebraic k-curves is that X is of "finite type" over k.

   The following exercise illustrates that type 2, 3 and 4 points are sort of "generic points" of the curves, so (informally) one can imagine them as points of dimension 1.

Exercise 3.5.0.4 Let x be a point of $\mathbf{A}_k^1$.

(i) Show that x is of type 1 if and only if the fiber over x in any ground field extension $\mathbf{A}_K^1$ is profinite.
(ii) Show that x is not of type 1 if and only if the fiber over x in some $\mathbf{A}_K^1$ contains a closed K-disc.

4 Analytic Spaces

4.1 The Category of k-Analytic Spaces

4.1.1 Nets

Definition 4.1.1.1 Let X be a topological space with a set of subsets T.

(i) T is a *quasi-net* if any point $x \in X$ has a neighborhood of the form $\cup_{i=1}^n V_i$ with $x \in V_i \in T$ for $1 \leq i \leq n$.
(ii) A quasi-net T is a *net* if for any choice of $U, V \subset T$ the restriction $T|_{U \cap V} = \{W \in T \mid W \subseteq U \cap V\}$ is a quasi-net of subsets of the topological space $U \cap V$.

We remark that the definition of nets axiomatizes properties an affinoid atlas should satisfy, and the definition of quasi-nets axiomatizes the properties of admissible coverings by analytic domains (see Remark 4.1.2.3 below). Now, let us assume that X is locally Hausdorff and the sets of T are compact (as will be the case with analytic spaces and their atlases).

Exercise 4.1.1.2 (i) With the above assumptions, X is locally compact.
(ii) A subset $Y \subseteq X$ is open if and only if $Y \cap U$ is open in U for any $U \in T$.
(iii) X is Hausdorff if and only if for any pair $U, V \in T$ the intersection $U \cap V$ is compact.
(iv) A subset $Y \subseteq X$ is compact if and only if it is covered by compact intersections of the form $Y \cap U$ with $U \in T$ (but there may exist non-compact intersections with other elements of T).

Remark 4.1.1.3 Nets in our sense are slightly analogous to ε-nets on metric spaces, and they should not be confused with the nets that generalize sequences in the definition of limits.

4.1.2 Analytic Spaces

We will freely view a net as a category with morphisms being the inclusions.

Definition 4.1.2.1 A k_H-*analytic space* is a locally Hausdorff topological space X with an *atlas of k_H-affinoid domains*, where the atlas consists of a net τ_0 on X, a functor $\phi : \tau_0 \to k_H$-*Aff* taking morphisms in τ_0 to embeddings of affinoid domains, and an isomorphism i of the two natural topological realization functors from τ_0 to the category of topological spaces, Top and $Top \circ \phi$. In concrete terms, we will write $\phi(U) = \mathcal{M}(\mathcal{A}_U)$, and i reduces to giving a homeomorphism $i_U : U \xrightarrow{\sim} \phi(U)$ for any $U \in \tau_0$ such that for any inclusion $j : U \hookrightarrow V$ in τ these homeomorphisms are compatible with $\phi(j) : \phi(U) \to \phi(V)$.

By Exercise 4.1.1.2(i) X is locally compact.

Definition 4.1.2.2 (i) A k_H-*analytic domain* in X is a subset $V \subseteq X$ that admits a covering $V = \cup_{i \in I} V_i$ such that each V_i is a k_H-affinoid domain in some element of τ_0 and $\{V_i\}_{i \in I}$ is a quasi-net on V.

(ii) By τ_H (resp. τ_H^c) we denote the sets of all (resp. compact) k_H-analytic domains. A covering of an element $V \in \tau_H$ by elements $V_i \in \tau_H$ is *admissible* if V_i's form a quasi-net on V.

Remark 4.1.2.3 Note that a closed unit disc X is a disjoint union of the open unit disc $D(0, 1)$ and the closed annulus $A(0; 1, 1)$. However, the covering $X = D(0, 1) \coprod A(0; 1, 1)$ is not admissible thanks to the condition $x \in \cap_{i=1}^n V_i$ in the definition of quasi-nets. In particular, one can easily show that X is τ_H-connected. This explains the role of the first condition in the definition of quasi-nets, and the second condition is needed to ensure that analytic spaces are locally compact.

Exercise 4.1.2.4 (i) Show that τ_H with admissible coverings is a G-topology and give an example where it is not closed under finite unions. In the sequel we will refer to τ_H as the G-*topology of* X.

Hint: already the union of the open polydisc of polyradius $(1, 2)$ with the closed polydisc of polyradius $(2, 1)$ is not locally compact.

(ii) Show that although τ_H^c does not have products in general (e.g. X is not in τ_H^c if it is not compact and τ_H^c is not closed under intersections of pairs for a non-Hausdorff space), it is closed under fibred products (i.e. if $U, V \subset W$ are three elements of τ_H^c then $U \cap V$ is in τ_H^c). Use this to define τ_H^c-sheaves.

(iii) Show that any admissible covering of a compact k_H-analytic domain by compact k_H-analytic domains possesses a finite subcovering. Deduce that $V \in \tau_H^c$ if and only if $V = \cup_{i=1}^n V_i$ with each V_i being k_H-affinoid in an element of τ_0 and all intersections $V_i \cap V_j$ being compact.

(iv) Deduce that the correspondence $V \to \mathcal{A}_V$ on τ_0 extends uniquely to a τ_H^c-sheaf of Banach k-algebras.

Hint: find an affinoid covering as in (iii), set $\mathcal{A}_V = \mathrm{Ker}(\prod_i \mathcal{A}_{V_i} \to \prod_{i,j} \mathcal{A}_{V_{ij}})$, and use Tate's acyclity theorem to establish independence of the covering.

The sheaf from (iv) will be called the *structure sheaf* and denoted $\mathcal{O}_{X_H}$. Any k_H-analytic space given by an atlas will automatically be provided with this additional structure.

4.1.3 Morphisms Between Analytic Spaces

Intuitively, a morphism between k_H-analytic spaces should be a continuous and G-continuous map $f:Y \to X$ (i.e. the preimage of an analytic domain is an analytic domain) provided with a bounded homomorphism $f^\#:\mathcal{O}_{X_H} \to f_*\mathcal{O}_{Y_H}$. However, the direct image $f_*\mathcal{O}_{Y_H}$ does not really make sense for non-compact morphisms, including $\mathbf{A}_k^1 \to \mathcal{M}(k)$. Therefore, we suggest the following definition.

Definition 4.1.3.1 A morphism $f:Y \to X$ between k_H-analytic spaces consists of a continuous and G-continuous map $f:Y \to X$ and a compatible family of bounded homomorphisms $f_{U,V}^\#:\mathcal{O}_{X_H}(U) \to \mathcal{O}_{Y_H}(V)$ for any pair of compact k_H-analytic domains $U \subseteq X$ and $V \subseteq f^{-1}(U)$.

A priori it is not clear how to compose such morphisms because the image of a compact set does not have to be Hausdorff. This forces us to show that a morphism is determined already by its restriction to atlases. (Note that the atlas definition of morphisms is used in [Ber2, §1.2] and [Ber3, 3.1].)

Exercise 4.1.3.2 (i) Assume that Y and X are provided with affinoid atlases τ_Y and τ_X such that for any $U \in \tau_X$ the restriction $\tau_Y|_{f^{-1}(U)}$ is an atlas of $f^{-1}(U)$. Then to give a morphism $Y \dashrightarrow X$ is equivalent to give a similar data $(g, g^\#)$ but with $g_{U,V}^\#$ defined only for $U \in \tau_X$ and $V \in \tau_Y$ with $f(V) \subseteq U$.
(ii) Use this to define composition of morphisms.
(iii) Show that any k_H-analytic domain $V \subseteq X$ possesses a canonical structure of a k_H-analytic space. Moreover, the inclusion underlies the *analytic domain embedding* morphism $i_V:V \to X$ which possesses the universal property that any morphism $Y \to X$ with set-theoretical image in V factors through V (in the analytic category).

Thanks to the claim of (ii) we have now introduced the category of k_H-analytic spaces, which will be denoted k_H-*An*. The particular case of (resp. *strictly*) k-*analytic spaces* corresponding to $H = \mathbf{R}_+^\times$ (resp. $H = |k^\times|$) will be denoted k-*An* (resp. *st-k-An*).

Remark 4.1.3.3 (i) Note that an isomorphism between two analytic spaces X and Y is a homeomorphism $f:Y \to X$ which induces a bijection $\tau_{H,Y} \xrightarrow{\sim} \tau_{H,X}$ and an isomorphism of the structure sheaves $f^\#:\mathcal{O}_{X_H} \to f_*(\mathcal{O}_{Y_H})$. In particular, this allows to consider k_H-analytic spaces without any fixed affinoid atlas.
(ii) We gave the usual definition of analytic spaces that follows [Ber2] but used a different definition of morphisms. Berkovich first defines a category of spaces with a fixed atlas (and morphisms between such objects) and then inverts morphisms corresponding to refinements of atlases.

One can also define analytic spaces in a more invariant way that does not involve atlases. This is worked out in the following exercise.

Exercise 4.1.3.4 Show that the following definition of analytic spaces is equivalent to the standard one: a k_H-analytic space X is a topological space $|X|$ provided with a G-topology τ_H^c and a τ_H^c-sheaf of Banach k-algebras $\mathcal{O}_{X_H}$ such that the elements of τ_H^c are compact (in the usual topology), τ_H^c is closed under finite unions, and X is G-locally isomorphic to k_H-affinoid sets, in the sense that there is a quasi-net $\{X_i\}$ on X such that each triple $(X_i, \tau_H|_{X_i}, \mathcal{O}_{X_H}|_{X_i})$ is isomorphic to a k_H-affinoid space.

Hint: use Fact 3.3.3.3.

The following useful result is surprisingly difficult (for non-separated spaces). It was proved in [Tem2] for the strictly analytic category and was generalized to general H-strict spaces in [CT].

Fact 4.1.3.5 Assume that $H' \subseteq \mathbf{R}_+^\times$ is a subgroup containing H. The natural embedding functor $k_H\text{-}An \to k_{H'}\text{-}An$ is fully faithful. In particular, any analytic space admits at most one (up to an isomorphism) structure of an H-strict analytic space.

An equivalent way to reformulate this fact is by saying that any k-analytic morphism between H-strict k-analytic spaces can be described using H-strict atlases. The fact implies that even when studying k_H-analytic spaces we can (and in the sequel will) safely work within the category of all k-analytic spaces. In the sequel, our default G-topology is the G-topology τ of all k-analytic domains, and τ^c denotes the G-topology of compact k-analytic domains.

Exercise 4.1.3.6 Show that the τ^c-sheaf $\mathcal{O}_{X_{\mathbf{R}_+^\times}}$ uniquely extends to a G-sheaf of rings $\mathcal{O}_{X_G}$ and, moreover, $\mathcal{O}_{X_G}(V) = \mathrm{Mor}_{k-An}(V, \mathbf{A}_k^1)$.

From now on, the *structure sheaf* of X refers to the sheaf $\mathcal{O}_{X_G}$.

Remark 4.1.3.7 (i) We aware the reader that there is an abuse of language in our notion of the structure sheaf because $\mathcal{O}_{X_G}$ is not a part of the definition of X, but only an additional structure X is provided with. Moreover, in a sharp contrast with the τ^c-sheaf $\mathcal{O}_{X_{\mathbf{R}_+^\times}}$, the G-sheaf $\mathcal{O}_{X_G}$ is not a sheaf of Banach rings and it does not contain enough information to define the analytic space (at least when the valuation is trivial).

(ii) One can provide $\mathcal{O}_{X_G}$ with a natural structure of a *pluri-normed* k-algebra, i.e. k-algebra with a family of k-bounded seminorms. If the family is countable, then this agrees with the usual notion of a Frechet k-algebra. Probably, one can develop a theory that includes also spectra of pluri-normed algebras, but this was not done so far (to the best of my knowledge).

4.1.4   Gluing of Analytic Spaces

There are three main constructions of new k-analytic spaces: by gluing (or using atlases), by analytification of algebraic k-varieties, and as the generic fiber of formal k°-schemes. Here we consider only the first construction because the other two will be studied later.

Exercise 4.1.4.1 Assume that $\{X_i\}_{i \in I}$ is a family of k-analytic spaces provided with analytic domains $X_{ij} \hookrightarrow X_i$ and isomorphisms $\phi_{ij}: X_{ij} \overset{\sim}{\to} X_{ji}$ that satisfy the usual cocycle compatibility condition on the intersections $X_{ijk} = X_{ij} \cap X_{ik}$. Show that in the following cases they glue to a k-analytic space X covered by domains isomorphic to X_i so that $X_i \cap X_j \overset{\sim}{\to} X_{ij}$.

Case 1:   The domains $X_{ij} \subseteq X_i$ are open.

Case 2;   The domains $X_{ij} \subseteq X_i$ are closed and for each i only finitely many domains X_{ij} are non-empty.

Exercise 4.1.4.2   (i) Let X be glued from annuli $X_1 = A(0; r, s)$ and $X_2 = A(0; s, t)$ along $X_{12} = A(0; s, s)$ so that the orientation of the annuli is preserved. The latter means that the gluing homomorphism $k\{s^{-1}T_1, rT_1^{-1}\} \to k\{s^{-1}T, sT^{-1}\}$ takes T_1 to an element $\sum_{i=-\infty}^{\infty} a_i T^i$ with $|a_1 T| = s > |a_i T^i|$ for $i \neq 1$, and similarly for the second chart. Show that X is isomorphic to the annulus $A(0; r, t)$.

   Hint: show that the intersection of $k\{s^{-1}T_1, rT_1^{-1}\}$ and $k\{t^{-1}T_2, sT_2^{-1}\}$ inside of $k\{s^{-1}T, sT^{-1}\}$ is isomorphic to $k\{t^{-1}R, rR^{-1}\}$.

(ii) In the same way show that if X_1 is the disc $E(0, s)$ and X_2, X_{12} are as in (i) then $X \overset{\sim}{\to} E(0, t)$.

(iii) We define $\mathbf{P}_k^1$ as the obvious gluing of $\mathcal{M}(k\{T\})$ and $\mathcal{M}(k\{T^{-1}\})$ along $\mathcal{M}(k\{T, T^{-1}\})$. Show that any other gluing of two discs with the same choice of orientation is isomorphic to $\mathbf{P}_k^1$. A wrong choice of orientation leads to a space that we call a closed unit disc with doubled open unit disc. This space is Hausdorff, but we will later see that it is not locally separated at the maximal point of the disc.

(iv) Define $\mathbf{P}_k^n$ with homogeneous coordinates $T_0 \ldots T_n$ in two different ways: (a) as a gluing of $n + 1$ unit polydiscs, (b) as a gluing of $n + 1$ affine spaces $\mathbf{A}_k^n$.

   Hint: in both cases, it is convenient to symbolically denote coordinates on the i-th chart as $\frac{T_j}{T_i}$ for $0 \le j \le n, j \neq i$.

Definition 4.1.4.3   (i) A seminorm on a graded ring $A = \oplus_{d \in \mathbf{N}} A_d$ is *homogeneous* if it is determined by its values on the homogeneous elements via the max formula $|\sum_{d \in \mathbf{N}} a_d| = \max_{d \in \mathbf{N}} |a_d|$.

(ii) Seminorms $|\ |$ and $\|\ \|$ on A are *homothetic* if there exists a number $C > 0$ such that $|a_d| = C^d \|a_d\|$ for any $a_d \in A_d$.

(iii) Let k be a non-archimedean field with a graded k-algebra A. The projective spectrum $\mathcal{M}\text{Proj}(A)$ is the set of all homothety equivalence classes of homogeneous semivaluations on A that extend the valuation of k and do not vanish on the whole A_1.

Exercise 4.1.4.4   (i) Show that $\mathbf{P}_k^n = \mathcal{M}\text{Proj}(k[T_0 \ldots T_n])$ by a direct computation.

(ii) Alternatively, show that two points $x, y \in \mathbf{A}^{n+1} \setminus \{0\} = \mathcal{M}\text{Spec}(k[\underline{T}]) \setminus \{0\}$ are mapped to the same point by the projection $\mathbf{A}^{n+1} \setminus \{0\} \to \mathbf{P}_k^n$ if and only if there exists $C > 0$ such that $|f_d(x)| = C^d |f_d(y)|$ for any homogeneous

$f_d(\underline{T}) \in k[\underline{T}]$ of degree d. Show that the set of homogeneous semivaluations in any fiber of the projection is a single homothety equivalence class.

4.1.5 Fibred Products

Fact 4.1.5.1 (i) The category k_H-*An* possesses a fibred product $Y \times_X Z$ which
agrees with the fibred product in any category $k_{H'}$-*An* for $H \subseteq H'$ and in the
category of k-affinoid spaces. In particular, if $X = \mathcal{M}(A)$, $Y = \mathcal{M}(B)$ and
$Z = \mathcal{M}(C)$ then $\mathcal{M}(B \hat{\otimes}_A C) \xrightarrow{\sim} Y \times_X Z$
(ii) Let $f\colon Y \to X$ and $g\colon Z \to X$ be morphisms of k-analytic spaces, and assume
that $X = \cup_i(X_i)$, $f^{-1}(X_i) = \cup_j Y_{ij}$ and $g^{-1}(X_i) = \cup_k Z_{ik}$ are admissible
coverings by affinoid domains. Then $Y \times_X Z$ admits an admissible covering by
affinoid domains $Y_{ij} \times_{X_i} Z_{ik}$.

Actually, the second part of this result indicates how the fibred product is
constructed.

4.1.6 The Category An-k

Often one also needs to consider morphisms between analytic spaces defined
over different fields. For example, to define fibers of morphisms or ground field
extensions. For this Berkovich introduces the category of analytic k-spaces.

Definition 4.1.6.1 (i) An *analytic k-space* is a pair (X, K) where K is a non-
archimedean k-field and X is a K-analytic space.
(ii) A morphism $(Y, L) \to (X, K)$ consists of a bounded homomorphism $\phi :
K \hookrightarrow L$, a continuous and G-continuous map $f : Y \to X$, affinoid nets
(or atlases) $Y = \cup_{i \in I} Y_i$ and $X = \cup_{i \in I} X_i$ such that $f(Y_i) \subseteq X_i$, and
bounded homomorphisms $\phi_i : \mathcal{O}_{X_G}(X_i) \to \mathcal{O}_{Y_G}(Y_i)$ that extend ϕ, agree with
$f : Y_i \to X_i$ after applying $\mathcal{M}$, and pairwise agree on intersections (i.e. ϕ_i
and ϕ_j agree on any $Y_k \subseteq Y_i \cap Y_j$). One identifies morphisms via refinement
of atlases pretty similar to the definition of morphisms of k-analytic spaces.
(iii) The category of *analytic k-spaces* is denoted An-k.

4.1.7 Fibers of Morphisms and Base Change

Definition/Exercise 4.1.7.1 (i) It follows from fact 3.2.3.2 that for any point
$x \in X$ the field $\mathcal{H}(x)$ is well defined: one takes any affinoid domain $x \in V$
and defines $\mathcal{H}(x)$ using that domain. Obviously, $\mathcal{H}(x)$ is preserved when we
replace X with any analytic domain containing x.
(ii) Assume that $f\colon Y \to X$ is a morphism of k-analytic spaces and $x \in X$ is a
point. Define the fibred product $Y_x = Y \times_X \mathcal{M}(\mathcal{H}(x))$ as an $\mathcal{H}(x)$-analytic

space. Show that $Y_x = f^{-1}(x)$ set-theoretically. The space Y_x is called the *fiber* of f over x.

(iii) Assume that X is a k-analytic space and K is a non-archimedean k-field. Then one can define a universal K-analytic space $X_K = X \hat{\otimes}_k K$ such that any morphism $Y \to X_k$ factors uniquely through X_K. One simply takes an atlas $\{X_i = \mathcal{M}(\mathcal{A}_i)\}$ of X, shows that $\{X_{K,i} = \mathcal{M}(\mathcal{A}_i \hat{\otimes}_k K)\}$ is an atlas of a K-analytic space, and calls that space $X \hat{\otimes}_k K$. The construction does not depend on the atlas and provides a functor $k\text{-}An \to K\text{-}An$ called the *ground field extension* functor.

Actually, one can unify these two constructions by saying that a fibred product $Y \times_X Z$ in $An\text{-}k$ exists whenever X and Y are k-analytic and $Z = \mathcal{M}(K)$ with K a non-archimedean field.

Remark 4.1.7.2 The definition of analytic k-space $X = (X, K)$ fixes a field of definition K. Sometimes this may be too restrictive because the "same" X may admit many different fields of definition. (This is analogous to the fact that complete local rings or even non-reduced affine varieties usually admit different fields of definitions.) For example, $K_r\{T\}$ admits many bounded k-automorphisms ϕ that do not preserve K_r, and one may wish to view $\mathcal{M}(\phi)$ as an automorphism of the analytic k-space $\mathcal{M}(K_r\{T\})$. There is a simple way to extend the category $An\text{-}k$ accordingly: in Definition 4.1.6.1, one omits a morphism between fields of definitions and allows all morphisms $Y_i \to X_i$ corresponding to a k-bounded homomorphism $\mathcal{O}_{X_G}(X_i) \to \mathcal{O}_{Y_G}(Y_i)$.

4.2 Basic Classes of Analytic Spaces and Morphisms

4.2.1 Good Spaces

Definition 4.2.1.1 (i) An analytic space X is called *good at* a point x if x has an affinoid neighborhood. The space X is *good* if it is good at all its points.

(ii) The sheaf $\mathcal{O}_X$ is defined as the restriction of $\mathcal{O}_{X_H}$ to the usual topology of X.

(iii) For any point $x \in X$ we define $\kappa(x)$ as the residue field of the local ring $\mathcal{O}_{X,x}$. Note that $\kappa(x)$ may change when we replace X with an analytic domain containing x (and this happens already in the affinoid case).

Good spaces are often more convenient to work with. In particular, the definition of $\mathcal{O}_X$ applies to any X, but it can play the role of a structure sheaf only for the class of good spaces (see Example 4.2.1.3). That is why we will only consider $\mathcal{O}_X$ and $\kappa(x)$ on good spaces.

Fact 4.2.1.2 For a good k_H-analytic space X the category of coherent $\mathcal{O}_{X_H}$-modules is equivalent to the category of *coherent* $\mathcal{O}_X$-modules (i.e. $\mathcal{O}_X$-modules locally isomorphic to a quotient of $\mathcal{O}_X^n$).

Example/Exercise 4.2.1.3 (i) If an analytic space X is good at a point $x \in X$ then $\mathcal{H}(x)$ is the completion of $\kappa(x)$.

(ii) Let X be a closed unit disc with doubled open unit disc as in Exercise 4.1.4.2, and let x be its maximal point. Show that in some but not all cases $k \xrightarrow{\sim} \mathcal{O}_{X,x}$. Thus, the usual topology is too crude to allow non-constant functions in a neighborhood of x. Clearly, X is not good and $\widehat{\kappa(x)} \subsetneq \mathcal{H}(x)$ in this case.

Here are simplest separated examples of non-good spaces. They will be studied further in Sect. 5.7 (and, at least, one has to use Raynaud's theory from Sect. 5.3 to show that these examples are non-good spaces.)

Example 4.2.1.4 (i) Assume that $\underline{r}$ is a tuple of $n > 1$ positive numbers. A closed polydisc $E(0, \underline{r})$ of polyradius $\underline{r}$ with removed open polydisc of polyradius $\underline{r}$ is a compact not good analytic domain in $X \subset E(0, \underline{r})$.

(ii) In the affine plane $X = \mathcal{M}\mathrm{Spec}(k[S, T])$ consider the affinoid domains $V_1 = X\{r \leq |S| \leq 1, |ST| \leq 1\}$ and $V_2 = X\{1 \leq |S| \leq r^{-1}, |S^{-1}T| \leq 1\}$ with $0 < r < 1$. One can show that the compact analytic domain $V = V_1 \cup V_2$ is not good at the maximal point of the unit polydisc $X\{S, T\}$. Moreover, if we consider the natural projection $f : V \to \mathbf{A}_k^1 = \mathcal{M}\mathrm{Spec}(k[S])$ then V_1 and V_2 are the preimages in V of the affinoid domains $A(0; r, 1)$ and $A(0; 1, r^{-1})$. Since the union of these two annuli is an affinoid space (it is $A(0; r, r^{-1})$)), we see that no reasonable notion of good or affinoid morphisms exists.

4.2.2   Finite Morphisms, Closed Immersions and Zariski Topology

Definition 4.2.2.1 A morphism $f : Y \to X$ is called a *closed immersion* (resp. *finite*) if there exists an admissible covering by affinoid domains $X = \cup_i X_i$ such that $Y_i = X_i \times_X Y$ are k-affinoid and the homomorphisms of Banach algebras $\mathcal{O}_{X_G}(X_i) \to \mathcal{O}_{Y_G}(Y_i)$ are surjective (resp. finite) and admissible.

We have already discussed in Exercise 3.1.4.4 why admissibility is essential when the valuation on k is trivial.

Exercise 4.2.2.2 (i) Assume that $f : Y \to X$ is finite. Then for any affinoid domain $\mathcal{M}(\mathcal{A}) = X' \subseteq X$ its preimage $Y' = Y \times_X X'$ is affinoid, say $Y' = \mathcal{M}(\mathcal{B})$, and the homomorphism $\mathcal{A} \to \mathcal{B}$ is surjective (resp. finite) and admissible.

Hint: $\mathcal{O}_{Y'_G}$ is a coherent $\mathcal{O}_{X'_G}$-algebra.

(ii) The class of closed immersions (resp. finite morphisms) is closed under compositions, base changes and ground field extensions.

Definition 4.2.2.3 Any subset $Z \subseteq X$ that is the image of a closed immersion is called *Zariski closed*. The complement of such set is called *Zariski open*.

Exercise 4.2.2.4 A point $x \in X$ is Zariski closed if and only if $[\mathcal{H}(x) : k] < \infty$. (Zariski closed points of X are precisely its classical rigid points. The set of all such points is denoted X_0.)

When working with Zariski topology one must be very careful because it becomes stronger when passing to analytic domains (even open ones). In other words, coherent ideals on an open subspace do not have to extend to the whole space. Such phenomenon does not occur for algebraic varieties (for obvious reasons) but does occur for formal varieties.

Example 4.2.2.5 (i) Give an example of a k-analytic space X with an open subspace U and a closed subspace $Z \subset U$ which does not extend to the whole X.

Hint: take $X = \mathcal{M}(k\{T, S\})$ the unit polydisc, U an open polydisc of polyradius $(1, r)$ with $r < 1$ and Z given by $T - f(S)$, where $f(S)$ has radius of convergence between r and 1.

(ii) An example of Ducros. Fix $r \notin \sqrt{|k^\times|}$ with $0 < r < 1$. Consider a polydisc $X = \mathcal{M}(k\{T, S\})$ with an affinoid domain $V = \mathcal{M}(k\{r^{-1}T, rT^{-1}, S\})$, which is a unit K_r-disc and a non-strict k-surface (the product of the unit k-disc with the irrational k-annulus $\mathcal{M}(K_r)$). Using the hint from (i) find a Zariski closed point $x \in V$ with $\mathcal{H}(x) \xrightarrow{\sim} K_r$ such that the ideal of x in V does not extend to any neighborhood of x in X. (In a sense, x is a k-curve in the k-surface X which cannot be extended, so x is Zariski closed only in a sufficiently small domain in X.) Show that in this case $\mathcal{O}_{X,x}$ is a dense subfield of K_r and hence the character $\chi_{X,x} : k\{T, S\} \to \mathcal{H}(x)$ is injective and hence flat. On the other hand, its base change with respect to the homomorphism $k\{T, S\} \to k\{r^{-1}T, rT^{-1}, S\}$ is not flat because the character $\chi_{V,x}$ has a non-trivial kernel. In particular, one cannot define a reasonable class of flat morphisms between k-affinoid spaces just by saying that $\mathcal{M}(f)$ is flat whenever f is flat. (One can show that this approach works well for strictly k-affinoid spaces; see also Remark 4.2.5.5(ii).)

4.2.3 Separated Morphisms

Definition 4.2.3.1 (i) A morphism $f : Y \to X$ is *separated* if the diagonal morphism $\Delta_{Y/X} : Y \to Y \times_X Y$ is a closed immersion.

(ii) A k-analytic space is *separated* if its morphism to $\mathcal{M}(k)$ is so.

Exercise 4.2.3.2 (i) Formulate and prove the basic properties of separated morphisms analogous to the properties of separated morphisms of schemes. In particular, show that in a separated k-analytic space X the intersection of two k-affinoid domains is k-affinoid.

(ii) Prove Fact 4.1.3.5 for the subcategories of separated objects in k_H-An and k-An.

Hint: if X and Y are H-strict then any closed subspace in $X \times Y$ is affinoid by Exercise 4.2.2.2 and hence H-strict.

Non-separatedness of a space can be of two sorts, as is illustrated by the following example.

Example/Exercise 4.2.3.3 (i) Let X be the closed unit disc with doubled open unit disc. Show that X is not locally separated at its maximal point x (i.e. any k-analytic domain which is a neighborhood of x is not separated). In particular, X is a non-good k-analytic Hausdorff space.

(ii) Show that the closed unit disc Y with doubled origin is not separated but is locally separated at all its points. Moreover, Y is a good non-Hausdorff k-analytic space.

4.2.4 Boundary and Proper Morphisms

Definition 4.2.4.1 (i) The *relative interior* $\mathrm{Int}(Y/X)$ of a morphism $f:Y \to X$ is the set of all points $y \in Y$ such that for any affinoid domain $U \subseteq X$ containing $x = f(y)$ there exists an affinoid domain $V \subseteq f^{-1}(U)$ such that V is a neighborhood of x in $f^{-1}(U)$ and $x \in \mathrm{Int}(V/U)$. The complement $\partial(Y/X) = Y \setminus \mathrm{Int}(Y/X)$ is called the *relative boundary* and we say that f is *boundaryless* or without boundary if $\partial(Y/X)$ is empty (in [Ber2] such morphisms are called "closed").

(ii) A morphism $f:Y \to X$ is *proper* if it is boundaryless and compact (i.e. the preimage of a compact domain is compact).

As usual, the absolute analogs of these notions are defined relatively to $\mathcal{M}(k)$. Note that (part (i) of) this definition agrees with our earlier definitions from Sect. 3.4.2.

Example/Exercise 4.2.4.2 (i) A k-analytic space has no boundary if and only if any its point x possesses an affinoid neighborhood U such that $x \in \mathrm{Int}(U)$. For example, an open polydisc, $\mathbf{P}_k^n$ and $\mathbf{A}_k^n$ have no boundary, and a closed polydisc has a boundary. So far, $\mathbf{P}_k^n$ is the only example of a proper non-discrete k-analytic space we have considered. It follows that any *projective k-analytic space*, i.e. a closed subspace of $\mathbf{P}_k^n$ is also proper.

(ii) Any boundaryless k-analytic space X is good. If the valuation on k is non-trivial then X is also strict.

(iii) A morphism between affinoid spaces is proper if and only if it is finite. Any finite morphism is proper.

(iv) A boundaryless morphism is separated if and only if the preimage of any Hausdorff domain is Hausdorff. In particular, proper morphisms are separated.

(v) There is a theory of analytic tori which is parallel in some part to the classical theory of complex tori. An *analytic torus* is defined as the quotient $T_\Lambda = \mathbf{G}_m^d/\Lambda$ where $\Lambda = \oplus_{i=1}^d \lambda_i^{\mathbf{Z}}$ is a multiplicative lattice such that $|\Lambda|$ is a lattice in $(\mathbf{R}_+^\times)^d$. It is easy to see that T_Λ is a proper analytic space. There also is an analog of Riemann's positivity conditions on Λ that are necessary and sufficient for T_Λ to be algebraic (and even projective). Similarly to the complex case, if $d = 1$ then T_Λ is always algebraic, but a generic two dimensional torus is not algebraic.

Fact 4.2.4.3   (i) If Y is an analytic domain in X then $\text{Int}(Y/X)$ is the topological interior of Y in X.

(ii) Boundaries are G-local on the base, i.e. given an admissible covering of X by affinoid domains X_i one has that $\partial(Y/X) = \cup_i \partial(X_i \times_X Y/X_i)$.

(iii) The classes of proper morphisms and morphisms without boundary are G-local on the base and are preserved by compositions, base changes and ground field extensions.

(iv) If $f : Y \to X$ is a separated boundaryless morphism and X is k-affinoid then for any affinoid domain $U \subseteq Y$ there exists a larger affinoid domain $V \subseteq Y$ such that $U \subseteq \text{Int}(V/X)$ and U is a Weierstrass domain in V.

Remark 4.2.4.4 Surprisingly enough, already (ii) is really difficult. It turns out that when one wants to show that various morphisms have no boundary, the difficult part of the proof is to show that the preimage of an affinoid domain under these morphisms is a good domain. Once this is established, one can use the theory of boundaries for affinoid spaces as outlined in Sect. 3.4.2. See also Remark 4.2.4.6 below.

Similarly to algebraic and complex analytic geometries, coherence is preserved by higher direct images with respect to proper morphisms.

Fact 4.2.4.5 (Kiehl's theorem on direct images) If $f : Y \to X$ is a proper morphism between k-analytic spaces and $\mathcal{F}$ is a coherent $\mathcal{O}_{Y_G}$-modules then the $\mathcal{O}_{X_G}$-modules $R^i f_*(\mathcal{F})$ arc coherent.

Note that we use here that f is a compact map because otherwise $f_*(\mathcal{F})$ is not a sheaf of Banach $\mathcal{O}_{X_G}$-modules.

Remark 4.2.4.6 Kiehl introduced the notion of proper morphisms and proved the above result (for rigid spaces) in [Ki]. One can easily show that our definition of proper morphisms (in the strict case) is equivalent to the original Kiehl's definition. The definition of proper morphisms is designed so that the theorem on direct images can be proved rather easily and naturally (one computes Čech complexes and shows that certain differentials are compact operators). As was already remarked, it is very difficult to establish some other properties, that one might expect to be more foundational. For example, the fact that proper morphisms are preserved by compositions was open for more than twenty years (for a discretely valued k this was proved in [Lüt] and the general case was established in [Tem1] and [Tem2]).

4.2.5   Smooth and étale Morphisms

Definition 4.2.5.1   (i) A finite morphism $f : Y \to X$ is *étale* if for any affinoid domain $U \subseteq X$ and its preimage $V = f^{-1}(U)$ the finite homomorphism of

k-affinoid algebras $\mathcal{O}_{X_G}(U) \to \mathcal{O}_{Y_G}(V)$ is étale. (Recall that V is affinoid and finite over U by Example 4.2.2.2(i).)

(ii) In general, a morphism $f : Y \to X$ is *étale* if locally (on Y) it is finite étale. Namely, for any point $y \in Y$ there exist neighborhoods V of y and U of $f(y)$ such that f restricts to a finite étale morphism $V \to U$.

(iii) A morphism $f : Y \to X$ is *smooth* if it can be represented as an étale morphism $Y \to \mathbf{A}_X^n$ followed by the projection.

Exercise 4.2.5.2   (i) Any smooth morphism (e.g. an étale morphism) is boundaryless.

  (ii) If V is an analytic domain in X then the embedding $V \hookrightarrow X$ is étale if and only if it is an open immersion.

(iii)* Any smooth morphism is an open map.

Fact 4.2.5.3 The classes of étale and smooth morphisms are closed under compositions, base changes and ground field extensions. Also, it follows from Kiehl's theorem that étaleness is G-local on the base. Probably, smoothness is not G-local on the base.

This definition of étale and smooth morphisms is analogous to a complex analytic definition but it does not apply to nice morphisms with boundaries. For example, the closed unit disc is not smooth at its maximal point. The following definition is a natural generalization to the case when there are boundaries. We give it for the sake of completeness, but do not discuss all results one should prove to show that it really makes sense.

Definition 4.2.5.4   (i) A morphism $f : Y \to X$ between strictly k-analytic spaces is *rig-smooth* if the restriction of f on $\mathrm{Int}(Y/X)$ is smooth.

 (ii) In general, a morphism $f : Y \to X$ is *rig-smooth* if so is some (and then any) ground field extension $f_{\underline{r}} := f \hat{\otimes}_k K_{\underline{r}}$ such that $Y_{\underline{r}}$ and $X_{\underline{r}}$ are strictly K_r-analytic.

(iii) A rig-smooth morphism with discrete fibers is called *quasi-étale*.

Remark 4.2.5.5   (i) Alternatively, one can define quasi-étale morphisms directly and then rig-smooth morphisms are the morphisms that G-locally split into the composition of a quasi-étale morphism with the projection $\mathbf{A}_X^n \to X$.

 (ii) We have to extend the ground field in the general case because $\mathrm{Int}(Y/X)$ can be too small to test (any sort of) smoothness. A good example of such situation was studied in Exercise 4.2.2.5(ii).

(iii) The same problem as in (ii) happens when one wants to introduce flatness. A reasonable theory of flatness was developed very recently by Ducros. In the strict case one gives a naive definition, and in general f is called *flat* if so is its strictly analytic ground field extension.

(iv) Ducros also proves that f is rig-smooth if and only if it is flat, the coherent $\mathcal{O}_{Y_G}$-module of continuous differentials Ω_{Y_G/X_G}^1 is locally free, and locally the rank of Ω_{Y_G/X_G}^1 equals the relative dimension. (The sheaf Ω_{Y_G/X_G}^1 admits the following G-local description: if $X = \mathcal{M}(\mathcal{A})$ and $Y = \mathcal{M}(\mathcal{B})$ then $\Omega^1_{Y_G/X_G}$ corresponds to the module I/I^2 where $I = \mathrm{Ker}(\mathcal{B}\hat{\otimes}_{\mathcal{A}}\mathcal{B} \to \mathcal{B})$.

4.3 Topological Properties

4.3.1 Basic Properties

Fact 4.3.1.1 (i) Any connected k-analytic space is pathwise connected.
(ii) Any point has a fundamental family of neighborhoods which are compact and pathwise connected analytic domains.
(iii) The topological dimension of a paracompact X is at most $\dim(X)$ and both are equal in the strict case.

This fact was proved in [Ber1] (and the argument is correct, although by a misunderstanding some mathematicians thought that part (i) was not proved). Let us say few words about the proof of (i). We have already checked this fact for an affine line and one easily deduces the case of a polydisc. Studying finite covers of polydiscs one obtains the case of strictly k-analytic spaces. The general case is deduced by descent from an appropriate $X\hat{\otimes}_k K_r$. Next let us consider examples of analytic spaces with bad topological properties.

Example 4.3.1.2 (i) Assume that $\tilde{k}$ is uncountable (e.g. $k = \mathbf{C}((T))$). Let U be the open subset of a closed two-dimensional polydisc E obtained by removing the maximal point of E. Then U is not paracompact, i.e. it possesses open covers that do not admit locally finite refinements.
(ii) There exist (see Exercise 6.1.3.5) examples of k-analytic curves C such that C is a double covering of an open unit disc and C is a closed subspace in a two-dimensional open unit polydisc, but the first Betti number of C is infinite. In particular, one can construct C so that it can be retracted onto its subset Δ, which is a graph with infinitely many loops.

4.3.2 Contractions

In some cases one can construct by hands a retraction of a k-analytic space X onto a subset S which is of topologically finite type. One such method is to find an action of a k-affinoid group G on X with a continuous family of affinoid subgroups $\{G_t\}_{t\in[0,1]}$ such that $G_0 = \{e\}$, $G_1 = G$ and for each point $x \in X$ each orbit $G_t x$ is affinoid and possesses exactly one maximal point x_t. Then $(x,t) \mapsto x_t$ defines a deformational retraction of X onto some its subset, which is very small in some examples. Two good examples of such G are as follows: a closed unit polydisc $\mathcal{M}(k\{\underline{T}\})$ with an additive group structure, and a product of unit annuli $G^n_{m,1} = \mathcal{M}(k\{T_1, T_1^{-1} \dots T_n, T_n^{-1}\})$ with the multiplicative group structure. The groups G_t with $t < 1$ are the polydiscs of polyradius $(t \dots t)$ with center at 0 or

1, respectively. The action of the torus is much more important because tori play important role in the theory of reductive groups (see [Ber1, §5] for the connection to Bruhat-Tits buildings). So, we give the most fundamental example of a contraction by a torus action.

Example/Exercise 4.3.2.1   (i) Show that $\mathbf{R}_+^n$ embeds into $\mathbf{A}_k^n$ so that $\underline{r} = (r_1 \dots r_n)$ goes to the semivaluation $\| \ \|_{\underline{r}}$ (the maximal point of $E(0, \underline{r})$).
(ii)* Show that the action of $G_{m,1}^n$ on $\mathbf{A}_k^n$ contracts it onto $\mathbf{R}_+^n$. Moreover, the retraction can be explicitly described by the formula $|f(x_t)| = \max_{i \in \mathbf{N}^n} |\partial_i f(x)| t^i$ where $f(\underline{T}) \in k[T_1 \dots T_n]$ and $\partial_i : k[\underline{T}] \to k[\underline{T}]$ for $i \in \mathbf{N}^n$ is the logarithmic differential operator $\frac{T^i}{i!} \frac{\partial^i}{d\underline{T}^i}$.

Note also that Berkovich devoted a separate paper [Ber4] to proving the following very difficult result.

Fact 4.3.2.2 Any analytic domain in a smooth k-analytic space is locally contractible.

4.3.3   Topological Type of Analytic Spaces

We saw in Sect. 2.3.3 that the affine line is a sort of an infinite tree. The topological structure of general analytic spaces is much more complicated and somewhat mysterious. A major progress in its understanding was achieved very recently by Hrushovski-Loeser in [HL] via model-theoretic methods, and here is one of the main applications of their theory to analytic spaces.

Fact 4.3.3.1 Let $\overline{X}$ be a projective n-dimensional k-analytic space, and assume that $X = V \cap Y$ is the intersection of a Zariski open subspace $V \subset \overline{X}$ with a compact analytic domain $Y \subset \overline{X}$. Then X contains a family of topological spaces $\{S_i\}_{i \in I}$ filtered by inclusion such that each S_i is homeomorphic to a finite simplicial complex of dimension at most n and there exists a projective family of maps $f_{ij} : S_i \to S_j$ (for each pair $S_j \subseteq S_i$) such that $X \overset{\sim}{\to} \operatorname{proj} \lim_{i \in I} S_i$. Moreover, this family extends to a compatible family of deformational retractions $\Phi_{ij} : S_i \times [0, 1] \to S_j$ with $\Phi_{ij}(x, 1) = f_{ij}(x)$ which induce deformational retractions of X onto each of S_i's.

Since any rig-smooth analytic space can be locally embedded as a subdomain in a projective variety, this result immediately implies Fact 4.3.2.2. Moreover, this result is of global nature and it treats most types of singularities as well (although, there exist non-algebraizable singularities that cannot be locally embedded into varieties). It seems natural to expect that the projective variety $\overline{X}$ in Fact 4.3.3.1 can be replaced with an arbitrary compact k-analytic space, but in such generality this is a widely open conjecture. (Note that in order to exclude bad spaces discussed in Example 4.3.1.2, some compactness assumption should be present in the formulation.)

5 Relation to Other Categories

This section contains various material, and some of its subsections are rather advanced. We place it before section Sect. 6 on analytic curves because some results of sections Sects. 5.1–5.3 will be used to study curves. So, the reader can look through the first three subsections and go directly to Sect. 6.

5.1 Analytification of Algebraic k-Varieties

5.1.1 The Analytification Functor

Let k-Var be the category of algebraic k-varieties (i.e. schemes of finite type over k). We are going to describe a construction of an analytification functor k-$Var \rightarrow k$-An. The analytification of a morphism $f: \mathcal{Y} \rightarrow \mathcal{X}$ will be denoted $f^{an}: \mathcal{Y}^{an} \rightarrow \mathcal{X}^{an}$. For $\mathcal{X} = \mathrm{Spec}(k[T_1 \ldots T_n])$ we set $\mathcal{X}^{an} = \mathbf{A}_k^n = \mathcal{M}\mathrm{Spec}(k[\underline{T}])$. For any quotient $A = k[\underline{T}]/I$ the analytification of $\mathcal{Y} = \mathrm{Spec}(A)$ is the closed subspace of $\mathcal{X}^{an}$ defined by vanishing of $I\mathcal{O}_{\mathcal{X}^{an}}$.

Exercise 5.1.1.1 (i) Prove that this definition is independent of choices. Also, show that $\mathcal{Y}^{an} = \mathcal{M}\mathrm{Spec}(A)$ is the set of all real semivaluations on A bounded on k.

Hint: two embeddings of $\mathcal{Y}$ into affine spaces are dominated by a third such embedding.

(ii) Extend this to a functor from the category of affine k-varieties to the category of boundaryless k-analytic spaces.

(iii) Show that the latter functor takes open immersions to open immersions and hence extends (by gluing) to an analytification functor k-$Var \rightarrow k$-An. Show that any analytification is a boundaryless space.

(iv) Show that $(\mathrm{Proj}(A))^{an} \xrightarrow{\sim} \mathcal{M}\mathrm{Proj}(A)$ for any graded finitely generated k-algebra A. In particular, $\mathcal{M}\mathrm{Proj}(A)$ is a projective analytic variety.

Fact 5.1.1.2 The analytification functor can be described via the following universal property. For any good k-analytic space Y let $F_{\mathcal{X}}(Y)$ be the set of morphisms of locally ringed spaces $(Y, \mathcal{O}_Y) \rightarrow (\mathcal{X}, \mathcal{O}_{\mathcal{X}})$. Then $X = \mathcal{X}^{an}$ is the k-analytic space that represents $F_{\mathcal{X}}$.

In particular, a morphism $\pi_{\mathcal{X}}: (\mathcal{X}^{an}, \mathcal{O}_{\mathcal{X}^{an}}) \rightarrow (\mathcal{X}, \mathcal{O}_{\mathcal{X}})$ arises.

Fact 5.1.1.3 $\mathcal{X}^{an}(K) \xrightarrow{\sim} \mathcal{X}(K)$ for any non-archimedean k-field K, in particular, $\pi_{\mathcal{X}}$ is surjective.

Definition 5.1.1.4 For a coherent $\mathcal{O}_{\mathcal{X}}$-module $\mathcal{F}$ the module $\mathcal{F}^{an} = \pi_{\mathcal{X}}^*(\mathcal{F})$ is a coherent $\mathcal{O}_{\mathcal{X}^{an}}$-module called the *analytification* of $\mathcal{F}$.

5.1.2 GAGA

The analytification functor preserves almost all properties of varieties and their morphisms, and here is a (partial) list.

Fact 5.1.2.1 Let $f : \mathcal{Y} \to \mathcal{X}$ be a morphism between algebraic k-varieties. Then f satisfies one of the following properties if and only if so does f^{an}: smooth, étale, finite, closed immersion, open immersion, isomorphism, proper, separated.

Fact 5.1.2.2 (i) For a proper variety $\mathcal{X}$ the analytification functor induces an equivalence $\mathrm{Coh}(\mathcal{O}_{\mathcal{X}}) \xrightarrow{\sim} \mathrm{Coh}(\mathcal{O}_{\mathcal{X}^{\mathrm{an}}})$.
(ii) The functor $\mathcal{X} \mapsto \mathcal{X}^{\mathrm{an}}$ is fully faithful on the category of proper varieties

Exercise 5.1.2.3 (i) The assertions of Fact 5.1.2.2 do not hold for general algebraic varieties.
(ii) For a proper variety $\mathcal{X}$, the analytification functor induces an equivalence between the categories of finite (resp. finite étale) $\mathcal{X}$-schemes and $\mathcal{X}^{\mathrm{an}}$-spaces.
(iii) Any projective k-analytic space X is algebraizable by a projective k-variety $\mathcal{X}$ (i.e. $X \xrightarrow{\sim} \mathcal{X}^{\mathrm{an}}$).

When the valuation on k is trivial, the properness assumption can be eliminated. Let $X_t \subset X$ be the set of points $x \in X$ with trivially valued completed residue field $\mathcal{H}(x)$.

Fact 5.1.2.4 Assume that the valuation on k is trivial.

(i) $\mathcal{X}_t^{\mathrm{an}} \xrightarrow{\sim} \mathcal{X}$.
(ii) The analytification functor is fully faithful.
(iii) For a variety $\mathcal{X}$ the analytification functor induces an equivalence of categories $\mathrm{Coh}(\mathcal{O}_{\mathcal{X}}) \xrightarrow{\sim} \mathrm{Coh}(\mathcal{O}_{\mathcal{X}^{\mathrm{an}}})$.

5.2 Generic Fibers of Formal k°-Schemes

5.2.1 Reminds on Formal Schemes

Definition 5.2.1.1 Let A be a ring with an ideal I.

(i) The *I-adic topology* on A is generated by the cosets $a + I^n$.
(ii) The *separated I-adic completion* is defined as $\hat{A} = \mathrm{proj} \lim_n A/I^n$.
(iii) A is *I-adic* if $A \xrightarrow{\sim} \hat{A}$.
(iv) Any ideal J with $I^n \subseteq J$ and $J^n \subseteq I$ for large enough n is called *ideal of definition* of A. It generates the same topology and can be used instead of I in all definitions.

Example/Exercise 5.2.1.2 (i) If $k°$ is a real valuation ring with fraction field k and $\pi \in k°°$ is any non-zero element then the (π)-adic completion of $k°$ is the ring of integers $\hat{k}°$ of the completion of k.

(ii) The separated $(k°°)$-completion of k is either $\hat{k}°$ or $\tilde{k}$. Moreover, the first possibility occurs only when k is discrete or trivially valued.

Definition 5.2.1.3 (i) The *formal spectrum* $\mathfrak{X} = \mathrm{Spf}(A)$ of an I-adic ring A is the set of open prime ideals of A with the topology generated by the sets $D(f)$, where $D(f)$ is the non-vanishing locus of an element $f \in A$.

(ii) Each $D(f)$ is homeomorphic to $\mathrm{Spf}(A_{\{f\}})$, where the *formal localization* $A_{\{f\}}$ is the universal I-adic A-algebra with inverted f.

(iii) The structure sheaf $\mathcal{O}_{\mathfrak{X}}$ is the sheaf of topological rings determined by the condition $\mathcal{O}_{\mathfrak{X}}(D(f)) = A_{\{f\}}$. The topologically ringed space $(\mathfrak{X}, \mathcal{O}_{\mathfrak{X}})$ is called the *affine formal scheme* associated with A.

(iv) The *closed fiber* (or special fiber) $\mathfrak{X}_s$ of $\mathfrak{X}$ is the reduction of $\mathrm{Spec}(A/J)$ for any ideal of definition J.

Exercise 5.2.1.4 (i) Show that $A_{\{f\}} \xrightarrow{\sim} \mathrm{proj}\lim_n (A/I^n)_f$ and $A\{T\}/(Tf - 1) \xrightarrow{\sim} A_{\{f\}}$, where $A\{T\} = \mathrm{proj}\lim_n (A/I^n)[T]$ is the ring of convergent power series over A.

(ii) If $\pi^n \in I$ for some n then $A_{\{\pi\}} = 0$. Thus, formal localization at a topologically nilpotent element has the same effect as inverting a nilpotent element in a ring.

(iii) Show that $\mathfrak{X}_s$ does not depend on the ideal of definition and $|\mathfrak{X}_s| \xrightarrow{\sim} |\mathfrak{X}|$. Actually, $\mathfrak{X}$ can be viewed as the inductive limit of schemes $(\mathfrak{X}_s, \mathcal{O}_{\mathfrak{X}}/J)$ where J runs through the ideals of definition.

Definition 5.2.1.5 A general *formal scheme* is a topologically ringed space $(\mathfrak{X}, \mathcal{O}_{\mathfrak{X}})$ which is locally isomorphic to affine formal schemes. Morphisms of such creatures are morphisms of topologically ringed spaces that induce local homomorphisms on the ring-theoretical stalks. The notions of ideals of definitions and of the closed fiber are extended to the general formal schemes in the obvious way.

Definition/Exercise 5.2.1.6 (i) The n-dimensional affine space over an adic ring A is defined as $\mathbf{A}_A^n = \mathrm{Spf}(A\{T_1 \ldots T_n\})$.

(ii) A formal scheme over an I-adic ring A is of *(topologically) finite presentation* (resp. *special*) if it is locally of the form $\mathrm{Spf}(A\{T_1 \ldots T_n\}/(f_1 \ldots f_m))$ (resp. $\mathrm{Spf}(A\{T_1 \ldots T_n\}[[S_1 \ldots S_l]]/(f_1 \ldots f_m)))$.

5.2.2 Generic Fibers of Formal $k°$-Schemes of Finite Type

In this section we are going to define a *generic fiber functor* η which assigns to a formal $k°$-scheme $\mathfrak{X}$ of locally finite type a Hausdorff strictly k-analytic space $\mathfrak{X}_\eta$ (even when the valuation is trivial). Intuitively, $\mathfrak{X}_\eta$ is the "missing generic fiber of $\mathfrak{X}$"

and when k is non-trivially valued it is defined by inverting a non-zero element $\pi \in k^{\circ\circ}$. (By Exercise 5.2.1.4(ii) we kill any formal k°-scheme by such an operation, so it is not surprising that $\mathfrak{X}_\eta$ is not a formal scheme but leaves in another category.) If k is trivially valued then we set $\pi = 0$ to uniformize the exposition.

In general, the definition of η is very similar to the definition of the analytification. One defines $(\mathbf{A}_{k^\circ}^n)_\eta$ to be the closed unit polydisc $E^n(0, 1)$. An affine scheme given by vanishing of $f_1 \ldots f_n$ in $\mathbf{A}_{k^\circ}^n$ is defined as the closed subspace in $E^n(1, 0)$ given by vanishing of f_i's. In general, η is defined via gluing. Let us realize this program with some details.

Definition 5.2.2.1 If $A = k^\circ\{T_1 \ldots T_n\}/I$ is a π-adic ring with a finitely generated I then $\mathcal{A} = A_\pi = A \otimes_{k^\circ} k$ is a k-affinoid algebra isomorphic to $k\{\underline{T}\}/Ik\{\underline{T}\}$. For the affine formal scheme $\mathfrak{X} = \mathrm{Spf}(A)$ we set $\mathfrak{X}_\eta = \mathcal{M}(\mathcal{A})$.

Exercise 5.2.2.2 Assume that A has no π-torsion, and so A embeds into $\mathcal{A}$.

 (i) The integral closure of A in $\mathcal{A}$ is $\mathcal{A}^\circ$.
 (ii) If the valuation on k is non-trivial and A is reduced then A is the unit ball for a Banach norm $| \ |$ on $\mathcal{A}$. This means that $| \ |$ is equivalent to $\rho_\mathcal{A}$ or, equivalently, $\pi^n \mathcal{A}^\circ \subseteq A \subseteq \mathcal{A}^\circ$ for large enough n.
     Hint: use Fact 3.1.2.1(iii).
(iii) Formal localization is compatible with inverting π. Namely,
     $$(A_{\{f\}})_\pi \xrightarrow{\sim} A_\pi\{f^{-1}\}.$$
     Hint: use Fact 3.1.2.1.

Part (iii) of the above exercise implies that η (defined for affine formal schemes) takes open immersions to embeddings of affinoid domains. Now we can define the functor η in general.

Definition/Exercise 5.2.2.3 (i) If a separated formal scheme $\mathfrak{X}$ of finite type over k° is glued from open subschemes $\mathfrak{X}_i$ along the intersections $\mathfrak{X}_{ij}$ then the gluing of $(\mathfrak{X}_i)_\eta$ along $(\mathfrak{X}_{ij})_\eta$ is possible by Exercise 4.1.4.2(ii) and the obtained k-analytic space is set to be $\mathfrak{X}_\eta$.
 (ii) For a general formal scheme $\mathfrak{X}$ we repeat the same construction but with $\mathfrak{X}_{ij}$ being separated formal schemes now.
(iii) Check that this construction defines the promised generic fiber functor (in particular, it extends to morphisms).

As one might expect, η preserves (or naturally modifies) almost all properties of morphisms.

Fact 5.2.2.4 Let $f:\mathfrak{Y} \to \mathfrak{X}$ be a morphism between formal k°-schemes without π-torsion. If f is an isomorphism, separated, proper, a closed immersion, finite étale, then f_η is so. If f is an open immersion, étale, or smooth, then f_η is a compact analytic domain embedding, quasi-étale, or rig-smooth, respectively.

Excluding preservation of properness, all these claims are simple. The remaining claim is really difficult, though this is not so surprising in view of other problems with properness discussed in Sect. 4.2.4. Actually, to prove this claim is essentially equivalent to prove the other difficult properties of properness listed in Sect. 4.2.4.

As for the opposite implications, at first glance, one cannot expect that something can be proved in that direction. For example, a generically finite morphism does not have to be finite, etc. However, the following result holds true and its proof is relatively simple.

Fact 5.2.2.5 If f_η is proper or separated then so is f.

Finally, there exists an anti-continuous *reduction map* $\pi_{\mathfrak{X}}:\mathfrak{X}_\eta \to \mathfrak{X}_s$ defined similarly to the affinoid reduction map.

Definition/Exercise 5.2.2.6 Check that for any point $x \in \mathfrak{X}_\eta$ its character $\chi_x:\mathcal{M}(\mathcal{H}(x)) \to \mathfrak{X}_\eta$ is induced by a morphism $\chi_x^\circ:\mathrm{Spf}(\mathcal{H}(x)^\circ) \to \mathfrak{X}$. This induces a point $\tilde{\chi}_x:\mathrm{Spec}(\widetilde{\mathcal{H}(x)}) \to \mathfrak{X}_s$ on the closed fiber and hence gives rise to a map $\pi_{\mathfrak{X}}$.

Remark 5.2.2.7 The reduction map is an analog of the following specialization construction. If X is a scheme over a henselian valuation ring (e.g. k°), $(X_\eta)_0$ is the set of closed points of the generic fiber and $(X_s)_0$ is the set of closed points of the closed fiber then specialization induces a map $(X_\eta)_0 \to (X_s)_0$.

5.2.3 Relation to the Analytification

If k is trivially valued then the analytification and the generic fiber constructions provide two functors from the category of k-varieties to the category of k-analytic spaces. More generally, for any k we have two functors $\mathcal{F}$ and $\mathcal{G}$ from the category of k°-schemes of finite type to the category of k-analytic spaces: $\mathcal{G}(X) = (X_\eta)^{\mathrm{an}}$ and $\mathcal{F}(X) = (\hat{X})_\eta$. In the first case, we first pass to the generic fiber of the morphism $X \to \mathrm{Spec}(k^\circ)$ and then analytify the obtained k-variety. In the second case, we first complete X and then take the generic fiber of the obtained formal k°-scheme of finite type.

Exercise 5.2.3.1   (i) Assume that $X = \mathrm{Spec}(A)$ and $f_1 \ldots f_n$ generate A over k°. Show that $(\hat{X})_\eta$ can be naturally identified with the affinoid domain in $(X_\eta)^{\mathrm{an}}$ defined by the conditions $|f_i| \leq 1$.
(ii) Let $\mathcal{F}^{\mathrm{sep}}$ and $\mathcal{G}^{\mathrm{sep}}$ be the restrictions of the functors $\mathcal{F}$ and $\mathcal{G}$ onto the category of separated k°-schemes of finite type. Extend the construction of (i) to a morphism of functors $\phi:\mathcal{F}^{\mathrm{sep}} \to \mathcal{G}^{\mathrm{sep}}$ which is a compact embedding of a strictly analytic domain (i.e. each morphism $\phi(X):\mathcal{F}(X) \to \mathcal{G}(X)$ is embedding of a compact analytic domain).
(iii) Show that when restricted to proper k°-schemes ϕ induces an isomorphism of functors, i.e. the embedding of the analytic domain $\phi(X):(\hat{X})_\eta \hookrightarrow (X_\eta)^{\mathrm{an}}$ is an isomorphism for a k°-proper X.
(iv) Show that ϕ extends to non-separated k° schemes but then it does not have to be embedding of an analytic domain.

Hint: if X is the relative affine line over k° with doubled origin then $\mathcal{F}(X)$ is not locally separated and hence cannot be embedded into the good (although not Hausdorff) space $\mathcal{G}(X)$.

5.2.4 Generic Fibers of k°-Special Formal Schemes

For completeness, we discuss briefly how the generic fiber functor extends to all k°-special formal schemes in the case of a discretely valued (or trivially valued) ground field k. (The non-discretely valued case was not studied in the literature because the rings $k^\circ[[T_1 \ldots T_n]]$ are rather pathological, e.g. they contain non-closed ideals.)

The general idea of defining $\mathfrak{X}_\eta$ is actually the same: for

$$\mathfrak{X} = \mathrm{Spf}(k^\circ[T_1 \ldots T_n][[T_{n+1} \ldots T_m]])$$

one defines $\mathfrak{X}_\eta \subseteq \mathcal{M}(k\{T_1 \ldots T_m\})$ as the unit polydisc given by the conditions $|T_i| < 1$ for $n < i \le m$. In particular, the polydisc is open when $n = 0$ and is closed when $n = m$. For an affine $\mathfrak{Y}$ one defines $\mathfrak{Y}_\eta$ using a closed embedding into $\mathfrak{X}$ as above, and for a general special formal scheme the functor is defined using gluing. An anti-continuous reduction map $\mathfrak{X}_\eta \to \mathfrak{X}_s$ is defined as earlier.

Exercise 5.2.4.1   (i) $\mathfrak{X}_\eta$ is a good k-analytic space, and it is strict when the valuation is non-trivial.

(ii) In the case of affine $\mathfrak{X} = \mathrm{Spf}(A)$, the generic fiber can be identified with the set of real semivaluations on A that extend the valuation on k, are bounded and are strictly smaller than one on the elements of an ideal of definition of A.

(iii) In the affine case, $\mathfrak{X}_\eta$ is an increasing union of affinoid domains X_n such that X_n is Weierstrass in each X_m for $m \ge n$ (e.g. an open polydisc is an increasing union of smaller closed polydiscs).

One of motivations to introduce generic fibers of special fibers is the following result of Berkovich.

Fact 5.2.4.2 Let $\mathfrak{X}$ be a k°-special formal scheme (e.g. a formal scheme of finite type over k°) and let $Z \hookrightarrow \mathfrak{X}_s$ be a closed subscheme. Then the preimage of Z under the reduction map $\pi_\mathfrak{X}:\mathfrak{X}_\eta \to \mathfrak{X}_s$ depends only on the formal completion of $\mathfrak{X}$ along Z. Moreover, this preimage is precisely the generic fiber $(\hat{\mathfrak{X}}_Z)_\eta$ of the formal completion of $\mathfrak{X}$ along Z.

The following conjecture in the opposite direction describes the precise information about the formal scheme that is kept in the generic fiber $(\hat{\mathfrak{X}}_Z)_\eta$.

Conjecture 3 If $\mathfrak{X}$ is locally of the form $\mathrm{Spf}(\mathcal{A}^\circ)$ for a k-affinoid algebra $\mathcal{A}$ (i.e. $\mathfrak{X}$ is of finite type over k° and is normal in its generic fiber), then the henselization of $\mathfrak{X}$ along a closed subscheme $Z \hookrightarrow \mathfrak{X}_s$ is completely determined by the generic fiber $(\hat{\mathfrak{X}}_Z)_\eta$.

A partial evidence in favor of this conjecture is provided by the following result that was proved in [Tem5] and applied to resolution of singularities in positive characteristic.

Fact 5.2.4.4 Assume that $\mathfrak{X}'$ and $\mathfrak{X}$ are locally of the form $\mathrm{Spf}(\mathcal{A}^\circ)$ for a k-affinoid algebra $\mathcal{A}$, $Z \hookrightarrow \mathfrak{X}_s$ and $Z' \hookrightarrow \mathfrak{X}'_s$ are closed subschemes, and $f : \mathfrak{X}' \to \mathfrak{X}$ is a morphism that induces isomorphism between the preimages of Z' and Z in the generic fibers. Then the restriction of f onto a small neighborhood of Z' is strictly étale over Z.

5.3 Raynaud's Theory

5.3.1 An Overview

Assume that the valuation is non-trivial. We constructed a functor η whose source is the category $k^\circ\text{-}Fsch$ of formal k°-schemes of finite type and whose target is the category $st\text{-}k\text{-}An^c$ of compact strictly k-analytic spaces. Raynaud's theory completely describes this functor in the following terms: η is the localization of the source by an explicitly given family of morphisms $\mathcal{B}$. In particular, one can view a compact strictly k-analytic space $\mathfrak{X}_\eta$ as its *formal model* $\mathfrak{X}$ given up to a morphism from $\mathcal{B}$. (A possible analogy is to think about $\mathfrak{X}$ as a particular atlas of a manifold $\mathfrak{X}_\eta$ with morphisms from $\mathcal{B}$ being the refinements of the atlases.)

Clearly, the central part of the theory should be to describe the morphisms $f : \mathfrak{Y} \to \mathfrak{X}$ that are rig-isomorphisms (or generic isomorphisms). Although this is not so easy, we will find a nice cofinal family $\mathcal{B}$ among all rig-isomorphisms. To guess what such $\mathcal{B}$ can be, let us consider a very similar problem in the theory of schemes. Given a scheme X with a schematically dense open subscheme U (which will play the role of the generic fiber) by a U-modification of X we mean a proper morphism $f : X' \to X$ such that $f^{-1}(U)$ is schematically dense in X' and is mapped isomorphically onto U. A strong version of Chow lemma states that the family of U-admissible blow ups, i.e. blow ups whose center is disjoint from U, form a cofinal family among all U-modifications. Now, it is natural to expect that in our situation one can take $\mathcal{B}$ to be the family of all formal blow ups along open ideals (i.e. ideals supported on $\mathfrak{X}_s$).

5.3.2 Admissible Blow Ups

Definition 5.3.2.1 Recall that the blow up $\mathrm{Bl}_{\mathcal{J}}(X)$ of a scheme X along an ideal $\mathcal{J} \subseteq \mathcal{O}_X$ is defined as $\mathbf{Proj}(\oplus_{n=0}^\infty \mathcal{J}^n)$. If A is an I-adic ring then the formal blow up of the affine formal scheme $\mathrm{Spf}(A)$ along an ideal $J \subseteq A$ is defined as the I-adic completion of $\mathrm{Bl}_J(\mathrm{Spec}(A))$. This definition is local on the base and hence globalizes to a definition of *formal blow up* $\widehat{\mathrm{Bl}}_{\mathcal{J}}(\mathfrak{X})$ of a formal scheme $\mathfrak{X}$ along an ideal $\mathcal{J} \subseteq \mathcal{O}_{\mathfrak{X}}$. If $\mathcal{J}$ is open (i.e. contains an ideal of definition) then the formal blow up is called *admissible*.

Fact 5.3.2.2 (i) Any composition of (admissible) formal blow ups is an (admissible) formal blow up.
(ii) (Admissible) blow ups form a filtered family.

Exercise 5.3.2.3 If $\mathfrak{X}$ is of finite type over k° then any admissible formal blow up is a rig-isomorphism.

5.3.3 The Main Results

Fact 5.3.3.1 (Raynaud) The family $\mathcal{B}$ of formal blow ups in the category $k^{\circ}\text{-}F\,sch$ admits a calculus of right fractions and the localized category is equivalent to $st\text{-}k\text{-}An^{c}$. The localization functor is isomorphic to the generic fiber functor.

Remark 5.3.3.2 This fact implies the following two corollaries:

(i) The family of admissible blow ups of a formal scheme $\mathfrak{X}$ from $k^{\circ}\text{-}\mathcal{F}sch$ is cofinal in the family of all rig-isomorphisms $\mathfrak{X}' \to \mathfrak{X}$.
(ii) Each compact strictly k-analytic space X admits a *formal model* $\mathfrak{X}$, i.e. a formal scheme $\mathfrak{X}$ in $k^{\circ}\text{-}\mathcal{F}sch$ with an isomorphism $\mathfrak{X}_{\eta} \xrightarrow{\sim} X$.

Actually, these two statements serve as intermediate steps while proving Fact 5.3.3.1. Moreover, one proves that if $\{X_i\}$ is a finite family of compact strictly analytic domains in X then there exists a model $\mathfrak{X}$ with open subschemes $\mathfrak{X}_i$ such that $(\mathfrak{X}_i)_{\eta} \xrightarrow{\sim} X_i$.

Let us say a couple of words on the proof of Fact 5.3.3.1. The functor η takes morphisms of $\mathcal{B}$ to isomorphisms hence it induces a functor $\mathcal{F}\colon k^{\circ}\text{-}F\,sch/\mathcal{B} \to st\text{-}k\text{-}An^{c}$. One easily sees that $\mathcal{F}$ is faithful. The proof that $\mathcal{F}$ is full reduces to proving claim (i) of the above remark. This is essentially a strong version of the Chow lemma and the main ingredient in its proof is Gerritzen-Grauert theorem. Once we know that $\mathcal{F}$ is fully faithful, it remains to show that it is essentially surjective, i.e. to prove claim (ii) of the remark. For a strictly k-affinoid space it is very easy to find a formal model, and in general one chooses an affinoid covering $X = \cup_i X_i$, finds models $\mathfrak{X}_i$ and then uses that $\mathcal{F}$ is full to find formal blow ups $\mathfrak{X}'_i \to \mathfrak{X}$ so that the formal schemes $\mathfrak{X}'_i$ glue to a formal model $\mathfrak{X}'$ of X.

5.4 Rigid Geometry

A naive attempt to construct the generic fiber of an affine formal scheme $\mathfrak{X} = \mathrm{Spf}(A)$ is to declare that $\mathfrak{X}_{\eta}$ is the set $\mathrm{Spec}(A) \setminus \mathrm{Spf}(A)$ of all non-open ideals. Such definition is not compatible with formal localizations, because the Zariski topology becomes stronger on localizations. In particular, this definition cannot be globalized.

The situation improves, however, if one only considers the set of closed points of $\operatorname{Spec}(A) \setminus \operatorname{Spf}(A)$. In some cases such spectrum can be globalized, though the usual Zariski topology should be replaced with a certain G-topology in order to achieve this. We will not develop this point of view, but we will show how a similar approach gives rise to the rigid geometry of Tate.

Let k be a non-archimedean field with a non-trivial valuation. Strictly k-affinoid algebras are the basic objects of rigid geometry over k. An affinoid space $X_0 = \operatorname{Sp}(\mathcal{A})$ is defined as the set of maximal ideals of A provided with the G-topology of finite unions of affinoid domains. By Hilbert Nullstellensatz, the residue field of any point $x \in X_0$ is finite over k and hence X_0 is the set of Zariski closed points of $X = \mathcal{M}(\mathcal{A})$. The theory of rigid affinoid spaces and general rigid analytic spaces is developed similarly to the theory of strictly k-analytic spaces from §§2–3. Some intermediate results are slightly easier to prove because we only worry for Zariski closed points, but in the end one has less tools to solve problems. For example, Shilov boundaries and the class of good spaces are not seen in rigid geometry. Another example of an application where generic points are very important is the theory of étale cohomology of analytic spaces. There exist non-zero étale sheaves which have zero stalks at all rigid points but they necessarily have a non-zero stalk at a point of X. (Rigid points form a conservative family for coherent sheaves, and so the latter are easily tractable in the framework of Tate's rigid geometry.)

5.5 Adic Geometry

Adic geometry replaces formal schemes with more general objects that have a honest generic fiber (as an adic space). Let us recall why formal schemes have no generic fiber. Let k be a non-archimedean field with a non-trivial valuation and non-zero $\pi \in k^{\circ\circ}$ and let A be a π-adic k°-algebra of finite type over. Formal inverting of π produces the zero ring in two stages: first we invert π obtaining a strictly k-affinoid algebra $\mathcal{A}$ and then we have to factor over the unit ideal because the π-adic topology on $\mathcal{A}$ is trivial (and so the π-adic separated completion of $\mathcal{A}$ is 0). This suggests to extend the category of adic rings so that topological rings like $\mathcal{A}$ (with its Banach topology) are included. R. Huber suggested a way to do that, and it is very natural if we recall how the topology of k is actually defined.

Definition 5.5.0.1 An f-adic ring is a topological ring that contains an open adic ring A_0 with a finitely generated ideal of definition. Any such A_0 is called a ring of definition (because it can be used to define the topology of A).

Note that ring of definition is an analog of a unit ball for a norm.

Exercise 5.5.0.2 (i) Any k-affinoid algebra $\mathcal{A}$ is f-adic.
(ii) $\mathcal{A}$ is reduced if and only if $\mathcal{A}^{\circ}$ is a ring of definition.

Adic spectrum is defined analogously to analytic spectrum but using all continuous semivaluations. This forces one to modify the notion of a basic ring as follows.

Definition 5.5.0.3 (i) An *affinoid ring* is a pair $A = (A^\triangleright, A^+)$ where $A^\triangleright$ is an f-adic ring and A^+ is an open subring which is integrally closed in $A^\triangleright$ and is contained in the ring of power-bounded elements A°. The ring A^+ is called the *ring of integers* of A.

(ii) The adic spectrum $\mathrm{Spa}(A)$ is the set of all equivalence classes of continuous semivaluations on $A^\triangleright$ such that $|a| \le 1$ for any $a \in A^+$.

The (usual!) topology and the structure sheaf on $\mathrm{Spa}(A)$ are defined using rational domains. This is done similarly to our definitions, so we omit the details.

Example/Exercise 5.5.0.4 (i) The k-affinoid rings in adic geometry are the pairs $(\mathcal{A}, \mathcal{A}^\circ)$ with (k, k°) playing the role of the ground field. Show that the only k°-ring of integers of a strictly k-affinoid algebra $\mathcal{A}$ is $\mathcal{A}^\circ$.

Hint: use Exercise 5.2.2.2.

(ii) For any adic k°-algebra A of finite type, (A, A) is an affinoid ring. Check that $\mathfrak{X} = \mathrm{Spa}(A, A)$ is an adic space and set-theoretically $\mathfrak{X}$ is the disjoint union of the *generic fiber* $\mathfrak{X}_\eta = \mathrm{Spa}(\mathcal{A}, \mathcal{A}^\circ)$ where $\mathcal{A} = A_\pi$ and the closed fiber $\mathfrak{X}_s$ which consists of all semivaluations that vanish on π. Note that $\mathrm{Spf}(A)$ naturally embeds into $\mathfrak{X}_s$ as the set of points of $\mathfrak{X}$ with trivial valuation on the residue field.

(iii) An *affinoid field* is an affinoid ring k such that $k^\triangleright$ is a valued field of height $h^\triangleright \le 1$ (with the induced topology) and k^+ is a valuation ring of k contained in k°. Let h^+ be the height of k^+. Show that $\mathrm{Spa}(k)$ is a chain (under specialization) of $h^+ + 1 - h^\triangleright$ points.

Remark 5.5.0.5 (i) Spectra of affinoid fields are "atomic objects" in the sense that they do not admit non-trivial monomorphisms from other spaces such that the closed point is contained in the image. In particular, a point of height at least two or a fiber of a morphism over such point is not an adic space.

(ii) Points of height at least two are very different from the usual analytic points. For example, their local rings are usually not henselian (because there is no reasonable completion for valued fields of height more than one).

Exercise 5.5.0.6 Describe all adic points of the affine line over k.

Hint: show that the only new points are height two points contained in the closures of type two points $x \in \mathbf{A}_k^{\mathrm{an}}$. Each connected component of $\mathbf{A}_k^{\mathrm{an}} \setminus \{x\}$ contains one such point.

5.6 Comparison of Categories and Spaces

In principle, all approaches to non-archimedean analytic geometry produce the same category of spaces of finite type (with rational radii of convergence).

Fact 5.6.0.1 The following categories are naturally equivalent: (a) the category of compact strictly k-analytic spaces, (b) the category of formal k°-schemes of finite type localized by admissible blow ups, (c) the category of quasi-compact and quasi-separated rigid k-analytic space, (d) the category of quasi-compact and quasi-separated adic $\text{Spa}(k, k^{\circ})$-spaces of locally finite type.

Let $\mathfrak{X}$ be a formal k°-scheme of finite type and let $\mathfrak{X}_{\eta}^{\text{rig}}$, $\mathfrak{X}_{\eta}^{\text{an}}$ and $\mathfrak{X}_{\eta}^{\text{ad}}$ be its generic fibers in the three categories of non-archimedean spaces over k. On the level of topological spaces these objects are related as follows.

Fact 5.6.0.2 One has that $\text{proj}\lim_{f:\mathfrak{X}'\to\mathfrak{X}} |\mathfrak{X}'| \xrightarrow{\sim} |\mathfrak{X}_{\eta}^{\text{ad}}| \supset |\mathfrak{X}_{\eta}^{\text{an}}| \supset |\mathfrak{X}_{\eta}^{\text{rig}}|$, where the limit is taken over all admissible formal blow ups f. Furthermore, $\mathfrak{X}_{\eta}^{\text{an}}$ is the set of all points of height one in $\mathfrak{X}_{\eta}^{\text{ad}}$ and also it is homeomorphic to the maximal Hausdorff quotient of $\mathfrak{X}_{\eta}^{\text{ad}}$, and $\mathfrak{X}_{\eta}^{\text{rig}}$ is the set of Zariski closed points of $\mathfrak{X}_{\eta}^{\text{an}}$.

Finally, the sheaves on these spaces are connected as follows.

Fact 5.6.0.3 The topoi (i.e. the categories of sheaves of sets) of the following sites are equivalent: $\mathfrak{X}_{\eta}^{\text{an}}$ with the G-topology of compact strictly k-analytic domains, $\mathfrak{X}_{\eta}^{\text{rig}}$ with the topology of compact rigid domains, and $\mathfrak{X}_{\eta}^{\text{ad}}$ with its usual topology. In particular, $\mathfrak{X}_{\eta}^{\text{ad}}$ is simply the set of (equivalence classes of) points of all these sites.

5.7 Reduction of Germs and Riemann-Zariski Spaces

The aim of this section is to use certain reduction data to study the local structure of an analytic space X at a point x. For simplicity we will assume that X is strictly analytic, and will discuss the general case in Sect. 5.7.5. The main tool we will develop and use is a reduction functor that associates to the germ (X, x) certain birational object $(X, x)^{\sim}$ over $\tilde{k}$. We will see that it has interesting connections to Raynaud's theory, adic geometry and the classical (though not very well known) Riemann-Zariski spaces. The reduction functor is used to establish fine local properties of analytic spaces, including Facts 4.2.4.3 and 4.1.3.5.

5.7.1 Reduction of Germs: Preliminary Version

To motivate the definition of $(X, x)^{\sim}$ let us first pursue the following mini-goal: find a simple criterion for an analytic space X to be not good at a point $x \in X$ and use it to justify the claim of Example 4.2.1.4. Note that an affinoid space $X = \mathcal{M}(A)$ possesses an affine formal model $\mathfrak{X}$: although $\mathcal{A}^{\circ}$ dos not have to be of finite type over k°, we can simply take an admissible surjection $f : k\{\underline{T}\} \to \mathcal{A}$ and set $\mathfrak{X} = \text{Spf}(f(k^{\circ}\{\underline{T}\}))$. By Raynaud's theory, for any other formal model $\mathfrak{X}'$ there exists a formal model $\mathfrak{X}''$ which is an admissible blow up of both $\mathfrak{X}$ and $\mathfrak{X}'$. It follows that the reduction $\tilde{X} = \mathfrak{X}'_s$ is a $\tilde{k}$-variety of a rather special form: there exists an $\tilde{X}$-proper variety which admits a proper morphism to an affine variety. It turns out that this observation can be localized to a criterion of goodness at a point.

Exercise 5.7.1.1

(i)* Let X be strictly analytic, let $x \in X$ be a point and let $\mathfrak{X}$ be a formal model with reduction $\tilde{X} = \mathfrak{X}_s$. Let Z denote the Zariski closure of the image of x under the reduction map $\pi_{\mathfrak{X}} : X \to \tilde{X}$, and assume that there is no proper morphism $Z' \to Z$ such that Z' admits a proper morphism to an affine variety. Show that X is not good at x.

   Hint: use Fact 3.4.2.3(vi) and Raynaud's theory.

(ii) Deduce that both spaces in Example 4.2.1.4 are non-good.

   Now, it is natural to expect that one can define an interesting reduction invariant of the germ (X, x) by considering varieties $Z = \overline{\pi_{\mathfrak{X}}(x)}$ as above modulo an equivalence relation that eliminates the freedom in the choice of the reduction. A straightforward attempt would be to factor the category of $\tilde{k}$-varieties by surjective proper morphisms, but we can slightly refine this by taking the residue field $\widetilde{\mathcal{H}(x)}$ into account (recall that it does not have to be finitely generated over $\tilde{k}$). This is based on the observation that $\pi_{\mathfrak{X}}(x)$ is the image of the reduction morphism $\mathrm{Spec}(\widetilde{\mathcal{H}(x)}) \to \tilde{X}$. Now we are ready to give a preliminary definition of the reduction functor.

Definition 5.7.1.2 (i) Let $\mathrm{Var}_{\tilde{k}}$ be the category of *pointed $\tilde{k}$-varieties* (i.e. morphisms $\mathrm{Spec}(K) \to Z$, where K is a $\tilde{k}$-field and Z is a $\tilde{k}$-variety) with morphisms f been compatible pairs of morphisms $f_{\eta} : \mathrm{Spec}(K') \to \mathrm{Spec}(K)$ and $f_s : Z' \to Z$. Let $\mathcal{B}$ denote the family of morphisms in which f_s is proper and f_{η} is an isomorphism, and let $\mathrm{Var}_{\tilde{k}}/\mathcal{B}$ denote the localization category.

(ii) For any compact strictly analytic X with a point x and a formal model $\mathfrak{X}$ consider the object $\mathrm{Spec}(\widetilde{\mathcal{H}(x)}) \to \mathfrak{X}_s$ of $\mathrm{Var}_{\tilde{k}}$. The same object viewed in the category $\mathrm{Var}_{\tilde{k}}/\mathcal{B}$ is denoted $(X, x)_{\tilde{\mathfrak{X}}}$

Fact 5.7.1.3 The isomorphism class of $(X, x)_{\tilde{\mathfrak{X}}}$ depends only on the germ (X, x) (and so is independent of the choice of $\mathfrak{X}$ and is preserved if we replace X with a neighborhood of x). Moreover, this construction is functorial on the category of germs of strictly analytic spaces at a point.

5.7.2   The Category $\mathrm{bir}_{\tilde{k}}$

The family $\mathcal{B}$ in Definition 5.7.1.2 is easily seen to admit the calculus of right fractions (i.e. any morphism in $\mathrm{Var}_{\tilde{k}}/\mathcal{B}$ is of the form $f \circ b^{-1}$ where $b \in \mathcal{B}$ and f is a morphism in $\mathrm{Var}_{\tilde{k}}$). In particular, the localization functor can be easily described. Nevertheless, it would be desirable to have a more geometric interpretation of $\mathrm{Var}_{\tilde{k}}/\mathcal{B}$, and it turns out that the latter is provided by classical RZ (or Riemann-Zariski) spaces.

Definition 5.7.2.1 For a $\tilde{k}$-field K let $\mathrm{RZ}_{\tilde{k}}(K)$ denote the set of all valuation rings $k \subseteq \mathcal{O} \subseteq K$ with $\mathrm{Frac}(\mathcal{O}) = K$. Provide $\mathrm{RZ}_{\tilde{k}}(K)$ with the topology whose basis is formed by the sets

$$\mathrm{RZ}_{\tilde{k}}(K[f_1 \ldots f_n]) = \{\mathcal{O} \in \mathrm{RZ}_{\tilde{k}}(K) | \ f_1 \ldots f_n \in \mathcal{O}\}$$

for any choice of $f_1 \ldots f_n \in K$.

Remark 5.7.2.2 Such spaces were introduced by Zariski in 1930ies. He called them Riemann spaces for their (very relative) analogy with Riemann surfaces. These spaces are sometimes used in birational geometry, for example, for resolution of singularities, and their modern name is Riemann-Zariski or Zariski-Riemann spaces (in both variants).

Fact 5.7.2.3 The spaces $\mathrm{RZ}_{\tilde{k}}(K)$ are quasi-compact and quasi-separated (qcqs).

Remark 5.7.2.4 This simple fact is due to Zariski. Funny enough, it was reproved in many works and often incorrectly (including Nagata's work on compactification and [Tem1]). The mistake is always the same—one assumes that quasi-compactness is preserved under projective limits, which is incorrect in general.

Definition 5.7.2.5 A *birational space over* $\tilde{k}$ is a qcqs topological space X with a local homeomorphism $\phi : X \to \mathrm{RZ}_{\tilde{k}}(K)$. A morphism between $\phi' : X' \to \mathrm{RZ}_{\tilde{k}}(K')$ and ϕ is a $\tilde{k}$-embedding $i : K \hookrightarrow K'$ and a continuous map $X' \to X$ compatible with the map $\mathrm{RZ}_{\tilde{k}}(K') \to \mathrm{RZ}_{\tilde{k}}(K)$ induced by i. The category of birational spaces over $\tilde{k}$ will be denoted $\mathrm{bir}_{\tilde{k}}$.

Remark 5.7.2.6 In this definition, ϕ plays the role of a structure sheaf. Such a naive structure sheaf suffices to define the category $\mathrm{bir}_{\tilde{k}}$. For the sake of completeness we note that one can provide X with a real structure sheaf $\mathcal{O}_X$ whose stalk at a point $x \in X$ is the corresponding valuation ring. This more refined approach becomes useful when one works with (much more general) relative Riemann-Zariski spaces.

Fact 5.7.2.7 For any pointed $\tilde{k}$-variety $f : \mathrm{Spec}(K) \to X$ of $\mathrm{Var}_{\tilde{k}}$ one can construct a birational space consisting of all pairs $(\mathcal{O}, g)$, where $\mathcal{O} \in \mathrm{RZ}_{\tilde{k}}(K)$ and $g : \mathrm{Spec}(\mathcal{O}) \to X$ is a morphism compatible with f (forgetting g defines an obvious projection onto $\mathrm{RZ}_{\tilde{k}}(K)$). This defines a functor $\mathrm{Val} : \mathrm{Var}_{\tilde{k}} \to \mathrm{bir}_{\tilde{k}}$ which induces an equivalence $\mathrm{Var}_{\tilde{k}}/\mathcal{B} \xrightarrow{\sim} \mathrm{bir}_{\tilde{k}}$.

Definition/Exercise 5.7.2.8 (i) Given an object (resp. morphism) X of $\mathrm{bir}_{\tilde{k}}$, by a *scheme model* of X we mean any object $\mathcal{X}$ (resp. morphism) of $\mathrm{Var}_{\tilde{k}}$ such that $\mathrm{Val}(\mathcal{X}) \xrightarrow{\sim} X$.

(ii) A birational space $\phi : X \to \mathrm{RZ}_{\tilde{k}}(K)$ is *affine* if ϕ is injective and its image is of the form $\mathrm{RZ}_{\tilde{k}}(K[f_1 \ldots f_n])$. Show that ϕ is affine if and only if it admits an affine scheme model.

(iii) A morphism $g : \phi' \to \phi$ is *separated* (resp. *proper*) if the map $X' \to X \times_{\mathrm{RZ}_{\tilde{k}}(K)} \mathrm{RZ}_{\tilde{k}}(K')$ is injective (resp. bijective). Show that this happens if and only if g admits a separated (reps. proper) scheme model, and then any scheme model is separated (resp. proper).

Hint: this is a slightly refined version of the valuative criteria.

(iv) A birational space $\phi : X \to \mathrm{RZ}_{\tilde{k}}(K)$ is *separated* (resp. *proper*) if so is its morphism to the final object $\mathrm{RZ}_{\tilde{k}}(\tilde{k})$. Show that this happens if and only if ϕ is injective (resp. bijective).

(v)* Show that there is no reasonable notion of affine morphisms in $\mathrm{bir}_{\tilde{k}}$.

Hint: if you have solved Exercise 5.7.1.1 then you may already know an appropriate example of a morphism in $\mathrm{bir}_{\tilde{k}}$.

The following fact is already due to Zariski.

Fact 5.7.2.9 If $X \to \mathrm{RZ}_{\tilde{k}}(K)$ is a birational space and $\mathrm{Spec}(K) \to \mathcal{X}_i$, $i \in I$ is the family of all its scheme models then the natural map $X \to \mathrm{proj\,lim}_{i \in I} \mathcal{X}_i$ is a homeomorphism.

Finally, we define germ reductions as birational spaces.

Definition 5.7.2.10 For any germ (X, x) of a strictly analytic space X take a compact neighborhood V of x with a formal model $\mathfrak{V}$ and define $(X, x)^\sim$ to be the image of $(V, x)^\sim_{\mathfrak{V}}$ in $\mathrm{bir}_{\tilde{k}}$. It is called *reduction of X at x* or *reduction of the germ (X, x)*.

Due to Facts 5.7.1.3 and 5.7.2.7, germ reduction is a functor from the category of germs to the category of birational spaces.

5.7.3 Relation to Other Theories

Remark 5.7.3.1 There is a strong analogy between formal k°-schemes, adic spaces and the functor $\mathfrak{X} \mapsto \mathfrak{X}^{\mathrm{ad}}_\eta$ on one side, and pointed varieties, birational spaces and the functor Val on the other hand. (Also, we stress this analogy by saying formal models and scheme models.) In particular:

(a) both adic and birational spaces are built from valuations of arbitrary height.

(b) Raynaud's theory says that η induces an equivalence of a category of formal schemes localized by admissible blow ups and a category of adic spaces. The equivalence $\mathrm{Var}_{\tilde{k}}/\mathcal{B} \xrightarrow{\sim} \mathrm{bir}_{\tilde{k}}$ induced by Val can be viewed as a baby version of Raynaud's theory.

(c) On the set-theoretic level, Raynaud's theory reduces to homeomorphism between adic (resp. birational) spaces and projective limits of their formal (resp. scheme) models.

An additional link between adic spaces and the germ reduction functor is provided by the following fact.

Fact 5.7.3.2 Let X be a strictly analytic space with associated adic space X^{ad}, let $x \in X$ be a point and let $\overline{x}$ be the closure of x in X^{ad}. Then $(X, x)^\sim$ is naturally homeomorphic to $\overline{x}$.

Remark 5.7.3.3 (i) This fact suggests an alternative definition of the reduction functor. It is more elegant, but not suited for computations.

(ii) In some sense, the role of the reduction functor is to make visible some local information contained in X^{ad} but not seen directly in the analytic geometry.

5.7.4 Main Properties and Applications

It turns out that many local properties of analytic spaces and their morphisms are reflected by the reduction functor, and here is the list of the main ones.

Fact 5.7.4.1 (i) The reduction functor establishes a bijection between germ sub-domains (Y, x) of (X, x) and birational subspaces of $(X, x)^{\sim}$. This bijection preserves intersections, finite unions and inclusions.

(ii) A strictly analytic space X is good at a point x if and only if the reduction $(X, x)^{\sim}$ is affine.

(iii) A morphism $f : Y \to X$ of strictly analytic spaces is separated at a point $y \in Y$ (resp. $y \in \text{Int}(Y/X)$) if and only if the reduction morphism $(Y, y)^{\sim} \to (X, f(y))^{\sim}$ is separated (resp. proper).

Exercise 5.7.4.2 Check that Fact 5.7.4.1(iii) reduces to Fact 3.4.2.3(vi) when X and Y are affinoid. In particular, show that if $X = \mathcal{M}(\mathcal{A})$ and $x \in X$ is a point with character $\chi_x : \mathcal{H}(x) \to \mathcal{A}$ then $\text{Spec}(\widetilde{\mathcal{H}(x)}) \to \text{Spec}(\tilde{\chi}_x(\tilde{\mathcal{A}}))$ is a scheme model of $(X, x)^{\sim}$.

Note that Fact 5.7.4.1(ii) gives a simple necessary and sufficient criterion of goodness. We have already used this criterion to show that certain spaces are not good. The opposite implication (i.e. the criterion of goodness) is the difficult one, and it is the deepest property asserted by the Fact. All other claims follow relatively easily. (We already mentioned in Remark 4.2.4.4 that the most difficult task in studying properness is to show that certain spaces are good.)

5.7.5 Germ Reduction of Non-Strict Spaces

Finally, let us discuss how the reduction functor can be extended to non-strict analytic spaces. There exists no generalization of formal models and Raynaud's theory for general H-strict k-analytic spaces (though affinoid reduction can be defined as in Remark 3.4.1.7). On the other hand, one can define a category $\text{bir}_{\tilde{k}_H}$ of H-graded birational spaces over $\tilde{k}_H$ analogous to the category $\text{bir}_{\tilde{k}}$ (this requires to define H-graded valuation rings, etc.). Then an H-graded germ reduction functor with values in $\text{bir}_{\tilde{k}_H}$ can be constructed as follows: for an affinoid $X = \mathcal{M}(\mathcal{A})$ take $(X, x)^{\sim}_H$ to be the H-graded birational space corresponding to the homomorphism $(\widetilde{\chi_x})_H : \tilde{\mathcal{A}}_H \to \widetilde{\mathcal{H}(x)}_H$, and in general we cover a germ (X, x) by good subdomains (X_i, x) and glue $(X, x)^{\sim}_H$ from $(X_i, x)^{\sim}_H$.

Fact 5.7.5.1 (i) All assertions of Fact 5.7.4.1 generalize verbatim to the H-graded setting.

(ii) There is a natural fully faithful embedding $\mathrm{bir}_{\widetilde{k}_H} \hookrightarrow \mathrm{bir}_{\widetilde{k}_G}$, where $G = \mathbf{R}_+^\times$. An analytic space X is H-strict locally at x if and only if the reduction $(X, x)_{\widetilde{G}}$ comes from the subcategory $\mathrm{bir}_{\widetilde{k}_H}$.

Fact 5.7.5.1(ii) provides a local criterion of H-strictness which was proved in [Tem2] for strictly analytic category and generalized to any H in [CT]. As a simple corollary, it was shown in loc.cit. that $k_H\text{-}An$ is a full subcategory of $k\text{-}An$.

6 Analytic Curves

Definition 6.1 (i) A k-*analytic curve* C is a k-analytic space of pure dimension one, i.e. $\dim(C) = 1$ and C does not contain discrete Zariski closed points.

(ii) In the same way as in Definition 2.3.3.3 we divide the points of C into four types accordingly to their completed residue fields.

6.1 Examples

6.1.1 Constructions

Let us first list some constructions that allow to create/enrich our list of k-analytic curves: (i) analytification of an algebraic curve, (ii) generic fibers of formal curves, (iii) an analytic domain in a curve, (iv) a finite covering of a curve (or, more generally, a covering with discrete fibers).

Example/Exercise 6.1.1.1 (i) The following curves can be obtained by the first method: affine line, projective curves, affine line with doubled origin.

(ii) Let $\mathcal{X}$ be an irreducible projective algebraic k-curve with $k(\mathcal{X}) = K$. The Zariski closed points of $X = \mathcal{X}^{\mathrm{an}}$ are in one-to-one correspondence with the closed points of $\mathcal{X}$, and other points of X are in one-to-one correspondence with the valuations on K that extend the valuation on k. In particular, $K \hookrightarrow \mathcal{H}(x)$ and $\hat{K} \xrightarrow{\sim} \mathcal{H}(x)$ for any non Zariski closed point $x \in X$.

(iii) The following curves can be obtained by the second method: compact k-analytic curves.

(iv) Most of Hausdorff curves admit a formal model of locally finite type over k°. For example, find such models for an affine line and for an open unit disc.

Now, let us study the other two methods with more details.

6.1.2 Domains in the Affine Line

A typical example of a compact domain X is a closed disc $E(a, r)$ with finitely many removed open discs $E(a_i, r_i)$.

Exercise 6.1.2.1 (i) Prove that X is a Laurent domain in $E(a, r)$.
(ii) Show that if r and r_1 are linearly independent over $|k^\times|$ then X is not a finite covering of a disc.
  Hint: if $X = \mathcal{M}(A)$ is finite over $E(b, s)$ then $\rho(A) \subset \{0\} \cup \sqrt{s^{\mathbf{Z}}|k^\times|}$.
(iii) Show that one can extend X a little bit so that $r_i \in \sqrt{r^{\mathbf{Z}}|k^\times|}$ and then X is a finite covering of $E(0, r)$.

Next we describe neighborhoods of an especially simple form.

Exercise 6.1.2.2 Assume that $k = k^a$. Show that a point $x \in \mathbf{A}_k^1$ admits a fundamental family of open neighborhoods X_i as follows:

(i) if x is of type 1 or 4 then X_i are open discs,
(ii) if x is of type 3 then X_i are open annuli,
(iii) if x is of type 2 then X_i are open discs with removed finitely many closed discs, and in addition one can achieve that $X_i \setminus \{x\}$ is a disjoint union of open discs and finitely many open annuli.

Finally, let us discuss some open domains in $\mathbf{A}_k^1$ with $k = k^a$.

Exercise 6.1.2.3 (i) Show that a separated gluing of two open annuli along an open annulus is an open annulus.
  Hint: use Exercise 4.1.4.2(i).
(ii) Show that a filtered union of a countable family of annuli does not have to be an annulus.
  Hint: take $\mathbf{P}_k^1$ and remove two type 4 points.
(iii) We say that a non-archimedean field k is *local* if $|k^\times| \xrightarrow{\sim} \mathbf{Z}$ and $\tilde{k}$ is finite. Show that either k is finite over $\mathbf{Q}_p$ or $k \xrightarrow{\sim} \mathbf{F}_p((t))$. Drinfel'd upper half-plane is defined as $\mathbf{P}_k^1 \setminus \mathbf{P}_k^1(k)$. Show that it is an open analytic domain in $\mathbf{P}_k^1$.

6.1.3 Finite Covers

First, we consider an inseparable cover giving rise to a pathology that cannot occur in the algebraic world.

Example/Exercise 6.1.3.1 (i) Construct a non-archimedean field k with $\mathrm{char}(k) = p$ and $[k : k^p] = \infty$.
(ii) Choose elements $a_i \in k$ which are linearly independent over k^p and such that $|a_i|$ tend to zero. Set $\mathcal{A} = k\{T, S\}/(S^p - \sum_{i=0}^\infty a_i T^{pi})$. Show that $X = \mathcal{M}(\mathcal{A})$ is a finite covering of $E(0, 1)$ of degree p, $X \otimes_k l$ is reduced for any finite field extension l/k but $X \hat{\otimes}_k k^{1/p}$ is not reduced.

Next, we study some quadratic covers. They will give rise to various interesting examples.

Exercise 6.1.3.2 Assume that $\text{char}(\tilde{k}) \neq 2$ and consider a power series of the form $f(T) = \sum_{i=0}^{\infty} a_i T^i \in k\{T\}$ with the affinoid algebra $\mathcal{A} = k\{T, S\}/(S^2 - f(T))$ and the quadratic covering $\phi : X = \mathcal{M}(\mathcal{A}) \to E(0, 1) = \mathcal{M}(k\{T\})$. Show that any type 4 point $x \in \mathbf{A}_k^1$ has two preimages, and the maximal point p_r of the disc $E(0, r)$ has two preimages if and only if there exists $i \in 2\mathbf{N}$ such that $|a_i| r^i > |a_j| r^j$ for any $i \neq j \in \mathbf{N}$. In particular, a Zariski closed point x has two preimages if and only if $f(x) \neq 0$.

Hint: show that the binomial expansion of $\sqrt{1 + z}$ has radius of convergence 1 over k.

Now let us study elliptic curves using double covers. For simplicity, we also assume that $k = k^a$.

Example/Exercise 6.1.3.3 It is known from algebraic geometry that any elliptic curve over k can be realized as the double covering $\phi : E \to \mathbf{P}_k^1$ given by $S^2 = T(T - 1)(T - \lambda)$.

(i) Assume that $|\lambda| > 1$.

    (a) Show that the points with one preimage are precisely the points of the disjoint intervals $[0, 1]$ and $[\lambda, \infty]$. In particular, X contains a cycle $\Delta(E)$ which is the preimage of the interval $I = [p_1, p_{|\lambda|}]$ and a contraction of $\mathbf{P}_k^1$ onto I lifts to the contraction of E onto $\Delta(E)$.

    (b) Show that the preimage of the disc $E(0, r)$ with $1 < r < |\lambda|$ is a closed annulus, and deduce that E is glued from two annuli.

    (c)* Show that E can be obtained from $A(0; 1, |\lambda|^2)$ by identifying $A(0; 1, 1)$ with $A(0; |\lambda|^2, |\lambda|^2)$. Moreover, the universal cover of E is isomorphic to G_m and $G_m/q^{\mathbf{Z}} \xrightarrow{\sim} E$ for an element $q \in k$ with $|q| = |\lambda|^2$.

(ii) Show that if $|\lambda - 1| < 1$ or $|\lambda| < 1$ then the structure of E is similar but with respect to other intervals connecting the four points.

(iii) Assume that $|\lambda| = |\lambda - 1| = 1$.

    (a) Show that the points with one preimage are p_1 and the points of the disjoint intervals $[0, p_1)$, $[1, p_1)$, $[\lambda, p_1)$ and $[\infty, p_1)$. Let $z = \Delta(E)$ be the preimage of p_1; then the contraction of $\mathbf{P}_k^1$ onto p_1 lifts to the contraction of E onto z.

    (b) Show that $E \setminus \{z\}$ is a disjoint union of open discs. Furthermore, $\widetilde{\mathcal{H}(z)}$ is of genus one over $\tilde{k}$ and the closed points of its projective model parameterize the open discs of $E \setminus \{z\}$. In particular, z is not locally embeddable into $\mathbf{A}_k^1$.

The curves from (i) and (ii) are called Tate curves, or elliptic curves with bad reduction. The curves from (iii) are called curves with good reduction. In the sequel by *genus* of a type two point z we mean the algebraic genus of $\widetilde{\mathcal{H}(z)}$ over $\tilde{k}$. Points of positive genus are very special and very informative.

Exercise 6.1.3.4 (i) Study curves C of genus two given by $S^2 = f(T)$ with $f(T)$ of degree five. Show that the first Betti number of C plus the sum of genera of its type two points equals to the genus of C.

(ii)* Prove the same for any C given by $S^2 = f(T)$ where $f(T)$ is a polynomial without multiple roots.

Finally, let us construct wild non-compact examples.

Exercise 6.1.3.5 Assume that k is not discretely valued.

(i) Show that if $|a_i|$ increase and tend to 1 then $f(T) = \sum_{i=0}^{\infty} a_i T^i$ is a bounded function with infinitely many roots on the open unit disc $D(0, 1)$.

(ii) Show that by an appropriate choice of $f(T)$ as above one can achieve that the corresponding double cover C of $D(0, 1)$ has infinitely many loops and infinitely many positive genus points.

It will follow from some further results that C is an example of a non-compactifiable space. In the sequel we will study compactifiable (usually compact) curves.

6.2 General Facts About Compact Curves

6.2.1 Algebraization

Fact 6.2.1.1 Any proper k-analytic curve X is projective. In particular, X is algebraizable.

Exercise 6.2.1.2 (i)* Prove the above fact.

Hint: take a Zariski closed point P and show that $H^1(X, \mathcal{O}_X(nP))$ vanishes for large enough n by Kiehl's theorem on direct images 4.2.4.5. Deduce that a linear system $\mathcal{O}_X(nP)$ with large enough n gives rise to a finite morphism from X to the projectivization of $H^0(X, \mathcal{O}_X(nP))$. Then use Fact 5.1.2.2 from GAGA.

(ii) Deduce that the curve from Exercise 6.1.3.1 is not a domain in a proper curve. Moreover, find k with $[k : k^p] < \infty$ and a finite extension K/K_r such that K is not isomorphic to the completion of a finitely generated k-field of transcendence degree one. Use this to construct a k-affinoid curve that cannot be embedded into a proper curve.

6.2.2 Compactification

We saw that if a separated compact curve C is not geometrically reduced then it does not have to be embeddable into a proper one. In the opposite direction we have the following result.

Fact 6.2.2.1 Any compact separated geometrically reduced k-analytic curve is isomorphic to a domain in a projective k-curve.

A general idea of the proof is as follows: we would like to patch the boundary of C, which consists of generic points (of types 2 and 3). One proves that a curve is geometrically reduced at a non Zariski closed point if and only if it is rig-smooth at such point. If C is rig-smooth at x then it admits a quasi-étale morphism $\phi:C \to \mathbf{A}_k^1$ locally around x. If we deform a quasi-étale morphism slightly then the isomorphism class of C does not change. Therefore we can define ϕ using only equations of the form $\sum_{i=0}^{n} a_i(T)y^i$ where the coefficients a_i are meromorphic. This allows to compactify C at all points of its boundary.

Using the Riemann-Roch theorem on a projective curve one deduces the following corollary.

Fact 6.2.2.2 A separated, compact, and geometrically reduced k-analytic curve is affinoid if and only if it does not contain proper irreducible components.

6.2.3 Formal Models

In the sequel we assume that C is a compact strictly k-analytic rig-smooth curve and the valuation is non-trivial. For any formal model $\mathfrak{C}$ of C let $\mathfrak{C}^0 \subset C$ be the preimage of the set of generic points of $\mathfrak{C}_s$ under the reduction map.

Fact 6.2.3.1 (i) The set $\mathfrak{C}_0$ determines the formal model $\mathfrak{C}$ up to a finite admissible blow up.

(ii) If C is separated then a finite set V of type 2 points is of the form $\mathfrak{C}_0$ for some formal model if and only if V contains the boundary of C and hits each proper irreducible component of C.

Exercise 6.2.3.2 (i) Show that (ii) above does not hold in the non-separated case
         Hint: take the closed disc with doubled open disc, and patch in an open annulus instead of the doubled open disc. Then there is no formal model with a single generic point (although such a model exists as a not locally separated formal algebraic space).
(ii)* Deduce Fact 6.2.3.1 from Fact 6.2.2.2.

6.3 Rig-Smooth Curves

Until the end of the paper, C is a compact rig-smooth k-analytic curve. For simplicity, we also assume that k is algebraically closed and C is connected. In general, all our results hold up to a finite ground field extension and obvious corrections needed to deal with disconnected curves.

6.3.1 Geometric Structure of Analytic Curves

Here is the main result about the structure of C in its geometric formulation. An equivalent approach via formal models will be discussed later.

Fact 6.3.1.1 There exists a finite set V of type two points such that $C \setminus V$ is a disjoint union of open discs and finitely many open annuli.

This claim is very strong and implies many other important results that we state as exercises.

Exercise 6.3.1.2 (i) C has finitely many points of positive genus and C can be contracted onto its subset $\Delta(V)$ homeomorphic to a finite graph.

Hint: take $\Delta(V)$ to be the the union of V and the open intervals through the annuli; then $C \setminus \Delta(V)$ is a disjoint union of open unit discs which can be easily contracted.

(ii)* If C is proper then its algebraic genus equals to the sum of the genera of type two points plus the first Betti number of $\Delta(V)$.

Hint: use the semistable formal model associated to V in the next section.

Next, let us study what is the freedom in the choice of V.

Exercise 6.3.1.3 (i) Cofinality: show that any finite set of type two points can be enlarged to a set V as above.

(ii) Fact 4.3.3.1 holds true for curves.

Hint: the sets $\Delta(V)$ form the required filtered family.

(iii)* Minimality: show that there exists a minimal such V unless $C \xrightarrow{\sim} \mathbf{P}^1_k$ or C is a Tate curve. (Note that the degenerate cases are proper curves that can be covered by annuli.)

Next, let us describe the local structure of C.

Exercise 6.3.1.4 Show that a point $x \in C$ has a fundamental family of open neighborhoods X_i such that

(i) X_i are open discs when x is of type 1 or 4,
(ii) X_i are open annuli when x is of type 2,
(iii) $X_i \setminus \{x\}$ are disjoint unions of open discs and finitely many open annuli when x is of type 2.

Using gluing of annuli and discs from Fact 4.1.4.2, it is easy to show that the above local description of C is equivalent to its global description. This local fact, in its turn, easily reduces to study of the field $\mathcal{H}(x)$. For example, for types 3 and 4 it suffices to show that $\mathcal{H}(x)$ is topologically generated by an element T (i.e. $\mathcal{H}(x) = \widehat{k(T)}$). Surprisingly, no simple proof of this fact is known. A shortest currently known proof can be found in [Tem4], where it is used to obtain a new proof of the semistable reduction theorem (we will see that the semistable reduction theorem is equivalent to Fact 6.3.1.1).

6.3.2 Semistable Formal Models

Definition 6.3.2.1 A formal k°-scheme is *semistable* if it is étale-locally isomorphic to the formal schemes of the form $\mathfrak{Z}_{n,a} = \mathrm{Spf}(k^\circ\{T_1 \ldots T_n\}/(T_1 \ldots T_n - a))$ with $a \in k^\circ$.

Let $\mathfrak{X}$ be normal in its generic fiber. Then $\mathfrak{X}$ is semistable if and only if it has the same formal fibers as the model schemes $\mathfrak{Z}_{n,a}$ (this particular case of Conjecture 3 is easily verified by a direct computation). In the case of curves this gives the following result.

Fact 6.3.2.2 A formal k°-curve $\mathfrak{C}$ with a rig-smooth generic fiber is semistable if and only if the formal fibers over its closed points are open discs (over the smooth points) and open annuli (over the double points of $\mathfrak{C}_s$).

This exercise and Fact 6.2.3.1 imply that the global description of C given by Fact 6.3.1.1 is equivalent to the following fundamental result, which can be proved alternatively by a classical but rather complicated algebraic theory that involves stable reduction over a discretely valued field and the theory of moduli spaces of curves.

Fact 6.3.2.3 (Semistable reduction theorem for analytic curves) Any compact rig-smooth strictly analytic curve over an algebraically closed field k admits a semistable formal model.

In the same way, Exercise 6.3.1.3 implies the following generalization of the above fact.

Exercise 6.3.2.4 (i) Cofinality: any formal model $\mathfrak{C}$ of C admits an admissible blow up $\mathfrak{C}' \to \mathfrak{C}$ such that $\mathfrak{C}'$ is semistable.
(ii) Stable reduction theorem: if C is not isomorphic to $\mathbf{P}_k^1$ or a Tate curve then it possesses a minimal semistable formal model (called the stable formal model).

6.3.3 Skeletons

The reader that solved Exercises in Sect. 6.3.1 is probably familiar with part of the ideas of this section. All facts of this section follow easily from the results of Sect. 6.3.1.

Definition 6.3.3.1 (i) Let V_0 be a finite set of type 1 and 2 points of C. The *skeleton* $\Delta(C, V_0)$ is defined as follows: its set of vertices V is the set of points $x \in C$ that are not contained in an open annulus $A \subset C \setminus V_0$, and its edges are formed by the points $x \in C$ that are not contained in an open disc $D \subset C \setminus V_0$. A vertex is *infinite* if it is of type 1.
(ii) The skeleton $\Delta = \Delta(C, V_0)$ is *degenerate* if the set V_f of finite vertices is empty.

For the sake of completeness, we make a remark about the more general situation that we do not study.

Remark 6.3.3.2 The definition makes sense for any curve C over $k = k^a$ with a finite set V. For a curve over an arbitrary ground field k, its skeleton is defined as the image of the skeleton after the ground field extension to $\widehat{k^a}$.

Exercise 6.3.3.3   (i) Show that Δ is a finite graph whose infinite vertices are the points of V_0 of type 1.

(ii) Show that the only degenerate cases are when C is a Tate curve and V_0 is empty, or $C \xrightarrow{\sim} \mathbf{P}_k^1$ and V_0 is a set of at most two type 1 points. Thus, in the degenerate cases Δ is empty, an infinite vertex, an interval with infinite ends, or a loop without vertices.

(iii) Show that in the non-degenerate case V_f is the minimal set of points such that $C \setminus V_f$ is a disjoint union of open discs and annuli such that annuli are disjoint from V_0 and each open disc contains at most one infinite point of V_0.

(iv)* Show that if C is not proper or V_0 has finite vertices then the above description of Δ implies (and is equivalent to) the following stable modification theorem: if $\mathfrak{C}$ is a formal rig-smooth k°-curve with a generically reduced Cartier divisor $\mathfrak{D}$ then there exists a minimal *modification* $\mathfrak{C}' \to \mathfrak{C}$ (i.e. a proper morphism whose generic fiber is an isomorphism) such that $\mathfrak{C}'$ is semistable and the strict transform of $\mathfrak{D}$ is étale over $\mathrm{Spf}(k^\circ)$.

(v)* Formulate and prove an analogous statement when C is proper and V_0 has no finite vertices.

   Hint: this is the stable reduction theorem for a formal curve with a divisor.

Acknowledgements I want to thank A. Ducros for careful reading of the notes and making many valuable comments.

References

[Ber1] V. Berkovich, *Spectral Theory and Analytic Geometry Over Non-Archimedean Fields*. Mathematical Surveys and Monographs, vol. 33 (American Mathematical Society, Providence, RI, 1990)

[Ber2] V. Berkovich, Étale cohomology for non-Archimedean analytic spaces. Publ. Math. IHES **78**, 7–161 (1993)

[Ber3] V. Berkovich, *Non-Archimedean analytic spaces*, lecture notes, Trieste, 2009, available at http://www.wisdom.weizmann.ac.il/~vova/Trieste_2009.pdf

[Ber4] V. Berkovich, Smooth p-adic analytic spaces are locally contractible. Invent. Math. **137**, 1–84 (1999)

[BGR] S. Bosch, U. Güntzer, R. Remmert, *Non-Archimedean Analysis. A Systematic Approach to Rigid Analytic Geometry* (Springer, Berlin-Heidelberg-New York, 1984)

[Con] B. Conrad, Several approaches to non-archimedean geometry, in *p-adic Geometry*, University Lecture Series, vol. 45 (American Mathematical Society, Providence, RI, 2008)

[CT] B. Conrad, M. Temkin, Descent for non-archimedean analytic spaces. Available at http://www.math.huji.ac.il/~temkin/papers/Descent.pdf

[Duc1] A. Ducros, Variation de la dimension d'un morphisme analytique p-adique. Compositio
        Math. **143**, 1511–1532 (2007)
[Duc2] A. Ducros, Les espaces de Berkovich sont excellents. Annales de l'institut Fourier **59**,
        1443–1552 (2009)
[HL]   E. Hrushovski, F. Loeser, Non-archimedean tame topology and stably dominated types.
        `arXiv:1009.0252v2`
[Ki]   R. Kiehl, Der Endlichkeitssatz für eigentliche Abbildungen in der nichtarchimedischen
        Funktionentheorie. Inv. Math. **2**, 191–214 (1967)
[Lüt]   W. Lütkebohmert, Formal-algebraic and rigid-analytic geometry. Math. Ann. **286**, 341–
        371 (1990)
[Tem1] M. Temkin, On local properties of non-Archimedean spaces. Math. Ann. **318**, 585–607
        (2000)
[Tem2] M. Temkin, On local properties of non-Archimedean spaces II. Isr. J. Math. **140**, 1–27
        (2004)
[Tem3] M. Temkin, A new proof of the Gerritzen-Grauert theorem. Math. Ann. **333**, 261–269
        (2005)
[Tem4] M. Temkin, Stable modifications of relative curves. J. Alg. Geom. **19**, 603–677 (2010)
[Tem5] M. Temkin, Inseparable local uniformization. J. Algebra **373**, 65–119 (2013)

Étale Cohomology of Schemes and Analytic Spaces

Antoine Ducros

Contents

2000 *Mathematics Subject Classification.* — 14F20, 14G22.

1 Introduction

This text corresponds essentially to the four hour mini-course that was given by the author in July 2010 at the university Paris 6 (Jussieu) during the *Summer school on Berkovich spaces* which was organized by the ANR Project *Espaces de Berkovich*.

The goal of those talks was to give a survey on scheme-theoretic and Berkovich-analytic étale cohomology theories for people coming from various mathematical areas (including model theory, complex dynamics, tropical geometry, etc.), not

A. Ducros (✉)
Institut de Mathématiques de Jussieu, Université Paris 6, 4 place Jussieu, 75252 Paris Cedex 05, France
e-mail: ducros@math.jussieu.fr, Url: www.math.jussieu.fr/~ducros/

© Springer International Publishing Switzerland 2015
A. Ducros et al. (eds.), *Berkovich Spaces and Applications*, Lecture Notes
in Mathematics 2119, DOI 10.1007/978-3-319-11029-5_2

necessarily familiar with algebraic geometry *à la* Grothendieck. That is the reason why we chose to rather insist on the motivation, the rough ideas behind the theory and the basic examples, than to give proofs—those are usually technically involved, and can be found in the literature. Nevertheless, the interested reader may wish to have some technical references about this subject. Let me now give some of them.

- The notion of a derived functor was fully developed by Grothendieck in a beautiful paper [7].
- Foundations of the scheme-theoretic étale cohomology were laid in [14–17], and [19]. More precisely, [14] introduces the notion of an étale morphism, and Grothendieck's theory of the fundamental group; the three books [15–17] are devoted to Grothendieck topologies, general properties of the associated cohomology theories, and the étale cohomology of schemes; and the ℓ-adic formalism is introduced in [19].

However all those books, though extremely interesting and useful, are quite difficult to read because of their high degree of generality. The reader may thus prefer having a look at [18], in which the theory is presented without (too much detailed) proofs, and with a lot of explanations about the ideas, the geometric interpretation, the analogy with usual topology and so on. He could as well read [11], which is certainly far more understandable than the SGA's, though it also requires some familiarity with the theory of schemes.

Let us also mention a beautiful text by Luc Illusie [10] about the étale cohomology from a double viewpoint, historical and mathematical.

- The main reference for analytic étale cohomology is the foundational article by Berkovich himself [1], which requires the reader to already master both analytic geometry and Grothendiecks' language (Grothendieck topologies, sheaves, derived categories and so on). The comparison theorem of [1] has been improved in [3], which is also quite technical.

2   The General Motivation

As well for schemes as for Berkovich analytic spaces, étale cohomology is a cohomology theory which behaves like singular cohomology of complex analytic spaces; that is to say, it satisfies the same general theorems (bounds on cohomological dimension, Künneth formula, Poincaré duality, etc.), and "takes the expected values" on several classes of algebraic varieties (projective curves, abelian varieties) or analytic spaces (polydiscs, polyannuli, analytic tori, etc.). Moreover, as we will see, this theory also encodes in a very natural way the Galois theory of fields, which allows to think of the latter, which is *a priori* purely algebraic, in a quite geometrical way.

In both scheme-theoretic and analytic contexts, the need for a "singular-like" cohomology theory came from deep *arithmetical* conjectures.

- As far as schemes are concerned, this theory appeared in connection with the Weil conjectures. Weil himself suggested to prove them by defining such a cohomology, which is for that reason often called a *Weil cohomology*, and by applying it to algebraic varieties defined over finite fields; it is for that purpose that Grothendieck et al. developed the theory in the sixties, and the Weil conjectures were eventually proven by Deligne in 1974.

- As far as analytic spaces are concerned, étale cohomology was developed for applications to the Langlands program. Drinfeld and Carayol had conjectured that the local Langlands correspondence should live in the étale cohomology vector spaces of some meaningful p-adic analytic moduli spaces. An archetype of such a space is the so-called *p-adic upper half plane* $\mathbb{P}_{\mathbb{Q}_p}^{1,an} \setminus \mathbb{P}^1(\mathbb{Q}_p)$. But at that time, the étale cohomology of a p-adic analytic space was not defined. This was done for the first time by Berkovich, who used it to prove a conjecture by Deligne on vanishing cycles. After that, it was shown to have the expected applications to the local Langlands program in a series of works by Boyer, Dat, Hausberger, Harris, Taylor, etc.

Remark 1 (The p-adic upper half-plane) Recall that Berkovich analytic spaces are defined over *any* (i.e., possibly Archimedean) complete valued field. In particular, $\mathbb{A}_{\mathbb{R}}^{1,an}$ does make sense. As a topological space, it is the set of multiplicative semi-norms on $\mathbb{R}[T]$ which extend the usual absolute value on $\mathbb{R}$, equipped with the topology induced by the product topology on $\mathbb{R}^{\mathbb{R}[T]}$. The map $z \mapsto (P \mapsto |P(z)|)$ is easily seen to induce a homeomorphism, and even an isomorphism of locally ringed spaces, between the quotient of $\mathbb{C}$ by the complex conjugation and $\mathbb{A}_{\mathbb{R}}^{1,an}$.

The space $\mathbb{P}_{\mathbb{R}}^{1,an}$ is obtained by glueing two copies of $\mathbb{A}_{\mathbb{R}}^{1,an}$ in the usual way; hence there is a canonical isomorphism between $\mathbb{P}_{\mathbb{R}}^{1,an}$ and the quotient of $\mathbb{P}^1(\mathbb{C})$ by the complex conjugation; therefore, $(\mathbb{P}_{\mathbb{R}}^{1,an} \setminus \mathbb{P}^1(\mathbb{R})) \simeq \mathbb{H}$, where $\mathbb{H}$ denotes the usual upper half-plane.

This is one of the numerous reasons why $\mathbb{P}_{\mathbb{Q}_p}^{1,an} \setminus \mathbb{P}^1(\mathbb{Q}_p)$ should be considered as the p-adic analog of $\mathbb{H}$. Note that it is a well-defined Berkovich space. Indeed, as $\mathbb{Q}_p$ is locally compact, $\mathbb{P}^1(\mathbb{Q}_p)$ is compact. As a consequence, $\mathbb{P}_{\mathbb{Q}_p}^{1,an} \setminus \mathbb{P}^1(\mathbb{Q}_p)$ is an open subset of $\mathbb{P}^1(\mathbb{Q}_p)$ (one checks that it is dense, and connected); thus it inherits a natural structure of a p-adic analytic space.

Remark 2 In both the scheme-theoretic and the analytic settings, there are natural cohomology theories, namely the ones associated with the *topological spaces* involved. But even if they are highly interesting, they definitely can *not* play the expected role. Let us give one reason for that : if X is a d-dimensional variety (resp. a d-dimensional paracompact analytic space) over an algebraically closed field, then its topological cohomology vanishes in rank $\geq d + 1$, while non-zero cohomology groups should be allowed to occur up to rank $2d$ in a singular-like cohomology theory.

Remark 3 (Serre) Let us explain why it is not possible to hope for a singular-like cohomology theory for schemes providing $\mathbb{Q}$-vector spaces (by GAGA principles, the same argument would also work *mutatis mutandis* for analytic spaces).

Assume that there exists such a theory $H^\bullet$. Let us fix a prime number p and let $\overline{\mathbb{F}}_p$ be an algebraic closure of $\mathbb{F}_p$. It is a fact that there exists an elliptic curve X over $\overline{\mathbb{F}}_p$ with the following property: *the ring* $(\mathsf{End}\ X) \otimes_\mathbb{Z} \mathbb{Q}$ *is a non-split* $\mathbb{Q}$*-quaternion algebra.*[1] In other words, it is a skew field with center $\mathbb{Q}$ and whose dimension over $\mathbb{Q}$ is equal to 4.

By our assumptions on $H^\bullet$, the group $H^1(X)$ is a 2-dimensional $\mathbb{Q}$-vector space, on which $\mathsf{End}\ X$ acts. The corresponding homomorphism

$$\mathsf{End}\ X \to \mathsf{End}\ H^1(X) \simeq \mathsf{M}_2(\mathbb{Q})$$

induces an arrow

$$(\mathsf{End}\ X) \otimes_\mathbb{Z} \mathbb{Q} \to \mathsf{M}_2(\mathbb{Q}).$$

As the source of this morphism is a skew field, this map is injective. By comparison of dimensions it is bijective. But $\mathsf{M}_2(\mathbb{Q})$ is not a skew field, contradiction.

This is the reason why étale cohomology usually deals with torsion or ℓ-adic coefficients, and only rarely with integral or rational ones.

3   The Notion of a Grothendieck Topology

As a general reference about this notion, we refer to [15] or to the more accessible [11].

Classical general topology axiomatizes the notion of an open subset (or of a neighborhood); in order to develop étale cohomology, Grothendieck slightly generalized this formalism, extending it to the categorical setting, and giving rise to what one calls a *Grothendieck topology*. In that approach, one "only" axiomatizes the notion of a *covering*—this is precisely what is needed to define sheaves and their cohomology groups.

3.1   *Grothendieck Topology*

Now let us start with a category C. We will not give here the precise definition of a category; let us simply say that one has *objects* of C, *morphisms* between them (including the identity map of every object) and that there is a notion of composition of morphisms which satisfies the usual rules.

---

[1]It can be shown that X satisfies this property if and only if it is *supersingular*, that is, if and only if its p-torsion subgroup consists of *one* $\overline{\mathbb{F}}_p$-point, with multiplicity p^2.

We will moreover assume that if $Y \to X$ and $Z \to X$ are arrows of C, the *fiber product* $Y \times_X Z$ exists in C. Let us give some explanations. One says that an object T of C, equipped with two morphisms $T \to Y$ and $T \to Z$ such that the diagram

commutes, is a *fiber product of Y and Z over X* if for every object S of C, and every couple of morphisms $(S \to Y, S \to Z)$ such that

commutes, there exists a unique morphism $S \to T$ making the diagram

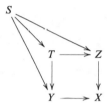

commutative.

As soon as such a fiber product exists, it is unique up to canonical isomorphism; it is therefore called *the* fiber product of Y and Z over X, and denoted by $Y \times_X Z$. Let us give a few examples of categories in which $Y \times_X Z$ always exists.

- The category of sets. In this case $Y \times_X Z$ is the set-theoretic fiber product, that is, the set of couples (y, z) with $y \in Y$ and $z \in Z$ such that y and z have the same image in X.
- The category of topological spaces. In this case $Y \times_X Z$ is the set-theoretic fiber product *endowed with the product topology*.
- The category of schemes. Let us assume for simplicity that X, Y and Z are affine, say respectively equal to Spec A, Spec B and Spec C. Then $Y \times_X Z$ is nothing but Spec $(B \otimes_A C)$.
- The category of analytic spaces over a non-Archimedean, complete field. Let us assume for simplicity that X, Y and Z are affinoid, say respectively equal to $\mathcal{M}(A), \mathcal{M}(B)$ and $\mathcal{M}(C)$. Then $Y \times_X Z$ is nothing but $\mathcal{M}(B \widehat{\otimes}_A C)$.
- The category of open subsets of a given topological space, in which morphisms are given by inclusions. In that case, $Y \times_X Z$ is nothing but the intersection $Y \cap Z$.
- The category of analytic domains of a given analytic space, in which morphisms are given by inclusions. In that case, $Y \times_X Z$ is nothing but the intersection $Y \cap Z$.

Definition 4 Let C be a category in which fiber products exist. A *Grothendieck topology* on C consists of the datum, for every object X of C, of a class of families of morphisms $(X_i \to X)_i$, which are called *coverings* of X, and which are subject to the following axioms:

1. If $Y \to X$ is an isomorphism, it is a covering of X;
2. if $(X_i \to X)_i$ is a covering of X, and if $Y \to X$ is any morphism,

$$(X_i \times_X Y)_i \to Y$$

   is a covering of Y;
3. if $(X_i \to X)_i$ is a covering of X and if $(X_{ij} \to X_i)_j$ is for every i a covering of X_i then $(X_{ij} \to X)_{i,j}$ is a covering of X.

Let us give two set-theoretic examples.

Example 5 On the category of open subsets of a given topological space, the usual open coverings define a Grothendieck topology.

Example 6 On the category of analytic domains of a given analytic space, the admissible coverings define a Grothendieck topology; it is called the *G-topology*.

Definition 7 A *site* is a category equipped with a Grothendieck topology.

3.2 Presheaves and Sheaves

We fix from now on a site C.

Remark 8 (Comments on the terminology) Though the behavior of C can be slightly different from that of an honest topological space, one will often use, while working with C, the vocabulary and, to some reasonable extent, the intuition of classical topology. Let us give a quite vague example.

If X is an objet of C and if P is a property, we will say that P is true *locally on* X if there exists a covering $(X_i \to X)$ such that P is true on every X_i. We will say that P is local if it is true on a given object of C as soon as it is true locally on it.

Definition 9 (Presheaf) A *presheaf* $\mathscr{F}$ on C is a *contravariant functor* from C to the category of sets. In other words it is an assignment which sends any object X of C to a set $\mathscr{F}(X)$, together with the datum, for every arrow $Y \to X$ in C, of a *restriction map* $\mathscr{F}(X) \to \mathscr{F}(Y)$ (note that the direction of the arrow is reversed). All these data have to behave well with respect to the composition of arrows.

The elements of $\mathscr{F}(X)$ for a given X will sometimes be called *sections of $\mathscr{F}$ on X*.

If there is no risk of confusion,[2] the restriction $\mathscr{F}(X) \to \mathscr{F}(Y)$ will be denoted $s \mapsto s_{|Y}$.

A *map* from a presheaf $\mathscr{F}$ to a presheaf $\mathscr{G}$ is a *morphism of functors* from $\mathscr{F}$ to $\mathscr{G}$, that is, a family of maps $\mathscr{F}(X) \to \mathscr{G}(X)$, where X runs through the collection of all objects of C, such that for every arrow $Y \to X$ in C the diagram

$$
\begin{array}{ccc}
\mathscr{F}(X) & \longrightarrow & \mathscr{F}(Y) \\
\downarrow & & \downarrow \\
\mathscr{G}(X) & \longrightarrow & \mathscr{G}(Y)
\end{array}
$$

commutes.

Definition 10 (Sheaf) A presheaf $\mathscr{F}$ on C is said to be a *sheaf* if it satisfies the following condition: *for every object X of* C, *for every covering* $(X_i \to X)$, *and for every family* $(s_i) \in \prod_i \mathscr{F}(X_i)$ *such that*

$$
s_{i|X_i \times_X X_j} = s_{j|X_i \times_X X_j} \in \mathscr{F}(X_i \times_X X_j) \text{ for all } (i, j),
$$

there exists a unique $s \in \mathscr{F}(X)$ such that $s_{|X_i} = s_i$ for every i.

Remark 11 If $i = j$, the notation $s_{i|X_i \times_X X_i}$ is confusing, because there are two arrows, namely, the two projections, from $X_i \times_X X_i$ to X. Therefore when $i = j$ the condition $s_{i|X_i \times_X X_j} = s_{j|X_i \times_X X_j}$ should be understood as follows: the images of s in $\mathscr{F}(X_i \times_X X_i)$ under the arrows induced by the two projections coincide; this is not automatic in general, as we will see later on.

Remark (Interpretation) The section s should be thought of as being obtained by *glueing* the s_i's, the equalities $s_{i|X_i \times_X X_j} = s_{j|X_i \times_X X_j}$ being the coincidence conditions that make the glueing process possible.

Remark (Abstract nonsense) If an objet X of C admits the *empty family* as a covering, then for every sheaf $\mathscr{F}$ on C, the set $\mathscr{F}(X)$ is a singleton. This follows formally from the definition, and the fact that an empty product of sets is a singleton. The reader who does not like such things can add this fact as an axiom.

Definition 12 (Maps) If $\mathscr{F}$ and $\mathscr{G}$ are two sheaves on C, a map from $\mathscr{F}$ to $\mathscr{G}$ is simply a map between the *presheaves* $\mathscr{F}$ and $\mathscr{G}$.

Remark (Link with old-fashioned sheaves) If C is the category of all open subsets of a given topological space T, equipped with the aforementioned Grothendieck topology, then a sheaf on C is nothing but a classical sheaf on the topological space T.

---

[2]There could be one because $\mathscr{F}(X) \to \mathscr{F}(Y)$ not only depends on Y, but also on the arrow $Y \to X$.

Remark (Presheaves and sheaves with some extra-structures) We have actually just defined presheaves and sheaves of *sets*. We will also use the notion of presheaves and sheaves of abelian groups.

The definitions are almost the same, except that $\mathscr{F}(X)$ comes for every object X of C with the structure of an abelian group and that restrictions map are required to be homomorphisms of groups; a morphism between two presheaves or sheaves of abelian groups $\mathscr{F}$ and $\mathscr{G}$ is a map from $\mathscr{F}$ to $\mathscr{G}$ such that $\mathscr{F}(X) \to \mathscr{G}(X)$ is a morphism of groups for every object X of C.

One can of course define in an analogous way presheaves and sheaves of rings, of k-vector spaces or k-algebras (for k a field), etc., and morphisms between them.

Definition 13 (Subsheaf of a sheaf) If $\mathscr{F}$ is a sheaf on C, a *subsheaf* of $\mathscr{F}$ is a sheaf $\mathscr{G}$ such that $\mathscr{G}(X)$ is for every object X of C a subset of $\mathscr{F}(X)$.

If $\mathscr{G}$ is a subsheaf of $\mathscr{F}$, the property for a section of $\mathscr{F}$ to be a section of $\mathscr{G}$ is a local property.

Remark (Presheaf associated with a sheaf) Every presheaf $\mathscr{F}$ on C can be sheafified in a natural way.

If A is a set and $\mathscr{F}$ is the constant presheaf $U \mapsto A$, its sheafification will be denoted by $\underline{A}_{\mathsf{C}}$ (or $\underline{A}$, or A_{C}, or A if there is no risk of confusion), and called the *constant sheaf* (associated with) A.

When C is the site of open subsets of a given topological space X, the sheaf $\underline{A}$ is the sheaf of locally constant A-valued functions on X.

If $\mathscr{O}$ is a sheaf of commutative rings on C and if $\mathscr{F}$ and $\mathscr{G}$ are two sheaves of $\mathscr{O}$-modules, the presheaf $U \mapsto \mathscr{F}(U) \otimes_{\mathscr{O}(U)} \mathscr{G}(U)$ is not a sheaf in general. Its sheafification is denoted by $\mathscr{F} \otimes_{\mathscr{O}} \mathscr{G}$.

3.3 Abelian Categories: The Example of the Sheaves of Abelian Groups

We will not give here the precise definition of an *abelian category*. Let us just say that this is a category A which "looks like" the category of modules over a ring. In particular, it satisfies the following properties:

- the set of morphisms between two given objects is an abelian group ;
- there is a zero object;
- one can define the direct sum of a family of objects;
- one can define the sub-objects of a given object, and the quotient of an object by one of its sub-objects;
- one can define the image, the kernel, and the cokernel of a map, and hence the notions of surjective and an injective maps;
- one can define exact sequences.

Moreover, all these notions behave in the usual way.

Example 14 The category of modules over a given ring (not necessarily commutative) is abelian.

Example 15 The category of sheaves of abelian groups on a site C is also abelian.

In particular one can define the notion of the image of a morphism, and of an exact sequence. Let us explain what they consist of.

Definition 16 (Image of a morphism of sheaves) Pick any map $f : \mathscr{F} \to \mathscr{G}$ of sheaves on a site C (the group structures are not needed). Consider the following assignment that sends an object X of C to the set of sections of $\mathscr{G}$ on X which belong *locally* to the image of f. More precisely, X is sent to the set of $t \in \mathscr{G}(X)$ satisfying the following condition: *there exist a covering (X_i) of X and, for every i, an element s_i of $\mathscr{F}(X_i)$ such that $f(s_i) = t_{|X_i}$.*

This assignment turns out to be a sheaf we will denote by $\mathsf{Im}\ f$.

Remark 17 The naive image of f, that is, the presheaf that sends X to the usual image of $f(X) : \mathscr{F}(X) \to \mathscr{G}(X)$, is not a sheaf in general: the fact for a section to belong to the image is not of local nature—see the counterexample below. It can be shown that $\mathsf{Im}\ f$ is precisely the sheafification of the naive image of f. This is also the smallest subsheaf of $\mathscr{G}$ containing it. We will say that f is *surjective* if $\mathsf{Im}\ f = \mathscr{G}$.

Remark (A fundamental example) Let $\mathscr{O}$ be the sheaf of holomorphic functions on $\mathbb{C}$. The fact for a holomorphic function to admit a primitive on a given open subset of $\mathbb{C}$ is not a local property. For instance, $1/z$ admits local primitives near every point in $\mathbb{C}^*$, but not a global primitive in $\mathbb{C}^*$. In particular the naive image of the derivation $d : \mathscr{O} \to \mathscr{O}$ is not a subsheaf of $\mathscr{O}$.

However, the sheaf-theoretic image $\mathsf{Im}\ d$ of d is very easy to describe: as every holomorphic function on an open subset of $\mathbb{C}$ has locally a primitive, $\mathsf{Im}\ d$ is nothing but the whole sheaf $\mathscr{O}$.

Definition 18 (Exact sequences of sheaves) Now let $\mathscr{F}, \mathscr{F}'$ and $\mathscr{F}''$ be three sheaves of abelian groups on C, and let $f : \mathscr{F}' \to \mathscr{F}$ and $g : \mathscr{F} \to \mathscr{F}''$ be two morphisms. The sequence

$$0 \longrightarrow \mathscr{F}' \overset{f}{\longrightarrow} \mathscr{F} \overset{g}{\longrightarrow} \mathscr{F}'' \longrightarrow 0$$

is exact if the following conditions hold:

i) for every object X of C, the sequence $0 \to \mathscr{F}'(X) \to \mathscr{F}(X) \to \mathscr{F}''(X)$ is exact (in the usual sense for a sequence of abelian groups);
ii) $\mathscr{F}'' = \mathsf{Im}\ g$ (in other words: g is surjective).

Remark 19 Observe that in i) we *do not* require $\mathscr{F}(X) \to \mathscr{F}''(X)$ to be surjective for every X. We simply demand in ii) that g is surjective, which is a weaker condition.

Example 20 Let $\mathcal{O}$ be the sheaf of holomorphic functions on $\mathbb{C}$ and let d be the derivation; recall that $\underline{\mathbb{C}}$ denotes the constant sheaf associated with $\mathbb{C}$, that is, the sheaf of locally constant complex-valued functions, which embeds naturally in $\mathcal{O}$. One $\mathcal{O}$. One immediately checks that the sequence

$$0 \longrightarrow \underline{\mathbb{C}} \longrightarrow \mathcal{O} \xrightarrow{\ d\ } \mathcal{O} \longrightarrow 0$$

is exact (the fact that $\mathsf{Im}\, d = \mathcal{O}$ has been mentioned above); Nevertheless, observe that the derivation $\mathcal{O}(\mathbb{C}^*) \to \mathcal{O}(\mathbb{C}^*)$ is *not* onto: again, consider the function $1/z$.

4 Cohomology of Sheaves

For this section, we refer to [11, 16], and to [7] for the notion of a derived functor.

4.1 Left-Exact Functors

Definition 21 Let $F : \mathsf{A} \to \mathsf{B}$ be a functor between two abelian categories. We say that F is *exact* (resp. *left-exact*) if for every exact sequence

$$0 \to X' \to X \to X'' \to 0$$

in A, the induced sequence

$$0 \to F(X') \to F(X) \to F(X'') \to 0 \ (\text{resp. } 0 \to F(X') \to F(X) \to F(X''))$$

is exact.

Observe that a left-exact functor is exact if and only if it carries surjections to surjections.

Remark 22 There is of course also a notion of a *right-exact* functor, but we will not use it here.

Example 23 Let C be a site and let X be an object of C. Let A be the category of sheaves of abelian groups on C, and let B be the category of abelian groups. Let Γ be the functor from A to B that sends a sheaf $\mathscr{F}$ to the group $\mathscr{F}(X)$.

By the very definition of an exact sequence of sheaves, Γ is left-exact. It is not exact in general, as we have seen above.

Example 24 Let X be a Hausdorff topological space, let A be the category of sheaves of abelian groups on X, and let B be the category of abelian groups. Let Γ_c be the functor from A to B that sends a sheaf $\mathscr{F}$ to the subgroup of $\mathscr{F}(X)$ that consists of sections with compact support. By the very definition of an exact sequence of sheaves, Γ_c is left-exact, but not exact in general.

As an example, let X be the open interval $]0; 1[$, let $\mathscr{F}$ be the sheaf of real analytic functions on X and let $\mathscr{G}$ be the sheaf on X that sends an open subset U to $\mathbb{R}$ if $\frac{1}{2} \in U$ and to 0 otherwise; evaluation at $\frac{1}{2}$ induces a natural map $\mathscr{F} \to \mathscr{G}$ which is easily seen to be surjective. Nevertheless, $\Gamma_c(\mathscr{F}) \to \Gamma_c(\mathscr{G}) = 0 \to \mathbb{R}$ is not onto; hence Γ_c is not exact.

Example 25 Let G be a group. Let A be the category of G-modules, that is, of abelian groups endowed with an action of G (or, equivalently, of modules over the ring $\mathbb{Z}[G]$), and let B be the category of abelian groups. Let F be the functor from A to B that sends a G-module M to its subgroup M^G consisting of G-invariant elements. It is an easy exercise to see that F is left-exact. Let us show by a counterexample that it is not exact in general.

Let us take for G the group $\mathbb{Z}/2\mathbb{Z}$, acting on $\mathbb{C}^*$ through complex conjugation. The morphism $\mathbb{C}^* \overset{z \mapsto z^2}{\longrightarrow} \mathbb{C}^*$ is equivariant and surjective; but after taking the G-invariants one gets $z \mapsto z^2$ from $\mathbb{R}^*$ to $\mathbb{R}^*$, which is not onto anymore.

Example 26 There is a topological avatar of the preceding example, which will play a central role in étale cohomology: let G be a *profinite group* (e.g. $G = \text{Gal}(\overline{k}/k)$, for k a field and $\overline{k}$ an algebraic closure of k). Let A be the category of *discrete G-modules*, that is, of G-modules such that the stabilizer of every element is an *open* subgroup of G; it is an abelian category. Let B be the category of abelian groups, and let F be the functor from A to B that sends a G-module M to its subgroup M^G consisting of G-invariant elements. It is left-exact, but not exact in general (the above counterexample fits in that setting: as $\mathbb{Z}/2\mathbb{Z}$ is finite, it is profinite and every $\mathbb{Z}/2\mathbb{Z}$-module is discrete).

4.2 Derived Functors

Derived functors of a given left-exact functor are a way to encode its lack of exactness, in the following sense.

Let $F : \mathsf{A} \to \mathsf{B}$ be a left-exact functor between abelian categories. A collection $(\mathbf{R}^i F)_{i \geqslant 0}$ of functors from A to B is said to be a *collection of derived functors of F* if the following hold:

- $\mathbf{R}^0 F = F$;
- to every short exact sequence $0 \to X' \to X \to X'' \to 0$ of objects of A one can associate in a functorial way a long exact sequence

$$0 \to F(X') \to F(X) \to F(X'') \to$$

$$\mathbf{R}^1 F(X') \to \mathbf{R}^1 F(X) \to \mathbf{R}^1 F(X'') \to \mathbf{R}^2 F(X') \to \dots$$

$$\dots \to \mathbf{R}^i F(X') \to \mathbf{R}^i F(X) \to \mathbf{R}^i F(X'') \to \mathbf{R}^{i+1} F(X') \to \dots$$

- the collection $(\mathbf{R}^i F)$ is universal for the preceding properties, in some straight-forward sense we will not make explicit.

If such a collection exists, it is unique up to a canonical system of isomorphisms of functors; hence if it is the case, we will speak of *the* collection $(\mathbf{R}^i F)$ of the derived functors of F, and we will say that $\mathbf{R}^i F$ is *the* i-th derived functor of F.

Remark (A trivial example) If F is an exact functor, then it admits derived functors: one has $\mathbf{R}^0 F = F$ and $\mathbf{R}^i F = 0$ for positive i.

Theorem 1 (Conditions for the existence of derived functors) *If the abelian category* A *has enough injective objects, then every left-exact functor from* A *to another abelian category* B *admits derived functors.*

We shall give below a precise definition of this rather technical assumption of having enough injective objects.

Remark 1. The following categories which we have already encountered have enough injective objects: the category of sheaves of abelian groups on a site; the category of modules over a ring, including that of G-modules for G a group; the category of *discrete* G-modules, for G a profinite group.[3] Hence on those categories, derived functors will always exist.
2. The proof of the existence of derived functors (provided there are enough injective objects) proceeds by *building* them, but in a highly abstract way which is usually not helpful at all when one needs to actually compute them.
3. If G is a (profinite) group, the derived functors of $M \mapsto M^G$, with M going through the category of (discrete) G-modules, can be computed explicitly using (continuous) G-cochains, cocycles and coboundaries with coefficients in M. They will be denoted by $\mathrm{H}^{\bullet}(G, \cdot)$.

Let A be an abelian category. An object X of A is said to be *injective* if for every injective arrow $Y \hookrightarrow Z$ in A, every morphism from Y to X can be extended to a morphism from Z to X. If X is injective, every injection from X into an object X' of A admits a retraction or, otherwise said, identifies X with a direct summand of X' (apply the definition to the injection $X \hookrightarrow X'$ and the identity map $X \to X$).

Example 27 In the category of vector spaces over a given field, every object is injective.

That is not the case for the category of abelian groups; indeed, $\mathbb{Z}$ is not an injective object of this category: simply note that the injection $n \mapsto 2n$ from $\mathbb{Z}$ into itself does not admit a retraction because $2\mathbb{Z}$ is not a direct summand of $\mathbb{Z}$.

In fact, one can prove (exercise !) that an abelian group is injective if and only if it is divisible.

---

[3]The latter category is in fact (equivalent to) the category of sheaves of abelian groups on a suitable site.

Definition 28 An abelian category A is said to have *enough injective objects* if every object of A admits an injective arrow into an injective object.

The category of abelian groups has enough injective objects: it is not difficult (again, exercise !) to embed a given abelian group in a divisible one ($\mathbb{Z} \hookrightarrow \mathbb{Q}$ is an example of such an embedding). This fact plays a crucial role in the proof that the category of modules over a given ring and the category of sheaves of abelian groups on a site have enough injectives.

4.3   The Cohomology Groups of a Sheaf

Let us now fix a site C. Let X be an object of C; let A be the category of sheaves of abelian groups on C and let B be the category of abelian groups. As we have seen, the functor from A to B that sends $\mathscr{F}$ to $\mathscr{F}(X)$ is left-exact; thus it admits derived functors (see above).

4.3.1   Notation and Definition

The functor $\mathscr{F} \mapsto \mathscr{F}(X)$ is also denoted by $\mathscr{F} \mapsto H^0(X, \mathscr{F})$, and its i-th derived functor by $\mathscr{F} \mapsto H^i(X, \mathscr{F})$ if $i > 0$. We say that $H^i(X, \mathscr{F})$ is the i-th cohomology group of X with coefficients $\mathscr{F}$, or the i-th cohomology group of $\mathscr{F}$ on X.

Remark (Links with other cohomology theories) Let us assume that C is the site of open subsets of a given topological space X. Recall that, for every abelian group A, we denote by $\underline{A}$ the sheaf of locally constant A-valued functions on X.

Then if X is a "reasonable" topological space (e.g. a metrizable locally finite CW-complex) it can be shown that for every i the group $H^i(X, \underline{A})$ in the above sense coincides with the i-th group of *singular* cohomology of X with coefficients in A.

If Y is a closed subset of X (resp. if X is Hausdorff), one can also derive the functor of global sections with support in Y (resp. compact support). The resulting groups are denoted by $H_Y^\bullet(X, \cdot)$ (resp. $H_c^\bullet(X, \cdot)$) and are called the *cohomology groups with support in Y (resp. with compact supports)* of the sheaf involved. Again, under reasonable assumptions on X and Y (resp. on X) the group $H_Y^i(X, \underline{A})$ (resp. $H_c^i(X, \underline{A})$) coincides with the i-th group of *singular* cohomology of X with support in Y (resp. with compact support) and with coefficients A.

4.3.2   Restrictions

Let us still assume that C is the site of open subsets of a topological space X, let U be an open subset of X and let V be an open subset of U; let $\mathscr{F}$ be a sheaf of abelian groups on C. It defines by restriction a sheaf $\mathscr{F}_{|U}$ on the topological space U.

Now one can define for every i the groups $H^i(V, \mathscr{F})$ and $H^i(V, \mathscr{F}_{|U})$ as before. For the first one, one computes the i-th derived functor of $\mathscr{G} \mapsto \mathscr{G}(V)$ for $\mathscr{G}$ running through the category of abelian sheaves *on* X, and one applies it to $\mathscr{F}$. For the second one, one computes the i-th derived functor of $\mathscr{G} \mapsto \mathscr{G}(V)$ for $\mathscr{G}$ running through the category of abelian sheaves *on* U, and one applies it to $\mathscr{F}_{|U}$.

But it turns out that there is no reason to worry about it: the groups $H^i(V, \mathscr{F})$ and $H^i(V, \mathscr{F}_{|U})$ are canonically isomorphic; hence we will most of the time only use the notation $H^i(V, \mathscr{F})$.

4.3.3   Restrictions in a More General Setting

The phenomenon we have just mentioned extends to every site C, as follows.

If X is an object of C, let C/X be the so-called category *of objects of* C *over* X, which is defined in the following way:

- its objects are couples (Y, f) where $Y \in C$ and where f is an arrow from Y to X;
- if (Y, f) and (Z, g) are two objects of C/X, an arrow from (Y, f) to (Z, g) is nothing but a morphism $h : Y \to Z$ such that $g \circ h = f$.

The category C/X inherits a Grothendieck topology: if (Y, f) is an object of C/X, a family $((Y_i, f_i) \to (Y, f))$ of arrows in C/X is said to be a covering if and only if $(Y_i \to Y)$ is a covering. If $\mathscr{F}$ is a sheaf on C, the assignment $(Y, f) \mapsto \mathscr{F}(Y)$ defines a sheaf on C/X, which is denoted by $\mathscr{F}_{|X}$.

If $(Y, f) \in C/X$, one can define for every integer i the groups $H^i(Y, \mathscr{F})$ and $H^i((Y, f), \mathscr{F}_{|X})$. For the first one, one computes the i-th derived functor of $\mathscr{G} \mapsto \mathscr{G}(Y)$ for $\mathscr{G}$ going through the category of abelian sheaves *on* C, and one applies it to $\mathscr{F}$. For the second one, one computes the i-th derived functor of $\mathscr{G} \mapsto \mathscr{G}(Y, f)$ for $\mathscr{G}$ going through the category of abelian sheaves *on* C/X, and one applies it to $\mathscr{F}_{|X}$.

Again, there is no reason to worry about it. Indeed, the groups $H^i(Y, \mathscr{F})$ and $H^i((Y, f), \mathscr{F}_{|X})$ are canonically isomorphic; hence we will most of the time only use the notation $H^i(Y, \mathscr{F})$.

4.3.4   Cup-Products

If $\mathscr{F}$ and $\mathscr{G}$ are two sheaves of abelian groups on a site C, one has for every (i, j) and every $X \in C$ a natural *cup-product pairing*

$$\cup : H^i(X, \mathscr{F}) \otimes H^j(X, \mathscr{G}) \to H^{i+j}(X, \mathscr{F} \otimes_{\underline{\mathbb{Z}}} \mathscr{G}).$$

It is "graded-commutative": if $h \in H^i(X, \mathscr{F})$ and $h' \in H^j(X, \mathscr{G})$ then one has $h' \cup h = (-1)^{ij} h \cup h'$.

4.4   What Is étale Cohomology?

In the scheme-theoretic and Berkovich's setting, one defines the notion of an *étale map*.

Now, for X being a scheme (resp. a Berkovich analytic space) we will consider the site $X_{\text{ét}}$ defined as follows:

- its objects are couples (U, f) where U is a scheme (resp. an analytic space) and $f : U \to X$ an étale map (although we often will simply denote such an object by U, one has to keep in mind that f is part of the data);
- a morphism from (V, g) to (U, f) is a map $h : U \to V$ such that $f \circ h = g$; *such a map is automatically étale.*
- a family $(g_i : U_i \to U)$ is a covering if and only if it is *set-theoretically* a covering, that is, if and only if $U = \bigcup g_i(U_i)$.

We call $X_{\text{ét}}$ the *étale site of X*, and its topology is the *étale topology* on X; an *étale sheaf on X* will be a sheaf on $X_{\text{ét}}$.

The *étale cohomology* is simply the cohomology theory of étale sheaves.

Remark (About the need of a Grothendieck topology) The reader could wonder whether one actually needs the frame of Grothendieck topologies to define étale cohomology, and if the latter could not been built (for schemes as well as for analytic spaces) using a suitable *usual* topological space. But that is definitely not the case: indeed, one can prove that the category of sheaves for the étale topology on a scheme (resp. analytic space) is not equivalent to the category of sheaves on a topological space.

5   Étale Morphisms of Schemes

For this section, we refer to [14] or [11].

5.1   Motivation, Definition and First Properties

If Y and X are topological spaces (resp. complex analytic spaces) and if $y \in Y$ we say that a continuous (resp. holomorphic) map $f : Y \to X$ is *a local homeomorphism at y* (resp. *a local isomorphism at y*) if there exist an open neighborhood V of y and an open neighborhood U of $f(y)$ such that f induces a homeomorphism (resp. a holomorphic diffeomorphism).

What Is an étale Map? *The property of a morphism of schemes to be étale at a point of the source should be thought of as an analog of the fact for a continuous map between topological spaces to be a local homeomorphism at a point.*

Before giving a precise definition, let us give an example of a local homeomorphism coming from differential geometry which we will essentially mimic in the scheme-theoretic context. Let X be a $\mathscr{C}^\infty$-manifold, let n be an integer and let $P_1, \ldots, P_n$ be polynomials in n variables $T_1, \ldots, T_n$ with coefficients in $\mathscr{C}^\infty(X)$; let us denote by Y the subset of $X \times \mathbb{R}^n$ which consists of those points $(x, t_1, \ldots, t_n)$ such that $P_i(x, t_1, \ldots, t_n) = 0$ for every i. Let $f : Y \to X$ be the map induced by the first projection.

Let y be a point of Y at which the Jacobian determinant $\left|\dfrac{\partial P_i}{\partial T_j}\right|_{i,j}$ does not vanish. It follows from the inverse function theorem that f is a local diffeomorphism at y.

Definition 29 Let $\varphi : Y \to X$ be a morphism of schemes, let y be a point of Y and let x be its image. We will say that φ is *étale* at y if there exist an affine neighborhood $U = \mathsf{Spec}\ A$ of x in X, a neighborhood V of y in $\varphi^{-1}(U)$, polynomials $P_1, \ldots, P_n$ in $A[T_1, \ldots, T_n]$, and a commutative diagram

$$
\begin{array}{ccc}
& & \mathsf{Spec}\ A[T_1, \ldots, T_n]/(P_1, \ldots, P_n) \\
& \overset{\psi}{\nearrow} & \big\downarrow \\
V & \xrightarrow{\ \ \varphi\ \ } & \mathsf{Spec}\ A
\end{array}
$$

with ψ an open immersion such that $\left|\dfrac{\partial P_i}{\partial T_j}\right|_{i,j}$ does not vanish on $\psi(V)$.

We will say that φ is *étale* if it is étale at every point of Y.

Remark 30 • It follows from the definition that the étale locus of a given map is an open subset of the source space.

• If $\varphi : Y \to X$ is a morphism which is étale at some point y of Y, one can always find a local presentation of φ as above *with $n = 1$* (and then one has only to ensure that the partial derivative $\partial P/\partial T$ does not vanish on $\psi(V)$).

• There is another equivalent, yet more technical, definition of an étale map[4]: a map $Y \to X$ is étale at y if and only if it is *flat* and *unramified* at y.

We are now going to give some basic properties of étale maps. For the sake of simplicity, we restrict ourselves to *global* étale morphisms. But most of these results also have a local counterpart.

Proposition 31 (First properties of étale maps)

• *Every étale map is an open map.*
• *If $Z \to Y$ and $Y \to X$ are étale, then the composite map $Z \to X$ is étale too.*
• *If $Y \to X$ is étale and if $X' \to X$ is any morphism, then $Y \times_X X' \to X'$ is étale.*

---

[4]In the classical literature ([11, 14],...), the étale maps are first defined in that technically involved way; the equivalence with the existence of a one-variable presentation with non-vanishing jacobian is then proven (it is not obvious).

- If $Y \to X$ and $Z \to X$ are étale, then any morphism $Y \to Z$ making the diagram

  commutative is étale.

Observe that the last statement is obvious for local homeomorphisms in topology.

Examples of étale maps 1) Every open immersion is étale.

2) If $Y \to X$ is a morphism between two complex algebraic varieties (*i.e* between two schemes of finite type over $\mathbb{C}$) then it is étale if and only if the induced holomorphic map $Y(\mathbb{C}) \to X(\mathbb{C})$ is a local isomorphism on $Y(\mathbb{C})$.

Examples of finite étale maps We are now going to give three fundamental examples of *finite* étale maps. Such a map is also called an étale cover, and is the analog of the notion of a topological cover, see the discussion below about the étale fundamental group.

1. Let A be a commutative ring, let $f \in A^*$, and let n be an integer which is invertible in A. The finite morphism

$$X := \mathsf{Spec}\ A[T]/(T^n - f) \to \mathsf{Spec}\ A$$

   is étale. This follows from the fact that $\dfrac{\partial(T^n - f)}{\partial T}$ does not vanish on X. Indeed, for every point $x \in X$ one has $T^n(x) = f(x) \neq 0$, whence $nT^{n-1}(x) \neq 0$. Here we used the fact that both f and n are invertible in A.

   Note that in the particular case where $A = k[X, X^{-1}]$ for k a field in which $n \neq 0$, and where $f = X$, the above finite étale map is nothing but $\mathbb{G}_m \xrightarrow{x \mapsto x^n} \mathbb{G}_m$.

2. Let p be a prime number, let A be an $\mathbb{F}_p$-algebra and let $f \in A$. The finite morphism

$$\mathsf{Spec}\ A[T]/(T^p - T - f) \to \mathsf{Spec}\ A$$

   is étale. This follows from the fact that $\dfrac{\partial(T^p - T - f)}{\partial T} = -1$, hence is invertible on $\mathsf{Spec}\ A[T]/(T^p - T - f)$.

   Note that in the particular case where $A = k[X]$ with k a field of characteristic p and $f = X$, the above finite étale map is nothing but $\mathbb{A}^1_k \xrightarrow{x \mapsto x^p - x} \mathbb{A}^1_k$. Hence it appears as a *non-trivial étale cover of the affine line* (of degree p, by itself). Note that this example is "geometric" and not "arithmetic" in the sense that it holds for *every* field of characteristic p, even algebraically closed.

We thus see that *the affine line over an algebraically closed field of* char p *is never simply connected.*

3. Let k be a field. A finite extension L of k is said to be *separable* if it is isomorphic to $k[X]/P$ for some irreducible $P \in k[X]$ with $P' \neq 0$ (this is equivalent to the fact that $(P, P') = 1$). If char $k = 0$, if k is algebraically closed or if k is finite, every finite extension of k is separable. A finite k-algebra is said to be separable if it is a product of finitely many finite separable extensions of k.

One can prove that if X is a scheme of finite type over k, it is étale over k if and only if it is isomorphic to the spectrum of a finite separable k-algebra (and X is then finite étale over k).

Note that if k is algebraically closed, a finite separable k-algebra is an algebra isomorphic to k^n for some n. Hence a k-scheme of finite type is étale over k if and only if it is a disjoint sum of finitely many copies of Spec k.

The latter assertion is a very simple illustration of the following principle: from the étale viewpoint, the spectrum of an algebraically (and, more generally, separably) closed field is analogous to a one-point space in topology. This is definitively false for the spectrum of an arbitrary field: for example, as $\mathbb{C}$ is a separable extension of $\mathbb{R}$, the map Spec $\mathbb{C} \to$ Spec $\mathbb{R}$ is a *non-trivial finite étale cover of degree 2* of Spec $\mathbb{R}$.

Remark (Fibers of an étale map) Let $\varphi : Y \to X$ be an étale morphism of schemes, and let $x \in X$. The fiber $\varphi^{-1}(x)$ is discrete, and finite as soon as φ is of finite type (*e.g.* if Y and X are two algebraic varieties over a given field).

Scheme-theoretically one can thus write $\varphi^{-1}(x) = \coprod$ Spec k_i, with each k_i a finite separable extension of $\kappa(x)$.

5.2   The étale Topology on a Scheme

Let X be a scheme. We have defined in the previous section what an étale morphism of schemes is. This allows us to define the étale site $X_{\text{ét}}$ and to speak about the étale topology on X, see Sect. 4.4. To illustrate the flexibility allowed by the étale topology, we now give an example of a property which is satisfied locally for the étale topology, but not necessarily for the Zariski topology.

Let $Y \to X$ be a morphism of schemes, and $y \in Y$. We will say that $Y \to X$ is *smooth at* y if there exist an integer n, an open neighborhood U of y and a factorization of $U \to X$ through an étale map $U \to \mathbb{A}^n_X$. Note that if $f : Y \to X$ is étale at y, it is smooth at y (take $n = 0$ and U equal to the étale locus of f).

Now let k be a field and let X be a non-empty, smooth k-scheme. By definition of smoothness, there exist a non-empty open subset U of X and an étale map $U \to \mathbb{A}^n_k$ for some n. As $U \to \mathbb{A}^n_k$ is étale, its image is a non-empty open subset V of $\mathbb{A}^n_k$.

Let k^s be a separable closure of k. It is an infinite field, therefore V admits a k^s-point. But this exactly means that there exists $x \in V$ such that $\kappa(x)$ embeds into k^s, hence is finite separable over k.

By the very definition of V, there exists $y \in U$ whose image on $\mathbb{A}_k^n$ is x. As the map $V \to U$ is étale, the field $\kappa(y)$ is finite separable over $\kappa(x)$, hence over k. We have shown that any smooth k-scheme admits a point whose residue field is separable over k. Let us rephrase this property in two different ways (use the universal property of the fiber product to see that they are equivalent).

First way.   There exist a finite separable extension F of k and a commutative diagram

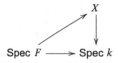

Second way.   There exist a finite separable extension F of k and a section of the morphism $(X \times_{\mathrm{Spec}\,k} \mathrm{Spec}\,F) \to \mathrm{Spec}\,F$.

For every finite separable extension F of k, the map $\mathrm{Spec}\,F \to \mathrm{Spec}\,k$ is étale. As it is trivially surjective ($\mathrm{Spec}\,k$ is a point !), it is an étale covering. In other words any non-empty smooth scheme over a field k has a section *locally for the étale topology on* $\mathrm{Spec}\,k$.

This statement can be generalized as follows. The proof is a little bit more involved but the idea is basically the same as above.

Proposition 32 *Any smooth,* surjective *morphism* $Y \to X$ *between two schemes admits a section locally for the étale topology on* X.

The above claim is definitely false for Zariski topology, which is too coarse. We give several counterexamples: two of 'arithmetic' nature (*i.e.* related to the fact that the ground field is not algebraically closed) and one of 'geometric' nature (*i.e.* over an algebraically closed field).

Example (Arithmetic example 1) Observe that since $\mathrm{Spec}\,\mathbb{R}$ is a point, any morphism $X \to \mathrm{Spec}\,\mathbb{R}$ with non-empty X is surjective, and to have a section locally for the Zariski topology on $\mathrm{Spec}\,\mathbb{R}$ simply means to have a section.

Now the surjective map $\mathrm{Spec}\,\mathbb{C} \to \mathrm{Spec}\,\mathbb{R}$ is smooth (even étale), but it does not have any section: the $\mathbb{R}$-scheme $\mathrm{Spec}\,\mathbb{C} = \mathrm{Spec}\,\mathbb{R}[T]/T^2 + 1$ has no real point. Of course, it admits two complex points, namely $T = i$ and $T = -i$.

Another less trivial arithmetic example is as follows.

Example (Arithmetic example 2) We denote by X the non-trivial real conic, that is, the projective $\mathbb{R}$-scheme defined in homogeneous coordinates by the equation

$$T_0^2 + T_1^2 + T_2^2 = 0 \,.$$

It can be easily shown that the surjective morphism $X \to \mathrm{Spec}\,\mathbb{R}$ is smooth, but it does not admit any section : $X(\mathbb{R}) = \emptyset$. Of course, we have $X(\mathbb{C}) \neq \emptyset$.

Example (**A geometric example**) Set

$$Y = \text{Spec } \mathbb{C}[U, V, S, T]/(SU^2 + TV^2 - 1),$$

and let $\varphi : Y \to \mathbb{A}^2_\mathbb{C}$ be the morphism induced by

$$\mathbb{C}[S, T] \to \mathbb{C}[U, V, S, T]/(SU^2 + TV^2 - 1).$$

At the level of points, the map φ is the projection sending (u, v, s, t) to (s, t) so that it is surjective. Let X be the complement of the origin in $\mathbb{A}^2_\mathbb{C}$. One can check that φ is smooth over X but it does not admit sections locally on X.

Suppose by contradiction that φ admits a section over some non-empty open subset X' of X. Then the restriction of this section to the generic point η of X would define a $\mathbb{C}(S, T)$-point on the generic fiber Y_η, that is, a solution of the equation $SU^2 + TV^2 = 1$ in $\mathbb{C}(S, T)$ (variables are U and V). We leave it as an exercise to check that there are no such solutions.

Let us now show concretely that φ has locally a section on X for the étale topology. For that purpose, denote by ψ_1 the map

$$\mathbb{A}^2_\mathbb{C} = \text{Spec } \mathbb{C}[S', T] \to \mathbb{A}^2_\mathbb{C} = \text{Spec } \mathbb{C}[S, T]$$

that is induced by $S \mapsto (S')^2$ and $T \to T$ and by ψ_2 the map

$$\mathbb{A}^2_\mathbb{C} = \text{Spec } \mathbb{C}[S, T'] \to \mathbb{A}^2_\mathbb{C} = \text{Spec } \mathbb{C}[S, T]$$

that is induced by $S \mapsto S$ and $T \to (T')^2$. At the level of points, ψ_1 and ψ_2 correspond respectively to the formulas $(s', t) \mapsto ((s')^2, t)$ and $(s, t') \mapsto (s, (t')^2)$.

Let X_1 (resp. X_2) be the open affine subscheme of $\mathbb{A}^2_\mathbb{C}$ defined as the invertibility locus of S (resp. T). It is easily seen that $\psi_i^{-1}(X_i) \to X_i$ is a finite étale cover of degree 2 for $i = 1, 2$. As $X = X_1 \cup X_2$, the family $(\psi_1^{-1}(X_1) \to X, \psi_2^{-1}(X_2) \to X)$ is an étale covering of X.

Now for every i the map $Y \times_{\mathbb{A}^2_\mathbb{C}} \psi^{-1}(X_i) \to \psi^{-1}(X_i)$ admits a section.

- If $i = 1$ one can consider the map defined at the level of rings by the formulas $U \mapsto (1/S')$, $V \mapsto 0$. At the level of points, this map corresponds to $(s', t) \mapsto (1/s', 0, s', t)$.
- If $i = 2$ one can consider the map defined at the level of rings by the formulas $U \mapsto 0$, $V \mapsto 1/T'$. At the level of points, this map corresponds to $(s, t') \mapsto (0, 1/t', s, t')$.

Remark 33 In the complex analytic case, it is obvious that a smooth surjective map admits sections locally on the target. This is so because any such map is locally on the source and on the target isomorphic to a projection $X \times \mathbb{D}^n \to X$ for some $n \geq 0$, where $\mathbb{D}$ is the open unit disc.

This explains why the étale topology is, for some purposes, a good algebraic analog of the classical topology over $\mathbb{C}$. It is at least far better than the Zariski topology!

5.3 The Fundamental Group of a Scheme

Grothendieck's theory of the fundamental group provides, through the notion of a finite étale map, a very natural framework containing both Galois theory and the theory of (finite) *topological* covers of complex algebraic varieties. This theory may thus help us to get a geometric intuition of some *a priori* purely algebraic field-theoretic notions.

Let X and Y be two connected non-empty schemes and let $Y \to X$ be a finite and étale map. It can be shown that $G := \mathsf{Aut}\,(Y/X)$ is a finite group.

One says that $Y \to X$ is *Galois* if X is *scheme-theoretically* equal to the quotient Y/G. This means that every arrow $Y \to Z$ which is G-invariant factors in a unique way through an arrow $X \to Z$. If $Y = \mathsf{Spec}\,A$, this is equivalent to the fact that $X = \mathsf{Spec}\,A^G$.

As an example, if k is a field and if L is a finite separable extension, then the finite étale map $\mathsf{Spec}\,L \to \mathsf{Spec}\,k$ is Galois if and only if L is a Galois extension of k in the usual sense.

To be consistent with the fact that the étale avatar of a one-point space is supposed to be the spectrum of a separably closed field, we define the notion of a *geometric point* on a scheme X to be a morphism $\mathsf{Spec}\,F \to X$ with F a separably closed field. In an equivalent way, a geometric point is defined as the datum of a (true) point x of the scheme X, together with an embedding $\kappa(x) \hookrightarrow F$ for some separably closed field F.

Let us choose a *geometric point* $\overline{x}$ on X (one should think of it as a 'basepoint'). One can associate to it a profinite group $\pi_1(X, \overline{x})$ which is called the *fundamental group* of $(X, \overline{x})$, which encodes the theory of finite étale covers of X. This quite vague claim can be given a precise meaning as follows.

Let C be the category of finite étale covers of X and let F be the functor from C to the category of finite sets that sends any finite étale cover $Y \to X$ to the finite set $Y \times_X \overline{x}$ (this is the *fiber functor* associated with $\overline{x}$). Then $\pi_1(X, \overline{x})$ is by definition the group of automorphisms of the functor F.

Pick any element $g \in \pi_1(X, \overline{x})$. Then for every finite Galois cover $Y \to X$ the permutation of $F(Y)$ induced by g arises from a unique automorphism of Y over X. One gets in this way a morphism from $\pi_1(X, \overline{x})$ to the inverse limit of all finite groups $\mathsf{Aut}\,(Y/X)$ where Y ranges over the collection of all finite Galois covers of X. This morphism turns out to be an isomorphism so that the group $\pi_1(X, \overline{x})$ inherits a natural topology, which makes it into a profinite group.

Grothendieck has proven that there is a natural identification between the category C and the category of finite, discrete $\pi_1(X, \overline{x})$-sets (that is, of finite sets equipped with an action of $\pi_1(X, \overline{x})$ with open stabilizers). Under this identification the functor F is the functor to sets that forgets the action of $\pi_1(X, \overline{x})$.

Let us list some concrete consequences of this rather abstract claim.

- To every open subgroup U of $\pi_1(X, \overline{x})$ is associated a connected, non-empty finite étale cover $\mathsf{Cov}\ (U)$ of X. This cover corresponds to the $\pi_1(X, \overline{x})$-set $\pi_1(X, \overline{x})/U$.

- If U and V are two open subgroups of $\pi_1(X, \overline{x})$ with $U \subset V$, the arrow $\mathsf{Cov}\ (U) \to X$ factors through a natural finite étale cover $\mathsf{Cov}\ (U) \to \mathsf{Cov}\ (V)$.

- If U is an open subgroup of $\pi_1(X, \overline{x})$, then the cover $\mathsf{Cov}\ (U) \to X$ is Galois if and only if U is normal in $\pi_1(X, \overline{x})$. In that case, there exists a natural isomorphism $\mathsf{Aut}\ (\mathsf{Cov}\ (U)/X) \simeq \pi_1(X, \overline{x})/U$.

- Every connected, non-empty, finite étale cover of X is (non-canonically) isomorphic to $\mathsf{Cov}\ (U)$ for some U.

- The group $\pi_1(X, \overline{x})$ is naturally isomorphic to

$$\varprojlim_{U \text{ open, normal in } \pi_1(X, \overline{x})} \mathsf{Aut}\ (\mathsf{Cov}\ U/X) = \varprojlim_{U \text{ open, normal in } \pi_1(X, \overline{x})} \pi_1(X, \overline{x})/U.$$

Let us mention that if $\overline{y}$ is another geometric point of X, then $\pi_1(X, \overline{y})$ is *non-canonically* isomorphic to $\pi_1(X, \overline{x})$. Therefore it may happen that one speaks of *the* fundamental group $\pi_1(X)$ of X, without mentioning any geometric point—this is however abusive and sometimes even dangerous!

Example 34 Suppose $X = \mathsf{Spec}\ k$, with k a field, and $\overline{x} = \mathsf{Spec}\ k^s$ for some separable closure k^s of x. Then $\pi_1(X, \overline{x})$ is nothing but $\mathrm{Gal}(k^s/k)$, and what is written above is (part of) the Galois correspondence.

Remark 35 Following our analogy, we see that choosing a separable closure of a given field corresponds to choosing a basepoint on a (connected, non-empty) topological space.

Functoriality Let $Y \to X$ be a morphism between two connected, non-empty schemes, and let $\overline{y}$ be a geometric point of Y. It defines by composition with $Y \to X$ a geometric point $\overline{x}$ of X. The map $Y \to X$ then induces in a natural way a continuous morphism $\pi_1(Y, \overline{y}) \to \pi_1(X, \overline{x})$.

5.4 Various Comparison Theorems

Let k be an algebraically closed field and let L be an algebraically closed extension of k. Recall that if X is a connected, non-empty algebraic variety over k, then the L-algebraic variety $X_L := X \times_k L$ is non-empty and connected too. If $k = \mathbb{C}$, then $X(\mathbb{C})$ is also non-empty and connected for the euclidean topology.

Our aim is two-fold. First we describe the behaviour of the fundamental group under an *algebraically closed* extension of an algebraically closed ground field.

Then we compare Grothendieck fundamental group and its classical counterpart when the ground field is $\mathbb{C}$.

Theorem 2 *Let k be an algebraically closed field and let X be a connected, non-empty algebraic variety over k. Let L be an algebraically closed extension of k. Pick any geometric point $\overline{\xi}$ of X_L and denote by $\overline{x}$ the geometric point on X defined by composition of $\overline{\xi}$ with $X_L \to X$.*

Assume moreover that k has characteristic 0 or that the variety X is proper (e.g. projective). Then the natural map $\pi_1(X_L, \overline{\xi}) \to \pi_1(X, \overline{x})$ is an isomorphism.

Remark 36 1. Concretely Theorem 2 says that if $\operatorname{char} k = 0$ or if X is proper then every finite étale cover of X_L can be defined over k.

2. Theorem 2 is definitely false if $\operatorname{char} k = p > 0$ and X is not proper. The fundamental group becomes strictly larger under an algebraically closed extension of the ground field. In other words, there are plenty of étale covers of X_L which are not definable over k as soon as $L \neq k$. For instance if $\lambda \in L \setminus k$, the étale cover

$$\operatorname{Spec} L[X,T]/(T^p - T - \lambda X) \to \operatorname{Spec} L[X] = \mathbb{A}_L^1$$

can not be defined over k.

3. Let us assume that $\operatorname{char} k = p > 0$. For every profinite group G, denote by $G_{\neq p}$ the largest profinite quotient of G such that all its finite quotients have prime to p order. One can then prove that $\pi_1(X_L, \overline{\xi})_{\neq p} \to \pi_1(X, \overline{x})_{\neq p}$ is an isomorphism even if X is not proper. This means that every *prime-to-p* Galois cover of X_L can be defined over k.

Let us come now to the comparison between the étale and topological fundamental groups.

Theorem 3 *Let X be a connected, non-empty complex algebraic variety. Pick $x \in X(\mathbb{C})$ and denote also by x the corresponding geometric point $\operatorname{Spec} \mathbb{C} \to X$ of the scheme X.*

Then there is a natural isomorphism between $\pi_1(X, x)$ and the profinite completion of the topological fundamental group $\pi_{1,\mathrm{top}}(X(\mathbb{C}), x)$.

Remark 37 Concretely Theorem 3 says that *every finite topological cover of $X(\mathbb{C})$ is algebraizable*. This statement is classically called 'Riemann's existence theorem'.

These comparison results have some striking consequences.

Corollary 38 *Let X be a connected, non-empty algebraic variety over $\mathbb{C}$. For any (possibly non-continuous) automorphism σ of $\mathbb{C}$, we denote by X_σ the corresponding twisted variety.*

Then the groups $\pi_{1,\mathrm{top}}(X(\mathbb{C}))$ and $\pi_{1,\mathrm{top}}(X_\sigma(\mathbb{C}))$ have isomorphic profinite completions.

This follows from Theorem 3 and the fact that X_σ coincides with X as a scheme since only its structure map to $\operatorname{Spec} \mathbb{C}$ has been changed (through σ). This assertion about profinite completions simply means that $X(\mathbb{C})$ and $X_\sigma(\mathbb{C})$ "have the same finite topological covers".

Remark 39 Serre has given an example in [12] of a complex algebraic variety X and an automorphism σ of $\mathbb{C}$ such that $\pi_1(X(\mathbb{C}))$ and $\pi_1(X_\sigma(\mathbb{C}))$ are *not* isomorphic. Roughly speaking, $X(\mathbb{C})$ and $X_\sigma(\mathbb{C})$ have the same finite topological covers, but not the same general topological covers.

Theorem 40 (The fundamental group of a projective curve) *Let k be an algebraically closed field and let X be a smooth, irreducible, projective k-curve.*

a) *If $X = \mathbb{P}^1_k$ then $\pi_1(X) = 0$.*
b) *If X has positive genus g and if* $\operatorname{char} k = 0$, *then $\pi_1(X)$ admits,* as a profinite group, *a presentation*

$$\langle a_1, b_1, \ldots, a_g, b_g \rangle / [a_1, b_1][a_2, b_2] \ldots [a_g, b_g].$$

c) *If X is of positive genus g and if* $\operatorname{char} k = p > 0$, *then $\pi_1(X)_{\neq p}$ admits,* as a profinite group with only prime-to-p finite quotients, *a presentation*

$$\langle a_1, b_1, \ldots, a_g, b_g \rangle / [a_1, b_1][a_2, b_2] \ldots [a_g, b_g].$$

Sketch of proof of Theorem 40 Claim a) follows easily from Hurwitz's formula. By Theorem 2, Claim b) can be reduced to the case where k is equal to $\mathbb{C}$. We may then apply Theorem 3 and use the standard description of the (topological) fundamental group of a Riemann surface.

Claim c) is far more subtle and was one of the first success of Grothendieck's approach to algebraic geometry.

First step.   One construct a complete, discrete valuation ring A with residue field k and with quotient field K of characteristic 0. This is the "Witt vectors construction".

Second step.   One extends X to a smooth, projective A-scheme $\mathscr{X}$, as follows. First, one makes a formal extension, which is possible because the obstruction to do so lives in a Zariski H^2-group of X, which is 0 by a dimension argument. Then, one algebraizes this formal extension, which is again possible since we are dealing with curves.

Third step.   This is the core of the proof. Let $\overline{K}$ be an algebraic closure of K. A comparison theorem which is due to Grothendieck says that

$$\pi_1(\mathscr{X} \times_A \overline{K})_{\neq p} \simeq \pi_1(X)_{\neq p} \, .$$

This result is the scheme-theoretic analog of the following classical fact. If $\varphi : Y \to X$ is a smooth, surjective and proper map with connected fibers between complex analytic spaces, then $x \mapsto \pi_{1,\mathrm{top}}(\varphi^{-1}(x))$ is locally constant on X.

Conclusion. We conclude by applying assertion b) to the curve $\mathscr{X} \times_A \overline{K}$. $\square$

6 Étale Morphisms of Analytic Spaces

We refer to [1].
In all this section, we fix a complete, non-Archimedean field k.

6.1 Definition of étale Maps

In order to simplify the exposition, we shall do a small twist to the standard presentation. First we define 'finite étale maps' and then 'étale maps'. But one can check that our terminology is consistent: the morphisms that are finite and étale according to our definition are precisely those that we have directly defined as 'finite étale'.

Definition 40 Let $\varphi : Y \to X$ be a morphism between two k-analytic spaces. We say that φ is *finite étale* if for every affinoid domain $U = \mathcal{M}(\mathscr{A})$ of X, the pre-image V of U on Y is an affinoid domain of Y equal to $\mathcal{M}(\mathscr{B})$ for some finite étale $\mathscr{A}$-algebra $\mathscr{B}$.

Remark 41 It is sufficient to check the above property on an affinoid G-covering of X.

Let us now come to the definition of general étale maps.

Definition 42 Let $\varphi : Y \to X$ be a morphism between two k-analytic spaces, let $y \in Y$ and let $x = \varphi(y)$. One says that φ is *étale at* y if there exist an open neighborhood U of x and an open neighborhood V of y such that φ induces a finite étale map (in the above sense) from V to U.

A morphism $Y \to X$ between two k-analytic spaces is *étale* if it is étale at every point of Y.

Remark 43 Beware that one can not, in general, take V being equal to $\varphi^{-1}(U)$.

Remark 44 Zariski topology is too coarse to make the above definition work in scheme theory. Indeed, let us assume that $p \neq 2$. Let Y be the invertibility locus of $T - 1$ on $\mathbb{G}_m$ and let φ be the map $Y \to \mathbb{G}_m$ induced by the finite étale arrow $\mathbb{G}_m \xrightarrow{z \mapsto z^2} \mathbb{G}_m$. Let y be the closed point of Y at which $T + 1 = 0$; its image x is the closed point of $\mathbb{G}_m$ at which $T = 1$, and one has $\kappa(y) = \kappa(x) = k$.

Then though φ is étale at y, it is not possible to find an open neighborhood V of y in Y and an open neighborhood U of x in $\mathbb{G}_m$ such that φ induces a finite étale map $V \to U$. Indeed, if it were possible, then by checking what happens at the generic point of the target, one sees that the degree of $V \to U$ would be equal to 2, contradicting the fact that $\varphi^{-1}(x) = \{y\}$ with $\kappa(y) = \kappa(x)$, hence is of degree *one* over $\kappa(x)$.

6.2 First Properties of étale Maps

Most of the results below have a counterpart for local étale maps.

Proposition 45

- Let $\varphi : Y \to X$ be any morphism of k-analytic spaces. Then φ is étale at a point $y \in Y$ if and only if φ is finite, flat and unramified at y.
- The étale locus of a map $\varphi : Y \to X$ is an open subset of x.
- Every étale map is open and boundaryless.
- If $Z \to Y$ and $Y \to X$ are étale, then the composite map $Z \to X$ is étale too.
- If $Y \to X$ is étale and if $X' \to X$ is any morphism, then $Y \times_X X' \to X'$ is étale.
- If $Y \to X$ and $Z \to X$ are étale, then any morphism $Y \to Z$ for which the diagram

is commutative is also étale.

6.3 Examples

- Every open immersion is étale. Be careful: the embedding of a non-open analytic domain (*e.g.* the closed unit ball in an affine space) is *not* étale because it has non-empty boundary.
- If $Y \to X$ is a morphism between algebraic k-varieties, then $Y^{\mathrm{an}} \to X^{\mathrm{an}}$ is étale if and only if $Y \to X$ is étale.
- A very important class of étale maps is that of *étale covers*. A morphism $\varphi : Y \to X$ is said to be an étale cover if X admits an open covering (X_i) such that for every i one can write

$$\varphi^{-1}(X_i) = \coprod_{j \in J_i} Y_{i,j}$$

with each $Y_{i,j}$ finite étale over X_i (and with the index set J_i possibly infinite). Let us now give three examples of étale covers.

1) *An example of finite étale cover.* Let r and R be two positive real numbers with $0 < r < R$. Let n be a positive integer which is non-zero in k, and let Y (resp. X) be the open subset of $\mathbb{A}_k^{1,an}$ defined by the conditions $r < |T| < R$

(resp. $r^n < |T| < R^n$). The map

$$Y \to X, z \mapsto z^n$$

is a finite étale cover of degree n.

2) *A topological cover.* Every topological cover of an analytic space inherits a structure of an analytic space for which it becomes an étale cover of the target.

Assume that k is algebraically closed and non-trivially valued, and let E be a k-elliptic curve with *bad* reduction. By a celebrated theorem of Tate,[5] there exist $q \in k^*$ with $|q| < 1$ and an isomorphism $E^{\mathrm{an}} \simeq \mathbb{G}_m^{\mathrm{an}}/q^{\mathbb{Z}}$.

The *uniformization* map $\mathbb{G}_m^{\mathrm{an}} \to E^{\mathrm{an}}$ is then a topological cover—this is even a universal cover of E^{an}.

3) *The logarithm.* Let $\mathbb{C}_p$ be the completion of an algebraic closure of $\mathbb{Q}_p$. Let X be the open unit $\mathbb{C}_p$-disc with center 1. Define the function $\log : X \to \mathbb{C}_p$ by the usual formula

$$\log(1 + z) = z - \frac{z^2}{2} + \frac{z^3}{3} - \dots$$

Then $\log$ induces an infinite étale cover $X \to \mathbb{A}_{\mathbb{C}_p}^{1,an}$. It is not a topological cover since $\mathbb{A}_{\mathbb{C}_p}^{1,an}$ is topologically simply connected.

Remark 46 Observe that the last two examples are relevant *infinite* étale covers between connected (and very basic) spaces.

Definition 47 (Galois cover) As in the algebraic setting, an étale cover between analytic spaces is said to be Galois if the target is the quotient of the source (in a suitable sense, analogous to the one we used in scheme-theory) by the automorphism group of the cover.

• Example 1) above is Galois as soon as k is algebraically closed, and its Galois group is then equal to $\mu_n(k)$—the latter being itself non-canonically isomorphic to $\mathbb{Z}/n\mathbb{Z}$.

• Example 2) above is Galois, with Galois group $q^{\mathbb{Z}}$, hence isomorphic to $\mathbb{Z}$.

• Example 3) above is Galois, with Galois group $\mu_{p^\infty}(\mathbb{C}_p) := \bigcup_n \mu_{p^n}(\mathbb{C}_p)$.

6.4   Fibers of étale Maps

Let $\varphi : Y \to X$ be an étale morphism between two k-analytic spaces and pick any point $x \in X$. Its fiber $\varphi^{-1}(x)$ is discrete: it is actually a disjoint (possibly infinite) union $\coprod \mathcal{M}(k_i)$ where every k_i is a finite separable extension of $\mathcal{H}(x)$.

---

[5]This result was written in the rigid-analytic language and was actually the first motivation for developing such a theory.

Moreover, if y is a preimage of x, the germ of morphism $(Y, y) \to (X, x)$ *only depends on the finite separable field extension* $\mathscr{H}(x) \hookrightarrow \mathscr{H}(y)$.

In particular, *an étale map φ is a local isomorphism at y if and only if* $\mathscr{H}(x) = \mathscr{H}(y)$.

If k is algebraically closed and if x is a k-point, then $\mathscr{H}(y) = \mathscr{H}(x)$ for every preimage y of x, hence φ is a local isomorphism at every pre-image of x.

However if k is not algebraically closed or if x is not a k-point then the latter fact is false. Let us give a simple explicit example. For every $s > 0$ denote by η_s the point of $\mathbb{A}^{1,an}_k$ defined by the semi-norm $\sum a_i T^i \mapsto \max |a_i| s^i$. Then in example 1) above for every $s \in]r; R[$ the pre-image of η_{s^n} is $\{\eta_s\}$, and the field extension $\mathscr{H}(\eta_{s^n}) \hookrightarrow \mathscr{H}(\eta_s)$ induced by our étale map is of degree n, hence is non-trivial as soon as $n > 0$.

6.5 The étale Topology of an Analytic Space

Since we have defined the notion of étale morphisms between k-analytic spaces, we may define the étale site $X_{\text{ét}}$ and speak about the étale topology of any k-analytic space X (see Sect. 5). Let us give an example of a property which is satisfied locally for the étale topology on X, but not in general for its usual topology.

Let $Y \to X$ be a morphism of analytic spaces, and let $y \in Y$. We say that $Y \to X$ is *smooth at y* if there exist an integer n, an open neighborhood U of y and a factorization of $U \to X$ through an étale map $U \to \mathbb{A}^n_X$. Note that if $Y \to X$ is étale at y, it is smooth at y (take $n = 0$ and U equal to the étale locus of $Y \to X$).

As in scheme theory, the following is true. *If $\varphi : Y \to X$ is a smooth, surjective morphism between two analytic spaces, then φ has a section locally for the étale topology on X.*

Remark 48 This is definitely false if one restricts to the usual topology on X, which is too coarse.

6.6 Fundamental Groups of Analytic Spaces

Let X be a non-empty, connected k-analytic space and let $\overline{x}$ be a geometric point of x, that is, a true point x of X together with an embedding of $\mathscr{H}(x)$ into a separably closed field. To the datum $(X, \overline{x})$ one can associate in a functorial way various topological groups, which are in general neither discrete nor profinite, each of which encoding part of the theory of étale covers of X.

As an example, the theory of finite étale covers of X is encoded in a profinite group which is denoted by $\pi_{1,f}(X, \overline{x})$. Similarly topological étale covers and (unrestricted) étale covers give rise respectively to discrete and topological groups $\pi_{1,\text{top}}(X, \overline{x})$ and $\pi_1(X, \overline{x})$). Yves André has introduced another one, which is denoted by $\pi_{1,\text{temp}}(X,\overline{x})$. It classifies the so-called *tempered* étale covers, that is, those covers which become topological after a finite étale base change. This group has turned out to be of deep arithmetical interest.

Remark 49 The profinite completions of $\pi_1(X,\overline{x})$ and $\pi_{1,\text{temp}}(X,\overline{x})$ are both isomorphic to $\pi_{1,f}(X,\overline{x})$.

Remark 50 The groups $\pi_{1,f}(X,\overline{x})$, $\pi_{1,\text{top}}(X,\overline{x})$, $\pi_1(X,\overline{x})$, and $\pi_{1,\text{temp}}(X,\overline{x})$ depend only on X *up to a non-canonical isomorphism*, and not on the geometric point $\overline{x}$. Therefore it may happen that one does not mention the latter.

Example 51

- If X is an algebraic variety over k and if $\overline{x}$ is a geometric point of X^{an}, then it induces a geometric point $\overline{\xi}$ of X, and there is a natural continuous homomorphism $\pi_{1,f}(X^{\text{an}},\overline{x}) \to \pi_1(X,\overline{\xi})$.

    If $\operatorname{char} k = 0$ or if X is proper, then $\pi_{1,f}(X^{\text{an}},\overline{x}) \to \pi_1(X,\overline{\xi})$ is an isomorphism. If $\operatorname{char} k = p > 0$ and X is not proper, then it only induces an isomorphism $\pi_{1,f}(X^{\text{an}},\overline{x})_{\neq p} \to \pi_1(X,\overline{\xi})_{\neq p}$. Roughly speaking, if $\operatorname{char} k = 0$ or X is proper (resp. if $\operatorname{char} k = p > 0$ and X is not proper) then every finite (resp. finite and prime-to-p) Galois cover of X^{an} is algebraic.

- Suppose that k is algebraically closed and that X is an open disc over k. Let p be the characteristic of the residue field $\tilde{k}$. If $p = 0$ (resp. if $p > 0$) then every finite (resp. finite and prime-to-p) Galois cover of X is trivial, hence $\pi_{1,f}(X) = 0$ (resp. $\pi_{1,f}(X)_{\neq p} = 0$).

    If $p > 0$, the fundamental group $\pi_{1,f}(X)$ is typically huge: indeed, one can show that the group $\operatorname{Hom}_{\text{cont}}(\pi_{1,f}(X),\mathbb{Z}/p\mathbb{Z})$ is infinite. We give a proof of this fact just before Sect. 9 in the case where the characteristic of k is zero. We also give, under this assumption, an example (due to Temkin) of a Galois cover of X with group $\mathbb{Z}/p\mathbb{Z}$ which is not isomorphic to an open subset of an algebraic curve.

- Suppose that the field k is algebraically closed of residual characteristic p and that X is an open annulus over k. If $p = 0$ (resp. if $p > 0$) then every finite (resp. finite and prime-to-p) Galois cover of X is Kummer, that is, as in example 1) above, hence $\pi_{1,f}(X) \simeq \hat{\mathbb{Z}}$ (resp. $\pi_{1,f}(X)_{\neq p} \simeq \hat{\mathbb{Z}}_{\neq p} = \prod_{\ell \neq p} \mathbb{Z}_\ell$).

- Suppose that k is algebraically closed and let E be an elliptic curve over k. If E has good reduction (in which case it is contractible) then

$$\pi_{1,f}(E^{\text{an}}) \simeq \hat{\mathbb{Z}}^2, \ \pi_{1,\text{top}}(E^{\text{an}}) = 0 \text{ and } \pi_{1,\text{temp}}(E^{\text{an}}) \simeq \hat{\mathbb{Z}}^2 .$$

If E has bad reduction (in which case it is homotopy equivalent to a circle) then

$$\pi_{1,f}(E^{\text{an}}) \simeq \hat{\mathbb{Z}}^2, \ \pi_{1,\text{top}}(E^{\text{an}}) \simeq \mathbb{Z} \text{ and } \pi_{1,\text{temp}}(E^{\text{an}}) \simeq \hat{\mathbb{Z}} \times \mathbb{Z} .$$

- The topological group $\pi_1(\mathbb{P}^{1,\,\mathrm{an}}_{\mathbb{C}_p})$ is huge, even though it has no non-trivial finite quotient. As an example, one can show, using a suitable 'period map', that $\mathrm{SL}_2(\mathbb{Q}_p)$ arises as a topological quotient of $\pi_1(\mathbb{P}^{1,\,\mathrm{an}}_{\mathbb{C}_p})$ ([6, Proposition 7.4]).

7 Étale Sheaves and étale Cohomology on Schemes

For this section, we refer to [11, 16–18], or to [19] for the ℓ-adic formalism.

Let us begin with a remark. As open immersions are étale, any Zariski-open covering of a scheme X is an étale covering of X. Therefore every sheaf on $X_{\mathrm{\acute{e}t}}$ is a usual sheaf when restricted to the category of open subsets of X.

7.1 Examples of Sheaves on $X_{\mathrm{\acute{e}t}}$

7.1.1 The Constant Sheaves

Let A be a set. The constant sheaf $\underline{A}$ associated with A admits the following concrete description: for every $U \in X_{\mathrm{\acute{e}t}}$, the set $\underline{A}(U)$ is the set of Zariski-locally constant functions from U to A. If U is locally connected (e.g if U is an algebraic variety over a field) then it can also be described as $A^{\pi_0(U)}$. If A is an abelian group, then $\underline{A}$ is a sheaf of abelian groups. As usual, we will most of the time write A instead of $\underline{A}$, and call it the *constant sheaf A*.

7.1.2 The Sheaf of Global Functions

The assignment $U \mapsto \mathscr{O}_U(U)$ is a sheaf on $X_{\mathrm{\acute{e}t}}$ (we already knew that it was a sheaf for the Zariski topology). This means that for every étale $U \in X_{\mathrm{\acute{e}t}}$, and for every étale covering $(U_i \to U)$, any family (s_i) with $s_i \in \mathscr{O}_{U_i}(U_i)$ for all i and $s_i|_{U_i \times_U U_j} = s_j|_{U_i \times_U U_j}$ for all (i, j) arises from a uniquely determined $s \in \mathscr{O}_U(U)$.

Let us see how this claim what this means in an interesting particular case. Suppose that $X = U = \mathrm{Spec}\,k$ for some field k and that $(U_i \to U)$ is a *single* map $\mathrm{Spec}\,L \to \mathrm{Spec}\,k$ for some finite Galois extension L of k with Galois group G. Let p_1 and p_2 be the two projections

$$\mathrm{Spec}\,L \times_{\mathrm{Spec}\,k} \mathrm{Spec}\,L \to \mathrm{Spec}\,L.$$

The claim says that every $l \in L$ such that $p_1^*(l) = p_2^*(l)$ belongs to k.

Now one has an isomorphism between $L \otimes_k L$ and L^G, which sends $(l \otimes \lambda)$ to $(lg(\lambda))_{g \in G}$; modulo this isomorphism, $p_1^*(l) = (l)_g$ and $p_2^*(l) = (g(l))_g$ for every $l \in L$.

Therefore one has $p_1^*(l) = p_2^*(l)$ if and only if $g(l) = l$ for every $g \in G$. Hence, our claim simply rephrases the basic fact $k = L^G$.

Some Comments We thus see that the theory of étale sheaves allows to think of the equality $k = L^G$ in a quite geometrical way: for an element of L, being G-invariant is the same as satisfying kind of a 'coincidence' condition; and the fact that such a G-invariant element comes from k is nothing but the result of a glueing process which was possible thanks to that coincidence condition.

This geometric vision of Galois phenomena has turned out to be very fruitful, especially in the context of *descent theory*. The typical problem addressed by this theory is: *does a given object defined on L and equipped with a Galois action come from k ?* It is very helpful to think of it as a glueing problem (of objects, now, not of functions), the existence of the Galois action corresponding precisely to the datum of 'a compatible system of isomorphisms on the intersections'.

7.1.3 Coherent Sheaves

More generally, let $\mathscr{F}$ be a coherent (Zariski) sheaf on X. Remember that by the very definition of $X_{\text{ét}}$ every object U of $X_{\text{ét}}$ comes equipped with an étale morphism $\pi : U \to X$; let us assign to such an U the $\mathscr{O}_U(U)$-module $(\pi^*\mathscr{F})(U)$. One can prove that this defines a sheaf on $X_{\text{ét}}$, which we denote by $\mathscr{F}_{\text{ét}}$; note that the sheaf we have considered above is nothing but $(\mathscr{O}_X)_{\text{ét}}$.

One can prove that for every coherent sheaf $\mathscr{F}$ on X and every integer i, the groups $\mathrm{H}^i(X_{\text{Zar}}, \mathscr{F})$ and $\mathrm{H}^i(X_{\text{ét}}, \mathscr{F}_{\text{ét}})$ are canonically isomorphic. Hence as far as cohomology of coherent sheaves is concerned, étale cohomology does not contain more information than Zariski cohomology.

Remark 52 One often writes $\mathscr{F}$ instead of $\mathscr{F}_{\text{ét}}$, when the context is sufficiently clear to indicate that one does not work with the Zariski sheaf $\mathscr{F}$ but with the étale sheaf naturally associated with it. We will sometimes make such an abuse—like writing $\mathscr{O}_X$ for the sheaf $U \mapsto \mathscr{O}_U(U)$ on $X_{\text{ét}}$; the latter is also sometimes denoted by $\mathbb{G}_a$.

7.1.4 The Multiplicative Group

The assignment $U \mapsto \mathscr{O}_U(U)^*$ defines a sheaf of abelian groups on $X_{\text{ét}}$. It is a subsheaf (of multiplicative monoids) of $\mathscr{O}_X$ which is often denoted by $\mathbb{G}_m$.

The generalization by Grothendieck of Hilbert's 90th theorem[6] says that there is a natural isomorphism $\mathrm{H}^1(X_{\text{Zar}}, \mathbb{G}_m) \simeq \mathrm{H}^1(X_{\text{ét}}, \mathbb{G}_m)$. This means that any 'étale line bundle'—that is, an étale sheaf of $\mathscr{O}_X$-modules locally isomorphic to $\mathscr{O}_X$ for the étale topology on X—is already locally isomorphic to $\mathscr{O}_X$ for the *Zariski* topology.

---

[6]This theorem asserts that if L/k is a finite Galois extension with cyclic Galois group $\langle \sigma \rangle$, every element of L whose norm is equal to one can be written $\sigma(x)/x$ for some $x \in L^*$.

More generally, for every n an étale sheaf of $\mathscr{O}_X$-modules which is locally isomorphic to $\mathscr{O}_X^n$ for the étale topology on X is already locally isomorphic to $\mathscr{O}_X^n$ for the *Zariski* topology.[7] When $X = \mathsf{Spec}\ k$ for k a field, the latter claim has a very concrete meaning: if L is a finite Galois extension of k and if G is the Galois group of L, then every finite dimensional L-vector space endowed with a semi-linear action of G is G-isomorphic to $L \otimes_k V$ for some k-vector space V (in other words, it admits a basis over L whose vectors are invariant under G). This is an example of the aforementioned descent theorems.

7.1.5   The Sheaf of Roots of Unity

Let n be an integer. Let μ_n be the presheaf on $X_{\text{ét}}$ defined by the assignment $U \mapsto \{f \in \mathbb{G}_m(U), f^n = 1\}$. It easily seen to be a subsheaf of $\mathbb{G}_m$, and we denote it by μ_n (the 'sheaf of n-th roots of unity').

Remark 53 If the ground scheme needs to be precised, one will write $A_X, \mathbb{G}_{m,X}$, $\mathsf{GL}_{n,X}$ and $\mu_{n,X}$ instead of $A, \mathbb{G}_m, \mathsf{GL}_n$ and μ_n.

7.1.6   An Important Example of a Locally Constant Sheaf

Let us assume that the integer n is invertible on X. We have seen that for every open affine subscheme $U = \mathsf{Spec}\ A$ of X the finite map $V := \mathsf{Spec}\ A[T]/(T^n - 1) \to U$ is an étale cover, and it is easily seen that the map $\overline{m} \mapsto \overline{T}^m$ induces an isomorphism of sheaves $(\mathbb{Z}/n\mathbb{Z})_V \simeq \mu_{n,V}$. The restriction of μ_n to V is thus constant, and we deduce that μ_n *is locally constant for the étale topology on* X.

7.1.7   Locally Constant Sheaves and Representations of π_1

Let A be a finite set (resp. a finite abelian group) and assume that X is a connected and non-empty scheme. Let $\overline{x}$ be a geometric point of X. There is exactly the same phenomenon as in usual topology: one can establish a natural dictionary between sheaves (resp. sheaves of abelian groups) on $X_{\text{ét}}$ which are locally isomorphic to the constant sheaf A and discrete (*i.e.* with open stabilizers) actions of $\pi_1(X, \overline{x})$ on A, up to conjugation by a permutation (resp. an automorphism) of A.

Let us give a very simple example of this in the arithmetical context. Let $X = \mathsf{Spec}\ k$ with k a field, and let $\overline{x}$ be given by a separable closure k^s of k whose Galois group is denoted by G; fix an integer n which is not zero in k, and let ζ be a primitive n-th root of 1 in k^s. The datum of ζ induces an isomorphism between the

---

[7]This can also be expressed as a comparison between étale and Zariski H^1, but with coefficients in GL_n—which is not an *abelian* group.

groups $\mathbb{Z}/n\mathbb{Z}$ and $\mu_n(k^s)$, whence a discrete action of G on $\mathbb{Z}/n\mathbb{Z}$. This is precisely the action that corresponds to the sheaf μ_n on $(\mathsf{Spec}\ k)_{\text{ét}}$ (we have seen above that μ_n is locally isomorphic to the constant sheaf $\mathbb{Z}/n\mathbb{Z}$).

7.2   Cohomology of a Locally Constant Sheaf

Let us return to a general, non-empty connected scheme X endowed with a geometric point $\overline{x}$. Let A be a finite abelian group and let $\mathscr{F}$ be an étale sheaf on X which is locally isomorphic to A. It corresponds to a discrete action of $\pi_1(X, \overline{x})$ on A.

It can then be proven that $\mathrm{H}^1(X_{\text{ét}}, \mathscr{F}) \simeq \mathrm{H}^1(\pi_1(X, \overline{x}), A)$, where A is endowed with the aforementioned $\pi_1(X, \overline{x})$-action (see Sect. 4 for the definition of $\mathrm{H}^1(\pi_1(X, \overline{x}), .)$).

The particular case where $\mathscr{F} = A$. In that case, the action of $\pi_1(X, \overline{x})$ on A is trivial, and $\mathrm{H}^1(\pi_1(X, \overline{x}), A)$ is nothing but $\mathsf{Hom}_{\text{cont}}(\pi_1(X, \overline{x}), A)$ (the topology on A being the discrete one).

The group $\mathsf{Hom}_{\text{cont}}(\pi_1(X, \overline{x}), A)$ is in bijection with the set of isomorphism classes of couples $(Y \to X, \iota)$, where $Y \to X$ is a connected Galois cover and ι an embedding $\mathsf{Aut}\ (Y/X) \hookrightarrow A$. In particular the same description holds for $\mathrm{H}^1(X_{\text{ét}}, A)$.

Notation If X is a scheme and if ℓ is a prime number, we will denote by $\mathsf{Sh}(X, \ell)$ the category of sheaves on $X_{\text{ét}}$ which are locally isomorphic to a constant sheaf associated with a finite abelian ℓ-group.

7.3   Pull-Backs

Let $f : Y \to X$ be a morphism of schemes. To every sheaf $\mathscr{F}$ on $X_{\text{ét}}$ one associates in a natural way a sheaf $f^*\mathscr{F}$ on $Y_{\text{ét}}$. If $\mathscr{F}$ is a sheaf of abelian groups, then there exists for every integer i a canonical morphism $\mathrm{H}^i(X_{\text{ét}}, \mathscr{F}) \to \mathrm{H}^i(Y_{\text{ét}}, f^*\mathscr{F})$.

If $\mathscr{F} = A_X$ for some set A then $f^*\mathscr{F} = A_Y$; one also has $f^*\mu_{n,X} = \mu_{n,Y}$ for every n invertible on X.

If $\mathscr{F}$ is locally isomorphic to a constant sheaf A_X, then $f^*\mathscr{F}$ is locally isomorphic to A_Y (note in particular that if $\mathscr{F} \in \mathsf{Sh}(X, \ell)$ for some prime number ℓ, then $f^*\mathscr{F} \in \mathsf{Sh}(Y, \ell)$). If one assumes moreover that A is finite and that Y and X are non-empty and connected, one has a nice description of $f^*\mathscr{F}$, as follows. Let $\overline{y}$ be a geometric point of Y and let $\overline{x}$ denote the corresponding geometric point of X. The locally constant sheaf $\mathscr{F}$ is given by an action of $\pi_1(X, \overline{x})$ on A. Its pull-back $f^*\mathscr{F}$ is then the locally constant sheaf on $Y_{\text{ét}}$ that corresponds to the action of $\pi_1(Y, \overline{y})$ on A obtained by composing that of $\pi_1(X, \overline{x})$ with $\pi_1(Y, \overline{y}) \to \pi_1(X, \overline{x})$.

If X is a scheme over a field k, if L is an extension of k and if $\mathscr{F}$ is an étale sheaf on X, its pull-back to X_L will be denoted by $\mathscr{F}_L$.

Assume now that X is a scheme of finite type over $\mathbb{C}$ and let $\mathscr{F}$ be an étale sheaf on X. There is a natural (topological) sheaf $\mathscr{F}_{\text{top}}$ on $X(\mathbb{C})$ associated with $\mathscr{F}$. If $\mathscr{F} = A_X$ for some set A then $\mathscr{F}_{\text{top}} = A_{X(\mathbb{C})}$.

If $\mathscr{F}$ is locally isomorphic to a constant sheaf A_X, then $\mathscr{F}_{\text{top}}$ is locally isomorphic to $A_{X(\mathbb{C})}$. Assume moreover that A is finite and that X is connected and non-empty. Then there is a nice description of $\mathscr{F}_{\text{top}}$ as follows. Let $x \in X(\mathbb{C})$ and let us still denote by x the corresponding geometric point of X. The locally constant sheaf $\mathscr{F}$ is given by an action of $\pi_1(X, x)$ on A. The sheaf $\mathscr{F}_{\text{top}}$ is then the locally constant locally constant sheaf that corresponds to the action of $\pi_1(X(\mathbb{C}), x)$ on A obtained by composing that of $\pi_1(X, x)$ with the profinite completion morphism $\pi_1(X(\mathbb{C}), x) \to \pi_1(X, x)$.

7.4 Two Fundamental Exact Sequences

7.4.1 The Kummer Exact Sequence

Let X be a scheme and let n be an integer which is invertible on X. The *Kummer sequence* of étale sheaves on X is the sequence

$$1 \to \mu_n \to \mathbb{G}_m \xrightarrow{z \mapsto z^n} \mathbb{G}_m \to 1.$$

To see that this is an exact sequence, the only point to check is that every section of $\mathbb{G}_m$ has locally for the étale topology a preimage by $z \mapsto z^n$, that is, an n-th square root.

So, let $U \in X_{\text{ét}}$ and let f be an invertible function on U. We have to prove that there exists an étale covering (U_i) of U such that $f_{|U_i}$ is an n-th power for every i. One immediately reduces to the case where U is affine, say equal to $\text{Spec } A$. But $\text{Spec } A[T]/(T^n - f) \to U$ is then an étale covering (see the examples in Sect. 5), and $\overline{T}$ is a n-th root of f in $A[T]/(T^n - f)$.

Remark 54 The Kummer exact sequence is often used as an algebraic substitute for the exponential sequence in complex analytic geometry.

7.4.2 The Artin-Schreier Exact Sequence

Let X be a scheme over $\mathbb{F}_p$. The map $\mathscr{O}_X \xrightarrow{z \mapsto z^p - z} \mathscr{O}_X$ is $\mathbb{F}_p$ linear, and it is not difficult to see that its kernel is precisely $\underline{\mathbb{F}_p}$. The *Artin-Schreier sequence* of étale sheaves on X is the sequence

$$0 \to \underline{\mathbb{F}_p} \to \mathscr{O}_X \xrightarrow{z \mapsto z^p - z} \mathscr{O}_X \to 0.$$

To see that this is an exact sequence, we proceed as above and check that every section of $\mathcal{O}_X$ has locally for the étale topology a preimage by $z \mapsto z^p - z$.

Pick $U = \operatorname{Spec} A \in X_{\text{ét}}$ and f a function on U. Observe that

$$\operatorname{Spec} A[T]/(T^p - T - f) \to U$$

is then an étale covering (see the examples in Sect. 5), and that $f = \overline{T}^p - \overline{T}$ in the ring $A[T]/(T^p - T - f)$. We conclude that there exists an étale covering (U_i) of U such that $f_{|U_i}$ is equal for every i to $z_i^p - z_i$ for a suitable function z_i on U_i as required.

7.5 Étale Cohomology with Support

We would like to mention here two important notions related to the étale cohomology of schemes.

7.5.1 Cohomology with Support

Let X be a scheme and let Y be a Zariski-closed subset of X. The functor that sends a sheaf of abelian groups $\mathcal{F}$ on $X_{\text{ét}}$ to the group of sections of $\mathcal{F}$ with support in Y (that is, which restrict to zero on $X - Y$) is left-exact, hence admits derived functors. The latter are denoted by $\mathrm{H}_Y^i(X_{\text{ét}}, .)$. The groups $\mathrm{H}_Y^i(X_{\text{ét}}, \mathcal{F})$'s are by definition the *étale cohomology groups with support in Y* of $\mathcal{F}$.

7.5.2 Cohomology with Proper Support

Let X be a scheme, and assume that *X is separated and of finite type over a field* k. One can then define for every sheaf $\mathcal{F}$ on $X_{\text{ét}}$ the *étale cohomology groups with proper (or compact) support of $\mathcal{F}$*, which are denoted by $\mathrm{H}_c^\bullet(X_{\text{ét}}, \mathcal{F})$.

The group $\mathrm{H}_c^0(X_{\text{ét}}, \mathcal{F})$ is precisely the group of sections of $\mathcal{F}$ with proper support (that is, whose restriction to $X - Z$ is zero for some suitable closed subset Z of X which is proper).

If X is proper then $\mathrm{H}_c^i(X_{\text{ét}}, \mathcal{F}) = \mathrm{H}^i(X_{\text{ét}}, \mathcal{F})$ for every i and every $\mathcal{F}$ (as expected!).

However we emphasize that in general, and contrary to the definition that is used in topology, the $\mathrm{H}_c^i(X_{\text{ét}}, .)$'s are *not* the derived functors of $\mathrm{H}_c^0(X_{\text{ét}}, .)$, though they give rise to long exact sequences starting from short ones.

Those derived functors do exist, but are in general simply irrelevant, because there are 'not enough' closed subsets of X which are proper—as an example, if X is affine every such closed subset is a union of finitely many closed points. That is why Grothendieck suggested a more *ad hoc* definition that we now present.

As X is of finite type and separated over a field k, there exists by Nagata's embedding theorem an open immersion j from X into a proper k-scheme $\overline{X}$. Now define $j_!\mathscr{F}$ as the 'extension by zero' of $\mathscr{F}$, that is, the sheafification on $\overline{X}_{\text{ét}}$ of the presheaf that sends any $U \in \overline{X}_{\text{ét}}$ to $\mathscr{F}(U)$ if $U \to \overline{X}$ factors through X and to zero otherwise.

One can prove that the groups $H^i(\overline{X}_{\text{ét}}, j_!\mathscr{F})$ do not depend on the chosen compactification $j : X \hookrightarrow \overline{X}$ of X, and for every i one *sets*

$$H_c^i(X_{\text{ét}}, \mathscr{F}) = H^i(\overline{X}_{\text{ét}}, j_!\mathscr{F}).$$

Remark 55 One can prove that if T is a compact topological space, if $\iota : U \hookrightarrow T$ is an open immersion, and if $\mathscr{G}$ is a sheaf on U one has natural isomorphisms

$$H_c^i(U, \mathscr{G}) \simeq H^i(T, \iota_!\mathscr{G})$$

where $\iota_!$ denotes the 'extension by zero' functor. This gives a motivation for the above definition.

7.5.3 Functoriality

If $f : Y \to X$ is a *proper* morphism between separated schemes of finite type over a given field k, and if $\mathscr{F}$ is a sheaf on $X_{\text{ét}}$, there is a natural morphism $H_c^i(X_{\text{ét}}, \mathscr{F}) \to H_c^i(Y_{\text{ét}}, f^*\mathscr{F})$ for every integer i.

If X is a separated scheme of finite type over a given field k, if U is an open subset of X and if $\mathscr{F}$ is a sheaf on $X_{\text{ét}}$, there is for every integer i a natural morphism $H_c^i(U_{\text{ét}}, \mathscr{F}_{|U}) \to H_c^i(X_{\text{ét}}, \mathscr{F})$.

7.6 Link Between étale and Zariski Topology

If X is a scheme and if $\mathscr{F}$ is an étale sheaf of abelian groups on X, we know that $\mathscr{F}$ is also a sheaf for the Zariski topology on X. It follows from the very definition of derived functors (which we didn't give in full detail) that there are for every i natural homomorphisms $H^i(X_{\text{Zar}}, \mathscr{F}) \to H^i(X_{\text{ét}}, \mathscr{F})$. The same also holds for cohomology with support in a given Zariski-closed subset of X.

If $\mathscr{F} = \mathbb{G}_m$ (resp. if $\mathscr{F}$ is the étale sheaf associated with a coherent sheaf), then the natural map $H^1(X_{\text{Zar}}, \mathscr{F}) \to H^1(X_{\text{ét}}, \mathscr{F})$ (resp. for every i, the natural map $H^i(X_{\text{Zar}}, \mathscr{F}) \to H^i(X_{\text{ét}}, \mathscr{F})$) is the comparison isomorphism mentioned in Sect. 7.1.3.

7.7　Étale Cohomology of an Algebraic Variety Over an Algebraically Closed Field

7.7.1　A Comparison Theorem

Theorem 5 *Let k be an algebraically closed field and let X be a separated k-scheme of finite type. Let ℓ be a prime number, and let $\mathscr{F} \in \mathsf{Sh}(X, \ell)$.*

A) *If the dimension of X is d, then $\mathrm{H}^i(X_{\text{ét}}, \mathscr{F}) = 0$ for $i > 2d$, and if X is affine then $\mathrm{H}^i(X_{\text{ét}}, \mathscr{F}) = 0$ for $i > d$.*

B) *For every i the group $\mathrm{H}^i_c(X_{\text{ét}}, \mathscr{F})$ is finite, and if L is an algebraically closed extension of k then $\mathrm{H}^i_c(X_{\text{ét}}, \mathscr{F}) \to \mathrm{H}^i_c(X_{L.\text{ét}}, \mathscr{F}_L)$ is an isomorphism.*

C) *Assume that X is proper or that $\ell \neq 0$ in k. Then $\mathrm{H}^i(X_{\text{ét}}, \mathscr{F})$ is finite, and if L is an algebraically closed extension of k then $\mathrm{H}^i(X_{\text{ét}}, \mathscr{F}) \to \mathrm{H}^i(X_{L.\text{ét}}, \mathscr{F}_L)$ is an isomorphism.*

D) *If $k = \mathbb{C}$ one has for every i natural isomorphisms*

$$\mathrm{H}^i(X_{\text{ét}}, \mathscr{F}) \simeq \mathrm{H}^i(X(\mathbb{C}), \mathscr{F}_{\text{top}}) \text{ and } \mathrm{H}^i_c(X_{\text{ét}}, \mathscr{F}) \simeq \mathrm{H}^i_c(X(\mathbb{C}), \mathscr{F}_{\text{top}}).$$

Remark 56 There is a complex-analytic analog of assertion A): if X is a complex analytic space of dimension d, then $\mathrm{H}^i(X, \mathbb{Z}) = 0$ for $i > 2d$ (because the topological dimension of X is $2d$), and if X is Stein, then $\mathrm{H}^i(X, \mathbb{Z}) = 0$ for $i > d$ (because one can prove that the homotopy type of such an X is that of a d-dimensional CW-complex); it is a general fact that affine schemes play a role which is analogous to that of Stein spaces.

7.7.2　Good Behaviour

Assertion D) ensures that if one works with a scheme of finite type over $\mathbb{C}$, then the étale cohomology groups will be exactly those one expects. But this also remains true over an arbitrarily algebraically closed field, at least for some broad classes of varieties.

Let us be more specific by describing the situation is the case of a curve, then of a torus and finally of an abelian variety. We keep the notations k and ℓ and we assume that $\ell \neq 0$ in k. We fix an integer n and we set $\Lambda = \mathbb{Z}/\ell^n\mathbb{Z}$

- If X is a smooth, projective, irreducible curve of genus g, then $\mathrm{H}^1(X_{\text{ét}}, \Lambda)$ is isomorphic to Λ^{2g}. If $n > 0$, if $P_1, \ldots, P_n$ are distinct k-points on X and if $U = X - \{P_1, \ldots, P_n\}$, one has an exact sequence

$$0 \to \mathrm{H}^1(X_{\text{ét}}, \Lambda) \simeq \Lambda^{2g} \to \mathrm{H}^1(U_{\text{ét}}, \Lambda) \to \bigoplus_{i=1}^{n} \Lambda \xrightarrow{\Sigma} \Lambda \to 0.$$

- Let $n \in \mathbb{N}$ and set $X = \mathbb{G}_{m,k}^n$. The group $\mathrm{H}^1(X_{\mathrm{\acute{e}t}}, \Lambda)$ is isomorphic to Λ^n. The connected Galois covers of X which are associated with elements of $\mathrm{H}^1(X_{\mathrm{\acute{e}t}}, \Lambda)$ are exactly the isogenies $X' \to X$ (with connected X') whose kernel embeds in Λ.

  The algebra $\mathrm{H}^\bullet(X_{\mathrm{\acute{e}t}}, \Lambda)$ where multiplication is given by the cup-product is isomorphic to the free exterior Λ-algebra generated by $\mathrm{H}^1(X_{\mathrm{\acute{e}t}}, \Lambda) \simeq \Lambda^n$.

- Let X be an abelian variety of dimension g over k. The group $\mathrm{H}^1(X_{\mathrm{\acute{e}t}}, \Lambda)$ is isomorphic to Λ^{2g}. As before, the connected Galois covers of X which are associated with elements of $\mathrm{H}^1(X_{\mathrm{\acute{e}t}}, \Lambda)$ are exactly the isogenies $X' \to X$ (with connected X') whose kernel embeds in Λ.

  The algebra $\mathrm{H}^\bullet(X_{\mathrm{\acute{e}t}}, \Lambda)$ is isomorphic to the free exterior Λ-algebra generated by $\mathrm{H}^1(X_{\mathrm{\acute{e}t}}, \Lambda) \simeq \Lambda^{2g}$.

7.7.3 Bad Behaviour

We now show that C) definitely fails if $\ell = 0$ in k and if X is not proper. For that purpose, assume that char $k = p > 0$ and that $\ell = p$. For the time being, we do not make any particular assumption on X. On $X_{\mathrm{\acute{e}t}}$ the Artin-Schreier exact sequence

$$0 \to \mathbb{F}_p \to \mathscr{O}_X \xrightarrow{z \mapsto z^p - z} \mathscr{O}_X \to 0$$

induces a long exact sequence

$$0 \to \mathrm{H}^0(X_{\mathrm{\acute{e}t}}, \mathbb{F}_p) \to \mathrm{H}^0(X_{\mathrm{\acute{e}t}}, \mathscr{O}_X) \to \mathrm{H}^0(X_{\mathrm{\acute{e}t}}, \mathscr{O}_X) \to \mathrm{H}^1(X_{\mathrm{\acute{e}t}}, \mathbb{F}_p) \to \mathrm{H}^1(X_{\mathrm{\acute{e}t}}, \mathscr{O}_X) \to \dots$$

Assume that X is connected and non-empty, and pick $f \in \mathrm{H}^0(X_{\mathrm{\acute{e}t}}, \mathscr{O}_X)$. With the image of f in $\mathrm{H}^1(X_{\mathrm{\acute{e}t}}, \mathbb{F}_p)$ is associated a connected Galois cover Y of X, together with an embedding of $\mathsf{Aut}\ (Y/X)$ into $\mathbb{F}_p$. Let us describe Y concretely and the embedding.

- If $f = z^p - z$ for some $z \in \mathrm{H}^0(X_{\mathrm{\acute{e}t}}, \mathscr{O}_X)$ then $Y = X$ (it is the case where the image of f in $\mathrm{H}^1(X_{\mathrm{\acute{e}t}}, \mathbb{F}_p)$ is zero), and the embedding is of course the trivial one.

- If not, for any open affine subset $U = \mathsf{Spec}\ A$ of X, set $Y_U = \mathsf{Spec}\ A[T]/(T^p - T - f)$. This is a natural étale cover of U, and the Galois cover $Y \to X$ is then obtained by patching the Y_U's together (for varying U). The embedding $\mathsf{Aut}\ (Y/X) \hookrightarrow \mathbb{F}_p$ then induces an isomorphism $\mathsf{Aut}\ (Y/X) \simeq \mathbb{F}_p$ as follows. If φ is an X-automorphism of Y, there exists a (unique) $\lambda_\varphi \in \mathbb{F}_p$ such that $\varphi^* \overline{T} = \overline{T} + \lambda_\varphi$. The assignment $\varphi \mapsto \lambda_\varphi$ is the required isomorphism.

Suppose now that $X = \mathbb{A}_k^1$. By the comparison theorem for coherent sheaves, one has $\mathrm{H}^1(X_{\mathrm{\acute{e}t}}, \mathscr{O}_X) \simeq \mathrm{H}^1(X_{\mathrm{Zar}}, \mathscr{O}_X)$. Since X is affine, the latter is zero. We thus have, in view of the above long exact sequence, a natural isomorphism

$$\mathrm{H}^1(X_{\mathrm{\acute{e}t}}, \mathbb{F}_p) \simeq k[T]/\{z^p - z, z \in k[T]\}.$$

We thus see that $H^1(X_{\text{ét}}, \mathbb{F}_p)$ is infinite, and that it actually increases by extension of the algebraically closed ground field. Whence both assertions of C) are false in this setting.

7.7.4 A Few Words About the Proof of Assertion D)

Let us begin with a general remark. Let C be a site, let $\mathscr{G}$ be a sheaf on C and let U be an object of C. Fix a cohomology class $h \in H^\bullet(U, \mathscr{G})$. One can then prove that, exactly like what happens in the classical case, the class h is locally trivial on U (for the given Grothendieck topology of C).

Now observe that in view of the aforementioned fact assertion D) implies the following:

Claim If $h \in H^\bullet(X(\mathbb{C}), \mathscr{F}_{\text{top}})$, then there exists an étale covering $(X_i \to X)$ such that the pull-back of h to $X_i(\mathbb{C})$ is zero for all i.

By very general 'abstract nonsense', it turns out that to prove assertion D) (at least for cohomology without support), it is *sufficient* to prove the claim. That is the way one does it when X is smooth, by using Riemann's existence theorem. The general case follows from Hironaka's resolution of singularities.

7.8 The ℓ-adic Formalism

Let us keep the notations k, X and ℓ of Theorem 5, and assume that $\ell \neq 0$ in k.

7.8.1 Definition

Though they are perfectly well-defined, the groups $H^i(X_{\text{ét}}, \mathbb{Z}_\ell)$ and $H^i(X_{\text{ét}}, \mathbb{Q}_\ell)$ in the sense of derived functors are irrelevant. Let us explain why through a basic example.

If X is connected and non-empty then $H^1(X_{\text{ét}}, \mathbb{Q}_\ell) \simeq \mathsf{Hom}_{\text{cont}}(\pi_1(X), \mathbb{Q}_\ell)$. As $\pi_1(X)$ is profinite and as $\mathbb{Q}_\ell$ has no torsion, the latter is zero, therefore $H^1(X_{\text{ét}}, \mathbb{Q}_\ell) = 0$. Recall that this statement is true for *any* algebraic k-variety! Following Grothendieck one rather *sets*

Definition 57

$$H^i(X_{\text{ét}}, \mathbb{Z}_\ell) := \varprojlim_n H^i(X_{\text{ét}}, \mathbb{Z}/\ell^n\mathbb{Z})$$

$$H^i(X_{\text{ét}}, \mathbb{Q}_\ell) := \left(\varprojlim_n H^i(X_{\text{ét}}, \mathbb{Z}/\ell^n\mathbb{Z}) \right) \otimes_{\mathbb{Z}_\ell} \mathbb{Q}_\ell$$

Observe that there is **an obvious abuse of notation** here! We can define similarly l-adic cohomology with compact support.

These definitions are justified by the fact that Theorem 5and the three examples which follow its statement also hold with $\mathscr{F} = \Lambda = \mathbb{Z}_\ell$ or $\mathbb{Q}_\ell$. We only need to suppose that $\ell \neq 0$ in k and replace the notion of a finite group by that of a finitely generated $\mathbb{Z}_\ell$-module or $\mathbb{Q}_\ell$-vector space in the assertion C).

7.8.2   An Interesting Consequence for ℓ-adic Coefficients

Assume that $k = \mathbb{C}$ and let σ be any (that is, possibly non-continuous) automorphism of $\mathbb{C}$. The corresponding twisted variety X_σ coincides with X as a scheme. Only its structure map to $\mathrm{Spec}\ \mathbb{C}$ has been changed (through σ).

Applying D) to X and X_σ and using the fact that they coincide as schemes, we get the existence of natural isomorphism of graded algebras

$$H^\bullet(X(\mathbb{C}), \mathbb{Z}_\ell) \simeq H^\bullet(X_\sigma(\mathbb{C}), \mathbb{Z}_\ell)$$

and

$$H^\bullet(X(\mathbb{C}), \mathbb{Q}_\ell) \simeq H^\bullet(X_\sigma(\mathbb{C}), \mathbb{Q}_\ell).$$

Note that the second isomorphism immediately implies that $X(\mathbb{C})$ and $X_\sigma(\mathbb{C})$ have the same Betti numbers.

When X is smooth and projective,[8] one can prove it directly without using étale cohomology. Serre observed that it is a consequence of GAGA and of the fact that the Betti numbers of a complex projective (or, more generally, compact Kähler) manifold Y can be computed, through Hodge theory, from the dimensions of the $H^q(Y, \Omega^p)$'s.

7.8.3   A Counterexample for Real Coefficients

Recently, François Charles has given in [5] an example of a complex, smooth projective variety X and of an automorphism σ of $\mathbb{C}$ such that the graded *algebras* $H^\bullet(X(\mathbb{C}), \mathbb{R})$ and $H^\bullet(X_\sigma(\mathbb{C}), \mathbb{R})$ are not isomorphic. But note that the graded *vector spaces* $H^\bullet(X(\mathbb{C}), \mathbb{R})$ and $H^\bullet(X_\sigma(\mathbb{C}), \mathbb{R})$ are isomorphic, because $X(\mathbb{C})$ and $X_\sigma(\mathbb{C})$ have the same Betti numbers: the problem comes from the cup-product.

---

[8]If X is only assumed to be smooth, this is still possible using Grothendieck's GAGA results for de Rham cohomology, which involve resolution of singularities.

8   Étale Sheaves and étale Cohomology of Analytic Spaces

For general references, see [1] and [3], the latter being used only for the comparison theorem—that is, Theorem 7.

In all this section we fix a non-Archimedean, complete field k.

8.1   Examples of étale Sheaves

Let us begin with a remark: as open immersions are étale, any open covering of an analytic space X is an étale covering of X. Therefore every sheaf on $X_{\text{ét}}$ is, when restricted to the category of open subsets of X, a usual sheaf.

We now present several important étale sheaves. There is a great similarity between these examples and those which were considered in the scheme-theoretic context.

8.1.1   Constant Sheaves

If A is any set, the sheaf on $X_{\text{ét}}$ which is associated with the presheaf $U \mapsto A$ sends every $U \in X_{\text{ét}}$ to the set of locally constant A-valued functions on U—which coincides with $A^{\pi_0(U)}$. As usual, it will be most of the time simply denoted by A, when there is no risk of confusion.

As for schemes, for every prime number ℓ we will denote by $\text{Sh}(X, \ell)$ the category of étale sheaves on X that are locally isomorphic to a constant sheaf associated with a finite abelian ℓ-group.

8.1.2   Coherent Sheaves

Let us assume that X is *good* (i.e. every point of X has an affinoid neighborhood) and let $\mathscr{F}$ be a coherent sheaf of $\mathscr{O}_X$-modules on X. The assignment that sends every $\pi : U \to X$ in $X_{\text{ét}}$ to $\pi^* \mathscr{F}(U)$ is a sheaf on $X_{\text{ét}}$, which is denoted by $\mathscr{F}_{\text{ét}}$ or by $\mathscr{F}$ if this does not cause any trouble. For this kind of sheaves, étale cohomology does not carry more information than the topological one: one has for every integer i a canonical isomorphism $\text{H}^i(X_{\text{ét}}, \mathscr{F}) \simeq \text{H}^i(X_{\text{top}}, \mathscr{F})$.

8.1.3   What Happens in the Non-Good Case?

What we have just said above remains essentially true, up to the following: coherent sheaves and their cohomology have to be considered with respect to the G-topology (see the definition at Sect. 3) instead of the usual topology on X, which is too coarse.

As we have already mentioned, this is a Grothendieck topology, but far more set-theoretic than the étale one.

In the good case, the theories of coherent sheaves (including their cohomology groups) for the usual topology on X and for the G-topology turn out to be the same.

8.1.4 The Multiplicative Group

The assignment $U \mapsto \mathscr{O}_U(U)^*$ is a sheaf on $X_{\text{ét}}$, which is usually denoted by $\mathbb{G}_m$. If X is good, one has a natural isomorphism $\mathrm{H}^1(X_{\text{ét}}, \mathbb{G}_m) \simeq \mathrm{H}^1(X_{\text{top}}, \mathbb{G}_m)$. In the general case, as above, one has to replace the topological H^1 with the H^1 related to the G-topology on X.

8.1.5 The Sheaf of Roots of Unity

For every $n \in \mathbb{N}$ the assignment

$$U \mapsto \{f \in \mathscr{O}_U(U)^*, f^n = 1\}$$

is a subsheaf of $\mathbb{G}_m$ which is denoted by μ_n. If $n \neq 0$ in k, this sheaf is locally isomorphic to the constant sheaf $\mathbb{Z}/n\mathbb{Z}$, and the *Kummer sequence*

$$1 \to \mu_n \to \mathbb{G}_m \xrightarrow{z \mapsto z^n} \mathbb{G}_m \to 1$$

is exact.

8.1.6 Locally Constant Sheaves

Let A be a finite set (resp. a finite abelian group) and assume that X is connected and non-empty. Let $\overline{x}$ be a geometric point of X. As for schemes, one can establish a natural dictionary between sheaves (resp. sheaves of abelian groups) on $X_{\text{ét}}$ which are locally isomorphic to the constant sheaf A and discrete (*i.e.* with open stabilizers) actions of $\pi_{1,\mathrm{f}}(X, \overline{x})$ on A, up to conjugation by a permutation (resp. an automorphism) of A.

If A is an abelian group and if $\mathscr{F}$ is the locally constant sheaf associated with a given action of $\pi_{1,\mathrm{f}}(X, \overline{x})$ on A then $\mathrm{H}^1(X_{\text{ét}}, \mathscr{F}) \simeq \mathrm{H}^1(\pi_{1,\mathrm{f}}(X, \overline{x}), A)$. In the particular case where $\mathscr{F} = A$ this is nothing but $\mathsf{Hom}_{\text{cont}}(\pi_{1,\mathrm{f}}(X, \overline{x}), A)$, which classifies up to isomorphism the couples $(Y \to X, \iota)$ where $Y \to X$ is a connected Galois cover and ι an embedding $\mathsf{Aut}\,(Y/X) \hookrightarrow A$.

8.1.7　Pull-Backs

If $f : Y \to X$ is a morphism of analytic spaces, and $\mathscr{F}$ is a sheaf on $X_{\text{ét}}$, then one can define in a natural way a sheaf $f^*\mathscr{F}$ on $Y_{\text{ét}}$. If $\mathscr{F}$ is a sheaf of abelian groups, there exists for every integer i a canonical morphism $\mathrm{H}^i(X_{\text{ét}}, \mathscr{F}) \to \mathrm{H}^i(Y_{\text{ét}}, f^*\mathscr{F})$.

If $\mathscr{F} = A_X$ for some set A then $f^*\mathscr{F} = A_Y$. One also has $f^*\mu_{n,X} = \mu_{n,Y}$ for every n invertible on X.

If $\mathscr{F}$ is locally isomorphic to a constant sheaf A_X, then $f^*\mathscr{F}$ is locally isomorphic to A_Y (note in particular that if $\mathscr{F} \in \mathsf{Sh}(X, \ell)$ for some prime number ℓ, then $f^*\mathscr{F} \in \mathsf{Sh}(Y, \ell)$). If one assumes moreover that A is finite and that Y and X are non-empty and connected, one has a nice description of $f^*\mathscr{F}$, as follows. Let $\overline{y}$ be a geometric point of Y and let $\overline{x}$ denote the corresponding geometric point of X. The locally constant sheaf $\mathscr{F}$ is given by an action of $\pi_{1,\mathrm{f}}(X, \overline{x})$ on A. Its pull-back $f^*\mathscr{F}$ is then the locally constant sheaf on $Y_{\text{ét}}$ that corresponds to the action of $\pi_{1,\mathrm{f}}(Y, \overline{y})$ on A obtained by composing that of $\pi_{1,\mathrm{f}}(X, \overline{x})$ with $\pi_{1,\mathrm{f}}(Y, \overline{y}) \to \pi_{1,\mathrm{f}}(X, \overline{x})$.

If X is a k-analytic space, if L is a complete extension of k and if $\mathscr{F}$ is an étale sheaf on X, its pull-back to X_L will be denoted by $\mathscr{F}_L$.

If $\mathscr{X}$ is a k-scheme of finite type, then with every sheaf $\mathscr{F}$ on $\mathscr{X}_{\text{ét}}$ is associated in natural way a sheaf $\mathscr{F}^{\text{an}}$ on $\mathscr{X}_{\text{ét}}^{\text{an}}$. If $\mathscr{F}$ is a sheaf of abelian groups, then there exists for every integer i a canonical morphism $\mathrm{H}^i(\mathscr{X}_{\text{ét}}, \mathscr{F}) \to \mathrm{H}^i(\mathscr{X}_{\text{ét}}^{\text{an}}, \mathscr{F}^{\text{an}})$.

If $\mathscr{F} = A_{\mathscr{X}}$ for some set A one has then $\mathscr{F}^{\text{an}} = A_{\mathscr{X}^{\text{an}}}$; if $\mathscr{F} = \mu_{n,\mathscr{X}}$ for some n invertible on $\mathscr{X}$ then $\mathscr{F}^{\text{an}} = \mu_{n,\mathscr{X}^{\text{an}}}$.

If $\mathscr{F}$ is locally isomorphic to a constant sheaf $A_{\mathscr{X}}$, then $\mathscr{F}^{\text{an}}$ is locally isomorphic to $A_{\mathscr{X}^{\text{an}}}$ (note in particular that if $\mathscr{F} \in \mathsf{Sh}(\mathscr{X}, \ell)$ for some prime number ℓ, then $\mathscr{F}^{\text{an}} \in \mathsf{Sh}(\mathscr{X}^{\text{an}}, \ell)$). If one assumes moreover that A is finite and that $\mathscr{X}$, and hence $\mathscr{X}^{\text{an}}$, are non-empty and connected, one has a nice description of $\mathscr{F}^{\text{an}}$, as follows. Let $\overline{x}$ be a geometric point of $\mathscr{X}^{\text{an}}$ and let $\overline{\xi}$ denote the corresponding geometric point of $\mathscr{X}$. The locally constant sheaf $\mathscr{F}$ is given by an action of $\pi_1(\mathscr{X}, \overline{\xi})$ on A; its pull-back $\mathscr{F}^{\text{an}}$ is then the locally constant sheaf on $\mathscr{X}_{\text{ét}}^{\text{an}}$ that corresponds to the action of $\pi_{1,\mathrm{f}}(\mathscr{X}^{\text{an}}, \overline{x})$ on A obtained by composing that of $\pi_1(\mathscr{X}, \overline{\xi})$ with $\pi_{1,\mathrm{f}}(\mathscr{X}^{\text{an}}, \overline{x}) \to \pi_1(\mathscr{X}, \overline{\xi})$.

8.2　Étale Cohomology with Support

As for schemes, one can define on X cohomology groups with support in a given closed subset (the definition will be the same), and also, if X is Hausdorff, cohomology groups with compact support. The definition of the latter will be simpler than in the algebraic case.

8.2.1 Cohomology with Support

If T is a closed subset of X, the functor that sends a sheaf of abelian groups $\mathscr{F}$ on $X_{\text{ét}}$ to the group of sections of $\mathscr{F}$ with support in T (that is, which restrict to zero on $X - T$) is left-exact, hence admits derived functors. The latter are denoted by $H_T^i(X_{\text{ét}}, .)$. We say that the $H_T^i(X_{\text{ét}}, \mathscr{F})$'s are the *étale cohomology groups with support in T* of $\mathscr{F}$.

8.2.2 Cohomology with Compact Support

One can also define cohomology with compact support for étale sheaves on a Hausdorff analytic space. Contrary to what happens in scheme theory, it is done like in topology, that is, by deriving the functor 'global sections with compact support'.

8.2.3 Functoriality

Let k be a complete, non-Archimedean field. If $f : Y \to X$ is a morphism between Hausdorff k-analytic spaces, if f is *topologically* proper (the pre-image of a compact subset is compact), and if $\mathscr{F}$ is a sheaf on $X_{\text{ét}}$, there is for every integer i a natural morphism $H_c^i(X_{\text{ét}}, \mathscr{F}) \to H_c^i(Y_{\text{ét}}, f^*\mathscr{F})$.

If X is a Hausdorff k-analytic space, if U is an open subset of X and if $\mathscr{F}$ is a sheaf on $X_{\text{ét}}$, then there is for every integer i a natural morphism $H_c^i(U_{\text{ét}}, \mathscr{F}_{|U}) \to H_c^i(X_{\text{ét}}, \mathscr{F})$.

If $\mathscr{X}$ is a separated k-scheme of finite type and if $\mathscr{F}$ is a sheaf on $\mathscr{X}_{\text{ét}}$, there is for every integer i a natural morphism $H_c^i(\mathscr{X}_{\text{ét}}, \mathscr{F}) \to H_c^i(\mathscr{X}_{\text{ét}}^{\text{an}}, \mathscr{F}^{\text{an}})$.

8.3 Link Between the étale and the Usual Topology

If X is an analytic space and if $\mathscr{F}$ is an étale sheaf of abelian groups on X, we know that $\mathscr{F}$ is also a sheaf for the usual topology on X. It follows from the very definition of derived functors (which we did not give in full detail) that there are for every i natural homomorphisms $H^i(X_{\text{top}}, \mathscr{F}) \to H^i(X_{\text{ét}}, \mathscr{F})$. The same also holds for cohomology with support in a given closed subset of X.

If X is good and if $\mathscr{F} = \mathbb{G}_m$ (resp. if $\mathscr{F}$ is the étale sheaf associated with a coherent sheaf), then the natural map $H^1(X_{\text{top}}, \mathscr{F}) \to H^1(X_{\text{ét}}, \mathscr{F})$ (resp. for every i, the natural map $H^i(X_{\text{top}}, \mathscr{F}) \to H^i(X_{\text{ét}}, \mathscr{F})$) is the comparison isomorphism mentioned above.

8.4 Étale Cohomology of Analytic Spaces: Two General Theorems

Theorem 6 (Over an algebraically closed ground field) *Assume that k is algebraically closed and let ℓ be a prime number which is non-zero in the residue field $\tilde{k}$. Let X be a Hausdorff k-analytic space and let $\mathscr{F} \in \mathsf{Sh}(X, \ell)$.*

1. *If X is paracompact of dimension d, then $\mathrm{H}^i(X_{\text{ét}}, \mathscr{F}) = 0$ for $i > 2d$.*
2. *If L is an algebraically closed complete extension of k, then for every i the natural arrows $\mathrm{H}^i(X_{\text{ét}}, \mathscr{F}) \to \mathrm{H}^i(X_{L.\text{ét}}, \mathscr{F}_L)$ and $\mathrm{H}^i_c(X_{\text{ét}}, \mathscr{F}) \to \mathrm{H}^i_c(X_{L.\text{ét}}, \mathscr{F}_L)$ are isomorphisms.*
3. *If X is compact then the group $\mathrm{H}^i(X_{\text{ét}}, \mathscr{F})$ is finite.*

Let us make some comments.

- Assertion 3) has been proved very recently by Berkovich ([4]). It was known till now only in some particular cases: if char. $k = 0$ (Huber, [9] and [8]); and if X is G-locally algebraizable (Berkovich, [2]).
- If X is affinoid of dimension d, it seems likely that $\mathrm{H}^i(X_{\text{ét}}, \mathscr{F}) = 0$ for $i > d$; but to the author's knowledge, this has yet not been proved.

It is possible to compare algebraic and analytic étale cohomology groups as follows.

Theorem 7 (Comparison theorem) *Let $\mathscr{X}$ be a separated k-scheme of finite type, let ℓ be a prime number and let $\mathscr{F} \in \mathsf{Sh}(\mathscr{X}, \ell)$.*

1. *One has $\mathrm{H}^i_c(\mathscr{X}^{an}_{\text{ét}}, \mathscr{F}^{an}) \simeq \mathrm{H}^i_c(\mathscr{X}_{\text{ét}}, \mathscr{F})$.*
2. *If moreover the prime ℓ is non-zero in k (but possibly zero in $\tilde{k}$) then*

$$\mathrm{H}^i(\mathscr{X}^{an}_{\text{ét}}, \mathscr{F}^{an}) \simeq \mathrm{H}^i(\mathscr{X}_{\text{ét}}, \mathscr{F}) \, .$$

Remark 58 Let k be any field (without absolute value) and let $\mathscr{X}$ be a separated k-scheme of finite type. By endowing k with the trivial absolute value and by applying assertion 1) above, one can define the cohomology with compact support of a torsion étale sheaf on $\mathscr{X}$ *without using any compactification of $\mathscr{X}$.*

Remark (About the ℓ-adic formalism in analytic geometry) If ℓ is a prime number which is non-zero in k then on a compact k-analytic space one can define relevant étale cohomology groups with coefficients $\mathbb{Z}_\ell$ or $\mathbb{Q}_\ell$ exactly in the same way as in scheme theory. For non-compact spaces it is slightly more complicated; the only reference on that topic are, as far as we know, unpublished notes by Berkovich.

8.5   The Cohomology of Some Analytic Spaces

We give some examples which show that the cohomology of analytic spaces over an algebraically closed field 'takes the expected values'. This means that these values are the same as for the analogous complex analytic space.

Throughout this section, we assume that k is algebraically closed, we fix a prime number ℓ which is non zero in $\tilde{k}$ and an integer n, and we set $\Lambda = \mathbb{Z}/\ell^n\mathbb{Z}$.

- If X in an open polydisc, then $\mathrm{H}^i(X_{\text{ét}}, \Lambda) = 0$ for every $i > 0$.
- If X is an open poly-annulus whose dimension is d, then $\mathrm{H}^1(X_{\text{ét}}, \Lambda) \simeq \Lambda^d$. The connected Galois covers of X which are associated with elements of $\mathrm{H}^1(X_{\text{ét}}, \Lambda)$ are exactly the connected Kummer covers $X' \to X$ whose group embeds in Λ.

  The algebra $\mathrm{H}^\bullet(X_{\text{ét}}, \Lambda)$ is isomorphic to the free exterior Λ-algebra generated by $\mathrm{H}^1(X_{\text{ét}}, \Lambda) \simeq \Lambda^d$.

- If X is a connected, proper, smooth analytic group of dimension g (*e.g.* the analytification of an abelian variety, or $\mathbb{G}_m^{g,an}/\mathsf{L}$ for some lattice $\mathsf{L} \subset (k^*)^g$) then $\mathrm{H}^1(X, \Lambda) \simeq \Lambda^{2g}$. The connected Galois covers of X which are associated with elements of $\mathrm{H}^1(X_{\text{ét}}, \Lambda)$ are exactly the isogenies $X' \to X$ (with connected X') whose kernel embeds in Λ.

  The algebra $\mathrm{H}^\bullet(X_{\text{ét}}, \Lambda)$ is isomorphic to the free exterior Λ-algebra generated by $\mathrm{H}^1(X_{\text{ét}}, \Lambda) \simeq \Lambda^{2g}$.

8.6   About the p-Primary Cohomology of the Open Unit Disc

In the discussion of the previous section, the assumption that $\ell \neq 0$ in $\tilde{k}$ is essential.

If $\tilde{k}$ is of characteristic $p > 0$ and if X is the open unit disc, then $\mathrm{H}^1(X, \mathbb{Z}/p\mathbb{Z})$ is infinite. Let us explain why this in the case when k is of characteristic zero. Note that $\mathbb{Z}/p\mathbb{Z}$ is then non-canonically isomorphic to μ_p. We are going to construct non-trivial elements in $\mathrm{H}^1(X, \mu_p)$.

Let us begin with some general facts. Let X be for the moment *any* k-analytic space. On $X_{\text{ét}}$ the Kummer exact sequence

$$1 \to \mu_p \to \mathbb{G}_m \xrightarrow{z \mapsto z^p} \mathbb{G}_m \to 1$$

induces a long exact sequence

$$0 \to \mathrm{H}^0(X_{\text{ét}}, \mu_p) \to \mathrm{H}^0(X_{\text{ét}}, \mathbb{G}_m) \to \mathrm{H}^0(X_{\text{ét}}, \mathbb{G}_m) \to \mathrm{H}^1(X_{\text{ét}}, \mu_p) \to \cdots$$

Assume that X is connected and non-empty, and let $f \in \mathrm{H}^0(X_{\text{ét}}, \mathbb{G}_m)$. Its image in $\mathrm{H}^1(X_{\text{ét}}, \mu_p)$ induces a connected Galois cover Y of X, together with an embedding of $\mathsf{Aut}\,(Y/X)$ into μ_p. Let us describe concretely Y and the embedding.

- If $f = z^p$ for some $z \in \mathrm{H}^0(X_{\text{ét}}, \mathbb{G}_m)$ then $Y = X$ (it is the case where the image of f in $\mathrm{H}^1(X_{\text{ét}}, \mu_p)$ is zero), and the embedding is of course the trivial one.
- If not, for any affinoid domain $U = \mathscr{M}(\mathscr{A})$ of the open disc X, denote by Y_U the affinoid space $\mathscr{M}(\mathscr{A}[T]/(T^p - f))$. This is an étale cover of U. The Galois cover $Y \to X$ is then obtained by patching the Y_U's together (for varying U). And the embedding $\mathsf{Aut}\,(Y/X) \hookrightarrow \mu_p$ is then an isomorphism $\mathsf{Aut}\,(Y/X) \simeq \mu_p$, defined as follows: if φ is an X-automorphism of Y, there exists a (unique) $\lambda_\varphi \in \mu_p$ such that $\varphi^* \overline{T} = \lambda_\varphi \overline{T}$; the assignment $\varphi \mapsto \lambda_\varphi$ is the required isomorphism.

Suppose now that the analytic space X is the open unit disc. The group $\mathscr{O}_X(X)^* = \mathrm{H}^0(X_{\text{ét}}, \mathbb{G}_m)$ is then the group of power series $\sum a_i T^i$ with $a_0 \neq 0$ and $|a_i| \leq |a_0|$ for every i. The map $f \mapsto a_1/a_0 = f'(0)/f(0)$ is a group homomorphism from $(\mathscr{O}_X(X)^*, \times)$ to $(k^\circ, +)$, which is surjective.

It induces a surjection $\mathscr{O}_X(X)^*/(\mathscr{O}_X(X)^*)^p \to k^\circ/pk^\circ$. As k°/pk° is infinite (there is a surjection $k^\circ/pk^\circ \to \tilde{k}$), the group $\mathscr{O}_X(X)^*/(\mathscr{O}_X(X)^*)^p$ is infinite. By the exact sequence above, it embeds into the cohomology group $\mathrm{H}^1(X, \mu_p)$. Therefore, $\mathrm{H}^1(X, \mathbb{Z}/p\mathbb{Z}) \simeq \mathrm{H}^1(X, \mu_p)$ is infinite.

Example 59 (A non-algebraizable finite Galois cover of the unit disc) Let I be the set of prime-to-p integers, and let $(a_i)_{i \in I}$ be a family of elements of k such that $|p| < |a_0| < 1$ and such that $(|a_i|)$ is strictly increasing and has limit 1 when $i \to \infty$. Let f be the invertible function $1 + \sum a_i T^i$ on the open unit disc, and let $Y \to X$ be the Galois cover associated with f by the procedure described above. It can be shown that Y is non-trivial (hence its Galois group is μ_p). It is non-algebraizable, in the following sense: it is not isomorphic to an open subset of an algebraic curve (the point is that Y has infinitely many type 2 points x with $\widetilde{\mathscr{H}(x)}$ of positive genus).

This example was communicated to the author by Michael Temkin.

9 Poincaré Duality

We refer to [11, 17, 18] in the case of schemes and [1] for analytic spaces.

Poincaré duality theorems look formally the same in the algebraic setting and in the analytic setting. We will therefore use a common formulation.

9.1 Cup-Products with Supports

Let us begin with a general remark, which holds for topological spaces as well as for étale topology of schemes and analytic spaces. In those theories, cup-product pairings are also defined for cohomology with a given closed support, and for cohomology with compact support when the latter makes sense. They give rise to pairings

$$H_Y^i(\cdot,\cdot) \times H_Z^j(\cdot,\cdot) \to H_{Y\cap Z}^{i+j}(\cdot,\cdot)$$

and

$$H_c^i(\cdot,\cdot) \times H^j(\cdot,\cdot) \to H_c^{i+j}(\cdot,\cdot)$$

which behave as expected.

9.2 Poincaré Duality: Context and Statement

Let us denote by k an algebraically closed field (resp. an algebraically closed non-Archimedean field). Let X be a separated, smooth k-scheme of finite type (resp. a Hausdorff smooth k-analytic space) of pure dimension d. We fix a prime number ℓ which is non zero in k (resp. $\tilde{k}$) and an integer n; set $\Lambda = \mathbb{Z}/\ell^n\mathbb{Z}$.

If $\mathscr{F} \in \mathsf{Sh}(X,\ell)$ is annihilated by ℓ^n and if $i \in \mathbb{N}$, we will denote by $\mathscr{F}(i)$ the sheaf $\mathscr{F} \otimes \mu_{\ell^n}^{\otimes i}$, and by $\mathscr{F}^\vee$ the sheaf that sends U to $\mathrm{Hom}(\mathscr{F}_{|U}, \Lambda_U)$ (it is called the *dual* sheaf of $\mathscr{F}$); if $i \in \mathbb{Z} \setminus \mathbb{N}$ the sheaf $\mathscr{F}(-i)^\vee$ will be denoted by $\mathscr{F}(i)$. The formulas $\mathscr{F}(i+j) = \mathscr{F}(i) \otimes \mathscr{F}(j)$ and $\mathscr{F}(i)^\vee = \mathscr{F}(-i)$ then hold for all i, j in $\mathbb{Z}$.

Theorem 8 (Poincaré duality for schemes and analytic spaces) *We use the notation above.*

1. *There is a natural* trace map $\mathsf{Tr} : H_c^{2d}(X_{\text{ét}}, \Lambda(d)) \to \Lambda$.
2. *For every* $\mathscr{F} \in \mathsf{Sh}(X,\ell)$ *which is annihilated by* ℓ^n *and every* $i \in \{0, \ldots, 2d\}$ *the map*

$$H^i(X_{\text{ét}}, \mathscr{F}) \times H_c^{2d-i}(X_{\text{ét}}, \mathscr{F}^\vee(d)) \to \Lambda,$$

induced by the cup-product pairing and the trace map, identifies $H^i(X_{\text{ét}}, \mathscr{F})$ *with the dual* Λ-*module of* $H_c^{2d-i}(X_{\text{ét}}, \mathscr{F}^\vee(d))$, *that is, with*

$$\mathsf{Hom}_\Lambda(H_c^{2d-i}(X_{\text{ét}}, \mathscr{F}^\vee(d)), \Lambda).$$

Remark (A fundamental case) If X is connected and non-empty, we have $H^0(X_{\text{ét}}, \Lambda) \simeq \Lambda$. We thus deduce from 2) that the trace map induces an isomorphism $H_c^{2d}(X_{\text{ét}}, \Lambda(d)) \simeq \Lambda$.

Remark (About duality) In the algebraic case, duality assertion 2) only involves *finite* Λ-modules. But that is not the case in the analytic context. Let us give a very simple example of that phenomenon in the *zero-dimensional* case (which is, maybe surprisingly, not completely empty): if X is a disjoint union of one-point spaces $\mathcal{M}(k)$ indexed by some set I (possibly infinite) then $H^0(X, \Lambda) = \Lambda^I$ and $H^0_c(X, \Lambda) = \Lambda^{(I)}$; classical duality between these two groups is then a particular case of Poincaré duality!

Remark (Transcendental Poincaré duality) In general topology, Poincaré duality between the cohomology without support and that with compact support on a topological variety involves the *orientation sheaf* of the variety (which is locally isomorphic to the constant sheaf $\mathbb{Z}$). In étale cohomology, the role of the latter is, in view of the theorem above, played by the sheaf of roots of unity.

However when one works with smooth complex analytic spaces, the orientation sheaf is usually not mentioned. This is because every smooth complex analytic space X comes with a natural orientation (defined through multiplication by i), whence an isomorphism between the orientation sheaf of X and $\mathbb{Z}_X$. We emphasize that this identification is related to the choice of one of the two square-roots of -1 in $\mathbb{C}$; such a choice also induces for every N an isomorphism $\mu_N \simeq \mathbb{Z}/N\mathbb{Z}$ (by distinguishing $e^{2i\pi/N}$), modulo which the isomorphisms of Theorem 5 D) are compatible with algebraic and transcendental Poincaré dualities for N-torsion finite coefficients.

9.3 Cycle Classes

We keep the notation of Theorem 8. Let D be an irreducible and reduced divisor on X, that is, an irreducible and reduced closed subscheme (resp. analytic subspace) of dimension $d - 1$. As X is smooth, D is Cartier (*i.e.* locally defined by one non-zero equation). The sheaf of rational (or meromorphic) functions defined on $X \setminus D$ and having at most simple poles along D is then a line bundle $\mathcal{O}(D)$ with a canonical section 1, which generates $\mathcal{O}(D)$ on $X \setminus D$. To such a datum is associated in a quite formal way a cohomology class of $H^1_D(X_{\text{top}}, \mathbb{G}_m)$ which one can push to $H^1_D(X_{\text{ét}}, \mathbb{G}_m)$, and then to $H^2_D(X_{\text{ét}}, \mu_{\ell^n}) = H^2_D(X_{\text{ét}}, \Lambda(1))$ through the long-exact cohomology sequence associated with the Kummer exact sequence of étale sheaves

$$1 \to \mu_{\ell^n} \to \mathbb{G}_m \xrightarrow{z \mapsto z^{\ell^n}} \mathbb{G}_m \to 1.$$

This construction extends to an assignment which sends every Zariski closed subset (resp. every closed analytic subset) Y of X of pure codimension i to a cohomology class in $H^{2i}_Y(X_{\text{ét}}, \Lambda(i))$, which can be pushed to $H^{2i}_c(X_{\text{ét}}, \Lambda(i))$ if Y is proper. The formation of those cohomology classes associated with Zariski-closed (resp. closed analytic) subsets of X makes *transverse* intersections commute with cup-products.

Now let x be a k-point on X. By the above, one can associate to it a class $h(x) \in H^{2d}_c(X_{\text{ét}}, \Lambda(d))$. One then has $\text{Tr}(h(x)) = 1 \in \Lambda$, and the remark following Theorem 8 can be rephrased as follows: $h(x)$ does not depend on x

and $H_c^{2d}(X_{\text{ét}}, \Lambda(d))$ is a free Λ-module generated by $h(x)$. One says that $h(x)$ is the *fundamental class* of X.

10   The Case of a Non-Algebraically Closed Ground Field

We refer to [16] and [11] for algebraic varieties, and to [1] for analytic spaces.

Up to now most of our theorems about the cohomology of algebraic varieties or analytic spaces concerned the case of an *algebraically closed* field. We would like to give some explanations about what happens if one removes that assumption.

10.1   *Étale Cohomology of a Field and Galois Cohomology*

Let k be a field, let k^s be a separable closure of k and let G be the profinite group $\text{Gal}(k^s/k)$. If $\mathscr{F}$ is an étale sheaf on $\textsf{Spec}\ k$ and if L is a finite sub-extension of k^s/k, then $\textsf{Spec}\ L \to \textsf{Spec}\ k$ is étale; therefore $\mathscr{F}(\textsf{Spec}\ L)$ does make sense, and will be written $\mathscr{F}(L)$ for the sake of simplicity. The direct limit of the $\mathscr{F}(L)$'s, for L going through the set of finite sub-extensions of k^s/k, is in a natural way a discrete G-set (discrete means that the stabilizers are open); we denote it by $\mathscr{F}(k^s)$; if $\mathscr{F}$ is a sheaf of abelian groups, then $\mathscr{F}(k^s)$ is a discrete G-module.

Theorem 9 *We use the above notation.*

1. *The assignment $\mathscr{F} \mapsto \mathscr{F}(k^s)$ induces an* equivalence *(i.e. a categorical 1-to-1 correspondence) between the category of étale sheaves on $\textsf{Spec}\ k$ and that of discrete G-sets.*
2. *If $\mathscr{F}$ is an étale sheaf on $\textsf{Spec}\ k$ and if L is a finite sub-extension of k^s/k corresponding to an open subgroup H of G then $\mathscr{F}(L) = \mathscr{F}(k^s)^H$.*
3. *If $\mathscr{F}$ is a sheaf of abelian groups on $\textsf{Spec}\ k$ then one has for every i an isomorphism $H^i((\textsf{Spec}\ k)_{\text{ét}}, \mathscr{F}) \simeq H^i(G, \mathscr{F}(k^s))$.*

In particular the étale cohomology theory of $\textsf{Spec}\ k$ is nothing but the cohomology theory of discrete $\text{Gal}(k^s/k)$-modules, which is also called the *Galois cohomology* of the field k. Galois cohomology plays a crucial role in modern arithmetic. We suggest the interested reader to look at Serre's book [13].

10.2   *A Topological Analog*

Let X be a connected, non-empty, "reasonable" topological space, admitting a universal cover $Y \to X$; let G be the automorphism group of Y over X. Let A be an abelian group equipped with an action of G. It corresponds to a local system

$\mathscr{F}$ on X which is locally isomorphic to A; the pull-back of $\mathscr{F}$ on Y is *globally* isomorphic to A. The group G acts in a natural way on $H^\bullet(Y, A)$.

There is a natural spectral sequence

$$H^p(G, H^q(Y, A)) \Rightarrow H^{p+q}(X, \mathscr{F}).$$

Now assume that $H^q(Y, A) = 0$ for every $q > 0$ (*e.g* Y is contractible). Then this spectral sequence simply furnishes isomorphisms $H^p(X, \mathscr{F}) \simeq H^p(G, A)$ for every p. Therefore we see that, as far as étale cohomology is concerned, the spectrum of a field k behaves, to some extent, like a connected, non-empty topological space having a contractible universal cover. The spectrum of a separable closure k^s of k can be thought of as such a cover.

10.3   Varieties or Analytic Spaces Over a Field

Our purpose is now to explain how the cohomology of a variety or an analytic space over a field k mixes, in some sense, Galois cohomology of the ground field and cohomology of the given variety over an algebraic closure of k.

Let us begin with some remarks. Fix a field k and an algebraic closure k^a of k, and let G be the absolute Galois group $\mathrm{Gal}(k^a/k)$.

1. If X is an algebraic variety over k and if $\mathscr{F}$ is an étale sheaf of abelian groups on X, there is a natural discrete action of G on $H^\bullet(X_{k^a, \text{ét}}, \mathscr{F}_{k^a})$.

2. If k is non-Archimedean and complete, and if $\widehat{k^a}$ denotes the completion of k^a, then for every *compact*[9] analytic space X and every étale sheaf of abelian groups $\mathscr{F}$ on X, there is a natural discrete action of G on $H^\bullet(X_{\widehat{k^a}, \text{ét}}, \mathscr{F}_{\widehat{k^a}})$.

Theorem 10 *Let k be a field and let k^a be an algebraic closure of k; let G be the absolute Galois group $\mathrm{Gal}(k^a/k)$.*

1. If X is a k-scheme of finite type and if $\mathscr{F}$ is an étale sheaf of abelian groups on X, one has a natural spectral sequence

$$H^p(G, H^q(X_{k^a, \text{ét}}, \mathscr{F}_{k^a})) \Rightarrow H^{p+q}(X_{\text{ét}}, \mathscr{F}).$$

2. Assume that k is non-Archimedean and complete, and let $\widehat{k^a}$ be the completion of k^a. If X is a compact analytic space over k and if $\mathscr{F}$ is an étale sheaf of abelian groups on X, one has a natural spectral sequence

$$H^p(G, H^q(X_{\widehat{k^a}, \text{ét}}, \mathscr{F}_{\widehat{k^a}})) \Rightarrow H^{p+q}(X_{\text{ét}}, \mathscr{F}).$$

---

[9]The fact for the action of G on the cohomology of $\mathscr{F}_{\widehat{k^a}}$ to be discrete essentially means that every cohomology class of $\mathscr{F}_{\widehat{k^a}}$ is already defined on a finite separable extension of k; one needs compactness to ensure this finiteness result.

Acknowledgements I would like to thank warmly: Mathilde Herblot, who has written detailed notes during the lectures, and has allowed me to use them for the redaction of this text; and Johannes Nicaise, who has read carefully several versions of this work and has made a lot of remarks and suggestions, which greatly helped me to improve the initial presentation.

References

1. V. Berkovich, Étale cohomology for non-archimedean analytic spaces. Inst. Hautes Études Sci. Publ. Math. **78**, 5–161 (1993)
2. V. Berkovich, Vanishing cycles for formal schemes. Invent. Math. **115**(3), 539–571 (1994)
3. V. Berkovich, On the comparison theorem for étale cohomology of non-Archimedean analytic spaces. Isr. J. Math. **92**, 45–60 (1995)
4. V. Berkovich, Vanishing cycles for formal schemes III. Preprint
5. F. Charles, Conjugate varieties with distinct real cohomology algebras. J. Reine Angew. Math. **630**, 125–139 (2009)
6. A.J. de Jong, Étale fundamental groups of non-Archimedean analytic spaces. Compositio Math. **97**, 89–118 (1995)
7. A. Grothendieck, Sur quelques points d'algèbre homologique. Tohoku Math. J. **9**, 119–221 (1957)
8. R. Huber, *Étale Cohomology of Rigid Analytic Varieties and Adic Spaces.* Aspects of Mathematics, vol. 30 (Friedr. Vieweg & Sohn, Braunschweig, 1996). x+450 pp.
9. R. Huber, A finiteness result for direct image sheaves on the étale site of rigid analytic varieties. J. Algebraic Geom. **7**(2), 359–403 (1998)
10. L. Illusie, *Grothendieck et la topologie étale*, available on Luc Illusie's home page (2008)
11. J.S. Milne, *Étale Cohomology* (Princeton University Press, Princeton, NJ, 1980)
12. J.-P. Serre, Exemple de variétés projectives conjuguées non homéomorphes. C. R. Acad. Sci. Paris **258**, 4194–4196 (1964)
13. J.-P. Serre, *Cohomologie Galoisienne*, 5th edn. Lecture Notes in Mathematics, vol. 5 (Springer, New York, 1997)
14. SGA 1, *Revêtements étales et groupe fondamental.* Lecture Notes in Mathematics, vol. 224 (Springer, New York, 1971)
15. SGA 4-1, *Théorie des topos et cohomologie étale des schÃl'mas*, vol. 1. Lecture Notes in Mathematics, vol. 269 (Springer, New York, 1972)
16. SGA 4-2, *Théorie des topos et cohomologie étale des schÃl'mas*, vol. 2. Lecture Notes in Mathematics, vol. 270 (Springer, New York, 1972)
17. SGA 4-3, *Théorie des topos et cohomologie étale des schÃl'mas*, vol. 3. Lecture Notes in Mathematics, vol. 305 (Springer, New York, 1972)
18. SGA $4\frac{1}{2}$, *Cohomologie étale.* Lecture Notes in Mathematics, vol. 569 (Springer, New York, 1977)
19. SGA5, *Cohomologie ℓ-adique et Fonctions L.* Lecture Notes in Mathematics, vol. 589 (Springer, New York, 1977)

Countability Properties of Some Berkovich Spaces

Charles Favre

Contents

1 Introduction

Our aim is to prove some facts on the topology of a particular class of non-Archimedean analytic spaces in the sense of Berkovich. One of the main feature of Berkovich analytic spaces is that their natural topology makes them both locally arcwise connected and locally compact. However, in general these spaces are far from being separable.[1] Nevertheless, we shall prove that analytic spaces retain some countability properties that reflect their algebraic nature. We expect these properties to be useful in a dynamical context.

Supported by the ERC-project 'Nonarcomp' no.307856., by the ANR project Berko, and by the ECOS project C07E01.

[1]A *separable* space is a topological space admitting a countable dense subset. We shall never use the notion of separable field extension so that no confusion should occur.

C. Favre (✉)
CNRS, CMLS, École Polytechnique, Palaiseau, 91128 Cedex, France
e-mail: charles.favre@polytechnique.edu, Url: http://www.math.polytechnique.fr/~favre/

© Springer International Publishing Switzerland 2015
A. Ducros et al. (eds.), *Berkovich Spaces and Applications*, Lecture Notes
in Mathematics 2119, DOI 10.1007/978-3-319-11029-5_3

We shall work in the following setting. We consider a normal algebraic variety X defined over a field[2] k. We fix an effective Cartier divisor $D \subset X$, and we let $\hat{X}$ be the formal completion of X along D. We then consider the generic fibre $\hat{X}_\eta$ of this formal scheme as defined in [Thu]: this is an analytic space in the sense of Berkovich [Ber] over k endowed with the trivial norm. When $X = \mathrm{Spec}\, A$ is affine, and $D = \{f = 0\}$ with $f \in A$, then $\hat{X}_\eta$ coincides with the set of multiplicative semi-norms $|\cdot| : A \to \mathbb{R}_+$ that are trivial on k, ≤ 1 on A and satisfy $0 < |f| < 1$. Note that replacing $|\cdot|$ by $v := -\log |\cdot|$, we can define in an equivalent way $\hat{X}_\eta$ as the set of valuations $v : A \to \mathbb{R}_+ \cup \{+\infty\}$ (possibly taking the value $+\infty$ on a non-zero element) that are trivial on k, ≥ 0 on A, and such that $\infty > v(f) > 0$. In the general case, X is covered by affine open subsets U_i, and $\hat{X}_\eta$ is obtained by patching together the $\hat{U}_{i,\eta}$'s in a natural way.

Multiplying a valuation by a positive constant yields an action of $\mathbb{R}_+^*$ on $\hat{X}_\eta$. It is therefore natural to introduce the *normalized generic fiber* $]X[\subset \hat{X}_\eta$ of valuations normalized by the condition $v(f) = +1$. In this way, $]X[$ can be identified with the quotient space of $\hat{X}_\eta$ by $\mathbb{R}_+^*$, and the projection map $\hat{X}_\eta \to]X[$ is a $\mathbb{R}_+^*$-fibration. Since X is of finite type, the space $]X[$ is always compact, see Lemma 4.3 below.

Our main result reads as follows.

Theorem A *Let X be a normal algebraic variety over k and D be an effective Cartier divisor. Let A be any subset of the normalized generic fiber $]X[$ of the formal completion of X along D.*

For any x in the closure of A, there exists a sequence of points $x_n \in A$ such that $x_n \to x$.

Recall that a topological space is *angelic* if any relatively ω-compact set[3] is relatively compact, and for any relatively ω-compact subset A, any point in $\bar{A}$ is the limit of a *sequence* of points in A, see [Flo] for more information. This property plays an important role in the study of Banach spaces. A first consequence of the previous result is the following

Corollary A *Let X be a normal algebraic variety over k and D be an effective Cartier divisor. Then $]X[$ is compact and angelic, hence sequentially compact.*

In order to state yet another consequence of Theorem A, we need to introduce some terminology. When X is affine, a divisorial valuation on the ring of regular functions $k[X]$ is a discrete valuation of rank 1 and transcendence degree $\dim(X) - 1$. Geometrically, it is given by the order of vanishing along a prime divisor in a suitable birational model of X. A point $x \in \hat{X}_\eta$ is called divisorial if it is given by a divisorial valuation in the ring of regular functions of some affine chart. If we interpret x as a semi-norm, then x is divisorial iff it is a norm, and the residue

---

[2]This means X is irreducible reduced of finite type and k is algebraically closed. The last assumption is however inessential in our paper.

[3]That is any sequence of points in this set has a cluster point in the ambient space.

field of the completion of $k[X]$ w.r.t. the induced norm by x has transcendence degree $\dim(X) - 1$ over k.

Finally recall that we have a natural reduction map $r_X : \hat{X}_\eta \to D$ defined in the affine case by sending a semi-norm $|\cdot|$ to the prime ideal $\{|\cdot| < 1\} \subset k[X]$. This reduction map sends a divisorial valuation to its center in X when it is non empty. Note that the center is automatically included in D if the valuation lies in $\hat{X}_\eta$.

We shall also obtain

Corollary B *Let X be a normal algebraic variety over k and D be an effective Cartier divisor. Then for any point $x \in \hat{X}_\eta$, there exists a sequence of divisorial points $x_n \in \hat{X}_\eta$ that converges to x, and such that $r_X(x_n) \in \overline{r_X(x)}$ for all n.*

Theorem A will be deduced from its analog on Riemann-Zariski spaces, see Theorem 3.1 below for details. In this case, it essentially boils down to the noetherianity of the Zariski topology on varieties.

Let us now explain how one can transfer the previous results to non-Archimedean analytic spaces. The normalized generic fiber $]X[$ is not a Berkovich analytic space in a canonical way. However in the special situation where we have a map $T : X \to \mathbb{A}_k^1$ such that $D = \{T = 0\}$, then $]X[$ turns out to be an analytic space over the non-archimedean field $k((T))$ (with the norm $\exp(-\mathrm{ord}_0)$). In the case $X = \mathrm{Spec}\,(k[x_i]/\mathfrak{a})$ is affine, then $\mathcal{A} = \varprojlim_n (k[x_i]/\mathfrak{a})/(T^n)$ is a $k[[T]]$-algebra topologically of finite type, and $]X[$ is the set of semi-norms ≤ 1 on $\mathcal{A}$ whose restriction on $k[[T]]$ is $\exp(-\mathrm{ord}_0)$. In general, $]X[$ coincides with the generic fiber of the T-adic completion of X along D in the sense of Berkovich (its construction in rigid geometry was previously given by Raynaud).

We shall prove:

Corollary C *Any compact Berkovich analytic space that is defined over the field $k((T))$ is angelic. It is in particular sequentially compact, and divisorial points are sequentially dense.*

Note that this result also holds over any (non-Archimedean) local field for simple reasons since any affinoid over such a field is separable. However when the residue field of k is not countable, strictly affinoid spaces in the sense of Berkovich of positive dimension are not metrizable.

For curves, the sequential compactness is a consequence of the semi-stable reduction theorem, and of the fact that any complete $\mathbb{R}$-tree (in the sense of [FJ]) is sequentially compact. We refer to [Mai] for a proof.

We note that any compact analytic space over *any* non-Archimedean complete field is angelic by the work of J. Poineau, [Poi]. Even if our Corollary C is much more restrictive, our method based on Riemann-Zariski spaces is substantially different from the approach of Poineau, and we hope it will prove useful in other contexts too.

We also refer to [Gub, Theorem 7.12] for a proof of the density of divisorial points in an arbitrary Berkovich space (see also [YZ, Lemma 2.3.1]).

The following natural question is related to the above results.

Question 1 Let X be any Berkovich analytic space. Is the support of any Radon measure separable?

In dimension 1, Question 1 has a positive answer, see [FJ, §7.3.6] and [BR, Lemma 5.6] that Borel measures on an arbitrary $\mathbb{R}$-tree are Radon measures. This would imply Question 1 in dimension 1. However both proofs are wrong, so that Question 1 remains open in any dimension. We refer to [Jon, §2.4] for a discussion on this problem.

2 Riemann-Zariski Spaces

Our basic references are [ZS, Vaq].

A non-trivial ring R is said to be a valuation ring if it has no divisors of zero, and for any non-zero element x in the fraction field of R, either x or x^{-1} belong to R. Any valuation ring is local, with maximal ideal $\mathfrak{m}_R$ consisting of those $x \in R$ such that x^{-1} does not belong to R.

A valuation on a domain R is a function $\nu : R \setminus \{0\} \to \Gamma$ to a totally ordered abelian group Γ such that $\nu(ab) = \nu(a) + \nu(b)$; and $\nu(a + b) \geq \max\{\nu(a), \nu(b)\}$. Any valuation extends in a unique way to a valuation on the fraction field K of R, and the set $R_\nu = \{a \in K, \nu(a) \geq 0\}$ is a valuation ring in K.

Conversely, to a valuation ring $R \subset K$ is associated a unique valuation $\nu : R \to \Gamma$ up to isomorphism, where Γ is the group obtained by moding out K^* by the multiplicative set $R \setminus \mathfrak{m}_R$.

In the sequel, we shall make no difference between valuation rings and valuations.

Let X be a *projective normal* variety defined over a field k. Abhyankhar's inequality for a valuation ν on $k(X)$ that is trivial on k states that

$$\text{rat.rk}(\nu) + \text{deg.tr}(\nu) \leq \dim(X) ,$$

where $\text{rat.rk}(\nu)$ denotes the dimension of the $\mathbb{Q}$ vector space $\Gamma \otimes_{\mathbb{Z}} \mathbb{Q}$; and $\text{deg.tr}(\nu)$ is the degree of transcendence of the residue field over k. In particular, $\Gamma \otimes_{\mathbb{Z}} \mathbb{Q}$ is countable. We thus have

Lemma 2.1 *The value group of any valuation on $k(X)$ that is trivial on k is countable.*

Define the Riemann-Zariski space $\mathfrak{X}$ as the set of valuation rings in $k(X)$ containing k. We endow it with the topology for which the sets

$$U(A) = \{R, \ A \subset R\}$$

where A ranges over all subrings of finite type of $k(X)$ that contains k form a basis of open sets. A theorem of Zariski states:

Theorem 2.2 *The Riemann-Zariski space is quasi-compact.*

Note that $\mathfrak{X}$ is never Hausdorff except in dimension 0.

Suppose $v : k(X) \to \Gamma$ is a valuation with valuation ring R_v. A projective birational model of X consists of a birational map $\phi : X' \dashrightarrow X$ from a normal projective variety to X. The map ϕ induces an isomorphism between $k(X)$ and $k(X')$, so that v can be viewed as a valuation on $k(X')$. The set of points $x' \in X'$ such that the local ring $\mathcal{O}_{X',x'}$ is included in R_v forms an irreducible subvariety called the center of v in X'. We denote it by $C(v, X')$. We shall view $C(v, X')$ scheme-theoretically as a (non necessarily closed) point in X'. A birational model $\psi : X'' \dashrightarrow X$ dominates another one $\phi : X' \dashrightarrow X$ iff $\mu := \phi^{-1} \circ \psi$ is regular. If X'' dominates X', then $\mu(C(v, X'')) = C(v, X')$.

Consider the category $\mathfrak{B}$ of all projective birational models of X up to natural isomorphism, each model endowed with the Zariski topology. It is an inductive set for the relation of domination introduced before. We may thus consider the projective limit of all birational models of X, that is $\varprojlim_{X' \in \mathfrak{B}} X'$ endowed with the projective limit topology. Concretely, a point in $\varprojlim_{X' \in \mathfrak{B}} X'$ is a collection of irreducible subvarieties $Z_{X'} \subset X'$ for each projective birational model X' such that $\mu(Z_{X''}) = Z_{X'}$ if X'' dominates X'.

For a given valuation v, we may attach the collection $\{C(v, X')\}_{X' \in \mathfrak{B}}$. This defines a map from $\mathfrak{X}$ to $\varprojlim_{X' \in \mathfrak{B}} X'$. Conversely, given a point $Z = \{Z_{X'}\}_{X' \in \mathfrak{B}}$ in the projective limit, we define the subset R_Z of $k(X)$ of those meromorphic functions that are regular at the generic point of $Z_{X'}$ for any birational model X' of X. It is not difficult to check that R_Z is a valuation ring.

These two maps are inverse one to the other. More precisely, one has the following fundamental result again due to Zariski:

Theorem 2.3 *The natural map* $\mathfrak{X} \to \varprojlim_{X' \in \mathfrak{B}} X'$ *given by* $v \mapsto \{C(v, X')\}_{X' \in \mathfrak{B}}$ *induces a homeomorphism.*

This result has the following useful consequence. For any model X', and any Weil divisor D' in X', let $\mathcal{U}(X', D')$ be the set of all valuation rings whose center is not included in D'. Then the collection of all sets of the form $\mathcal{U}(X', D')$ gives a basis for the topology of $\mathfrak{X}$.

Note that if we cover $X' \setminus D'$ by affine charts Y_i, then $X' \setminus Y_i := D_i$ is a divisor since X' is normal, and $\mathcal{U}(X', D') = \cup \mathcal{U}(X', D_i)$. Thus the collection of all $\mathcal{U}(X', D')$'s such that $X' \setminus D'$ is affine also forms a basis for the topology of $\mathfrak{X}$. The next lemma follows from the very definition of the topology on $\mathfrak{X}$. We include a proof for convenience.

Lemma 2.4 *A sequence of valuations* v_n *converges to* v *in* $\mathfrak{X}$ *iff for any* $f \in R_v$, *we have* $f \in R_{v_n}$ *for n large enough.*

Proof Suppose first $\nu_n \to \nu$, and pick $f \in R_\nu$. Choose a model X' such that f is regular on X', and let D' be the set of poles of f. Since f belongs to R_ν, the center of ν cannot be included in D', and $\nu \in \mathcal{U}(X', D')$. By assumption $\nu_n \to \nu$ hence $\nu_n \in \mathcal{U}(X', D')$ for n large enough. This implies the center of ν_n not to be included in D', and we conclude that $f \in R_{\nu_n}$.

Conversely, let us assume that for any $f \in R_\nu$, we have $f \in R_{\nu_n}$ for n large enough. The collection of open sets $\mathcal{U}(X', D')$ with X' a birational model of X, and $D' \subset X'$ a divisor forms a basis for the topology on $\mathfrak{X}$. Therefore proving the convergence of ν_n to ν is equivalent to show $\nu_n \in \mathcal{U}(X', D')$ for n large enough if $\nu \in \mathcal{U}(X', D')$. As noted above, we may assume $Y' = X' \setminus D'$ is affine. Choose a finite set of regular functions f_i on the affine space Y' that generate $k[Y']$. Since the center of ν in X' is not included in D', we have $\nu(f_i) \geq 0$ for all i. By assumption, we get $\nu_n(f_i) \geq 0$ for all i and all large enough n. But then the center of ν_n cannot be in D', hence $\nu_n \in \mathcal{U}(X', D')$. This concludes the proof. $\qquad\square$

Finally we shall use several times the

Lemma 2.5 *Suppose $\mathfrak{I}$ is a coherent sheaf of ideals on a normal variety X. Then there exists a regular birational map $\phi : X' \to X$ such that X' is normal and $\mathfrak{I} \cdot \mathcal{O}_{X'}$ is locally principal (i.e. invertible).*

Proof Take X' to be the normalization of the blow-up of $\mathfrak{I}$, and use the universal property of blow-ups. $\qquad\square$

3 Countability Properties in Riemann-Zariski Spaces

In this section, X is a *normal projective algebraic variety* defined over a (non necessarily algebraically closed) field k. Our aim is to prove

Theorem 3.1 *Let A be any subset of $\mathfrak{X}$. Then for any ν in the closure of A, one can find a sequence of valuations $\nu_n \in A$ such that $\nu_n \to \nu$.*

The proof relies on the following lemma. For any ordered group Γ, let Γ_+ be the set of non-negative elements. Recall that given an affine variety Y, a valuation $\nu : k[Y] \to \Gamma_+$ and $\gamma \in \Gamma$ positive, the valuation ideal is defined by

$$I(\nu, \gamma) := \{f, \ \nu(f) \geq \gamma\} \subset k[Y] \,.$$

Note that the condition on ν to be non-negative on $k[Y]$ is equivalent to have a center in Y.

Lemma 3.2 *Let Y be an irreducible affine variety. Suppose $\nu : k[Y] \to \Gamma$ is a valuation whose center in Y is non-empty. Pick $\gamma \in \Gamma$, with $\gamma > 0$, and consider a proper modification $\mu : Y' \to Y$ with Y' normal such that $I(\nu, \gamma) \cdot \mathcal{O}_{Y'} = \mathcal{O}_{Y'}(-D)$ for some effective Cartier divisor D.*

Then $C(v, Y')$ is included in the support of D, and for any $f \in k[Y]$ such that $v(f) = \gamma$, the divisor $\mathrm{div}(f \circ \mu) \subset Y'$ is equal to D at the generic point of $C(v, Y')$.

Proof Suppose there exists a closed point $p \in Y'$, and a regular function $f \in k[Y]$ such that $v(f) = \gamma$, and $f \circ \mu$ does not vanish at p. Then the lift of f^{-1} is regular at p but does not belong to the valuation ring at v. Whence $p \notin C(v, Y')$. This proves the first claim.

Take a local function h defining D at the generic point $p \in C(v, Y')$. Since $I(v, \gamma) \cdot \mathcal{O}_{Y'} = \mathcal{O}_{Y'}(-D)$, one can find elements $g_i \in I(v, \gamma)$ and germs of regular functions h_i at p such that $h = \sum h_i (g_i \circ \mu)$.

By taking a suitable linear combination of the g_i, we thus get an element g in the valuation ideal such that $v(g) = \gamma$, and the divisor $\{g \circ \mu = 0\}$ is equal to D at p. Define $\tilde{f} = (f \circ \mu)/(g \circ \mu)$. Since $v(f) = v(g) = \gamma$, we get $v(\tilde{f}) = 0$ which shows that $\tilde{f}$ is not vanishing at p, and proves the claim. $\qquad\square$

Proof of Theorem 3.1 There is no loss of generality in assuming $v \notin \overline{\{\mu\}}$ for any $\mu \in A$. Otherwise the constant sequence μ converges to v. Let $Y \subset X$ be an affine chart intersecting the center of v, and write $v : k[Y] \to \Gamma$ for some totally ordered group Γ. By Lemma 2.1, $v(k[Y])$ is countable. By Lemma 2.5, we may produce a sequence $Y_{n+1} \xrightarrow{\pi_n} Y_n \xrightarrow{\pi_{n-1}} \dots \xrightarrow{\pi_1} Y$ of birational models such that for any $\gamma \in \Gamma$, the valuation ideal sheaf $I(v, \gamma) \cdot \mathcal{O}_{Y_n}$ is locally principal for all n large enough.

For any subset $B \subset A$, we introduce the Zariski closed set

$$C_n(B) := \overline{\cup_{\mu \in B} C(\mu, Y_n)}.$$

Lemma 3.3 *There exists a countable subset $A' \subset A$ and irreducible subvarieties $Z_n \subset Y_n$ such that*

1. *$\pi_{n-1}(Z_n) = Z_{n-1}$ for all n;*
2. *the restriction maps $\pi_{n-1} : Z_n \to Z_{n-1}$ are birational;*
3. *for all n, $C(v, Y_n) \subset Z_n$;*
4. *for all n, $C_n(A') = Z_n$.*

Pick any enumeration $\{v_m\}$ of the elements of A', and consider an algebraically closed countable field K such that all varieties $Z_n, C(v_m, Y_n), C(v, Y_n)$ are defined over K.

Let us introduce the following terminology. A pro-divisor $W = \{W_n\}$ is a collection of (possibly zero) reduced divisors W_n in Z_n defined over K, such that $\pi_n(W_{n+1}) \subset W_n$ for all n, there exists an N for which W_N is irreducible, and $W_n = (\pi_N \circ \dots \circ \pi_{n-1})^{-1}(W_N)$ for all $n \geq N$. We call the minimal N having this property the height of W and denote it by $h(W)$. Since for each N the set of prime divisors of Z_N and defined over K is countable, we may find a sequence W^j enumerating all pro-divisors.

For any valuation $\mu \in A'$ and any pro-divisor W, we write $\mu \subset W$ if $C(\mu, Y_n) \subset W_n$ for all n. This condition is equivalent to impose $C(\mu, Y_N) \subset W_N$ for $N = h(W)$.

Lemma 3.4 *Suppose $\mu \in A'$. Then one can find an integer j such that $\mu \subset W^j$.*

We now define by induction a sequence of valuations $\tilde{v}_m$ and of integers $j_m < j_{m+1}$ such that

- $\tilde{v}_l \subset W^{j_l}$;
- $\tilde{v}_l \not\subset W^j$ for all $j < j_l$.

To do so we proceed as follows. Set $A_j = \{\mu \in A', \mu \subset W^j\}$. Set $j_1 = \min\{j, A_j \neq \emptyset\}$, and define recursively $j_{m+1} = \min\{j > j_m, A_j \setminus \cup_{l<j} A_l \neq \emptyset\}$.

Let us justify the existence of j_m for all m. By contradiction, assume that $A_j \setminus \cup_{l<j} A_l = \emptyset$ for all $j > j_m$. Since by Lemma 3.4 any valuation in A' belongs to some W^j, we have $A' = \cup_{j \geq 0} A_j$. As a consequence, we conclude that $A' \subset \cup_{j \leq j_m} A_j$ which forces all valuations in A' to have a center included in some fixed divisor of Z_1. This contradicts property (4) of Lemma 3.3.

Finally we pick any sequence $\tilde{v}_l \in A_{j_l}$.

We now prove that $\tilde{v}_n$ converges to v. Pick $f \in R_v$, write $f = \frac{g}{h}$ with $g, h \in k[Y]$. Pick N sufficiently large such that $I(v, v(g))$ and $I(v, v(h))$ are locally principal in Y_N. By Lemma 3.2, the lift of f to Y_N is regular at the generic point of $C(v, Y_N)$.

Let Z be the set of poles of f in Y_N (which is defined over k). Since f is regular at the generic point of $C(v, Y_N) \subset Z_N$, and Z_N is irreducible, $Z \cap Z_N$ is a divisor over k of Z_N that is possibly empty. The set of K-rational points of $Z \cap Z_N$ is included in the K-rational points of a hypersurface W of Z_N defined over K. Then W determines a unique pro-divisor say W^J for some J such that $W_N^J = Z \cap Z_N$, and $W_n^J = (\pi_N \circ \ldots \circ \pi_{n-1})^{-1}(W_N^J)$ for all $n \geq N$.

By construction, for any integer l such that $j_l > J$, we have $\tilde{v}_l \not\subset W^J$. In other words, we have $C(\tilde{v}_l, Y_N) \not\subset W_N^J$. Since both varieties are defined over K which is algebraically closed, we can find a K-rational point $x \in C(\tilde{v}_l, Y_N)(K) \setminus W(K)$. In particular, x does not belong to $Z \cap Z_N$ which shows $C(\tilde{v}_l, Y_N) \not\subset Z \cap Z_N$. We have thus proved that f is regular at the generic point of $C(\tilde{v}_l, Y_N)$ for l large enough. This implies $f \in R_{\tilde{v}_l}$ for l large enough, and concludes the proof. $\square$

Proof of Lemma 3.4 Pick any valuation $\mu \in A'$. Assume that the center of μ in Y_n is equal to Z_n for all n. We need to prove that $v \in \overline{\mu}$, i.e. $R_v \subset R_\mu$.

Pick any $f \in R_v$. As in the proof above, we can find an N such that f is regular at the generic point of the center of v in Y_N. Hence f is regular at the generic point of Z_N. Since the latter is the center of μ, we conclude that $f \in R_\mu$. $\square$

Proof of Lemma 3.3 We first construct varieties Z_n in Y_n satisfying the first three properties. For that purpose, let us introduce the set $\mathcal{Z}$ of all sequences $Z_\bullet := \{Z_n\}_n$ of subvarieties of Y_n such that Z_n is an irreducible component of $C_n(A)$, and $\pi_{n-1}(Z_n) \subset Z_{n-1}$.

For any fixed n, the set of irreducible components of $C_n(A)$ is finite of cardinality $d(n)$. The set of sequences of irreducible components of $C_n(A)$ is thus in natural bijection with $\Sigma = \prod_{n=1}^{\infty} \{1, \ldots, d(n)\}$. For the product topology, Σ is a compact (totally disconnected) space.

Since $\pi_{n-1}C(\mu, Y_n) = C(\mu, Y_{n-1})$ for any valuation, it follows that $\pi_{n-1}C_n(A) = C_{n-1}(A)$. From this we infer that $\mathcal{Z}$ is a non empty closed (hence compact) subset of Σ.

Now consider

$$\mathcal{K}_n := \{Z_\bullet \in \mathcal{Z}, \ Z_n \supset C(v, Y_n)\}$$

If $Z_\bullet$ belongs to $\mathcal{K}_n$, then we have $Z_{n-1} \supset \pi_{n-1}(Z_n) \supset \pi_{n-1}(C(v, Y_n)) = C(v, Y_{n-1})$. Hence $\mathcal{K}_n$ forms a decreasing sequence of non empty compact subsets of $\mathcal{Z}$. The intersection $\cap_n \mathcal{K}_n$ is thus non empty, and we may pick $Z_\bullet \in \cap_n \mathcal{K}_n$. By construction (3) is verified.

Since $\pi_{n-1}(Z_n) \subset Z_{n-1}$, the dimension of Z_n is increasing hence stationary. Replacing the sequence Y_n by Y_{n+N} with N large enough, we may thus assume $\pi_{n-1} : Z_n \to Z_{n-1}$ is birational for all n, so that (2) also holds. By the irreducibility of Z_n, it follows that (1) is also satisfied.

For each n, we now define $A_n := \{\mu \in A, \ C(\mu, Y_n) \subset Z_n\}$. We claim that for all $k \leq n$, we have $C_k(A_n) = Z_n$.

Since Z_n is an irreducible component of $C_n(A)$, it is clear that $C_n(A_n) = Z_n$. Now π_{n-1} is proper hence closed, which implies $\pi_{n-1}(\bar{S}) = \overline{\pi_{n-1}(S)}$ for any subset $S \subset Y_n$. We infer

$$Z_{n-1} = \pi_{n-1}(Z_n) = \pi_{n-1}(C_n(A_n)) = \overline{\pi_{n-1}(\cup_{\mu \in A_n} C(\mu, Y_n))} =$$

$$\overline{\cup_{\mu \in A_n} C(\mu, Y_{n-1})} = C_{n-1}(A_n) \, ,$$

and we conclude by a descending induction.

Finally we construct the subset A'.

First we shall construct countable subsets of A with special properties. To any countable subset $\mathcal{N}$ of A_n, we attach the integer $d_n(\mathcal{N}) = \dim C_n(\mathcal{N})$, and let $r_n(\mathcal{N})$ be the number of irreducible components of $C_n(\mathcal{N})$. Suppose one can find a sequence of countable sets $\mathcal{N}^j$ such that $d_n(\mathcal{N}^j)$ is independent of j, and $r_n(\mathcal{N}^j) \to_{j \to \infty} \infty$. Then $d_n(\mathcal{N}) > d_n(\mathcal{N}^j)$ for $\mathcal{N} := \cup_j \mathcal{N}^j$. This shows that there exists a countable subset $\mathcal{N}$ of A_n maximizing the pair $(d_n(\mathcal{N}), r_n(\mathcal{N}))$ for the lexicographic order on $\mathbb{N}^2$. We claim that $d_n(\mathcal{N}) = \dim Z_n$.

Indeed if it were not the case, and since $\cup_{A_n} C(\mu, Y_n)$ is Zariski dense in Z_n, then we could find a valuation $\mu \in A_n$ such that $C(\mu, Y_n) \not\subset C_n(\mathcal{N})$ which would contradict the maximality of $(d_n(\mathcal{N}), r_n(\mathcal{N}))$. Since Z_n is irreducible, for each n we have found a countable subset $\mathcal{N}^n \subset A_n$ such that $C_n(\mathcal{N}^n) = Z_n$. We conclude the proof by setting $A' := \cup_n \mathcal{N}^n$. $\qquad \square$

4 Proof of the Main Results

In this last section, we explain how Theorem A can be deduced from Theorem 3.1. Proofs of Corollaries A and B are given at the end of this section.

4.1 The Projection of the Riemann-Zariski Space to the Generic Fiber

Proposition 4.1 *Pick any normal variety Y, and any effective Cartier divisor E in Y. Denote by $\mathfrak{Y}(E)$ the subset of the Riemann-Zariski space $\mathfrak{Y}$ of Y consisting of those valuations whose center in Y is included in E.*

Then there exists a surjective and continuous map $\Pi : \mathfrak{Y}(E) \to]Y[$ such that for any v, the center of v in Y is included in $r_Y(\Pi(v))$ (with equality when v is divisorial), and any divisorial point in $]Y[$ has a preimage in $\mathfrak{Y}(E)$ which is divisorial too.

This result is well-known but we give a proof for sake of completeness.

Proof To construct Π, pick a finite collection of affine charts $Y_j \subset Y$, each intersecting E, and such that $\cup_j Y_j \supset E$. For each j, pick an equation $f_j \in k[Y_j]$ of E. Recall that $]Y_j[$ is the set of bounded multiplicative semi-norms $|\cdot| : k[Y_j] \to \mathbb{R}_+$ such that $-\log|f_j| = +1$. Then $]Y[$ is the disjoint union of the $]Y_j[$'s patched together in a natural way. For any valuation $v \in \mathfrak{Y}(E)$ with valuation ring R_v, we take j such that the center of v in Y_j is non-empty, and define the function $\Pi(v) \equiv |\cdot|_v : k[Y_j] \to \mathbb{R}_+$ by setting

$$\Pi(v)(f) = \exp\left(-\sup\{p/q \in \mathbb{Q}_+, \text{ such that } f^q/f_j^p \in R_v, \, p \in \mathbb{N}, \, q \in \mathbb{N}^*\}\right)$$

for any $f \in k[Y_j]$. It is clearly ≤ 1. We claim this is a multiplicative semi-norm. To simplify notation, we shall work with $\mu(f) = -\log(\Pi(v)(f))$. We shall use repeatedly the fact that $g \in k(Y)$ belongs to the valuation ring R_v iff $g^n \in R_v$ for some $n \in \mathbb{N}^*$.

Fix $f \in k[Y_j]$, and let $I(f) = \{p/q \in \mathbb{Q}_+, \text{ such that } f^q/f_j^p \in R_v\}$. Then $I(f)$ is a segment containing 0. Indeed pick $p/q > p'/q' \in I(f)$. Then $(f^q/f_j^p)^{q'} = (f^{q'}/f_j^{p'})^q \times f_j^{qp'-pq'} \in R_v$, and $(f^q/f_j^p) \in R_v$.

Next assume $f^q/f_j^p, g^q/f_j^p \in R_v$. The same argument as before shows that for all $0 \leq i \leq q$ we have $(f^{q-i}g^i)/f_j^p \in R_v$. Whence $(f+g)^q/f_j^p \in R_v$. This shows $\mu(f+g) \geq \min\{\mu(f), \mu(g)\}$.

Finally if $f^q/f_j^p, g^{q'}/f_j^{p'} \in R_v$, then $(fg)^{qq'}/f_j^{pq'+p'q} \in R_v$ so that $\mu(fg) \geq \mu(f) + \mu(g)$. Conversely, if $p_0/q_0 > \mu(f) + \mu(g)$ then we may find two rational numbers $p/q > \mu(f)$, $p'/q' > \mu(g)$ such that $p_0/q_0 = p/q + p'/q'$. Since

f_j^p/f^q, $f_j^{p'}/g^{q'} \in R_\nu$, we get $f_j^{p_0}/(fg)^{q_0} \in R_\nu$. This proves $\mu(fg) = \mu(f) + \mu(g)$, and concludes the proof that $\Pi(\nu)$ is a multiplicative semi-norm.

To see the continuity of Π we pick a (net) ν_n converging to a valuation ν in the Riemann-Zariski space, and we pick $f \in k[Y_j]$. Take $p/q \in \mathbb{Q}_+$ such that $f^q/f_j^p \in R_\nu$, i.e. $\Pi(\nu)(f) \leq \exp(-p/q)$. Then for an index n large enough $f^q/f_j^p \in R_{\nu_n}$ too, hence $\limsup_n \Pi(\nu_n)(f) \leq \Pi(\nu)(f)$. Conversely, suppose $f_j^p/f^q \in R_\nu$ (i.e. $\Pi(\nu)(f) \geq \exp(-p/q)$). The same argument shows $\liminf_n \Pi(\nu_n)(f) \leq \Pi(\nu)(f)$.

Let $Z \subset E$ be the center of a valuation $\nu \in \mathfrak{Y}(E)$, and suppose $Z \cap Y_j \neq \emptyset$. Then $r_Y(\Pi(\nu))$ is described by the prime ideal of functions $f \in k[Y_j]$ such that $\Pi(\nu)(f) < 1$, or in an equivalent way such that $f^q/f_j^p \in R_\nu$ for some $p/q > 0$. But $f_j \in \mathfrak{m}_{R_{\nu_j}}$, hence f too, and $r_Y(\Pi(\nu)) \supset Z$.

Observe that by construction, a rank 1 valuation $\nu \in \mathfrak{Y}(E)$ is mapped to the unique norm $|\cdot|$ on $k(Y)$ such that $R_\nu = \{f, |f| \leq 1\}$ and $-\log|f_j| = +1$ in some chart.

Now pick any multiplicative seminorm $|\cdot| \in]Y_j[$. Its kernel $\mathfrak{p}$ is a prime ideal of $k[Y_j]$ not containing f_j, and $|\cdot|$ induces a norm $|\cdot|_\mathfrak{p}$ on the quotient ring $k[Y_j]/\mathfrak{p}$. Pick any divisorial valuation ν associated to any exceptional divisor in the normalized blow-up of $\mathfrak{p}$. Then the composition of ν with the valuation associated to $|\cdot|_\mathfrak{p}$ is mapped to $|\cdot|$ by Π. This map is therefore surjective.

We complete the proof by noting that Π maps divisorial valuations to divisorial points. $\qquad\square$

We shall also need the following proposition.

Proposition 4.2 *Pick any normal variety Y, and any effective Cartier divisor E in Y as above, and pick any $y \in]Y[$.*

Then there exists $\nu \in \Pi^{-1}(y)$ such that for any open neighborhood $\mathcal{U} \subset \mathfrak{Y}(E)$ of ν, there exists an open neighborhood $U \subset]Y[$ of y such that $\Pi^{-1}(U) \subset \mathcal{U}$.

Proof We may assume Y is affine, and we fix $f \in k[Y]$ such that $E = \{f = 0\}$.

Assume first that the kernel of y is trivial and denote by ν its associated rank 1 valuation on $k(Y)$. We may take $\mathcal{U}$ as the set of valuation rings on $k(Y)$ whose center in some birational model Y' dominating Y is included in some Zariski open set $V \subset Y'$. Observe that $F := Y' \setminus V$ is Zariski closed.

Denote by C the center of ν in Y'. By assumption its generic point does not belong to F.

Pick $g_1, \ldots, g_n \in k(Y)$ such that $g_i^{-1} \in \mathfrak{m}_{C,Y'}$, and $g_i \in \mathfrak{m}_{F,Y'}$ for all i. By the noetherianity of the Zariski topology we can also assume that for any point $p \in F$ one can find an index i such that g_i is regular at p. Write $g_i = h_i/\tilde{h}_i$ with $h_i, \tilde{h}_i \in k[Y]$ and define

$$U := \{x \in]Y[, \ |h_i(x)| > |\tilde{h}_i(x)| \text{ for all } i = 1, \ldots, n\}.$$

This is an open neighborhood of y. We claim that $\Pi^{-1}(U) \subset \mathcal{U}$.

To see this pick any valuation $\mu \in \mathfrak{Y}(E)$ whose center in Y' is not included in V. Then we may find an index i such that g_i is vanishing at the generic point of the center of μ, hence $g_i \in \mathfrak{m}_\mu$. To conclude the proof it is sufficient to show that $|h_i|(\Pi(\mu)) \leq |\tilde{h}_i|(\Pi(\mu))$ since this implies $\Pi(\mu) \notin U$.

We proceed by contradiction assuming that $|h_i|(\Pi(\mu)) > |\tilde{h}_i|(\Pi(\mu))$. Since by definition $-\log|h|(\Pi(\mu)) = \sup\{p/q \in \mathbb{Q}_+, h^p/f^q \in R_\mu\}$, we may suppose there exists $p/q \in \mathbb{Q}_+$ such that $(h_i^p/f^q)^{-1} \in \mathfrak{m}_\mu$ and $\tilde{h}_i^p/f^q \in \mathfrak{m}_\mu$. This would imply $(\tilde{h}_i/h_i)^p \in \mathfrak{m}_\mu$ hence $g_i^{-1} \in \mathfrak{m}_\mu$ which is absurd.

In the case y has a non trivial kernel $\mathfrak{p}$, we may assume that Y' dominates the normalized blow-up of $\mathfrak{p}$. We pick any irreducible Cartier divisor D in Y' that is contracted onto the variety associated to $\mathfrak{p}$ in Y. We let v be the valuation obtained by composing the divisorial valuation associated to this divisor with the valuation induced by y on $k[Y]/\mathfrak{p}$, as in the proof of the previous proposition. Observe that the center C of v in Y' is included in the intersection of D with the preimage of E and is strictly included in both hypersurfaces. Also the generic point of D necessarily lies in V.

Choose functions g_i as above that are regular and non vanishing at the generic point of D. The key observation is that for such a choice the open set U defined above still contains y. One can now repeat the previous proof to conclude. $\square$

We also collect the following result for later reference

Lemma 4.3 *Pick any normal algebraic variety Y, and any effective Cartier divisor E in Y then $]Y[$ is compact.*

Proof Since Y is of finite type it is covered by finitely many affine charts, and we can suppose Y is affine. Choose an equation f of E. Then $]Y[$ is the set of seminorms $|\cdot| : k[Y] \to [0, 1]$ such that $|f| = 1/e$ hence is compact by Tychonoff. $\square$

4.2 Proof of Theorem A

Recall our assumption: X is a normal algebraic variety, D is an effective Cartier divisor in X, and A is any subset of $]X[$.

We first cover X by finitely many affine open sets X_i, and write $D_i = D \cap X_i$. We let $]X_i[$ be the normalized generic fiber of the formal completion of X_i along D_i. This is a compact subset of $]X[$, and $\cup_i]X_i[=]X[$.

For each i, we fix a projective (normal) compactification $X_i \subset \bar{X}_i$. We write $\bar{D}_i$ for the closure of D in $\bar{X}_i$. It is a priori only a Weil divisor, but we may suppose it is Cartier by taking the normalized blow up associated to the ideal defining $\bar{D}_i$.

Now pick any point x in the closure of A in $]X[$, and set $A_i = A \cap]X_i[$. Then $x \in \bar{A} = \overline{\cup A_i} \subset \cup \bar{A}_i$, so that $x \in \bar{A}_i$ for some i. Since $]X_i[$ is closed, it follows that x belongs to $]X_i[$ too.

Apply Proposition 4.1 to $Y = \bar{X}_i$, and $E = \bar{D}_i$. This yields a continuous and surjective map $\Pi : \mathfrak{V}(E) \to]\bar{X}_i[\supset]X_i[$. By Proposition 4.2, we can find a valuation $\nu \in \Pi^{-1}(x) \subset \mathfrak{V}(E)$ that lies in the closure of $\Pi^{-1}(A_i)$. Theorem 3.1 implies the existence of a sequence of valuations $\nu_n \in \Pi^{-1}(A_i)$ converging to ν so that $\Pi(\nu_n) \to x$ by continuity.

4.3 Proofs of Corollaries A and B

As in the previous section, X is a normal algebraic variety, and D is an effective Cartier divisor.

We first prove Corollary A. Pick any subset A of $]X[$. First take x in the closure of A. By Theorem A, there exists a sequence $x_n \in A$ such that $x_n \to x$. Since $]X[$ is compact, any of its subset is relatively compact. This implies $]X[$ is angelic and concludes the proof of Corollary A.

We now prove Corollary B. Pick any $x \in \hat{X}_\eta$. Note that for any closed subset C of D the preimage $r_X^{-1}(C)$ is open. In particular, $r_X^{-1}(\overline{r_X(x)})$ is an open neighborhood of x in $\hat{X}_\eta$. We claim that divisorial norms are dense in $\hat{X}_\eta$ (hence in $]X[\cap r_X^{-1}(\overline{r_X(x)}))$. We conclude by applying Theorem A.

To justify our claim, we proceed as follows. Just as in the proof of Theorem A, we cover X by affine charts X_i and take projective compactifications $\bar{X}_i$ of X_i.

For any subset $E \subset \bar{X}_i$, denote by $\mathfrak{X}_i(E)$ the subset of the Riemann-Zariski space of $\bar{X}_i$ consisting of valuations centered in E. The set of divisorial valuations on $k(\bar{X}_i)$ that are centered in $\bar{D}_i$ is dense in $\mathfrak{X}_i(\bar{D}_i)$. On the other hand the set $\mathfrak{X}_i(\bar{D}_i \setminus D_i)$ is closed in $\mathfrak{X}_i(\bar{D}_i)$, hence divisorial valuations are dense in $\mathfrak{X}_i(\bar{D}_i)$. By Proposition 4.1, this shows divisorial norms are dense in $]X_i[$, hence in $]X[$ and in $\hat{X}_\eta$ as required.

4.4 Proofs of Corollary C

Since a compact analytic space over $k((T))$ is covered by finitely many affinoids, and any affinoid is a closed subset in a ball of a suitable dimension, it is sufficient to treat the case of the unit ball in $\mathbb{A}^n_{k((T))}$.

Let D be the hyperplane $\{x_1 = 0\}$ in the affine space $(x_1, \ldots, x_{n+1}) \in X := \mathbb{A}^{n+1}_k$. The normalized generic fiber $]X[$ of the formal completion of X along D is the set of multiplicative semi-norms $|\cdot| : k[x_1, \ldots, x_{n+1}] \to \mathbb{R}_+$ trivial on k, and such that $|x_1| = e^{-1}$ and $|x_i| \leq 1$ for all $i \geq 2$. This is precisely the unit ball in $X = \mathbb{A}^n_{k((T))}$. By Theorem A, the unit ball in any dimension is thus angelic.

Acknowledgements We thank T. De Pauw, A. Ducros, M. Jonsson, J. Kiwi, J. Nicaise, and R. Menares for useful discussions on the material presented in this paper. Also we deeply thank

J. Poineau for kindly informing the author about his proof of the sequential compactness of compact analytic spaces over an arbitrary field.

References

[BR]   M. Baker, R. Rumely, *Potential Theory and Dynamics on the Berkovich Projective Line*. Mathematical Surveys and Monographs, vol. 159 (American Mathematical Society, Providence, RI, 2010)

[Ber]   V.G. Berkovich, *Spectral Theory and Analytic Geometry Over Non-Archimedean Fields*. Math. Surveys Monographs, vol. 33 (American Mathematical Society, Providence, RI, 1990)

[FJ]   C. Favre, M. Jonsson, *The Valuative Tree*. Lecture Notes in Mathematics, vol. 1853 (Springer, Berlin, 2004)

[Flo]   K. Floret, *Weakly Compact Sets*. Lecture Notes in Mathematics, vol. 801 (Springer, Berlin, 1980)

[Gub]   W. Gubler, Local heights of subvarieties over non-archimedean fields. J. Reine Angew. Math. **498**, 61–113 (1998)

[Jon]   M. Jonsson, *Dynamics on Berkovich Spaces in Low Dimensions*, this volume

[Mai]   N. Maïnetti, Sequential compactness of some analytic spaces. J. Anal. **8**, 39–54 (2000)

[Poi]   J. Poineau, Les espaces de Berkovich sont angéliques. Bull. de la Soc. Math. France **141**(2), 267-297 (2013)

[Thu]   A. Thuillier, Géométrie toroïdale et géométrie analytique non archimédienne. Application au type d'homotopie de certains schémas formels. Manuscripta Math. **123**(4), 381–451 (2007)

[Vaq]   M. Vaquié, *Valuations*, in *Resolution of Singularities (Obergurgl, 1997)*. Progress in Math., vol. 181 (Birkhäuser Verlag, Basel, 2000), pp. 539–590

[YZ]   X. Yuan, S.-W. Zhang, Calabi-Yau theorem and algebraic dynamics. Preprint (2009) www.math.columbia.edu/~szhang/papers/Preprints.htm

[ZS]   O. Zariski, P. Samuel, *Commutative Algebra*, vol. 2. Graduate Texts in Mathematics, No. 29 (Springer, New York-Heidelberg-Berlin, 1975)

Part II
Applications to Geometry

Cohomological Finiteness of Proper Morphisms in Algebraic Geometry: A Purely Transcendental Proof, Without Projective Tools

Antoine Ducros

Contents

1   Introduction

If $f : \mathscr{Y} \to \mathscr{X}$ is a proper morphism between locally noetherian schemes, and if $\mathscr{F}$ is a coherent $\mathscr{O}_{\mathscr{Y}}$-module, then $\mathrm{R}^q f_* \mathscr{F}$ is a coherent $\mathscr{O}_{\mathscr{X}}$ module for every integer q. In order to prove this result, one proceeds usually in two steps; first, one supposes that f is projective, and one uses tools which are specific to this situation: tensorisation with $\mathscr{O}(n)$, explicit computations of the Čech cohomology groups associated to the standard affine covering of $\mathbb{P}^r$, and so on; then one reduces the general case to the projective one using Chow's lemma.

The purpose of this note is to give, when $\mathscr{Y}$ and $\mathscr{X}$ are locally of finite type over a field k, a new proof of that assertion, which uses neither projective tools nor Chow's lemma. The principle is very simple.

- This theorem is true in the setting of Berkovich analytic spaces. One sees it by reducing, thanks to a suitable ground field extension, to the case of a proper morphism between two rigid-analytic spaces, in which the cohomological finiteness has been proved by Kiehl, of course without Chow's lemma (a proper

A. Ducros (✉)

Institut de Mathématiques de Jussieu, Université Paris 6, 4 place Jussieu, 75252 Paris Cedex 05, France

e-mail: ducros@math.jussieu.fr, Url: www.math.jussieu.fr/~ducros/

© Springer International Publishing Switzerland 2015

A. Ducros et al. (eds.), *Berkovich Spaces and Applications*, Lecture Notes in Mathematics 2119, DOI 10.1007/978-3-319-11029-5__4 rigid-analytic space is in general far from being bi-meromorphically isomorphic to a projective variety): Kiehl uses strong refinements of affinoid covers, and the fact that such a refinement induces completely continuous maps at the level of Čech complexes.

- To any algebraic variety $\mathscr{X}$ over k can be associated an analytic space $\mathscr{X}^{\beth}$ over the field k *endowed with the trivial absolute value*. If $\mathscr{Y} \to \mathscr{X}$ is proper, then $\mathscr{Y}^{\beth} \to \mathscr{X}^{\beth}$ is proper too (this can be proven using only valuative tools).
- Thanks to an *explicit* computation of Čech complexes, one proves in this very particular case a GAGA comparison theorem for the cohomology of coherent sheaves, which allows to conclude.

2 Reminders on Coherent Sheaves in Analytic Geometry

Let k be a complete, non-Archimedean field, let Z be a k-analytic space in the sense of Berkovich [1, 2]. Let us assume that Z is paracompact and separated (*ie.*, the diagonal map is a closed immersion). Let $\mathscr{F}$ be a coherent sheaf on the site Z_G; for a definition of this ringed site and basic facts[1] about coherent sheaves on it, see [2, §1.3]. Let us fix a locally finite affinoid G-covering of Z. Thanks to Tate's acyclicity theorem, and because the intersection of two affinoid domains of Z is an affinoid domain of Z (this is due to our separatedness assumption), there is a natural isomorphism $H^\bullet(Z_G, \mathscr{F}) \simeq H^\bullet \mathscr{C}(\mathfrak{U}, \mathscr{F})$, where $\mathscr{C}$ denotes the Čech complex.

Let $f : Y \to X$ be a morphism between k-analytic spaces which is proper, that is, topologically proper and without boundary; and let $\mathscr{F}$ be a coherent sheaf on Y_G. Let q be an integer. Let us recall how one proves ([1, Proposition 3.3.5]) using in a crucial way Kiehl's corresponding theorem in rigid geometry, that $R^q f_* \mathscr{F}$ is coherent.

One immediately reduces to the case where X is affinoid, say $X = \mathscr{M}(\mathscr{A})$. As $Y \to X$ has no boundary, there exists, for every $y \in X$, two affinoid neighborhoods U_x and V_x of x in X, such that U_x is included in the *relative interior* of V_x over Y, in the sense of [1, Definition 2.5.7]. Note that we used here the fact that Y is good, which comes from the goodness of X (which is affinoid) and the fact that a boundaryless morphism between general analytic spaces is good by its very definition ([2, §1.5.3]). The space Y being compact, there exists two finite affinoid covers $\mathfrak{U}$ and $\mathfrak{V}$ of Y, such that every affinoid domain of Y belonging to $\mathfrak{U}$ is included in the relative interior over Y of an affinoid domain belonging to $\mathfrak{V}$.

---

[1]One of them is missing in [2]: the coherence of the sheaf $\mathscr{O}_{Z_G}$ itself. It is proven in [4, Lemma 0.1]. Note that it was pointed out to the author by Jérôme Poineau that there is a mistake in this proof: it establishes that a *surjection* $\mathscr{O}^n \to \mathscr{O}$ has a locally finitely generated kernel, though in order to get the coherence, this finiteness claim should be established for any, *i.e.* non necessarily surjective, map $\mathscr{O}^n \to \mathscr{O}$; but it turns out that the proof doesn't make any use of this inaccurate surjectivity assumption.

Let $\mathbf{r} = (r_1, \ldots, r_n)$ be a family of positive real numbers which is free in $\mathbb{Q} \otimes_{\mathbb{Z}} (\mathbb{R}_+^* / |k^*|)$, and which is such that $|k_{\mathbf{r}}^*| \neq \{1\}$ and that all the affinoid spaces involved become strict after extending the scalars to $k_{\mathbf{r}}$ (the notation $k_{\mathbf{r}}$ has its usual meaning in Berkovich theory, see for example [1, §2.1]). Every affinoid domain belonging to $\mathfrak{U}_{\mathbf{r}}$ is 'included in the relative interior over $Y_{\mathbf{r}}$ of an affinoid domain belonging to $\mathfrak{V}_{\mathbf{r}}$', in the sense of Berkovich; but in view of [1, Lemma 2.5.11], this remains true in the sense of Kiehl (see [5, Definition 2.1]). The two following facts are therefore consequences of the assertion preceding of [5, Theorem 2.6]:

i) the differential maps of $\mathscr{C}(\mathfrak{U}_{\mathbf{r}}, \mathscr{F}_{\mathbf{r}})$ are admissible, which implies, among other things, that each of its cohomology spaces has a natural structure of a Banach $\mathscr{A}_{\mathbf{r}}$-module;

ii) for every q, the Banach $\mathscr{A}_{\mathbf{r}}$-module $H^q \mathscr{C}(\mathfrak{U}_{\mathbf{r}}, \mathscr{F}_{\mathbf{r}})$ is a finitely generated $\mathscr{A}_{\mathbf{r}}$-module; as $|k_{\mathbf{r}}^*| \neq \{1\}$ this implies, by the open mapping theorem, that $H^q \mathscr{C}(\mathfrak{U}_{\mathbf{r}}, \mathscr{F}_{\mathbf{r}})$ is a finite Banach $\mathscr{A}_{\mathbf{r}}$-module, that is, a topological quotient of the Banach $\mathscr{A}_{\mathbf{r}}$-module $\mathscr{A}_{\mathbf{r}}^m$ for some m.

One deduces from i) that the differentials of $\mathscr{C}(\mathfrak{U}, \mathscr{F})$ are admissible; the space $H^q \mathscr{C}(\mathfrak{U}, \mathscr{F})$ inherits therefore for every q a natural structure of a Banach $\mathscr{A}$-module. By [1, Proposition 2.1.2] there is for all i a natural isomorphism $H^q \mathscr{C}(\mathfrak{U}, \mathscr{F})_{\mathbf{r}} \simeq H^q \mathscr{C}(\mathfrak{U}_{\mathbf{r}}, \mathscr{F}_{\mathbf{r}})$; by ii) above and [1, Proposition 2.1.11], it follows that $H^q \mathscr{C}(\mathfrak{U}, \mathscr{F})$ is a finite Banach $\mathscr{A}$-module, and in particular a finitely generated $\mathscr{A}$-module. In other words, $H^q(Y_G, \mathscr{F})$ is a finitely generated $\mathscr{A}$-module.

It remains to show (we will not need it in this paper) that for every affinoid domain Z of X and for every q, the natural arrow

$$H^q(Y_G, \mathscr{F}) \otimes \mathscr{A}_Z \dashrightarrow H^q(f^{-1}(Z)_G, \mathscr{F})$$

is an isomorphism. One can do it by extending the scalars to $k_{\mathbf{r}}$ for a suitable $\mathbf{r}$, and applying [5, Theorem 3.3].

3 The Analytic Space Associated to a Scheme and the Comparison Theorem

Let k be *any* field. To every k-scheme of finite type $\mathscr{X}$, there are two ways to associate an analytic space over the field k endowed with the trivial absolute value.

The first one consists in considering its analytification in the sense of [2, §2.6]. The second one, which has already been considered by Thuillier in [8], consists in viewing $\mathscr{X}$ as a locally topologically finitely presented formal scheme over the discrete ring k (in other words, one is interested in the topologically locally ringed space deduced from $\mathscr{X}$ by endowing $\mathscr{O}_{\mathscr{X}}(\mathscr{U})$ with the discrete topology for every open subset $\mathscr{U}$ of $\mathscr{X}$); one can then consider its *generic fiber* in the sense of [3]. This is what we will do here, and we will denote by $\mathscr{X}^{\beth}$ this generic fiber.

Let us recall how one builds it. If $\mathscr{X}$ is affine, say $\mathscr{X} = \text{Spec } A$, then $\mathscr{X}^{\beth}$ is the affinoid space $\mathscr{M}(A^{\beth})$, where $A^{\beth}$ is the affinoid algebra which is equal to A equipped with the trivial norm. If $f \in A$, the map $A^{\beth} \to A^{\beth}_f$ is obviously bounded, and $\mathscr{M}(A^{\beth}_f) \to \mathscr{M}(A)$ identifies $\mathscr{M}(A^{\beth}_f)$ with the affinoid domain of $\mathscr{M}(A)$ described by the equation $|f| = 1$. One uses this remark to build $\mathscr{X}^{\beth}$ by gluing in the general case.

There is a natural reduction map $\mathscr{X}^{\beth} \to \mathscr{X}$, through which the pre-image of an open subset $\mathscr{V}$ of $\mathscr{X}$ is a compact analytic domain of $\mathscr{X}^{\beth}$ naturally isomorphic to $\mathscr{V}^{\beth}$. The arrow r appears as a morphism of *locally ringed sites* from $(\mathscr{X}^{\beth}_G, \mathscr{O}_{\mathscr{X}^{\beth}_G})$ to $(\mathscr{X}, \mathscr{O}_{\mathscr{X}})$. If $\mathscr{F}$ is a coherent sheaf on $\mathscr{X}$, we denote by $\mathscr{F}^{\beth}$ the $\mathscr{O}_{\mathscr{X}^{\beth}_G}$-module $r^* \mathscr{F}$.

3.1 Description of $\mathscr{F} \mapsto \mathscr{F}^{\beth}$ in the Affine Case

Let us assume that $\mathscr{X}$ is affine, say $\mathscr{X} = \text{Spec } A$. Let M be a finitely generated A-module; we denote by $\tilde{M}$ (resp. $\tilde{M}_G$) the associated coherent sheaf on $\mathscr{X}$ (resp. $\mathscr{X}^{\beth}_G$).

- If $\mathscr{V}$ is an open affine subset of $\mathscr{X}$, then $\tilde{M}(\mathscr{V}) = M \otimes_A \mathscr{O}_{\mathscr{X}}(\mathscr{V})$; as a consequence, for every $\mathscr{O}_{\mathscr{X}}$-module $\mathscr{F}$, the natural map

$$\text{Hom}(\tilde{M}, \mathscr{F}) \to \text{Hom}(M, \mathscr{F}(\mathscr{X}))$$

  is an isomorphism.
- If V is an affinoid domain of $\mathscr{X}^{\beth}$, then $\tilde{M}_G(V) \simeq M \otimes_A \mathscr{O}_{\mathscr{X}_G}(V)$; as a consequence, for every $\mathscr{O}_{\mathscr{X}^{\beth}_G}$-module $\mathscr{G}$, the natural map

$$\text{Hom}(\tilde{M}_G, \mathscr{F}) \to \text{Hom}(M, \mathscr{G}(\mathscr{X}^{\beth}))$$

  is an isomorphism.
- If $\mathscr{G}$ is an $\mathscr{O}_{\mathscr{X}^{\beth}_G}$-module, it follows from the above that one has a sequence of canonical isomorphisms

$$\text{Hom}(\tilde{M}_G, \mathscr{G}) \simeq \text{Hom}(M, \mathscr{G}(\mathscr{X}^{\beth})) \simeq \text{Hom}(M, r_* \mathscr{G}(\mathscr{X})) \simeq \text{Hom}(\tilde{M}, r_* \mathscr{G}).$$

Since r^* is the left-adjoint of r_*, one deduces from the last point that for every finitely generated A-module M, there is a canonical isomorphism $(\tilde{M})^{\beth} \simeq \tilde{M}_G$; in particular, $\tilde{M}(\mathscr{X}) \to \tilde{M}^{\beth}(\mathscr{X}^{\beth})$ is an isomorphism.

3.2 Description of $\mathscr{F} \mapsto \mathscr{F}^{\beth}$ in the General Case

We do not assume anymore that the space $\mathscr{X}$ is affine. Let $\mathscr{F}$ be a coherent sheaf on $\mathscr{X}$. By the above, $\mathscr{F}(\mathscr{V}) \to \mathscr{F}^{\beth}(\mathscr{V}^{\beth})$ is an isomorphism for every affine open subset $\mathscr{V}$ of $\mathscr{X}$. This fact immediately extends to *all* open subsets of $\mathscr{X}$.

Let us assume now that $\mathscr{X}$ is separated, and let $\mathfrak{U}$ be a finite affine cover of $\mathscr{X}$. It induces a finite affinoid cover $\mathfrak{U}^{\beth}$ of $\mathscr{X}^{\beth}$. The canonical arrows

$$\mathrm{H}^{\bullet}\mathscr{C}(\mathfrak{U}, \mathscr{F}) \to \mathrm{H}^{\bullet}(\mathscr{X}, \mathscr{F}) \text{ and } \mathrm{H}^{\bullet}\mathscr{C}(\mathfrak{U}^{\beth}, \mathscr{F}^{\beth}) \to \mathrm{H}^{\bullet}(\mathscr{X}^{\beth}, \mathscr{F}^{\beth})$$

are bijections. Moreover, what we have just seen about $\mathscr{F}^{\beth}$ implies that

$$\mathscr{C}(\mathfrak{U}, \mathscr{F}) \to \mathscr{C}(\mathfrak{U}^{\beth}, \mathscr{F}^{\beth})$$

is an isomorphism.

As a consequence, the canonical map $\mathrm{H}^{\bullet}(\mathscr{X}, \mathscr{F}) \to \mathrm{H}^{\bullet}(\mathscr{X}_{\mathrm{G}}^{\beth}, \mathscr{F}^{\beth})$ is an isomorphism.

4 The Main Theorem

Theorem 1 *Let k be a field and let $f : \mathscr{Y} \to \mathscr{X}$ be a proper morphism between k-schemes locally of finite type. Let $\mathscr{F}$ be a coherent sheaf on $\mathscr{Y}$. For every integer q, the $\mathscr{O}_{\mathscr{X}}$-module $\mathrm{R}^q f_* \mathscr{F}$ is coherent.*

Proof One can assume that $\mathscr{X}$ is affine; say $\mathscr{X} = \mathsf{Spec}\ \mathsf{A}$. As $\mathrm{R}^q f_* \mathscr{F}$ is quasi-coherent, it is sufficient to ensure that $\mathrm{H}^q(\mathscr{Y}, \mathscr{F})$ is a finitely generated A-module. By the preceding paragraph, it is naturally isomorphic to $\mathrm{H}^q(\mathscr{Y}^{\beth}, \mathscr{F}^{\beth})$. By the following lemma, $\mathscr{Y}^{\beth} \to \mathscr{X}^{\beth}$ is proper. In view of what we have recalled in Sect. 2, this implies that $\mathrm{H}^q(\mathscr{Y}^{\beth}, \mathscr{F}^{\beth})$ is a finitely generated A-module, which ends the proof. $\qquad\square$

Lemma 2 *The morphism $\mathscr{Y}^{\beth} \to \mathscr{X}^{\beth}$ is proper.*

We will give three different proofs of this lemma.

First proof The morphism $\mathscr{Y}^{\beth} \to \mathscr{X}^{\beth}$ is topologically proper. Hence it is sufficient to prove that it has no boundary. Thanks to [1, Corollary 2.5.12], this can be checked after extending the scalars to k_r for any $r \in]0; 1[$; one can identify $\mathscr{Y}_r^{\beth}$ (resp. $\mathscr{X}_r^{\beth}$) with the generic fiber of $\mathscr{Y} \times \mathsf{Spf}\ k_r^{\circ}$ (resp. $\mathscr{X} \times \mathsf{Spf}\ k_r^{\circ}$).

As $\mathscr{Y} \to \mathscr{X}$ is proper, $\mathscr{Y} \times \mathsf{Spf}\ k_r^{\circ} \to \mathscr{X} \times \mathsf{Spf}\ k_r^{\circ}$ is proper too; since the absolute value of k_r is not trivial, $\mathscr{Y}_r^{\beth} \to \mathscr{X}_r^{\beth}$ is then proper (and in particular boundaryless): this is a result by Temkin (see [6, Corollary 4.4]). $\qquad\square$

Remark 3 The field k_r is nothing but $k((t))$, endowed with the t-adic absolute value that sends t to r. The formal scheme $\mathscr{Y} \times \mathrm{Spf}\, k_r^\circ$ (resp. $\mathscr{X} \times \mathrm{Spf}\, k_r^\circ$) is simply the completion of $\mathscr{Y} \times \mathsf{Spec}\, k[[t]]$ (resp. of $\mathscr{X} \times \mathsf{Spec}\, k[[t]]$) along its special fiber.

Second proof One can do the same as in the first proof, but with avoiding ground field extension. It suffices to remark that the aforementioned result by Temkin (see [6, Corollary 4.4]) can be easily extended to the case of a trivially valued ground field, using the formalism of *graded* reductions, which is developed by Temkin in [7]. □

Third proof This proof was kindly indicated to the author by Amaury Thuillier.

We may assume that $\mathscr{X}$ is affine; say $\mathscr{X} = \mathsf{Spec}\, A$. The $\mathscr{X}^{\beth}$-analytic space $\mathscr{Y}^{\beth}$ embeds in a natural way into the analytification $(\mathscr{Y}/\mathscr{X}^{\beth})^{an}$ of the $A^{\beth}$-scheme of finite type $\mathscr{Y}$—note that $(\mathscr{Y}/\mathscr{X}^{\beth})^{an}$ is nothing but $\mathscr{Y}^{an} \times_{\mathscr{X}^{an}} \mathscr{X}^{\beth}$.

More precisely, $\mathscr{Y}^{\beth}$ can be identified with an analytic domain of the proper $\mathscr{X}^{\beth}$-analytic space $(\mathscr{Y}/\mathscr{X}^{\beth})$. Indeed, if $\mathscr{Y} = \mathsf{Spec}\, B$, then $\mathscr{Y}^{\beth}$ is naturally isomorphic to the affinoid domain of $(\mathscr{Y}/\mathscr{X}^{\beth})^{an}$ defined by the inequalities $|b| \leq 1$ for b running through a finite system of generators of B over A; in general, these local identifications glue obviously.

To conclude, it is sufficient to prove that $\mathscr{Y}^{\beth} \to (\mathscr{Y}/\mathscr{X}^{\beth})^{an}$ is onto, because it will then be an isomorphism. Let $y \in (\mathscr{Y}/\mathscr{X}^{\beth})^{an}$. By the definition of $A^{\beth}$, the natural map $A \to \mathscr{H}(y)$ induced by $\mathscr{Y} \to \mathscr{X}$ takes its values in $\mathscr{H}(y)^\circ$, hence induces a morphism $\mathsf{Spec}\, \mathscr{H}(y)^\circ \to \mathscr{X}$. As $\mathscr{Y} \to \mathscr{X}$ is proper, the $\mathscr{X}$-map $\mathsf{Spec}\, \mathscr{H}(y) \to \mathscr{Y}$ extends to an $\mathscr{X}$-map $\mathsf{Spec}\, \mathscr{H}(y)^\circ \to \mathscr{Y}$. If x denotes the image of the closed point of $\mathsf{Spec}\, \mathscr{H}(y)^\circ$ on $\mathscr{Y}$ and if $\mathscr{V} = \mathsf{Spec}\, B$ is an affine neighborhood of x in $\mathscr{Y}$, then $|b(y)| \leq 1$ for every $b \in B$. This implies that $y \in \mathscr{V}^{\beth} \subset \mathscr{Y}^{\beth}$. □

References

[1] V. Berkovich, *Spectral Theory and Analytic Geometry Over Non-archimedean Fields*. Mathematical Surveys and Monographs, vol. 33 (AMS, Providence, RI, 1990)
[2] V. Berkovich, Étale cohomology for non-archimedean analytic spaces. Inst. Hautes Études Sci. Publ. Math. **78**, 5–161 (1993)
[3] V. Berkovich, Vanishing cycles for formal schemes. Invent. Math. **115**(3), 539–571 (1994)
[4] A. Ducros, Les espaces de Berkovich sont excellents. Ann. Inst. Fourier **59**(4), 1407–1516 (2009)
[5] R. Kiehl, Der Endlichkeitsatz für eingentliche Abbildungen in der nichtarchimedischen Funktionentheorie. Invent. Math. **2**, 191–214 (1967)
[6] M. Temkin, On local properties of non-Archimedean analytic spaces. Math. Annalen **318**, 585–607 (2000)
[7] M. Temkin, On local properties of non-Archimedean analytic spaces. II. Isr. J. Math. **140**, 1–27 (2004)
[8] A. Thuillier, Géométrie toroïdale et géométrie analytique non archimédienne. Application au type d'homotopie de certains schémas formels. Manuscripta Math. **123**(4), 381–451 (2007)

Bruhat-Tits Buildings and Analytic Geometry

Bertrand Rémy, Amaury Thuillier, and Annette Werner

Contents

1   Introduction

This paper is mainly meant to be a survey on two papers written by the same authors, namely [RTW10] and [RTW12]. The general theme is to explain what the theory of analytic spaces in the sense of Berkovich brings to the problem of compactifying Bruhat-Tits buildings.

B. Rémy (✉)
CMLS École Polytechnique Route de Saclay Palaiseau 91128 Cedex, France

Université de Lyon 1 Institut Camille Jordan 43 boulevard du 11 novembre 1918 F-69622 Villeurbanne cedex, France
e-mail: bertrand.remy@polytechnique.edu

A. Thuillier
Université de Lyon 1 Institut Camille Jordan, 43 Boulevard du 11 novemvre, 1918 F-69622, Villeurbanne cedex, France
e-mail: thuillier@math.univ-lyon1.fr

A. Werner
Institut für Mathematik, Goethe-Universität Frankfurt, Robert-Mayer-Str. 6-8, D-60325 Frankfurt a. M., Germany
e-mail: werner@math.uni-frankfurt.de

© Springer International Publishing Switzerland 2015
A. Ducros et al. (eds.), *Berkovich Spaces and Applications*, Lecture Notes in Mathematics 2119, DOI 10.1007/978-3-319-11029-5__5

1. *Bruhat-Tits buildings.* The general notion of a building was introduced by J. Tits in the 60ies [Tits74], [Bou07, Exercises for IV.2]. These spaces are cell complexes, required to have some nice symmetry properties so that important classes of groups may act on them. More precisely, it turned out in practice that for various classes of algebraic groups and generalizations, a class of buildings is adapted in the sense that any group from such a class admits a very transitive action on a suitable building. The algebraic counterpart to the transitivity properties of the action is the possibility to derive some important structure properties for the group.

This approach is particularly fruitful when the class of groups is that of simple Lie groups over non-Archimedean fields, or more generally reductive groups over non-Archimedean valued fields—see Sect. 4. In this case the relevant class of buildings is that of Euclidean buildings (2.1). *This is essentially the only situation in building theory we consider in this paper.* Its particularly nice features are, among others, the facts that in this case the buildings are (contractible, hence simply connected) gluings of Euclidean tilings and that deep (non-positive curvature) metric arguments are therefore available; moreover, on the group side, structures are shown to be even richer than expected. For instance, topologically the action on the buildings enables one to classify and understand maximal compact subgroups (which is useful to representation theory and harmonic analysis) and, algebraically, it enables one to define important integral models for the group (which is again useful to representation theory, and which is also a crucial step towards analytic geometry).

One delicate point in this theory is merely to prove that for a suitable non-Archimedean reductive group, there does exist a nice action on a suitable Euclidean building: this is the main achievement of the work by F. Bruhat and J. Tits in the 70ies [BrT72, BrT84]. Eventually, Bruhat-Tits theory suggests to see the Euclidean buildings attached to reductive groups over valued fields (henceforth called *Bruhat-Tits buildings*) as non-Archimedean analogues of the symmetric spaces arising from real reductive Lie groups, from many viewpoints at least.

2. *Some compactification procedures.* Compactifications of symmetric spaces were defined and used in the 60s; they are related to the more difficult problem of compactifying locally symmetric spaces [Sat60b], to probability theory [Fur63], to harmonic analysis... One group-theoretic outcome is the geometric parametrization of classes of remarkable closed subgroups [Moo64]. For all the above reasons and according to the analogy between Bruhat-Tits buildings and symmetric spaces, it makes therefore sense to try to construct compactifications of Euclidean buildings.

When the building is a tree, its compactification is quite easy to describe [Ser77]. In general, and for the kind of compactifications we consider here, the first construction is due to E. Landvogt [Lan96]: he uses there the fact that the construction of the Bruhat-Tits buildings themselves, at least at the beginning of Bruhat-Tits theory for the simplest cases, consists in defining a suitable gluing equivalence relation for infinitely many copies of a well-chosen Euclidean tiling. In Landvogt's approach, the equivalence relation is extended so that it glues together infinitely many compactified copies of the Euclidean tiling used to construct the building. Another approach is more group-theoretic and relies on the analogy with symmetric spaces: since the symmetric space of a simple real Lie group can be seen as the space of maximal compact subgroups of the group, one can compactify this space by taking its closure in the (compact) Chabauty space of all closed subgroups. This approach is carried out by Y. Guivarc'h and the first author [GR06]; it leads to statements in group theory which are analogues of [Moo64] (e.g., the virtual geometric classification of maximal amenable subgroups) but the method contains an intrinsic limitation due to which one cannot compactify more than the set of vertices of the Bruhat-Tits buildings.

The last author of the present paper also constructed compactifications of Bruhat-Tits buildings, in at least two different ways. The first way is specific to the case of the general linear group: going back to Bruhat-Tits' interpretation of Goldman-Iwahori's work [GI63], it starts by seeing the Bruhat-Tits building of $GL(V)$—where V is a vector space over a discretely valued non-Archimedean field—as the space of (homothety classes of) non-Archimedean norms on V. The compactification consists then in adding at infinity the (homothety classes of) non-zero non-Archimedean seminorms on V. Note that the symmetric space of $SL_n(\mathbf{R})$ is the set of normalized scalar products on $\mathbf{R}^n$ and a natural compactification consists in projectivizing the cone of positive nonzero semidefinite bilinear forms: what is done in [Wer04] is the non-Archimedean analogue of this; it has some connection with Drinfeld spaces and is useful to our subsequent compactification in the vein of Satake's work for symmetric spaces. The second way is related to representation theory [Wer07]: it provides, for a given group, a finite family of compactifications of the Bruhat-Tits building. The compactifications, as in E. Landvogt's monograph, are defined by gluing compactified Euclidean tilings but the variety of possibilities comes from exploiting various possibilities of compactifying equivariantly these tilings in connection with highest weight theory.

3. *Use of Berkovich analytic geometry.* The compactifications we would like to introduce here make a crucial use of Berkovich analytic geometry. There are actually two different ways to use the latter theory for compactifications.

The first way is already investigated by V. Berkovich himself when the algebraic group under consideration is split [Ber90, Chap. 5]. One intermediate step for it consists in defining a map from the building to the analytic space attached to the algebraic group: this map attaches to each point x of the building an affinoid subgroup G_x, which is characterized by a unique maximal point $\vartheta(x)$ in the ambient analytic space of the group. The map ϑ is a closed embedding when the ground field is local; a compactification is obtained when ϑ is composed with the (analytic map) associated to a fibration from the group to one of its flag varieties. One obtains in this way the finite family of compactifications described in [Wer07]. One nice feature is the possibility to obtain easily maps between compactifications of a given group but attached to distinct flag varieties. This enables one to understand in combinatorial

Lie-theoretic terms which boundary components are shrunk when going from a "big" compactification to a smaller one.

The second way mimics I. Satake's work in the real case. More precisely, it uses a highest weight representation of the group in order to obtain a map from the building of the group to the building of the general linear group of the representation space which, as we said before, is nothing else than a space of non-Archimedean norms. Then it remains to use the seminorm compactification mentioned above by taking the closure of the image of the composed map from the building to the compact space of (homothety classes of) seminorms on the non-Archimedean representation space.

For a given group, these two methods lead to the same family of compactifications, indexed by the conjugacy classes of parabolic subgroups. One interesting point in these two approaches is the fact that the compactifications are obtained by taking the closure of images of equivariant maps. The construction of the latter maps is also one of the main difficulties; it is overcome thanks to the fact that Berkovich geometry has a rich formalism which combines techniques from algebraic and analytic geometry (the possibility to use field extensions, or the concept of Shilov boundary, are for instance crucial to define the desired equivariant maps).

Structure of the Paper In Sect. 2, we define (simplicial and non-simplicial) Euclidean buildings and illustrate the notions in the case of the groups SL_n; we also show in these cases how the natural group actions on the building encode information on the group structure of rational points. In Sect. 3, we illustrate general notions thanks to the examples of spaces naturally associated to special linear groups (such as projective spaces); this time the notions are relevant to Berkovich analytic geometry and to Drinfeld upper half-spaces. We also provide specific examples of compactifications which we generalize later. In Sect. 4, we sum up quickly what we need from Bruhat-Tits theory, including the existence of integral models for suitable bounded open subgroups; following the classical strategy, we first show how to construct a Euclidean building in the split case by gluing together Euclidean tilings, and then how to rely on Galois descent arguments for non-necessarily split groups. In Sect. 5, we finally introduce the maps that enable us to obtain compactifications of Bruhat-Tits buildings (these maps from buildings to analytifications of flag varieties have been previously defined by V. Berkovich in the split case); a variant of this embedding approach, close to Satake's ideas using representation theory to compactify symmetric spaces, is also quickly presented. In the last section, we correct a mistake in the proof of an auxiliary lemma in [RTW10] which requires us to introduce an additional hypothesis for two results of [RTW12].

Conventions In this paper, as in [Ber90], valued fields are assumed to be non-Archimedean and complete, the valuation ring of such a field k is denoted by k°, its maximal ideal is by $k^{\circ\circ}$ and its residue field by $\tilde{k} = k^\circ / k^{\circ\circ}$. Moreover a *local field* is a non-trivially valued non-Archimedean field which is locally compact for the topology given by the valuation (i.e., it is complete, the valuation is discrete and the residue field is finite).

2   Buildings and Special Linear Groups

We first provide a (very quick) general treatment of Euclidean buildings; general references for this notion are [Rou09] and [Wei09]. It is important for us to deal with the simplicial as well as the non-simplicial version of the notion of a Euclidean building because compactifying Bruhat-Tits buildings via Berkovich techniques uses huge valued fields. The second part illustrates these definitions for special linear groups; in particular, we show how to interpret suitable spaces of norms to obtain concrete examples of buildings in the case when the algebraic group under consideration is the special linear group of a vector space. These spaces of norms will naturally be extended to spaces of (homothety classes of) seminorms when buildings are considered in the context of analytic projective spaces.

2.1   Euclidean Buildings

Euclidean buildings are non-Archimedean analogues of Riemannian symmetric spaces of the non-compact type, at least in the following sense: if G is a simple algebraic group over a valued field k, Bruhat-Tits theory (often) associates to G and k a metric space, called a Euclidean building, on which $G(k)$ acts by isometries in a "very transitive" way. This is a situation which is very close to the one where a (non-compact) simple real Lie group acts on its associated (non-positively curved) Riemannian symmetric space. In this more classical case, the transitivity of the action, the explicit description of fundamental domains for specific (e.g., maximal compact) subgroups and some non-positive curvature arguments lead to deep conjugacy and structure results—see [Mau09] and [Par09] for a modern account. Euclidean buildings are singular spaces but, by and large, play a similar role for non-Archimedean Lie groups $G(k)$ as above.

2.1.1   Simplicial Definition

The general reference for building theory from the various "discrete" viewpoints is [AB08]. Let us start with an affine reflection group, more precisely a *Coxeter group of affine type* [Bou07]. The starting point to introduce this notion is a locally finite family of hyperplanes—called *walls*—in a Euclidean space [**loc. cit.**, V §1 introduction]. An affine Coxeter group can be seen as a group generated by the reflections in the walls, acting properly on the space and stabilizing the collection of walls [**loc. cit.**, V §3 introduction]; it is further required that the action on each irreducible factor of the ambient space be via an infinite *essential* group (no non-zero vector is fixed by the group).

Example 2.1   1. The simplest (one-dimensional) example of a Euclidean tiling is provided by the real line tesselated by the integers. The corresponding affine

Coxeter group, generated by the reflections in two consecutive vertices (i.e., integers), is the infinite dihedral group D_∞.

2. The next simplest (irreducible) example is provided by the tessellation of the Euclidean plane by regular triangles. The corresponding tiling group is the Coxeter group of affine type $\widetilde{A_2}$; it is generated by the reflections in the three lines supporting the edges of any fundamental triangle.

Note that Poincaré's theorem is a concrete source of Euclidean tilings: start with a Euclidean polyhedron in which each dihedral angle between codimension 1 faces is of the form $\frac{\pi}{m}$ for some integer $m \geq 1$ (depending on the pair of faces), then the group generated by the reflections in these faces is an affine Coxeter group [Mas88, IV.H.11].

In what follows, Σ is a Euclidean tiling giving rise to a Euclidean reflection group by Poincaré's theorem (in Bourbaki's terminology, it can also be seen as the natural geometric realization of the Coxeter complex of an affine Coxeter group, that is the affinization of the Tits' cone of the latter group [Bou07]).

Definition 2.2 Let (Σ, W) be a Euclidean tiling and its associated Euclidean reflection group. A *(discrete) Euclidean building* of type (Σ, W) is a polysimplicial complex, say $\mathscr{B}$, which is covered by subcomplexes all isomorphic to Σ—called the *apartments*—such that the following incidence properties hold.

SEB 1   Any two cells of $\mathscr{B}$ lie in some apartment.
SEB 2   Given any two apartments, there is an isomorphism between them fixing their intersection in $\mathscr{B}$.

The cells in this context are called *facets* and the group W is called the *Weyl group* of the building $\mathscr{B}$. The facets of maximal dimension are called *alcoves*.

The axioms of a Euclidean building can be motivated by metric reasons. Indeed, once the choice of a W-invariant Euclidean metric on Σ has been made, there is a natural way the define a distance on the whole building: given any two points x and x' in $\mathscr{B}$, by *(SEB 1)* pick an apartment $\mathbb{A}$ containing them and consider the distance between x and x' taken in $\mathbb{A}$; then *(SEB 2)* implies that the so–obtained non-negative number doesn't depend on the choice of $\mathbb{A}$. It requires further work to check that one defines in this way a distance on the building (i.e., to check that the triangle inequality holds [Par00, Prop. II.1.3]).

Remark 2.3 The terminology "polysimplicial" refers to the fact that a building can be a direct product of simplicial complexes rather than merely a simplicial complex; this is why we provisionally used the terminology "cells" instead of "polysimplices" to state the axioms (as already mentioned, cells will henceforth be called facets— alcoves when they are top-dimensional).

Let us provide now some examples of discrete buildings corresponding to the already mentioned examples of Euclidean tilings.

Example 2.4 1. The class of buildings of type $(\mathbf{R}, D_\infty)$ coincides with the class of trees without terminal vertex (recall that a tree is a 1-dimensional simplicial complex—i.e., the geometric realization of a graph—without non-trivial loop [Ser77]).

2. A 2-dimensional $\widetilde{A_2}$-building is already impossible to draw, but roughly speaking it can be constructed by gluing half-tilings to an initial one along *walls* (i.e., fixed point sets of reflections) and by iterating these gluings infinitely many times provided a prescribed "shape" of neighborhoods of vertices is respected—see Example 2.7 for further details on the local description of a building in this case.

It is important to note that axiom (ii) does *not* require that the isomorphism between apartments extends to a global automorphism of the ambient building. In fact, it may very well happen that for a given Euclidean building $\mathscr{B}$ we have Aut($\mathscr{B}$) = {1} (take for example a tree in which any two distinct vertices have distinct valencies). However, J. Tits' classification of Euclidean buildings [Tit86] implies that in dimension $\geqslant 3$ any irreducible building comes—via Bruhat-Tits theory, see next remark—from a simple algebraic group over a local field, and therefore admits a large automorphism group. At last, note that there do exist 2-dimensional exotic Euclidean buildings, with interesting but unexpectedly small automorphism groups [Bar00].

Remark 2.5 In Sect. 4, we will briefly introduce Bruhat-Tits theory. The main outcome of this important part of algebraic group theory is that, given a semisimple algebraic group G over a local field k, there exists a discrete Euclidean building $\mathscr{B} = \mathscr{B}(G, k)$ on which the group of rational points $G(k)$ acts by isometries and *strongly transitively* (i.e., transitively on the inclusions of an alcove in an apartment).

Example 2.6 Let G as above be the group SL_3. Then the Euclidean building associated to SL_3 is a Euclidean building in which every apartment is a Coxeter complex of type $\widetilde{A_2}$, that is the previously described 2-dimensional tiling of the Euclidean space $\mathbf{R}^2$ by regular triangles. Strong transitivity of the $SL_3(k)$-action means here that given any alcoves (triangles) c, c' and any apartments $\mathbb{A}, \mathbb{A}'$ such that $c \subset \mathbb{A}$ and $c' \subset \mathbb{A}'$ there exists $g \in SL_3(k)$ such that $c' = g.c$ and $\mathbb{A}' = g.\mathbb{A}$.

The description of the apartments doesn't depend on the local field k (only on the Dynkin diagram of the semisimple group in general), but the field k plays a role when one describes the combinatorial neighborhoods of facets, or small metric balls around vertices. Such subsets, which intersect finitely many facets when k is a local field, are known to be realizations of some (spherical) buildings: these buildings are naturally associated to semisimple groups (characterized by some subdiagram of the Dynkin diagram of G) over the residue field $\tilde{k}$ of k.

Example 2.7 For G = SL_3 and $k = \mathbf{Q}_p$, each sufficiently small ball around a vertex is the flag complex of a 2-dimensional vector space over $\mathbf{Z}/p\mathbf{Z}$ and any edge in the associated Bruhat-Tits building is contained in the closure of exactly $p+1$ triangles. A suitably small metric ball around any point in the relative interior of an edge can be seen as a projective line over $\mathbf{Z}/p\mathbf{Z}$, that is the flag variety of SL_2 over $\mathbf{Z}/p\mathbf{Z}$.

2.1.2 Non-Simplicial Generalization

We will see, e.g. in Sect. 5.1, that it is often necessary to understand and use reductive algebraic groups over valued fields for *non-discrete* valuations even if in the initial situation the ground field is discretely valued. The geometric counterpart to this is the necessary use of non-discrete Euclidean buildings. The investigation of such a situation is already covered by the fundamental work by F. Bruhat and J. Tits as written in [BrT72] and [BrT84], but the intrinsic definition of a non-discrete Euclidean building is not given there—see [Tit86] though, for a reference roughly appearing at the same time as Bruhat-Tits' latest papers.

The definition of a building in this generalized context is quite similar to the discrete one (2.1.1) in the sense that it replaces an atlas by a collection of "slices" which are still called *apartments* and turn out to be maximal flat (i.e., Euclidean) subspaces once the building is endowed with a natural distance. What follows can be found for instance in A. Parreau's thesis [Par00].

Let us go back to the initial question.

Question 2.8 Which geometry can be associated to a group $G(k)$ when G is a reductive group over k, a (not necessarily discretely) valued field?

The answer to this question is a long definition to swallow, so we will provide some explanations immediately after stating it.

The starting point is again a d-dimensional Euclidean space, say Σ_{vect}, together with a finite group $\overline{W}$ in the group of isometries $\text{Isom}(\Sigma_{\text{vect}}) \simeq O_d(\mathbf{R})$. By definition, a *vectorial wall* in Σ_{vect} is the fixed-point set in Σ_{vect} of a reflection in $\overline{W}$ and a *vectorial Weyl chamber* is a connected component of the complement of the union of the walls in Σ_{vect}, so that Weyl chambers are simplicial cones.

Now assume that we are given an affine Euclidean space Σ with underlying Euclidean vector space Σ_{vect}. We have thus $\text{Isom}(\Sigma) \simeq \text{Isom}(\Sigma_{\text{vect}}) \ltimes \Sigma_{\text{vect}} \simeq O_d(\mathbf{R}) \ltimes \mathbf{R}^d$. We also assume that we are given a group W of (affine) isometries in Σ such that the vectorial part of W is $\overline{W}$ and such that there exists a point $x \in \Sigma$ and a subgroup $T \subset \text{Isom}(\Sigma)$ of translations satisfying $W = W_x \cdot T$; we use here the notation $W_x = \text{Stab}_W(x)$. A point x satisfying this condition is called *special*.

Definition 2.9 Let $\mathscr{B}$ be a set and let $\mathscr{A} = \{f : \Sigma \rightarrow \mathscr{B}\}$ be a collection of injective maps, whose images are called *apartments*. We say that $\mathscr{B}$ is a *Euclidean building* of type (Σ, W) if the apartments satisfy the following axioms.

EB 1   The family $\mathscr{A}$ is stable by precomposition with any element of W (i.e., for any $f \in \mathscr{A}$ and any $w \in W$, we have $f \circ w \in \mathscr{A}$).

EB 2   For any $f, f' \in \mathscr{A}$ the subset $\mathscr{C}_{f,f'} = f'^{-1}\big(f(\Sigma)\big)$ is convex in Σ and there exists $w \in W$ such that we have the equality of restrictions $(f^{-1} \circ f') \,|_{\mathscr{C}_{f,f'}} = w \,|_{\mathscr{C}_{f,f'}}$.

EB 3   Any two points of $\mathscr{B}$ are contained in a suitable apartment.

At this stage, there is a well-defined map $d : \mathscr{B} \times \mathscr{B} \rightarrow \mathbf{R}_{\geqslant 0}$ and we further require:

EB 4   Given any (images of) Weyl chambers, there is an apartment of X containing sub-Weyl chambers of each.

EB 5 Given any apartment $\mathbb{A}$ and any point $x \in \mathbb{A}$, there is a 1-lipschitz retraction map $r = r_{x,\mathbb{A}} : \mathcal{B} \to \mathbb{A}$ such that $r \mid_{\mathbb{A}} = \mathrm{id}_{\mathbb{A}}$ and $r^{-1}(x) = \{x\}$.

The above definition is taken from [Par00, II.1.2]; in these axioms a *Weyl chamber* is the affine counterpart to the previously defined notion of a *Weyl chamber* and a "sub-Weyl chamber" is a translate of the initial Weyl chamber which is completely contained in the latter.

Remark 2.10 A different set of axioms is given in G. Rousseau's paper [Rou09, §6]. It is interesting because it provides a unified approach to simplicial and non-simplicial buildings via incidence requirements on apartments. The possibility to obtain a non-discrete building with Rousseau's axioms is contained in the model for an apartment and the definition of a facet as a filter. The latter axioms are adapted to some algebraic situations which cover the case of Bruhat-Tits theory over non-complete valued fields—see [Rou09, Remark 9.4] for more details and comparisons.

Remark 2.11 In this paper we do not use the plain word "chamber" though it is standard terminology in abstract building theory. This choice is made to avoid confusion: alcoves here are chambers (in the abstract sense) in Euclidean buildings and parallelism classes of Weyl chambers here are chambers (in the abstract sense) in spherical buildings at infinity of Euclidean buildings [Wei09, Chap. 8], [AB08, 11.8].

It is easy to see that, in order to prove that the map d defined thanks to axioms (EB 1)–(EB 3) is a distance, it remains to check that the triangle inequality holds; this is mainly done by using the retraction given by axiom (EB 5). The previously quoted metric motivation (Remark 2.3) so to speak became a definition. Note that the existence of suitable retractions is useful to other purposes.

The following examples of possibly non-simplicial Euclidean buildings correspond to the examples of simplicial ones given in Example 2.4.

Example 2.12 1. Consider the real line $\Sigma = \mathbf{R}$ and its isometry group $\mathbf{Z}/2\mathbf{Z} \ltimes \mathbf{R}$. Then a Euclidean building of type $(\mathbf{R}, \mathbf{Z}/2\mathbf{Z} \ltimes \mathbf{R})$ is a real tree—see below.
2. For a 2-dimensional case extending simplicial $\tilde{A}_2$-buildings, a model for an apartment can be taken to be a maximal flat in the symmetric space of $\mathrm{SL}_3(\mathbf{R})/\mathrm{SO}(3)$ acted upon by its stabilizer in $\mathrm{SL}_3(\mathbf{R})$ (using the notion of singular geodesics to distinguish the walls). There is a geometric way to define the Weyl group and Weyl chambers (six directions of simplicial cones) in this differential geometric context—see [Mau09] for the general case of arbitrary symmetric spaces.

Here is a (purely metric) definition of real trees. It is a metric space (X, d) with the following two properties:

(i) it is *geodesic*: given any two points $x, x' \in X$ there is a (continuous) map $\gamma : [0; d] \to X$, where $d = d(x, x')$, such that $\gamma(0) = x$, $\gamma(d) = x'$ and $d(\gamma(s), \gamma(t)) = |s - t|$ for any $s, t \in [0; d]$;
(ii) any geodesic triangle is a tripod (i.e., the union of three geodesic segments with a common end-point).

Remark 2.13 Non-simplicial Euclidean buildings became more popular since recent work of geometric (rather than algebraic) nature, where non-discrete buildings appear as asymptotic cones of symmetric spaces and Bruhat-Tits buildings [KL97].

The remark implies in particular that there exist non-discrete Euclidean buildings in any dimension, which will also be seen more concretely by studying spaces of non-Archimedean norms on a given vector space—see Sect. 2.2.

Remark 2.14 Note that given a reductive group G over a valued field k, Bruhat-Tits theory "often" provides a Euclidean building on which the group $G(k)$ acts strongly transitively in a suitable sense (see Sect. 4 for an introduction to this subject).

2.1.3 More Geometric Properties

We motivated the definitions of buildings by metric considerations, therefore we must mention the metric features of Euclidean buildings once these spaces have been defined. First, a Euclidean building always admits a metric whose restriction to any apartment is a (suitably normalized) Euclidean distance [Rou09, Prop. 6.2]. Endowed with such a distance, a Euclidean building is always a geodesic metric space as introduced in the above metric definition of real trees (2.1.2).

Recall that we use the axioms (EB) *from Definition 2.9 to define a building; moreover we assume that the above metric is complete.* This is sufficient for our purposes since we will eventually deal with Bruhat-Tits buildings associated to algebraic groups over complete non-Archimedean fields.

Let $(\mathscr{B}, d)$ be a Euclidean building endowed with such a metric. Then $(\mathscr{B}, d)$ satisfies moreover a remarkable non-positive curvature property, called the CAT(0)-*property* (where "CAT" seems to stand for Cartan-Alexandrov-Toponogov). Roughly speaking, this property says that geodesic triangles are at least as thin as in Euclidean planes. More precisely, the point is to compare a geodesic triangle drawn in $\mathscr{B}$ with "the" Euclidean triangle having the same edge lengths. A geodesic space is said to have the CAT(0)-*property*, or to *be* CAT(0), if a median segment in each geodesic triangle is at most as long as the corresponding median segment in the comparison triangle drawn in the Euclidean plane $\mathbf{R}^2$ (this inequality has to be satisfied for all geodesic triangles). Though this property is stated in elementary terms, it has very deep consequences [Rou09, §7].

One first consequence is the uniqueness of a geodesic segment between any two points [BH99, Chap. II.1, Prop. 1.4].

The main consequence is a famous and very useful fixed-point property. The latter statement is itself the consequence of a purely geometric one: any bounded subset in a complete, CAT(0)-space has a unique, metrically characterized, circum-center [AB08, 11.3]. This implies that if a group acting by isometries on such a space (e.g., a Euclidean building) has a bounded orbit, then it has a fixed point. This is the *Bruhat-Tits fixed point lemma*; it applies for instance to any compact group of isometries.

Let us simply mention two very important applications of the Bruhat-Tits fixed point lemma (for simplicity, we assume that the building under consideration is discrete and locally finite—which covers the case of Bruhat-Tits buildings for reductive groups over local fields).

1. The Bruhat-Tits fixed point lemma is used to classify maximal bounded subgroups in the isometry group of a building. Indeed, it follows from the definition of the compact open topology on the isometry group $\text{Aut}(\mathscr{B})$ of a building $\mathscr{B}$, that a facet stabilizer is a compact subgroup in $\text{Aut}(\mathscr{B})$. Conversely, a compact subgroup has to fix a point and this point can be sent to a point in a given fundamental domain for the action of $\text{Aut}(\mathscr{B})$ on $\mathscr{B}$ (the isometry used for this conjugates the initial compact subgroup into the stabilizer of a point in the fundamental domain).
2. Another consequence is that any Galois action on a Bruhat-Tits building has "sufficiently many" fixed points, since a Galois group is profinite hence compact. These Galois actions are of fundamental use in Bruhat-Tits theory, following the general idea—widely used in algebraic group theory—that an algebraic group G over k is nothing else than a split algebraic group over the separable closure k^s, namely $G \otimes_k k^s$, together with a semilinear action of $\text{Gal}(k^s/k)$ on $G \otimes_k k^s$ [Bor91, AG § §11–14].

Arguments similar to the ones mentioned in 1. imply that, when k is a local field, there are exactly $d + 1$ conjugacy classes of maximal compact subgroups in $\text{SL}_{d+1}(k)$. They are parametrized by the vertices contained in the closure of a given alcove (in fact, they are all isomorphic to $\text{SL}_{d+1}(k^\circ)$ and are all conjugate under the action of $\text{GL}_{d+1}(k)$ by conjugation).

Remark 2.15 One can make 2. a bit more precise. The starting point of Bruhat-Tits theory is indeed that a reductive group G over any field, say k, splits—hence in particular is very well understood—after extension to the separable closure k^s of the ground field. Then, in principle, one can go down to the group G over k by means of suitable Galois action—this is one leitmotiv in [BT65]. In particular, Borel-Tits theory provides a lot of information about the group $G(k)$ by seeing it as the fixed-point set $G(k^s)^{\text{Gal}(k^s/k)}$. When the ground field k is a valued field, then one can associate a Bruhat-Tits building $\mathscr{B} = \mathscr{B}(G, k^s)$ to $G \otimes_k k^s$ together with an action by isometries of $\text{Gal}(k^s/k)$. The Bruhat-Tits building of G over k is contained in the Galois fixed-point set $\mathscr{B}^{\text{Gal}(k^s/k)}$, but this is inclusion is strict in general: the Galois fixed-point set is bigger than the desired building [Rou77, III]. Still, this may be a good first approximation of Bruhat-Tits theory to have in mind. We refer to Sect. 4.2.2 for further details.

2.2   The SL$_n$ Case

We now illustrate many of the previous notions in a very explicit situation, of arbitrary dimension. Our examples are spaces of norms on a non-Archimedean vector space. They provide the easiest examples of Bruhat-Tits buildings, and are also very close to spaces occurring in Berkovich analytic geometry. In this section, we denote by V a k-vector space and by $d + 1$ its (finite) dimension over k.

Note that until Remark 2.23 we assume that k is a local field.

2.2.1 Goldman-Iwahori Spaces

The materiel of this subsection is classical and could be find, for example, in [Wei74].

We are interested in the following space.

Definition 2.16 The *Goldman-Iwahori* space of the k-vector space V is the space of non-Archimedean norms on V; we denote it by $\mathcal{N}(V, k)$. We denote by $\mathcal{X}(V, k)$ the quotient space $\mathcal{N}(V, k)\big/ \sim$, where $\sim$ is the equivalence relation which identifies two homothetic norms.

To be more precise, let $\| \cdot \|$ and $\| \cdot \|'$ be norms in $\mathcal{N}(V, k)$. We have $\| \cdot \| \sim \| \cdot \|'$ if and only if there exists $c > 0$ such that $\|x\| = c \|x\|'$ for all $x \in V$. In the sequel, we use the notation $[\cdot]_\sim$ to denote the class with respect to the homothety equivalence relation.

Example 2.17 Here is a simple way to construct non-Archimedean norms on V. Pick a basis $\mathbf{e} = (e_0, e_1, \ldots, e_d)$ in V. Then for each choice of parameters $\underline{c} = (c_0, c_1, \ldots, c_d) \in \mathbf{R}^{d+1}$, we can define the non-Archimedean norm which sends each vector $x = \sum_i \lambda_i e_i$ to $\max_i\{\exp(c_i) \mid \lambda_i \mid\}$, where $\mid \cdot \mid$ denotes the absolute value of k. We denote this norm by $\| \cdot \|_{\mathbf{e},\underline{c}}$.

We also introduce the following notation and terminology.

Definition 2.18 (i) Let $\| \cdot \|$ be a norm and let $\mathbf{e}$ be a basis in V. We say that $\| \cdot \|$ is *diagonalized* by $\mathbf{e}$ if there exists $\underline{c} \in \mathbf{R}^{d+1}$ such that $\| \cdot \| = \| \cdot \|_{\mathbf{e},\underline{c}}$; in this case, we also say that the basis $\mathbf{e}$ is *adapted* to the norm $\| \cdot \|$.

(ii) Given a basis $\mathbf{e}$, we denote by $\widetilde{\mathbb{A}_{\mathbf{e}}}$ the set of norms diagonalized by $\mathbf{e}$:

$$\widetilde{\mathbb{A}_{\mathbf{e}}} = \{\| \cdot \|_{\mathbf{e},\underline{c}} : \underline{c} \in \mathbf{R}^{d+1}\}.$$

(iii) We denote by $\mathbb{A}_{\mathbf{e}}$ the quotient of $\widetilde{\mathbb{A}_{\mathbf{e}}}$ by the homothety equivalence relation: $\mathbb{A}_{\mathbf{e}} = \widetilde{\mathbb{A}_{\mathbf{e}}}\big/ \sim$.

Note that the space $\widetilde{\mathbb{A}_{\mathbf{e}}}$ is naturally an affine space with underlying vector space $\mathbf{R}^{d+1}$: the free transitive $\mathbf{R}^{d+1}$-action is by shifting the coefficients c_i which are the logarithms of the "weights" $\exp(c_i)$ for the norms $\| \cdot \|_{\mathbf{e},\underline{c}}$: $\sum_i \lambda_i e_i \mapsto \max_{0 \leqslant i \leqslant d}\{\exp(c_i)|\lambda_i|\}$. Under this identification of affine spaces, we have: $\mathbb{A}_{\mathbf{e}} \simeq \mathbf{R}^{d+1}/\mathbf{R}(1, 1, \ldots, 1) \simeq \mathbf{R}^d$.

Remark 2.19 The space $\mathscr{X}(V, k)$ will be endowed with a Euclidean building structure (Theorem 2.25) in which the spaces $\mathbb{A}_{\mathbf{e}}$—with $\mathbf{e}$ varying over the bases of V—will be the apartments.

The following fact can be generalized to more general valued fields than local fields but is *not* true in general (Remark 2.24).

Proposition 2.20 *Every norm of $\mathscr{N}(V, k)$ admits an adapted basis in* V.

Proof Let $\| \cdot \|$ be a norm of $\mathscr{N}(V, k)$. We prove the result by induction on the dimension of the ambient k-vector space. Let μ be any non-zero linear form on V. The map $V \setminus \{0\} \to \mathbf{R}_+$ sending y to $\frac{|\mu(y)|}{\|y\|}$ naturally provides, by homogeneity, a continuous map $\phi : \mathbf{P}(V)(k) \to \mathbf{R}_+$. Since k is locally compact, the projective space $\mathbf{P}(V)(k)$ is compact, therefore there exists an element $x \in V \setminus \{0\}$ at which ϕ achieves its supremum, so that

$$\frac{|\mu(z)|}{|\mu(x)|} \, \|x\| \leqslant \|z\| \tag{$*$}$$

for any $z \in V$.

Let z be an arbitrary vector of V. We write $z = y + \dfrac{\mu(z)}{\mu(x)}x$ according to the direct sum decomposition $V = \mathrm{Ker}(\mu) \oplus kx$. By the ultrametric inequality satisfied by $\| \cdot \|$, we have

$$\|z\| \leqslant \max\{\|y\|; \frac{|\mu(z)|}{|\mu(x)|} \, \|x\|\} \tag{$**$}$$

and

$$\|y\| \leqslant \max\{\|z\|; \frac{|\mu(z)|}{|\mu(x)|} \, \|x\|\} . \tag{$***$}$$

Inequality $(*)$ says that $\max\{\| z \|; \dfrac{|\mu(z)|}{|\mu(x)|} \, \| x \|\} = \| z \|$, so $(***)$ implies $\| z \| \geqslant \| y \|$. The latter inequality together with $(*)$ implies that $\| z \| \geqslant \max \{\| y \|; \dfrac{|\mu(z)|}{|\mu(x)|} \, \| x \|\}$. Combining this with $(**)$ we obtain the equality $\| z \| = \max\{\| y \|; \dfrac{|\mu(z)|}{|\mu(x)|} \, \| x \|\}$. Applying the induction hypothesis to $\mathrm{Ker}(\mu)$, we obtain a basis adapted to the restriction of $\| \cdot \|$ to $\mathrm{Ker}(\mu)$. Adding x we obtain a basis adapted to $\| \cdot \|$, as required (note that $\frac{\mu(z)}{\mu(x)}$ is the coordinate corresponding to the vector x in any such basis). $\qquad\square$

Actually, we can push a bit further this existence result about adapted norms.

Proposition 2.21 *For any two norms of $\mathcal{N}(V,k)$ there is a basis of V simultaneously adapted to them.*

Proof We are now given two norms, say $\| \cdot \|$ and $\| \cdot \|'$, in $\mathcal{N}(V,k)$. In the proof of Proposition 2.20, the choice of a non-zero linear form μ had no importance. In the present situation, we will take advantage of this freedom of choice. We again argue by induction on the dimension of the ambient k-vector space.

By homogeneity, the map $V \setminus \{0\} \to \mathbf{R}_+$ sending y to $\dfrac{\| y \|}{\| y \|'}$ naturally provides a continuous map $\psi : \mathbf{P}(V)(k) \to \mathbf{R}_+$. Again because the projective space $\mathbf{P}(V)(k)$ is compact, there exists $x \in V \setminus \{0\}$ at which ψ achieves its supremum, so that

$$\frac{\| y \|}{\| x \|} \leqslant \frac{\| y \|'}{\| x \|'} \text{ for any } y \in V.$$

Now we endow the dual space V^* with the operator norm $\| \cdot \|^*$ associated to $\| \cdot \|$ on V. Since V is finite-dimensional, by biduality (i.e. the normed vector space version of $V^{**} \simeq V$), we have the equality $\| x \| = \sup\limits_{\mu \in V^* \setminus \{0\}} \dfrac{|\mu(x)|}{\| \mu \|^*}$. By homogeneity and compactness, there exists $\lambda \in V^* \setminus \{0\}$ such that $\| x \| = \dfrac{|\lambda(x)|}{\| \lambda \|^*}$. For arbitrary $y \in V$ we have $|\lambda(y)| \leqslant \| y \| \cdot \| \lambda \|^*$, so the definition of x implies that

$$\frac{|\lambda(y)|}{|\lambda(x)|} \leqslant \frac{\| y \|}{\| x \|} \text{ for any } y \in V.$$

In other words, we have found $x \in V$ and $\lambda \in V^*$ such that

$$\frac{|\lambda(y)|}{|\lambda(x)|} \leqslant \frac{\| y \|}{\| x \|} \leqslant \frac{\| y \|'}{\| x \|'} \text{ for any } y \in V.$$

Now we are in position to apply the arguments of the proof of Proposition 2.20 to both $\| \cdot \|$ and $\| \cdot \|'$ to obtain that $\| z \| = \max\{\| y \|; \dfrac{|\lambda(z)|}{|\lambda(x)|} \| x \|\}$ and $\| z \|' = \max\{\| y \|'; \dfrac{|\lambda(z)|}{|\lambda(x)|} \| x \|'\}$ for any $z \in V$ decomposed as $z = x + y$ with $y \in \mathrm{Ker}(\lambda)$. It remains then to apply the induction hypothesis (i.e., that the desired statement holds in the ambient dimension minus 1). $\square$

2.2.2 Connection with Building Theory

It is now time to describe the connection between Goldman-Iwahori spaces and Euclidean buildings. As already mentioned, the subspaces $\mathbb{A}_e$ will be the apartments in $\mathscr{X}(V,k)$ (Remark 2.19).

Let us fix a basis $\mathbf{e}$ in V and consider first the bigger affine space $\widetilde{A_{\mathbf{e}}} = \{\| \cdot \|_{\mathbf{e},\underline{c}} : \underline{c} \in \mathbf{R}^{d+1}\} \simeq \mathbf{R}^{d+1}$. The symmetric group $\mathscr{S}_{d+1}$ acts on this affine space by permuting the coefficients c_i. This is obviously a faithful action and we have another one given by the affine structure. We obtain in this way an action of the group $\mathscr{S}_{d+1} \ltimes \mathbf{R}^{d+1}$ on $\widetilde{A_{\mathbf{e}}}$ and, after passing to the quotient space, we can see $A_{\mathbf{e}}$ as the ambient space of the Euclidean tiling attached to the affine Coxeter group of type $\widetilde{A_d}$ (the latter group is isomorphic to $\mathscr{S}_{d+1} \ltimes \mathbf{Z}^d$). The following result is due to Bruhat-Tits, elaborating on Goldman-Iwahori's investigation of the space of norms $\mathscr{N}(V, k)$ [GI63].

Theorem 2.22 *The space $\mathscr{X}(V, k) = \mathscr{N}(V, k)/ \sim$ is a simplicial Euclidean building of type $\widetilde{A_d}$, where $d + 1 = \dim(V)$; in particular, the apartments are isometric to $\mathbf{R}^d$ and the Weyl group is isomorphic to $\mathscr{S}_{d+1} \ltimes \mathbf{Z}^d$.*

Reference for the proof In [BrT72, 10.2] this is stated in group-theoretic terms, so one has to combine the quoted statement with [**loc. cit.**, 7.4] in order to obtain the above theorem. This will be explained in Sect. 4. □

The 0-skeleton (i.e., the vertices) for the simplicial structure corresponds to the k°-*lattices* in the k-vector space V, that is the free k°-submodules in V of rank $d +1$. To a lattice $\mathscr{L}$ is attached a norm $\| \cdot \|_{\mathscr{L}}$ by setting $\| x \|_{\mathscr{L}} = \inf\{|\lambda| : \lambda \in k^\times$ and $\lambda^{-1} x \in \mathscr{L}\}$. One recovers the k°-lattice $\mathscr{L}$ as the unit ball of the norm $\| \cdot \|_{\mathscr{L}}$.

Remark 2.23 Note that the space $\mathscr{N}(V, k)$ is an extended building in the sense of [Tit79]; this is, roughly speaking, a building to which is added a Euclidean factor in order to account geometrically for the presence of a center of positive dimension.

Instead of trying to prove this result, let us mention that Proposition 2.21 says, in our building-theoretic context, that any two points are contained in an apartment. In other words, this proposition implies axiom (SEB 1) of Definition 2.2: it is the non-Archimedean analogue of the fact that any two real scalar products are diagonalized in a suitable common basis (Gram-Schmidt).

Now let us skip the hypothesis that k is a local field. If k is a not discretely valued, then it is not true in general that every norm in $\mathscr{N}(V, k)$ can be diagonalized in some suitable basis. Therefore we introduce the following subspace:

$$\mathscr{N}(V, k)^{\text{diag}} = \{\text{norms in } \mathscr{N}(V, k) \text{ admitting an adapted basis}\}.$$

Remark 2.24 We will see (Remark 3.2) that the connection between Berkovich projective spaces and Bruhat-Tits buildings helps to understand why $\mathscr{N}(V, k) - \mathscr{N}(V, k)^{\text{diag}} \neq \varnothing$ if and only if the valued field k is *not* maximally complete (one also says spherically complete).

Thanks to the subspace $\mathscr{N}(V, k)^{\text{diag}}$, we can state the result in full generality.

Theorem 2.25 *The space $\mathscr{X}(V, k) = \mathscr{N}(V, k)^{\text{diag}}/ \sim$ is a Euclidean building of type $\widetilde{A_d}$ in which the apartments are isometric to $\mathbf{R}^d$ and the Weyl group is* *isomorphic to $\mathscr{S}_{d+1} \ltimes \Lambda$ where Λ is a translation group, which is discrete if and only if so is the valuation of k.*

Reference for the proof This is proved for instance in [Par00, III.1.2]; see also [BrT84] for a very general treatment.   □

Example 2.26 For $d = 1$, i.e. when $V \simeq k^2$, the Bruhat-Tits building

$$\mathscr{X}(V, k) = \mathscr{N}(V, k)^{\mathrm{diag}} / \sim$$

given by Theorem 2.25 is a tree, which is a (non-simplicial) real tree whenever k is not discretely valued.

2.2.3   Group Actions

After illustrating the notion of a building thanks to Goldman-Iwahori spaces, we now describe the natural action of a general linear group over the valued field k on its Bruhat-Tits building. We said that buildings are usually used to better understand groups which act sufficiently transitively on them. We therefore have to describe the $GL(V, k)$-action on $\mathscr{X}(V, k)$ given by precomposition on norms (that is, $g. \| \cdot \| = \| \cdot \| \circ g^{-1}$ for any $g \in GL(V, k)$ and any $\| \cdot \| \in \mathscr{N}(V, k)$). Note that we have the formula

$$g. \| \cdot \|_{\mathbf{e},\underline{c}} = \| \cdot \|_{g.\mathbf{e}.\underline{c}} .$$

We will also explain how this action can be used to find interesting decompositions of $GL(V, k)$. Note that the $GL(V, k)$-action on $\mathscr{X}(V, k)$ factors through an action by the group $PGL(V, k)$.

For the sake of simplicity, we assume that k is discretely valued until the rest of this section.

We describe successively: the action of monomial matrices on the corresponding apartment, stabilizers, fundamental domains and the action of elementary unipotent matrices on the buildings (which can be thought of as "foldings" of half-apartments fixing complementary apartments).

First, it is very useful to restrict our attention to apartments. Pick a basis $\mathbf{e}$ of V and consider the associated apartment $\mathbb{A}_{\mathbf{e}}$. The stabilizer of $\mathbb{A}_{\mathbf{e}}$ in $GL(V, k)$ consists of the subgroup of linear automorphisms g which are *monomial* with respect to $\mathbf{e}$, that is whose matrix expression with respect to $\mathbf{e}$ has only one non-zero entry in each row and in each column; we denote $N_{\mathbf{e}} = \mathrm{Stab}_{GL(V,k)}(\mathbb{A}_{\mathbf{e}})$. Any automorphism in $N_{\mathbf{e}}$ lifts a permutation of the indices of the vectors e_i ($0 \leqslant i \leqslant d$) in $\mathbf{e}$. This defines a surjective homomorphism $N_{\mathbf{e}} \twoheadrightarrow \mathscr{S}_{d+1}$ whose kernel is the group, say $D_{\mathbf{e}}$, of the linear automorphisms diagonalized by $\mathbf{e}$. The group $D_{\mathbf{e}} \cap SL(V, k)$ lifts the translation subgroup of the (affine) Weyl group $\mathscr{S}_{d+1} \ltimes \mathbf{Z}^d$ of $\mathscr{X}(V, k)$. Note that the latter translation group consists of the translations contained in the group generated by the reflections in the codimension 1 faces of a given alcove, therefore this group is (of finite index but) smaller than the "obvious" group given by translations with integral coefficients with respect to the basis $\mathbf{e}$. For any $\underline{\lambda} \in (k^\times)^n$, we have the following "translation formula":

$$\underline{\lambda}. \, \| \cdot \|_{\mathbf{e}.\underline{c}} = \| \cdot \|_{\mathbf{e}.(c_i - \log|\lambda_i|)_i},$$

Example 2.27 When $d = 1$ and when k is local, the translations of smallest displacement length in the (affine) Weyl group of the corresponding tree are translations whose displacement length along their axis is equal to twice the length of an edge.

The fact stated in the example corresponds to the general fact that the $\mathrm{SL}(\mathrm{V}, k)$-action on $\mathscr{X}(\mathrm{V}, k)$ is *type* (or *color*)-*preserving*: choosing $d + 1$ colors, one can attach a color to each *panel* (= codimension 1 facet) so that each color appears exactly once in the closure of any alcove; a panel of a given color is sent by any element of $\mathrm{SL}(\mathrm{V}, k)$ to a panel of the same color. Note that the action of $\mathrm{GL}(\mathrm{V}, k)$, hence also of $\mathrm{PGL}(\mathrm{V}, k)$, on $\mathscr{X}(\mathrm{V}, k)$ is not type-preserving since $\mathrm{PGL}(\mathrm{V}, k)$ acts transitively on the set of vertices.

It is natural to first describe the isotropy groups for the action we are interested in.

Proposition 2.28 *We have the following description of stabilizers:*

$$\mathrm{Stab}_{\mathrm{GL}(\mathrm{V}, k)}(\| \cdot \|_{\mathbf{e}.\underline{c}}) = \{g \in \mathrm{GL}(\mathrm{V}, k) : |\det(g)| = 1 \text{ and } \log(|g_{ij}|) \leqslant c_j - c_i\},$$

where $[g_{ij}]$ is the matrix expression of $\mathrm{GL}(\mathrm{V}, k)$ with respect to the basis $\mathbf{e}$.

Reference for the proof This is for instance [Par00, Cor. III.1.4]. ☐

There is also a description of the stabilizer group in $\mathrm{SL}(\mathrm{V}, k)$ as the set of matrices stabilizing a point with respect to a tropical matrix operation [Wer11, Prop. 2.4].

We now turn our attention to fundamental domains. Let x be a vertex in $\mathscr{X}(\mathrm{V}, k)$. Fix a basis $\mathbf{e}$ such that $x = [\| \cdot \|_{\mathbf{e}.\underline{0}}]_\sim$. Then we have an apartment $\mathbb{A}_\mathbf{e}$ containing x and the inequalities

$$c_0 \leqslant c_1 \leqslant \cdots \leqslant c_d$$

define a Weyl chamber with tip x (after passing to the homothety classes). The other Weyl chambers with tip x contained in $\mathbb{A}_\mathbf{e}$ are obtained by using the action of the spherical Weyl group $\mathscr{S}_{d+1}$, which amounts to permuting the indices of the c_i's (this action is lifted by the action of monomial matrices with coefficients ± 1 and determinant 1).

Accordingly, if we denote by ϖ a uniformizer of k, then the inequalities

$$c_0 \leqslant c_1 \leqslant \cdots \leqslant c_d \quad \text{and} \quad c_d - c_0 \leqslant -\log|\varpi|$$

define an alcove (whose boundary contains x) and any other alcove in $\mathbb{A}_\mathbf{e}$ is obtained by using the action of the affine Weyl group $\mathscr{S}_{d+1} \ltimes \mathbf{Z}^d$.

Proposition 2.29 *Assume k is local. We have the following description of fundamental domains.*

(i) *Given a vertex x, any Weyl chamber with tip x is a fundamental domain for the action of the maximal compact subgroup $\mathrm{Stab}_{\mathrm{SL}(\mathrm{V},k)}(x)$ on $\mathscr{X}(\mathrm{V},k)$.*

(ii) *Any alcove is a fundamental domain for the natural action of $\mathrm{SL}(\mathrm{V},k)$ on the building $\mathscr{X}(\mathrm{V},k)$.*

If we abandon the hypothesis that k is a local field and assume the absolute value of k is surjective (onto $\mathbf{R}_{\geqslant 0}$), then the $\mathrm{SL}(\mathrm{V},k)$-action on $\mathscr{X}(\mathrm{V},k)$ is transitive.

Sketch of proof Property (ii) follows from (i) and from the previous description of the action of the monomial matrices of $\mathrm{N}_\mathbf{e}$ on $\mathbb{A}_\mathbf{e}$ (note that $\mathrm{SL}(\mathrm{V},k)$ is type-preserving, so a fundamental domain cannot be strictly smaller than an alcove).

(i). A fundamental domain for the action of the symmetric group $\mathscr{S}_{d+1}$ as above on the apartment $\mathbb{A}_\mathbf{e}$ is given by a Weyl chamber with tip x, and the latter symmetric group is lifted by elements in $\mathrm{Stab}_{\mathrm{SL}(\mathrm{V},k)}(x)$. Therefore it is enough to show that any point of the building can be mapped into $\mathbb{A}_\mathbf{e}$ by an element of $\mathrm{Stab}_{\mathrm{SL}(\mathrm{V},k)}(x)$. Pick a point z in the building and consider a basis $\mathbf{e}'$ such that $\mathbb{A}_{\mathbf{e}'}$ contains both x and z (Proposition 2.21). We can write $x = \| \cdot \|_{\mathbf{e},0} = \| \cdot \|_{\mathbf{e}',\underline{c}}$, with weights $\underline{c}$ in $\log |k^\times|$ since x is a vertex. After dilation, if necessary, of each vector of the basis $\mathbf{e}'$, we may—and shall—assume that $\underline{c} = 0$. Pick $g \in \mathrm{SL}(\mathrm{V},k)$ such that $g.\mathbf{e} = \mathbf{e}'$. Since $\mathbf{e}$ and $\mathbf{e}'$ span the same lattice L over k°, which is the unit ball for x (see comment after Th. 2.22), we have $g.L = L$ and therefore g stabilizes x. We have therefore found $g \in \mathrm{Stab}_{\mathrm{SL}(\mathrm{V},k)}(x)$ with $g.\mathbb{A}_\mathbf{e} = \mathbb{A}_{\mathbf{e}'}$, in particular $g^{-1}.z$ belongs to $\mathbb{A}_\mathbf{e}$. □

Remark 2.30 Point (i) above is the geometric way to state the so-called Cartan decomposition: $\mathrm{SL}(\mathrm{V},k) = \mathrm{Stab}_{\mathrm{SL}(\mathrm{V},k)}(x) \cdot \overline{\mathrm{T}^+} \cdot \mathrm{Stab}_{\mathrm{SL}(\mathrm{V},k)}(x)$, where $\overline{\mathrm{T}^+}$ is the semigroup of linear automorphisms t diagonalized by $\mathbf{e}$ and such that $t.x$ belongs to a fixed Weyl chamber in $\mathbb{A}_\mathbf{e}$ with tip x. The Weyl chamber can be chosen so that $\overline{\mathrm{T}^+}$ consists of the diagonal matrices whose diagonal coefficients are powers of some given uniformizer with the exponents increasing along the diagonal. Let us recall how to prove this by means of elementary arguments [PR94, §3.4 p. 152]. Let $g \in \mathrm{SL}(\mathrm{V},k)$; we pick $\lambda \in k^\circ$ so that λg is a matrix of $\mathrm{GL}(\mathrm{V},k)$ with coefficients in k°. By interpreting left and right multiplication by elementary unipotent matrices as matrix operations on rows and columns, and since k° is a principal ideal domain, we can find $p, p' \in \mathrm{SL}_{d+1}(k^\circ)$ such that $p^{-1}\lambda g p'^{-1}$ is a diagonal matrix (still with coefficients in k°), which we denote by d. Therefore, we can write $g = p\lambda^{-1}dp'$; and since g, p and p' have determinant 1, so does $t = \lambda^{-1}d$. It remains to conjugate $\lambda^{-1}d$ by a suitable monomial matrix with coefficients ± 1 and determinant 1 in order to obtain the desired decomposition.

At the beginning of this subsection, we described the action of linear auto-morphisms on an apartment when the automorphisms are diagonalized by a basis defining the apartment. One last interesting point is the description of the action of elementary unipotent matrices (for a given basis). The action looks like a "folding" in the building, fixing a suitable closed half-apartment.

More precisely, let us introduce the elementary unipotent matrices $u_{ij}(v) = \mathrm{id} + vE_{ij}$ where $v \in k$ and E_{ij} is the matrix whose only non-zero entry is the (i, j)-th one, equal to 1.

Proposition 2.31 *The intersection $\widetilde{A_e} \cap u_{ij}(\lambda).\widetilde{A_e}$ is the half-space of $\widetilde{A_e}$ consisting of the norms $\| \cdot \|_{e,c}$ satisfying $c_j - c_i \geqslant \log |\lambda|$. The isometry given by the matrix $u_{ij}(\lambda)$ fixes pointwise this intersection and the image of the open half-apartment $\widetilde{A_e} - \{\| \cdot \|_{e,c} : c_j - c_i \geqslant \log |\lambda|\}$ is (another half-apartment) disjoint from $\widetilde{A_e}$.*

Proof In the above notation, we have $u_{ij}(v)(\sum_i \lambda_i e_i) = \sum_{k \neq i} \lambda_k e_k + (\lambda_i + v\lambda_j)e_i$ for any $v \in k$.

First, we assume that we have $u_{ij}(\lambda). \| \cdot \|_{e,c} = \| \cdot \|_{e,c}$. Then, applying this equality of norms to the vector e_j provides $e^{c_j} = \max\{e^{c_j}; e^{c_i} |\lambda|\}$, hence the inequality $c_j - c_i \geqslant \log |\lambda|$.

Conversely, pick a norm $\| \cdot \|_{e,c}$ such that $c_j - c_i \geqslant \log |\lambda|$ and let $x = \sum_i \lambda_i e_i$. By the ultrametric inequality, we have $e^{c_i} |\lambda_i - \lambda\lambda_j| \leqslant \max\{e^{c_i} |\lambda_i|; e^{c_i} |\lambda||\lambda_j|\}$, and the assumption $c_j - c_i \geqslant \log |\lambda|$ implies that $e^{c_i} |\lambda_i - \lambda\lambda_j| \leqslant \max\{e^{c_i} |\lambda_i|; e^{c_j} |\lambda_j|\}$, so that $e^{c_i} |\lambda_i - \lambda\lambda_j| \leqslant \max_{1 \leqslant \ell \leqslant d} e^{c_\ell} |\lambda_\ell|$. Therefore we obtain that $u_{ij}(\lambda). \| x \|_{e,c} \leqslant \| x \|_{e,c}$ for any vector x. Replacing λ by $-\lambda$ and x by $u_{ij}(-\lambda).x$, we finally see that the norms $u_{ij}(\lambda). \| \cdot \|_{e,c}$ and $\| \cdot \|_{e,c}$ are the same when $c_j - c_i \geqslant \log |\lambda|$. We have thus proved that the fixed-point set of $u_{ij}(\lambda)$ in $\widetilde{A_e}$ is the closed half-space $D_\lambda = \{\| \cdot \|_{e,c} : c_j - c_i \geqslant \log |\lambda|\}$.

It follows from this that $\widetilde{A_e} \cap u_{ij}(\lambda).\widetilde{A_e}$ contains D_λ. Assume that $\widetilde{A_e} \cap u_{ij}(\lambda).\widetilde{A_e} \supsetneq D_\lambda$ in order to obtain a contradiction. This would provide norms $\| \cdot \|$ and $\| \cdot \|'$ in $\widetilde{A_e} - D_\lambda$ with the property that $u_{ij}(\lambda). \| \cdot \| = \| \cdot \|'$. But we note that a norm in $\widetilde{A_e} - D_\lambda$ is characterized by its orthogonal projection onto the boundary hyperplane ∂D_λ and by its distance to ∂D_λ. Since $u_{ij}(\lambda)$ is an isometry which fixes D_λ we conclude that $\| \cdot \| = \| \cdot \|'$, which is in contradiction with the fact that the fixed-point set of $u_{ij}(\lambda)$ in $\widetilde{A_e}$ is exactly D_λ. $\qquad\square$

3 Special Linear Groups, Berkovich and Drinfeld Spaces

We ended the previous section by an elementary construction of the building of special linear groups over discretely valued non-Archimedean field. The general-ization to an arbitrary reductive group over such a field is significantly harder and requires the full development of Bruhat-Tits, which will be the topic of Sect. 4. Before diving into the subtleties of buildings construction, we keep for a moment the particular case of special linear groups and describe a realization of their buildings in the framework of Berkovich's analytic geometry, which leads very naturally to a compactification of those buildings. The general picture, namely Berkovich realizations and compactifications of general Bruhat-Tits buildings will be dealt with in Sect. 5).

Roughly speaking understanding the realization (resp. compactification) described below of the building of a special linear group amounts to understanding (homothety classes of) norms on a non-Archimedean vector space (resp. their degenerations), using the viewpoint of multiplicative seminorms on the corresponding symmetric algebra.

A useful reference for Berkovich theory is [Tem14]. *Unless otherwise indicated, we assume in this section that k is a local field.*

3.1 Drinfeld Upper Half Spaces and Berkovich Affine and Projective Spaces

Let V be a finite-dimensional vector space over k, and let $S^\bullet V$ be the symmetric algebra of V. It is a graded k-algebra of finite type. Every choice of a basis $v_0, \ldots, v_d$ of V induces an isomorphism of $S^\bullet V$ with the polynomial ring over k in $d + 1$ indeterminates. The affine space $\mathbf{A}(V)$ is defined as the spectrum $\mathrm{Spec}(S^\bullet V)$, and the projective space $\mathbf{P}(V)$ is defined as the projective spectrum $\mathrm{Proj}(S^\bullet V)$. These algebraic varieties give rise to analytic spaces in the sense of Berkovich, which we briefly describe below.

3.1.1 Drinfeld Upper Half-Spaces in Analytic Projective Spaces

As a topological space, the Berkovich affine space $\mathbf{A}(V)^{\mathrm{an}}$ is the set of all multiplicative seminorms on $S^\bullet V$ extending the absolute value on k together with the topology of pointwise convergence. The Berkovich projective space $\mathbf{P}(V)^{\mathrm{an}}$ is the quotient of $\mathbf{A}(V)^{\mathrm{an}} - \{0\}$ modulo the equivalence relation $\sim$ defined as follows: $\alpha \sim \beta$, if and only if there exists a constant $c > 0$ such that for all f in $S^n V$ we have $\alpha(f) = c^n \beta(f)$. There is a natural PGL(V)-action on $\mathbf{P}(V)^{\mathrm{an}}$ given by $g\alpha = \alpha \circ g^{-1}$. From the viewpoint of Berkovich geometry, Drinfeld upper half-spaces can be introduced as follows [Ber95].

Definition 3.1 We denote by Ω the complement of the union of all k-rational hyperplanes in $\mathbf{P}(V)^{\mathrm{an}}$. The analytic space Ω is called Drinfeld upper half space.

Our next goal is now to mention some connections between the above analytic spaces and the Euclidean buildings defined in the previous section.

3.1.2   Retraction Onto the Bruhat-Tits Building

Let α be a point in $\mathbf{A}(V)^{an}$, i.e. α is a multiplicative seminorm on $S^{\bullet}V$. If α is not contained in any k-rational hyperplane of $\mathbf{A}(V)$, then by definition α does not vanish on any element of $S^1 V = V$. Hence the restriction of the seminorm α to the degree one part $S^1 V = V$ is a norm. Recall that the Goldman-Iwahori space $\mathscr{N}(V, k)$ is defined as the set of all non-Archimedean norms on V, and that $\mathscr{X}(V, k)$ denotes the quotient space after the homothety relation (2.2.1). Passing to the quotients we see that restriction of seminorms induces a map

$$\tau : \Omega \longrightarrow \mathscr{X}(V, k).$$

If we endow the Goldman-Iwahori space $\mathscr{N}(V, k)$ with the coarsest topology, so that all evaluation maps on a fixed $v \in V$ are continuous, and $\mathscr{X}(V, k)$ with the quotient topology, then τ is continuous. Besides, it is equivariant with respect to the action of $PGL(V, k)$. We refer to [RTW12, §3] for further details.

3.1.3   Embedding of the Building (Case of the Special Linear Group)

Let now γ be a non-trivial norm on V. By Proposition 2.20, there exists a basis $e_0, \ldots, e_d$ of V which is adapted to γ, i.e. we have

$$\gamma\left(\sum_i \lambda_i e_i\right) = \max_i\{\exp(c_i)|\lambda_i|\}$$

for some real numbers $c_0, \ldots, c_d$. We can associate to γ a multiplicative seminorm $j(\gamma)$ on $S^{\bullet}V$ by mapping the polynomial $\sum_{I=(i_0,\ldots,i_d)} a_I e_0^{i_0} \ldots e_d^{i_d}$ to $\max_I\{|a_I| \exp(i_0 c_0 + \ldots + i_d c_d)\}$. Passing to the quotients, we get a continuous map

$$j : \mathscr{X}(V, k) \longrightarrow \Omega$$

satisfying $\tau\big(j(\alpha)\big) = \alpha$.

Hence j is injective and a homeomorphism onto its image. Therefore the map j can be used to realize the Euclidean building $\mathscr{X}(V, k)$ as a subset of a Berkovich analytic space. This observation is due to Berkovich, who used it to determine the automorphism group of Ω [Ber95].

Remark 3.2 In this remark, we remove the assumption that k is local and we recall that the building $\mathscr{X}(V, k)$ consists of homothety classes of *diagonalizable* norms on V (Theorem 2.25). Assuming $\dim(V) = 2$ for simplicity, we want to rely on analytic geometry to prove the existence of non-diagonalizable norms on V for some k.

The map $j : \mathscr{X}(V, k) \to \mathbf{P}^1(V)^{an}$ can be defined without any assumption on k. Given any point $x \in \mathscr{X}(V, k)$, we pick a basis $\mathbf{e} = (e_0, e_1)$ diagonalizing x and define $j(x)$ to be the multiplicative norm on $S^{\bullet}(V)$ mapping an homogeneous polynomial $f = \sum_\nu a_\nu e_0^{\nu_0} e_1^{\nu_1}$ to $\max_\nu \{|a_\nu| \cdot |e_0|(x)^{\nu_0} \cdot |e_1|(x)^{\nu_1}\}$. We do not distinguish between $\mathscr{X}(V, k)$ and its image by j in $\mathbf{P}(V)^{an}$, which consists only of points of types 2 and 3 (this follows from [Tem14, 3.2.11]).

Let us now consider the subset Ω' of $\Omega = \mathbf{P}(V)^{an} - \mathbf{P}(V)(k)$ consisting of multiplicative norms on $S^\bullet(V)$ whose restriction to V is diagonalizable. The map τ introduced above is well-defined on Ω' by $\tau(z) = z_{|V}$. This gives a continuous retraction of Ω' onto $\mathscr{X}(V, k)$. The inclusion $\Omega' \subset \Omega$ is strict in general, i.e. if k is not local. For example, assume that $k = \mathbf{C}_p$ is the completion of an algebraic closure of $\mathbf{Q}_p$; this non-Archimedean field is algebraically closed but not spherically complete. In this situation, Ω contains a point z of type 4 [Tem14, 2.3.13], which we can approximate by a sequence (x_n) of points in $\mathscr{X}(V, k)$ (this is the translation of the fact that z corresponds to a decreasing sequence of closed balls in k with empty intersection [Tem14, 2.3.11.(iii)]). Now, if $z \in \Omega'$, then $\tau(z) = \tau(\lim x_n) = \lim \tau(x_n) = \lim x_n$ and therefore z belongs to $\mathscr{X}(V, k)$. Since the latter set contains only points of type 2 or 3, this cannot happen and $z \notin \Omega'$; in particular, the restriction of z to V produces a norm which is not diagonalizable.

3.2 A First Compactification

Let us now turn to compactification of the building $\mathscr{X}(V, k)$. We give an outline of the construction and refer to [RTW12, §3] for additional details. The generalization to arbitrary reductive groups is the subject of Sect. 5.2. Recall that we assume that k is a local field.

3.2.1 The Space of Seminorms

Let us consider the set $\mathscr{S}(V, k)$ of non-Archimedean seminorms on V. Every non-Archimedean seminorm γ on V induces a norm on the quotient space $V/\ker(\gamma)$. Hence using Proposition 2.20, we find that there exists a basis $e_0, \ldots, e_d$ of V such that $\alpha(\sum_i \lambda_i e_i) = \max_i \{r_i \,|\lambda_i|\}$ for some non-negative real numbers $r_0, \ldots, r_d$. In this case we say that α is diagonalized by $\mathbf{e}$. Note that in contrast to Definition 2.18 we do no longer assume that the r_i are non-zero and hence exponentials.

We can extend γ to a seminorm $j(\gamma)$ on the symmetric algebra $S^\bullet V \simeq k[e_0, \ldots, e_d]$ as follows:

$$j(\gamma)\left(\sum_{I=(i_0,\ldots,i_d)} a_I e_0^{i_0} \ldots e_d^{i_d} \right) = \max\{|a_I| r_0^{i_0} \ldots r_d^{i_d}\}.$$

We denote by $\overline{\mathscr{X}(V, k)}$ the quotient of $\mathscr{S}(V, k) \setminus \{0\}$ after the equivalence relation $\sim$ defined as follows: $\alpha \sim \beta$ if and only if there exists a real constant c with $\alpha = c\beta$. We equip $\mathscr{S}(V, k)$ with the topology of pointwise convergence and

$\overline{\mathscr{X}}(V,k)$ with the quotient topology. Then the association $\gamma \mapsto j(\gamma)$ induces a continuous and PGL(V,k)-equivariant map

$$j : \overline{\mathscr{X}}(V,k) \to \mathbf{P}(V)^{\mathrm{an}}$$

which extends the map $j : \mathscr{X}(V,k) \to \Omega$ defined in the previous section.

3.2.2   Extension of the Retraction Onto the Building

Moreover, by restriction to the degree one part $S^1 V = V$, a non-zero multiplicative seminorm on $S^\bullet V$ yields an element in $\mathscr{S}(V,k) - \{0\}$. Passing to the quotients, this induces a map

$$\tau : \mathbf{P}(V)^{\mathrm{an}} \longrightarrow \overline{\mathscr{X}}(V,k)$$

extending the map $\tau : \Omega \to \mathscr{X}(V,k)$ defined in Sect. 3.1.

As in Sect. 3.1, we see that $\tau \circ j$ is the identity on $\overline{\mathscr{X}}(V,k)$, which implies that j is injective: it is a homeomorphism onto its (closed) image in $\mathbf{P}(V)^{\mathrm{an}}$. Since $\mathbf{P}(V)^{\mathrm{an}}$ is compact, we deduce that the image of j, and hence $\overline{\mathscr{X}}(V,k)$, is compact. As $\mathscr{X}(V,k)$ is an open subset of $\overline{\mathscr{X}}(V,k)$, the latter space is a compactification of the Euclidean building $\mathscr{X}(V,k)$; it was studied in [Wer04].

3.2.3   The Strata of the Compactification

For every proper subspace W of V we can extend norms on V/W to non-trivial seminorms on V by composing the norm with the quotient map $V \to V/W$. This defines a continuous embedding

$$\mathscr{X}(V/W,k) \to \overline{\mathscr{X}}(V,k).$$

Since every seminorm on V is induced in this way from a norm on the quotient space after its kernel, we find that $\overline{\mathscr{X}}(V,k)$ is the disjoint union of all Euclidean buildings $\mathscr{X}(V/W,k)$, where W runs over all proper subspaces of V. Hence our compactification of the Euclidean building $\mathscr{X}(V,k)$ is a union of Euclidean buildings of smaller rank.

3.3   Topology and Group Action

We will now investigate the convergence of sequences in $\overline{\mathscr{X}}(V,k)$ and deduce that it is compact. We also analyze the action of the group SL(V,k) on this space.

3.3.1 Degeneracy of Norms to Seminorms and Compactness

Let us first investigate convergence to the boundary of $\mathscr{X}(V,k)$ in $\overline{\mathscr{X}}(V,k) = (\mathscr{S}(V,k)\backslash\{0\})/\sim$. We fix a basis $\mathbf{e} = (e_0,\dots,e_d)$ of V and denote by $\mathbb{A}_{\mathbf{e}}$ the corresponding apartment associated to the norms diagonalized by $\mathbf{e}$ as in Definition 2.18. We denote by $\overline{\mathbb{A}}_{\mathbf{e}} \subset \overline{\mathscr{X}}(V,k)$ all classes of *seminorms* which are diagonalized by $\mathbf{e}$.

We say that a sequence $(z_n)_n$ of points in $\overline{\mathbb{A}}_{\mathbf{e}}$ is distinguished, if there exists a non-empty subset I of $\{0,\dots,d\}$ such that:

(a) For all $i \in I$ and all n we have $z_n(e_i) \neq 0$.

(b) for any $i, j \in \mathrm{I}$, the sequence $\left(\frac{z_n(e_j)}{z_n(e_i)}\right)_n$ converges to a positive real number;

(c) for any $i \in \mathrm{I}$ and $j \in \{0,\dots,d\} - \mathrm{I}$, the sequence $\left(\frac{z_n(e_j)}{z_n(e_i)}\right)_n$ converges to 0.

Here we define $\left(\frac{z_n(e_i)}{z_n(e_j)}\right)_n$ as $\left(\frac{x_n(e_i)}{x_n(e_j)}\right)_n$ for an arbitrary representative $x_n \in \mathscr{S}(V,k)$ of the class z_n. Note that this expression does not depend on the choice of the representative x_n.

Lemma 3.3 *Let $(z_n)_n$ be a distinguished sequence of points in $\overline{\mathbb{A}}_{\mathbf{e}}$. Choose some element $i \in I$. We define a point z_∞ in $\mathscr{S}(V,k)$ as the homothety class of the seminorm x_∞ defined as follows:*

$$
x_\infty(e_j) = \begin{cases} \lim_n \left(\frac{z_n(e_j)}{z_n(e_i)}\right) & \text{if } j \in \mathrm{I} \\ 0 & \text{if } j \notin \mathrm{I} \end{cases}
$$

and $x_\infty(\sum_j a_j e_j) = \max |a_j| x_\infty(e_j)$. Then z_∞ does not depend on the choice of i, and the sequence $(z_n)_n$ converges to z_∞ in $\overline{\mathscr{X}}(V,k)$.

Proof Let x_n be a representative of z_n in $\mathscr{S}(V,k)$. For i, j and ℓ contained in I we have

$$
\lim_n \left(\frac{x_n(e_j)}{x_n(e_\ell)}\right) = \lim_n \left(\frac{x_n(e_j)}{x_n(e_i)}\right) \lim_n \left(\frac{x_n(e_i)}{x_n(e_\ell)}\right),
$$

which implies that the definition of the seminorm class z_∞ does not depend on the choice of $i \in I$.

The convergence statement is obvious, since the seminorm x_n is equivalent to $(x_n(e_i))^{-1}x_n$.   □

Hence the distinguished sequence of norm classes $(z_n)_n$ considered in the Lemma converges to a seminorm class whose kernel W_I is spanned by all e_j with $j \notin I$. Therefore the limit point z_∞ lies in the Euclidean building $\mathscr{X}(V/W_I)$ at the boundary.

Note that the preceding Lemma implies that $\overline{\mathbb{A}}_{\mathbf{e}}$ is the closure of $\mathbb{A}_{\mathbf{e}}$ in $\overline{\mathscr{X}}(V,k)$. Namely, consider $z \in \overline{\mathbb{A}}_{\mathbf{e}}$, i.e. z is the class of a seminorm x on V which is diagonalizable by **e**. For every n we define a norm x_n on V by

$$x_n(e_i) = \begin{cases} x(e_i), & \text{if } x(e_i) \neq 0 \\ \frac{1}{n}, & \text{if } x(e_i) = 0 \end{cases}$$

and

$$x_n(\sum_i a_i e_i) = \max_i |a_i| x_n(e_i).$$

Then the sequence of norm classes $x_n = [z_n]_\sim$ in $\mathbb{A}_\mathbf{e}$ is distinguished with respect to the set $I = \{i : x(e_i) \neq 0\}$ and it converges towards z.

We will now deduce from these convergence results that the space of seminorms is compact. We begin by showing that $\overline{\mathbb{A}}_\mathbf{e}$ is compact.

Proposition 3.4 *Let $(z_n)_n$ be a sequence of points in $\overline{\mathbb{A}}_\mathbf{e}$. Then $(z_n)_n$ has a converging subsequence.*

Proof Let x_n be seminorms representing the points z_n. By the box principle, there exists an index $i \in \{0, \ldots, d\}$ such that after passing to a subsequence we have

$$x_n(e_i) \gtrsim x_n(e_j) \text{ for all } j = 0, \ldots, d, n \geq 0.$$

In particular we have $x_n(e_i) > 0$. For each $j = 0, \ldots, d$ we look at the sequence

$$\beta(j)_n = \frac{x_n(e_j)}{x_n(e_i)}$$

which lies between zero and one. In particular, $\beta(i)_n = 1$ is constant.

After passing to a subsequence of $(z_n)_n$ we may—and shall—assume that all sequences $\beta(j)_n$ converge to some $\beta(j)$ between zero and one. Let I be the set of all $j = 0, \ldots, n$ such that $\beta(j) > 0$. Then a subsequence of $(z_n)_n$ is distinguished with respect to I, hence it converges by Lemma 3.3. $\qquad\square$

Since $\overline{\mathbb{A}}_\mathbf{e}$ is metrizable, the preceding proposition shows that $\overline{\mathbb{A}}_\mathbf{e}$ is compact.

We can now describe the SL(V, k)-action on the seminorm compactification of the Goldman-Iwahori space of V. As before, we fix a basis $\mathbf{e} = (e_0, \ldots, e_n)$.

Let o be the homothety class of the norm on V defined by

$$\left| \sum_{i=0}^d a_i e_i \right| (o) = \max_{0 \leq i \leq d} |a_i|$$

and let

$$P_o = \{g \in \mathrm{SL}(V, k) \, ; \, g \cdot o \sim o\}$$

be the stabilizer of o. It follows from Proposition 2.28 that $P_o = SL_{d+1}(k^0)$ with respect to the basis $\mathbf{e}$.

Lemma 3.5 *The map* $P_o \times \overline{\mathbb{A}}_{\mathbf{e}} \rightarrow \overline{\mathscr{X}}(V, k)$ *given by the* $SL(V, k)$-*action is surjective.*

Proof Let $[x]_\sim$ be an arbitrary point in $\overline{\mathscr{X}}(V, k)$. The seminorm x is diagonalizable with respect to some basis $\mathbf{e}'$ of V. A similar argument as in the proof of Proposition 2.29 shows that there exists an element $h \in P_o$ such that hx lies in $\overline{\mathbb{A}}_{\mathbf{e}}$ (actually hx lies in the closure, taken in the seminorm compactification, of a Weyl chamber with tip o). $\qquad\square$

The group P_o is closed and bounded in $SL(V, k)$, hence compact. Since $\overline{\mathbb{A}}_{\mathbf{e}}$ is compact by Proposition 3.4, the previous Lemma proves that $\overline{\mathscr{X}}(V, k)$ is compact.

3.3.2   Isotropy Groups

Let z be a point in $\overline{\mathscr{X}}(V, k)$ represented by a seminorm x with kernel $W \subset V$. By $\overline{x}$ we denote the norm induced by x on the quotient space V/W. By definition, an element $g \in PGL(V, k)$ stabilizes z if and only if one (and hence any) representative h of g in $GL(V, k)$ satisfies $hx \sim x$, i.e. if and only if there exists some $\gamma > 0$ such that

$$x(h^{-1}(v)) = \gamma x(v) \text{ for all } v \in V. \qquad (*)$$

This is equivalent to saying that h preserves the subspace W and that the induced element $\overline{h}$ in $GL(V/W, k)$ stabilizes the equivalence class of the norm $\overline{x}$ on V/W. Hence we find

$$\text{Stab}_{PGL(V,k)}(z) = \{h \in GL(V, k) : h \text{ fixes the subspace } W \text{ and } \overline{h}\overline{x} \sim \overline{x}\}/k^\times.$$

Let us now assume that z is contained in the compactified apartment $\overline{\mathbb{A}}_{\mathbf{e}}$ given by the basis $\mathbf{e}$ of V. Then there are non-negative real numbers $r_0, r_1, \ldots, r_d$ such that

$$x\left(\sum_i a_i e_i\right) = \max_i\{r_i|a_i|\}.$$

The space W is generated by all vectors e_i such that $r_i = 0$. We assume that if r_i and r_j are both non-zero, the element r_j/r_i is contained in the value group $|k^*|$ of k. In this case, if h stabilizes z, we find that $\gamma = x(h^{-1}e_i)/r_i$ is contained in the value group $|k^*|$ of k, i.e. we have $\gamma = |\lambda|$ for some $\lambda \in k^*$. Hence $(\lambda h)x = x$. Therefore in this case the stabilizer of z in $PGL(V, k)$ is equal to the image of

$$\{h \in GL(V, k) : h \text{ fixes the subspace } W \text{ and } \overline{h}\overline{x} = \overline{x}\}$$

under the natural map from $GL(V, k)$ to $PGL(V, k)$.

Lemma 3.6 *Assume that z is contained in the closed Weyl chamber $\overline{\mathscr{C}} = \{[x]_\sim \in$ $\overline{\mathbb{A}}_{\mathbf{e}} : x(e_0) \leqslant x(e_1) \leqslant \ldots \leqslant x(e_d)\}$, i.e. using the previous notation we have $r_0 \leqslant r_1 \leqslant \ldots \leqslant r_d$. Let $d - \mu$ be the index such that $r_{d-\mu} = 0$ and $r_{d-\mu+1} > 0$. (If z is contained in $\mathbb{A}_{\mathbf{e}}$, then we put $\mu = d + 1$.) Then the space W is generated by the vectors e_i with $i \leqslant d - \mu$. We assume as above that r_j / r_i is contained in $|k^*|$ if $i > d - \mu$ and $j > d - \mu$. Writing elements in $GL(V)$ as matrices with respect to the basis $\mathbf{e}$, we find that $\mathrm{Stab}_{PGL(V,k)}(z)$ is the image of*

$$\left\{ \begin{pmatrix} A & B \\ 0 & D \end{pmatrix} \in GL_{d+1}(k) : D = (\delta_{ij}) \in GL_\mu(k), \right.$$

$$\left. with \ |\det(D)| = 1 \ and \ |\delta_{ij}| \leqslant r_j / r_i \ for \ all \ i, j \ \leqslant \mu. \right\}$$

in $PGL(V, k)$.

Proof This follows directly from the previous considerations combined with Proposition 2.28 which describes the stabilizer groups of norms. □

The isotropy groups of the boundary points can also be described in terms of tropical linear algebra, see [Wer11, Proposition 3.8].

4 Bruhat-Tits Theory

We provide now a very short survey of Bruhat-Tits theory. The main achievement of the latter theory is the existence, for many reductive groups over valued fields, of a combinatorial structure on the rational points; the geometric viewpoint on this is the existence of a strongly transitive action of the group of rational points on a Euclidean building. Roughly speaking, one half of this theory (the one written in [BrT72]) is of geometric and combinatorial nature and involves group actions on Euclidean buildings: the existence of a strongly transitive action on such a building is abstractly shown to come from the fact that the involved group can be endowed with the structure of a valued root group datum. The other half of the theory (the one written in [BrT84]) shows that in many situations, in particular when the valued ground field is local, the group of rational points can be endowed with the structure of a valued root group datum. This is proved by subtle arguments of descent of the ground field and the main tool for this is provided by group schemes over the ring of integers of the valued ground field. Though it concentrates on the case when the ground field is local, the survey article [Tit79] written some decades ago by J. Tits himself is still very useful. For a very readable introduction covering also the case of a non-discrete valuation, we recommend the recent text of Rousseau [Rou09].

4.1 Reductive Groups

We introduce a well-known family of algebraic groups which contains most classical groups (i.e., groups which are automorphism groups of suitable bilinear or sesquilinear forms, possibly taking into account an involution, see [Wei60] and [KMRT98]). The ground field here is not assumed to be endowed with any absolute value. The structure theory for rational points is basically due to C. Chevalley over algebraically closed fields [Che05], and to A. Borel and J. Tits over arbitrary fields [BT65] (assuming a natural isotropy hypothesis).

4.1.1 Basic Structure Results

We first need to recall some facts about general linear algebraic groups, up to quoting classical conjugacy theorems and showing how to exhibit a root system in a reductive group. Useful references are A. Borel's [Bor91], Demazure-Gabriel's [DG70] and W.C. Waterhouse's [Wat79] books.

Linear algebraic groups. By convention, unless otherwise stated, an "algebraic group" in what follows means a "linear algebraic group over some ground field"; being a linear algebraic group amounts to being a smooth affine algebraic group scheme (over a field). Any algebraic group can be embedded as a closed subgroup of some group GL(V) for a suitable vector space over the same ground field (see [Wat79, 3.4] for a scheme-theoretic statement and [Bor91, Prop. 1.12 and Th. 5.1] for stronger statements but in a more classical context).

Let G be such a group over a field k; we will often consider the group $G_{k^a} = G \otimes_k k^a$ obtained by extension of scalars from k to an algebraic closure k^a.

Unipotent and diagonalizable groups. We say that $g \in G(k^a)$ is *unipotent* if it is sent to a unipotent matrix in some (*a posteriori* any) linear embedding $\varphi : G \hookrightarrow GL(V)$: this means that $\varphi(g) - \mathrm{id}_V$ is nilpotent. The group G_{k^a} is called *unipotent* if so are all its elements; this is equivalent to requiring that the group fixes a vector in any finite-dimensional linear representation as above [Wat79, 8.3].

The group G is said to be a *torus* if it is connected and if G_{k^a} is *diagonalizable*, which is to say that the algebra of regular functions $\mathcal{O}(G_{k^a})$ is generated by the characters of G_{k^a}, i.e., $\mathcal{O}(G_{k^a}) \simeq k^a[X(G_{k^a})]$ [Bor91, §8]. Here, $X(G_{k^a})$ denotes the finitely generated abelian group of characters $G_{k^a} \to \mathbf{G}_{\mathbf{m},k^a}$ and $k^a[X(G_{k^a})]$ is the corresponding group algebra over k^a. A torus G defined over k (also called a *k-torus*) is said to be *split over* k if the above condition holds over k, i.e., if its coordinate ring $\mathcal{O}(G)$ is the group algebra of the abelian group $X^*(G) = \mathrm{Hom}_{k,\mathbf{Gr}}(G, \mathbf{G}_{m,k})$. In other words, a torus is a connected group of simultaneously diagonalizable matrices in any linear embedding over k^a as above, and it is k-split if it is diagonalized in any linear embedding defined over k [Wat79, §7].

Lie algebra and adjoint representation. One basic tool in studying connected real Lie groups is the Lie algebra of such a group, that is its tangent space at the identity element [Bor91, 3.5]. In the context of algebraic groups, the definition is the same but it is conveniently introduced in a functorial way [Wat79, §12].

Definition 4.1 Let G be a linear algebraic group over a field k. The *Lie algebra* of G, denoted by $\mathscr{L}(G)$, is the kernel of the natural map $G(k[\varepsilon]) \to G(k)$, where $k[\varepsilon]$ is the k-algebra $k[X]/(X)$ and ε is the class of X; in particular, we have $\varepsilon^2 = 0$.

We have $k[\varepsilon] = k \oplus k\varepsilon$ and the natural map above is obtained by applying the functor of points G to the map $k[\varepsilon] \to k$ sending ε to 0. The bracket for $\mathscr{L}(G)$ is given by the commutator (group-theoretic) operation [Wat79, 12.2–12.3].

Example 4.2 For $G = GL(V)$, we have $\mathscr{L}(G) \simeq End(V)$ where $End(V)$ denotes the k-vector space of all linear endomorphisms of V. More precisely, any element of $\mathscr{L}(GL(V))$ is of the form $id_V + u\varepsilon$ where $u \in End(V)$ is arbitrary. The previous isomorphism is simply given by $u \mapsto id_V + u\varepsilon$ and the usual Lie bracket for $End(V)$ is recovered thanks to the following computation in $GL(V, k[\varepsilon])$: $[id_V + u\varepsilon, id_V + u'\varepsilon] = id_V + (uu' - u'u)\varepsilon$—note that the symbol $[.,.]$ on the left hand-side stands for a commutator and that $(id_V + u\varepsilon)^{-1} = id_V - u\varepsilon$ for any $u \in End(V)$.

An important tool to classify algebraic groups is the adjoint representation [Bor91, 3.13].

Definition 4.3 Let G be a linear algebraic group over a field k. The *adjoint representation* of G is the linear representation $Ad : G \to GL(\mathscr{L}(G))$ defined by $Ad(g) = int(g)|_{\mathscr{L}(G)}$ for any $g \in G$, where $int(g)$ denotes the conjugacy $h \mapsto ghg^{-1}$—the restriction makes sense since, for any k-algebra R, both $G(R)$ and $\mathscr{L}(G) \otimes_k R$ can be seen as subgroups of $G(R[\varepsilon])$ and the latter one is normal.

In other words, the adjoint representation is the linear representation provided by differentiating conjugacies at the identity element.

Example 4.4 For $G = SL(V)$, we have $\mathscr{L}(G) \simeq \{u \in End(V) : tr(u) = 0\}$ and $Ad(g).u = gug^{-1}$ for any $g \in SL(V)$ and any $u \in \mathscr{L}(G)$. In this case, we write sometimes $\mathscr{L}(G) = \mathfrak{sl}(V)$.

Reductive and semisimple groups. The starting point for the definition of reductive and semisimple groups consists of the following existence statement [Bor91, 11.21].

Proposition/Definition 4.5 *Let G be a linear algebraic group over a field k.*

(i) *There is a unique connected, unipotent, normal subgroup in G_{k^a}, which is maximal for these properties. It is called the* unipotent radical *of G and is denoted by $\mathscr{R}_u(G)$.*

(ii) *There is a unique connected, solvable, normal subgroup in G_{k^a}, which is maximal for these properties. It is called the* radical *of G and is denoted by $\mathscr{R}(G)$.*

The statement for the radical is implied by a finite dimension argument and the fact that the Zariski closure of the product of two connected, normal, solvable subgroups is again connected, normal and solvable. The unipotent radical is also the unipotent part of the radical: indeed, in a connected solvable group (such as $\mathscr{R}(G)$), the unipotent elements form a closed, connected, normal subgroup [Wat79, 10.3]. Note that by their very definitions, the radical and the unipotent radical depend only on the k^a-group G_{k^a} and not on the k-group G.

Definition 4.6 Let G be a linear algebraic group over a field k.

(i) We say that G is *reductive* if we have $\mathscr{R}_u(G) = \{1\}$.
(ii) We say that G is *semisimple* if we have $\mathscr{R}(G) = \{1\}$.

Example 4.7 For any finite-dimensional k-vector space V, the group GL(V) is reductive and SL(V) is semisimple. The groups Sp_{2n} and $SO(q)$ (for most quadratic forms q) are semisimple.

If, taking into account the ground field k, we had used a rational version of the unipotent radical, then we would have obtained a weaker notion of reductivity. More precisely, it makes sense to introduce the *rational unipotent radical*, denoted by $\mathscr{R}_{u,k}(G)$ and contained in $\mathscr{R}_u(G)$, defined to be the unique maximal connected, unipotent subgroup in G *defined over k*. Then G is called k-*pseudo-reductive* if we have $\mathscr{R}_{u,k}(G) = \{1\}$. This class of groups is considered in the note [BT78], it is first investigated in some of J. Tits' lectures ([Tit92] and [Tit93]). A thorough study of pseudo-reductive groups and their classification are written in B. Conrad, O. Gabber and G. Prasad's book [CGP10] (an available survey is for instance [Rém11]).

In the present paper, we are henceforth interested in reductive groups.

Parabolic subgroups. The notion of a parabolic subgroup can be defined for any algebraic group [Bor91, 11.2] but it is mostly useful to understand the structure of rational points of reductive groups.

Definition 4.8 Let G be a linear algebraic group over a field k and let H be a Zariski closed subgroup of G. The subgroup H is called *parabolic* if the quotient space G/H is a complete variety.

It turns out *a posteriori* that for a parabolic subgroup H, the variety G/H is actually a projective one; in fact, it can be shown that H is a parabolic subgroup if and only if it contains a *Borel subgroup*, that is a maximal connected solvable subgroup [Bor91, 11.2].

Example 4.9 For $G = GL(V)$, the parabolic subgroups are, up to conjugacy, the various groups of upper triangular block matrices (there is one conjugacy class for each "shape" of such matrices, and these conjugacy classes exhaust all possibilities).

The completeness of the quotient space G/H is used to have fixed-points for some subgroup action, which eventually provides conjugacy results as stated below [DG70, IV, §4, Th. 3.2].

Conjugacy theorems. We finally mention a few results which, among other things, allow one to formulate classification results independent from the choices made to construct the classification data (e.g., the root system—see Sect. 4.1.2 below) [Bor91, Th. 20.9].

Theorem 4.10 *Let* G *be a linear algebraic group over a field* k. *We assume that* G *is reductive.*

(i) *Minimal parabolic* k-*subgroups are conjugate over* k, *that is any two minimal parabolic* k-*subgroups are conjugate by an element of* $G(k)$.
(ii) *Accordingly, maximal* k-*split tori are conjugate over* k.

For the rational conjugacy of tori, the reductivity assumption can be dropped and simply replaced by a connectedness assumption; this more general result is stated in [CGP10, C.2]. In the general context of connected groups (instead of reductive ones), one has to replace parabolic subgroups by *pseudo-parabolic* ones in order to obtain similar conjugacy results [CGP10, Th. C.2.5].

4.1.2 Root System, Root Datum and Root Group Datum

The notion of a root system is studied in detail in [Bou07, IV]. It is a combinatorial notion which encodes part of the structure of rational points of semisimple groups. It also provides a nice uniform way to classify semisimple groups over algebraically closed fields up to isogeny, a striking fact being that the outcome does not depend on the characteristic of the field [Che05].

In order to state the more precise classification of reductive groups up to isomorphism (over algebraically closed fields, or more generally of split reductive groups), it is necessary to introduce a more subtle notion, namely that of a *root datum*:

Definition 4.11 Let X be a finitely generated free abelian group; we denote by $X^{\vee}$ its **Z**-dual and by $\langle \cdot, \cdot \rangle$ the duality bracket. Let R and $R^{\vee}$ be two finite subsets in X and $X^{\vee}$, respectively. We assume we are given a bijection $^{\vee} : \alpha \mapsto \alpha^{\vee}$ from R onto $R^{\vee}$. We have thus, for each $\alpha \in R$, endomorphisms

$$s_{\alpha} : x \mapsto x - \langle x, \alpha^{\vee} \rangle \alpha \qquad \text{and} \qquad s_{\alpha}^{\vee} : x^{\vee} \mapsto x^{\vee} - \langle \alpha, x^{\vee} \rangle \alpha^{\vee}$$

of the groups X and $X^{\vee}$, respectively. The quadruple $\Psi = (X, R, X^{\vee}, R^{\vee})$ is said to be a *root datum* if it satisfies the following axioms:

RD 1   For each $\alpha \in R$, we have $\langle \alpha, \alpha^{\vee} \rangle = 2$.
RD 2   For each $\alpha \in R$, we have $s_{\alpha}(R) = R$ and $s_{\alpha}^{\vee}(R^{\vee}) = R^{\vee}$.

This formulation is taken from [Spr98]. The elements of R are called *roots* and the reflections s_{α} generate a finite group W of automorphisms of X, called the *Weyl group* of Ψ.

Let Q denote the subgroup of X generated by R. Up to introducing $V = Q \otimes_{\mathbf{Z}} \mathbf{R}$ and choosing a suitable W-invariant scalar product on V, we can see that R is a root system in the following classical sense:

Definition 4.12 Let V be a finite-dimensional real vector space endowed with a scalar product which we denote by $\langle \cdot, \cdot \rangle$. We say that a finite subset R of $V - \{0\}$ is a *root system* if it spans V and if it satisfies the following two conditions.

RS 1   To each $\alpha \in R$ is associated a reflection s_α which stabilizes R and switches α and $-\alpha$.

RS 2   For any $\alpha, \beta \in R$, we have $s_\alpha(\beta) - \beta \in \mathbf{Z}\alpha$.

The Weyl group of Ψ is identified with the group of automorphisms of V generated by the euclidean reflections s_α.

Let R be a root system. For any subset Δ in R, we denote by $R^+(\Delta)$ the set of roots which can be written as a linear combination of elements of Δ with non-negative integral coefficients. We say that Δ is a *basis* for the root system R if it is a basis of V and if we have $R = R^+(\Delta) \sqcup R^-(\Delta)$, where $R^-(\Delta) = -R^+(\Delta)$. Any root system admits a basis and any two bases of a given root system are conjugate under the Weyl group action [Bou07, VI.1.5, Th. 2]. When Δ is a basis of the root system R, we say that $R^+(\Delta)$ is a *system of positive roots* in R; the elements in Δ are then called *simple roots* (with respect to the choice of Δ). The *coroot* associated to α is the linear form $\alpha^\vee$ on V defined by $\beta - s_\alpha(\beta) = \alpha^\vee(\beta)\alpha$; in particular, we have $\alpha^\vee(\alpha) = 2$.

Example 4.13 Here is a well-known concrete construction of the root system of type A_n. Let $\mathbf{R}^{n+1} = \bigoplus_{i=0}^{n} \mathbf{R}\varepsilon_i$ be equipped with the standard scalar product, making the basis (ε_i) orthonormal. Let us introduce the hyperplane $V = \{\sum_i \lambda_i \varepsilon_i : \sum_i \lambda_i = 0\}$; we also set $\alpha_{i,j} = \varepsilon_i - \varepsilon_j$ for $i \neq j$. Then $R = \{\alpha_{i,j} : i \neq j\}$ is a root system in V and $\Delta = \{\alpha_{i,i+1} : 0 \leqslant i \leqslant n - 1\}$ is a basis of it for which $R^+(\Delta) = \{\alpha_{i,j} : i < j\}$. The Weyl group is isomorphic to the symmetric group $\mathscr{S}_{n+1}$; canonical generators leading to a Coxeter presentation are for instance given by transpositions $i \leftrightarrow i + 1$.

Root systems in reductive groups appear as follows. The restriction of the adjoint representation (Definition 4.3) to a maximal k-split torus T is simultaneously diagonalizable over k, so that we can write:

$$\mathscr{L}(G) = \bigoplus_{\varphi \in X^*(T)} \mathscr{L}(G)_\varphi \quad \text{where}$$

$$\mathscr{L}(G)_\varphi = \{v \in \mathscr{L}(G) : \mathrm{Ad}(t).v = \varphi(t)v \text{ for all } t \in T(k)\}.$$

The normalizer $N = N_G(T)$ acts on $X^*(T)$ via its action by (algebraic) conjugation on T, hence it permutes algebraic characters. The action of the centralizer $Z = Z_G(T)$ is trivial, so the group actually acting is the finite quotient $N(k)/Z(k)$ (finiteness follows from rigidity of tori [Wat79, 7.7], which implies that the identity component N° centralizes T; in fact, we have $N^\circ = Z$ since centralizers of tori in connected groups are connected).

$$R = R(T, G) = \{\varphi \in X^*(T) : \mathscr{L}(G)_\varphi \neq \{0\}\}.$$

It turns out that [Bor91, Th. 21.6]:

1. the **R**-linear span of R is $V = Q \otimes_{\mathbf{Z}} \mathbf{R}$, where $Q \subset X^*(T)$ is generated by R;
2. there exists an $N(k)/Z(k)$-invariant scalar product V;
3. the set R is a root system in V for this scalar product;
4. the Weyl group W of this root system is isomorphic to $N(k)/Z(k)$.

Moreover one can go further and introduce a root datum by setting $X = X^*(T)$ and by taking $X^{\vee}$ to be the group of all 1-parameter multiplicative subgroups of T. The roots α have just been introduced before, but distinguishing the coroots among the cocharacters in $X^{\vee}$ is less immediate (over algebraically closed fields or more generally in the split case, they can be defined by means of computation in copies of SL_2 attached to roots as in Example 4.15 below). We won't need this but, as already mentioned, in the split case the resulting quadruple $\Psi = (X, R, X^{\vee}, R^{\vee})$ characterizes, up to isomorphism, the reductive group we started with (see [SGA3] or [Spr98, Chap. 9 and 10]).

One of the main results of Borel-Tits theory [BT65] about reductive groups over arbitrary fields is the existence of a very precise combinatorics on the groups of rational points. The definition of this combinatorial structure—called a *root group datum*—is given in a purely group-theoretic context. It is so to speak a collection of subgroups and classes modulo an abstract subgroup T, all indexed by an abstract root system and subject to relations which generalize and formalize the presentation of SL_n (or of any split simply connected simple group) over a field by means of elementary unipotent matrices [Ste68]. This combinatorics for the rational points $G(k)$ of an isotropic reductive group G is indexed by the root system $R(T, G)$ with respect to a maximal split torus which we have just introduced; in that case, the abstract group T of the root group datum can be chosen to be the group of rational points of the maximal split torus (previously denoted by the same letter!). More precisely, the axioms of a root group datum are given in the following definition, taken from [BrT72, 6.1].[1]

Definition 4.14 Let R be a root system and let G be a group. Assume that we are given a system $\left(T, (U_{\alpha}, M_{\alpha})_{\alpha \in R}\right)$ where T and each U_{α} is a subgroup in G, and each M_{α} is a right congruence class modulo T. We say that this system is a *root group datum* of type R for G if it satisfies the following axioms:

RGD 1   For each $\alpha \in R$, we have $U_{\alpha} \neq \{1\}$.

RGD 2   For any $\alpha, \beta \in R$, the commutator group $[U_{\alpha}, U_{\beta}]$ is contained in the group generated by the groups U_{γ} indexed by roots γ in $R \cap (\mathbf{Z}_{>0}\alpha + \mathbf{Z}_{>0}\beta)$.

RGD 3   If both α and 2α belong to R, we have $U_{2\alpha} \subsetneq U_{\alpha}$.

RGD 4   For each $\alpha \in R$, the class M_{α} satisfies $U_{-\alpha}\text{-}\{1\} \subset U_{\alpha}M_{\alpha}U_{\alpha}$.

---

[1]Though the notion is taken from [BrT72], the terminology we use here is not the exact translation of the French "donnée radicielle" as used in [loc. cit.]: this is because we have already used the terminology "root datum" in the combinatorial sense of [SGA3]. Accordingly, we use the notation of [SGA3] instead of that of [BrT72], e.g. a root system is denoted by the letter R instead of Φ.

RGD 5   For any $\alpha, \beta \in R$ and each $n \in M_\alpha$, we have $n U_\beta n^{-1} = U_{s_\alpha(\beta)}$.

RGD 6   We have $TU^+ \cap U^- = \{1\}$, where $U^\pm$ is the subgroup generated by the groups U_α indexed by the roots α of sign $\pm$.

The groups U_α are called the *root groups* of the root group datum.

This list of axioms is probably a bit hard to swallow in one stroke, but the example of GL_n can help a lot to have clearer ideas. We use the notation of Example 4.13 (root system of type A_n).

Example 4.15 Let $G = GL_{n+1}$ and let T be the group of invertible diagonal matrices. To each root $\alpha_{i,j}$ of the root system R of type A_n, we attach the subgroup of elementary unipotent matrices $U_{i,j} = U_{\alpha_{i,j}} = \{I_n + \lambda E_{i,j} : \lambda \in k\}$. We can see easily that $N_G(T) = \{\text{monomial matrices}\}$, that $Z_G(T) = T$ and finally that $N_G(T)/Z_G(T) \simeq \mathscr{S}_{n+1}$. Acting by conjugation, the group $N_G(T)$ permutes the subgroups $U_{\alpha_{i,j}}$ and the corresponding action on the indexing roots is nothing else than the action of the Weyl group $\mathscr{S}_{n+1}$ on R. The axioms of a root group datum follow from matrix computation, in particular checking axiom *(RGD4)* can be reduced to the following equality in SL_2:

$$\begin{pmatrix} 1 & 0 \\ 1 & 1 \end{pmatrix} = \begin{pmatrix} 1 & 1 \\ 0 & 1 \end{pmatrix} \begin{pmatrix} 0 & -1 \\ 1 & 0 \end{pmatrix} \begin{pmatrix} 1 & 1 \\ 0 & 1 \end{pmatrix}.$$

We can now conclude this subsection by quoting a general result due to A. Borel and J. Tits (see [BrT72, 6.1.3 c)] and [BT65]).

Theorem 4.16 *Let G be a connected reductive group over a field k, which we assume to be k-isotropic. Let T be a maximal k-split torus in G, which provides a root system $R = R(T, G)$.*

  (i) *For every root $\alpha \in R$ the connected subgroup U_α with Lie algebra $\mathscr{L}(G)_\alpha$ is unipotent; moreover it is abelian or two-step nilpotent.*

 (ii) *The subgroups $T(k)$ and $U_\alpha(k)$, for $\alpha \in R$, are part of a root group datum of type R in the group of rational points $G(k)$.*

Recall that we say that a reductive group is *isotropic over k* if it contains a non-central k-split torus of positive dimension (the terminology is inspired by the case of orthogonal groups and is compatible with the notion of isotropy for quadratic forms [Bor91, 23.4]). Note finally that the structure of a root group datum implies that (coarser) of a Tits system (also called BN-pair) [Bou07, IV.2], which was used by J. Tits to prove, in a uniform way, the simplicity (modulo center) of the groups of rational points of isotropic simple groups (over sufficiently large fields) [Tit64].

4.1.3   Valuations on Root Group Data

Bruhat-Tits theory deals with isotropic reductive groups over valued fields. As for Borel-Tits theory (arbitrary ground field), a substantial part of this theory can also be summed up in combinatorial terms. This can be done by using the notion of a *valuation* of a root group datum, which formalizes among other things the fact that the valuation of the ground field induces a filtration on each root group. The definition is taken from [BrT72, 6.2].

Definition 4.17 Let G be an abstract group and let $(T, (U_\alpha, M_\alpha)_{\alpha \in R})$ be a root group datum of type R for it. A *valuation* of this root group datum is a collection $\varphi = (\varphi_\alpha)_{\alpha \in R}$ of maps $\varphi_\alpha : U_\alpha \to \mathbf{R} \cup \{\infty\}$ satisfying the following axioms.

V0   For each $\alpha \in R$, the image of φ_α contains at least three elements.

V1   For each $\alpha \in R$ and each $\ell \in \mathbf{R} \cup \{\infty\}$, the preimage $\varphi_\alpha^{-1}([\ell; \infty])$ is a subgroup of U_α, which we denote by $U_{\alpha,\ell}$; moreover we require $U_{\alpha,\infty} = \{1\}$.

V2   For each $\alpha \in R$ and each $n \in M_\alpha$, the map $u \mapsto \varphi_{-\alpha}(u) - \varphi_\alpha(nun^{-1})$ is constant on the set $U^*_{-\alpha} = U_{-\alpha} - \{1\}$.

V3   For any $\alpha, \beta \in R$ and $\ell, \ell' \in \mathbf{R}$ such that $\beta \notin -\mathbf{R}_+\alpha$, the commutator group $[U_{\alpha,\ell}, U_{\beta,\ell'}]$ lies in the group generated by the groups $U_{p\alpha+q\beta, p\ell+q\ell'}$ where $p, q \in \mathbf{Z}_{>0}$ and $p\alpha + q\beta \in R$.

V4   If both α and 2α belong to R, the restriction of $2\varphi_\alpha$ to $U_{2\alpha}$ is equal to $\varphi_{2\alpha}$.

V5   For $\alpha \in R$, $u \in U_\alpha$ and $u', u'' \in U_{-\alpha}$ such that $u'uu'' \in M_\alpha$, we have $\varphi_{-\alpha}(u') = -\varphi_\alpha(u)$.

The geometric counterpart to this list of technical axioms is the existence, for a group endowed with a valued root group datum, of a Euclidean building (called the *Bruhat-Tits building* of the group) on which it acts by isometries with remarkable transitivity properties [BrT72, §7]. For instance, if the ground field is discretely valued, the corresponding building is simplicial and a fundamental domain for the group action is given by a maximal (poly)simplex, also called an *alcove* (in fact, if the ground field is discretely valued, the existence of a valuation on a root group datum can be conveniently replaced by the existence of an affine Tits system [BrT72, §2]). As already mentioned, the action turns out to be strongly transitive, meaning that the group acts transitively on the inclusions of an alcove in an apartment (Remark 2.5 in Sect. 2.1.1).

4.2   Bruhat-Tits Buildings

The purpose of this subsection is to roughly explain how Bruhat-Tits theory attaches a Euclidean building to a suitable reductive group defined over a valued field. This Bruhat-Tits building comes equipped with a strongly transitive action by the group of rational points, which in turn implies many interesting decompositions of the group. The latter decompositions are useful for instance to doing harmonic analysis or studying various classes of linear representations of the group. We roughly explain the descent method used to perform the construction of the Euclidean buildings, and finally mention how some integral models are both one of the main tools and an important outcome of the theory.

4.2.1 Foldings and Gluing

We keep the (connected) semisimple group G, defined over the (now, complete valued non-Archimedean) field k but from now on, *we assume for simplicity that k is a local field (i.e., is locally compact) and we denote by ω its discrete valuation,* normalized so that $\omega(k^\times) = \mathbf{Z}$. Hence $\omega(\cdot) = -\log_q|\cdot|$, where $q > 1$ is a generator of the discrete group $|k^\times|$.

We also assume that G contains a k-split torus of positive dimension: this is an isotropy assumption over k already introduced at the end of Sect. 4.1.2 (in this situation, this algebraic condition is equivalent to the topological condition that the group of rational points $G(k)$ is non-compact [Pra82]). In order to associate to G a Euclidean building on which $G(k)$ acts strongly transitively, according to [Tit79] we need two things:

1. a model, say Σ, for the apartments;
2. a way to glue many copies of Σ altogether in such a way that they will satisfy the incidence axioms of a building (2.1.1).

Model for the apartment. References for what follows are [Tit79, §1] or [Lan96, Chapter I]. Let T be a maximal k-split torus in G and let $X_*(T)$ denote its group of 1-parameter subgroups (or *cocharacters*). As a first step, we set $\Sigma_{\text{vect}} = X_*(T) \otimes_{\mathbf{Z}} \mathbf{R}$.

Proposition 4.18 *There exists an affine space Σ with underlying vector space Σ_{vect}, equipped with an action by affine transformations $\nu : N(k) = N_G(T)(k) \to \text{Aff}(\Sigma)$ and having the following properties.*

(i) *There is a scalar product on Σ such that $\nu\big(N(k)\big)$ is an affine reflection group.*
(ii) *The vectorial part of this group is the Weyl group of the root system $R = R(T, G)$.*
(iii) *The translation (normal) subgroup acts cocompactly on Σ, it is equal to $\nu\big(Z(k)\big)$ and the vector $\nu(z)$ attached to an element $z \in Z(k)$ is defined by $\chi\big(\nu(z)\big) = -\omega\big(\chi(z)\big)$ for any $\chi \in X^*(T)$.*

If we go back to the example of GL(V) acting by precomposition on the space of classes of norms $\mathscr{X}(V, k)$ as described in Sect. 2.2, we can see the previous statement as a generalization of the fact, mentioned in Sect. 2.2.3, that for any basis **e** of V, the group $N_\mathbf{e}$ of monomial matrices with respect to **e** acts on the apartment $\mathbb{A}_\mathbf{e}$ as $\mathscr{S}_d \ltimes \mathbf{Z}^d$ where $d = \dim(V)$.

Filtrations and gluing. Still for this special case, we saw (Proposition refprop - folding) that any elementary unipotent matrix $u_{ij}(\lambda) = I_d + \lambda E_{ij}$ fixes pointwise a closed half-apartment in $\mathbb{A}_\mathbf{e}$ bounded by a hyperplane of the form $\{c_i - c_j = \text{constant}\}$ (the constant depends on the valuation $\omega(\lambda)$ of the additive parameter λ), the rest of the apartment $\mathbb{A}_\mathbf{e}$ associated to **e** being "folded" away from $\mathbb{A}_\mathbf{e}$.

In order to construct the Bruhat-Tits building in the general case, the gluing equivalence will impose this folding action for unipotent elements in root groups; this will be done by taking into account the "valuation" of the unipotent element under consideration. What formalizes this is the previous notion of a valuation for a root group datum (Definition 4.17), which provides a filtration on each root group. For further details, we refer to the motivations given in [Tit79, 1.1–1.4]. It is not straightforward to perform this in general, but it can be done quite explicitly when the group G is *split* over k (i.e., when it contains a maximal torus which is k-split). For the general case, one has to go to a (finite, separable) extension of the ground field splitting G and then to use subtle descent arguments. The main difficulty for the descent step is to handle at the same time Galois actions on the split group and on its "split" building in order to descend the ground field both for the valuation of the root group datum and at the geometric level (see Sect. 4.2.2 for slightly more details).

Let us provisionally assume that G is split over k. Then each root group $U_\alpha(k)$ is isomorphic to the additive group of k and for any such group $U_\alpha(k)$ we can use the valuation of k to define a decreasing filtration $\{U_\alpha(k)_\ell\}_{\ell \in \mathbf{Z}}$ satisfying:

$$\bigcup_{\ell \in \mathbf{Z}} U_\alpha(k)_\ell = U_\alpha(k) \quad \text{and} \quad \bigcap_{\ell \in \mathbf{Z}} U_\alpha(k)_\ell = \{1\},$$

and further compatibilities, namely the axioms of a valuation (Definition 4.17) for the root group datum structure on $G(k)$ given by Borel-Tits theory (Theorem 4.16)—the latter root group datum structure in the split case is easy to obtain by means of Chevalley bases [Ste68] (see remark below). For instance, in the case of the general linear group, this can be merely done by using the parameterizations

$$(k, +) \simeq U_{\alpha_{i,j}}(k) = \{\mathrm{id} + \lambda E_{i,j} : \lambda \in k\}.$$

Remark 4.19 Let us be slightly more precise here. For a split group G, each root group U_α is k-isomorphic to the additive group $\mathbf{G}_a$, and the choice of a Chevalley basis of $\mathrm{Lie}(G)$ determines a set of isomorphisms $\{p_\alpha : U_\alpha \to \mathbf{G}_a\}_{\alpha \in R}$. It is easily checked that the collection of maps

$$\varphi_\alpha : U_\alpha(k) \xrightarrow{\ p_\alpha\ } \mathbf{G}_a(k) \xrightarrow{\ \omega\ } \mathbf{R}$$

defines a valuation on the root group datum $(T(k), (U_\alpha(k), M_\alpha))$.

For each $\ell \in \mathbf{R}$, the condition $|p_\alpha| \leqslant q^{-s}$ defines an *affinoid* subgroup $U_{\alpha,s}$ in U_α^{an} such that $U_\alpha(k)_\ell = U_{\alpha,s}(k)$ for any $s \in (\ell - 1, \ell]$. The latter identity holds after replacement of k by any finite extension k', *as long as we normalize the valuation of k' in such a way that is extends the valuation on k.* This shows that Bruhat-Tits filtrations on root groups, in the split case at this stage, comes from a decreasing, exhaustive and separated filtration of U_α^{an} by affinoid subgroups $\{U_{\alpha,s}\}_{s \in \mathbf{R}}$.

Let us consider again the apartment Σ with underlying vector space $\Sigma_{\mathrm{vect}} = X_*(T) \otimes_{\mathbf{Z}} \mathbf{R}$. We are interested in the affine linear forms $\alpha + \ell$ ($\alpha \in R$, $\ell \in \mathbf{Z}$). We fix an origin, say o, such that $(\alpha + 0)(o) = 0$ for any root $\alpha \in R$. We have "level sets" $\{\alpha + \ell = 0\}$ and "positive half-spaces" $\{\alpha + \ell \geqslant 0\}$ bounded by them.

For each $x \in \Sigma$, we set $N_x = \mathrm{Stab}_{G(k)}(x)$ (using the action ν of Proposition 4.18) and for each root α we denote by $U_\alpha(k)_x$ the biggest subgroup $U_\alpha(k)_\ell$ such that $x \in \{\alpha + \ell \geqslant 0\}$ (i.e. ℓ is minimal for the latter property). At last, we define P_x to be the subgroup of $G(k)$ generated by N_x and by $\{U_\alpha(k)_x\}_{\alpha \in R}$. We are now in position to define a binary relation, say $\sim$, on $G(k) \times \Sigma$ by:

$$(g, x) \sim (h, y) \quad \Longleftrightarrow \quad \text{there exists } n \in N_G(T)(k) \text{ such that}$$
$$y = \nu(n).x \text{ and } g^{-1}hn \in P_x.$$

Construction of the Bruhat-Tits buildings. This relation is exactly what is needed in order to glue together copies of Σ and to finally obtain the desired Euclidean building.

Theorem 4.20 *The relation $\sim$ is an equivalence relation on the set $G(k) \times \Sigma$ and the quotient space $\mathscr{B} = \mathscr{B}(G, k) = (G(k) \times \Sigma) / \sim$ is a Euclidean building whose apartments are isomorphic to Σ and whose Weyl group is the affine reflection group $W = \nu(N(k))$. Moreover the $G(k)$-action by left multiplication on the first factor of $G(k) \times \Sigma$ induces an action of $G(k)$ by isometries on $\mathscr{B}(G, k)$.*

Notation According to Definition 2.9, copies of Σ in $\mathscr{B}(G, k)$ are called *apartments*; they are the maximal flat (i.e., euclidean) subspaces. Thanks to $G(k)$-conjugacy of maximal split tori 4.10, apartments of $\mathscr{B}(G, k)$ are in bijection with maximal split tori of G. Therefore, we will speak of the *apartment of a maximal split torus* S of G and write $A(S, k)$. By construction, this is an affine space under the **R**-vector space $\mathrm{Hom}_{\mathbf{Ab}}(X^*(S), \mathbf{R})$.

Reference for the proof As already explained, the difficulty is to check the axioms of a valuation (Def 4.17) for a suitable choice of filtrations on the root groups of a Borel-Tits root group datum (Th. 4.16). Indeed, the definition of the equivalence relation $\sim$, hence the construction of a suitable Euclidean building, for a valued root group datum can be done in this purely abstract context [BrT72, §7]. The existence of a valued root group datum for reductive groups over suitable valued (not necessarily complete) fields was announced in [BrT72, 6.2.3 c)] and was finally settled in the second IHÉS paper (1984) by F. Bruhat and J. Tits [BrT84, Introduction and Th. 5.1.20]. □

One way to understand the gluing equivalence relation $\sim$ is to see that it prescribes stabilizers. Actually, it can eventually be proved that *a posteriori* we have:

$$\Sigma^{U_{\alpha,\ell}(k)} = \{\alpha + \ell \geqslant 0\} \quad \text{and} \quad \mathrm{Stab}_{G(k)}(x) = P_x \text{ for any } x \in \mathscr{B}.$$

A more formal way to state the result is to say that to each valued root group datum on a group is associated a Euclidean building, which can be obtained by a gluing equivalence relation defined as above [BrT72, §7].

Example 4.21 In the case when G $=$ SL(V), it can be checked that the building obtained by the above method is equivariantly isomorphic to the Goldman-Iwahori space $\mathscr{X}(V, k)$ [BrT72, 10.2].

4.2.2 Descent and Functoriality

Suitable filtrations on root groups so that an equivalence relation $\sim$ as above can be defined do not always exist. Moreover, even when things go well, the way to construct the Bruhat-Tits building is not by first exhibiting a valuation on the root group datum given by Borel-Tits theory and then by using the gluing relation $\sim$. As usual in algebraic group theory, one has first to deal with the split case, and then to apply various and difficult arguments of descent of the ground field. Bruhat and Tits used a two-step descent, allowing a fine description of smooth integral models of the group associated with facets. A one-step descent was introduced by Rousseau in his thesis [Rou77], whose validity in full generality now follows from recent work connected to Tits' Center Conjecture ([Str11]).

Galois actions. More precisely, one has to find a suitable (finite) Galois extension k'/k such that G splits over k' (or, at least, *quasi-splits* over k', i.e. admits a Borel subgroup defined over k') and, which is much more delicate, which enables one:

1. to define a Gal(k'/k)-action by isometries on the "(quasi)-split" building $\mathscr{B}(G, k')$;
2. to check that a building for G(k) lies in the Galois fixed point set $\mathscr{B}(G, k')^{\mathrm{Gal}(k'/k)}$.

Similarly, the group G(k') of course admits a Gal(k'/k)-action.

Remark 4.22 Recall that, by completeness and non-positive curvature, once step 1 is settled we know that we have sufficiently many Galois-fixed points in $\mathscr{B}(G, k')$ (see the discussion of the Bruhat-Tits fixed point theorem in Sect. 2.1.3).

F. Bruhat and J. Tits found a uniform procedure to deal with various situations of that kind. The procedure described in [BrT72, 9.2] formalizes, in abstract terms of buildings and group combinatorics, how to exhibit a valued root group datum structure (resp. a Euclidean building structure) on a subgroup of a bigger group with a valued root group datum (resp. on a subspace of the associated Bruhat-Tits building). The main result [BrT72, Th. 9.2.10] says that under some sufficient conditions, the restriction of the valuation to a given sub-root group datum "descends" to a valuation and its associated Bruhat-Tits building is the given subspace. These sufficient conditions are designed to apply to subgroups and convex subspaces obtained as fixed-points of "twists" by Galois actions (and they can also be applied to non-Galois twists "à la Ree-Suzuki").

Two descent steps. As already mentioned, this needn't work over an arbitrary valued field k (even when k is complete). Moreover F. Bruhat and J. Tits do not perform the descent in one stroke, they have to argue by a two step descent.

The first step is the so-called *quasi-split* descent [BrT84, §4]. It consists in dealing with field extensions splitting an initially quasi-split reductive group. The Galois twists here (of the ambient group and building) are shown, by quite concrete arguments, to fit in the context of [BrT72, 9.2] mentioned above. This is possible thanks to a deep understanding of quasi-split groups: they can even be handled via a presentation (see [Ste68] and [BrT84, Appendice]). In fact, the largest part of the chapter about the quasi-split descent [BrT84, §4] is dedicated to another topic which will be presented below (4.2.3), namely the construction of suitable integral models (i.e. group schemes over $k°$ with generic fiber G) defined by geometric conditions involving bounded subsets in the building. The method chosen by F. Bruhat and J. Tits to obtain these integral models is by using a linear representation of G whose underlying vector space contains a suitable $k°$-lattice, but they mention themselves that this could be done by Weil's techniques of group chunks. Since then, thanks to the developments of Néron model techniques [BLR90], this alternative method has been written down [Lan96].

The second step is the so-called *étale* descent [BrT84, §5]. By definition, an étale extension, in the discretely valued case (to which we stick here), is unramified with separable residual extension; let us denote by k^{sh} the maximal étale extension of k. This descent step consists in considering situations where the semisimple k-group G is such that $G \otimes_k k^{sh}$ is quasi-split (so that, by the first step, we already have a valued root group datum and a Bruhat-Tits building for $G(k^{sh})$, together with integral structures). Checking that this fits in the geometric and combinatorial formalism of [BrT72, 9.2] is more difficult in that case. In fact, this is the place where the integral models over the valuation ring $k°$ are used, in order to find a suitable torus in G which become maximal split in $G \otimes_k k'$ for some étale extension k' of k [BrT84, Cor. 5.1.12].

Remark 4.23 In the split case, we have noticed that the Bruhat-Tits filtrations on rational points of root groups come from filtrations by affinoid subgroups (4.19). This fact holds in general and can be checked as follows: let k'/k be a finite Galois extension splitting G and consider a maximal torus T of G which splits over k' and contains a maximal split torus S. The canonical projection $X^*(T \otimes_k k') \rightarrow X^*(S \otimes_k k') \tilde{=} X^*(S)$ induces a surjective map

$$p : R(T \otimes_k k', G \otimes_k k') \longrightarrow R(S, G) \cup \{0\}$$

and there is a natural k'-isomorphism

$$\prod_{\beta \in p^{-1}(\alpha)} U_\beta \times \prod_{\beta \in p^{-1}(2\alpha)} U_\beta \simeq U_\alpha \otimes_k k'$$

for any ordering of the factor.

A posteriori, Bruhat-Tits two-step descent proves that any maximal split torus S of G is contained in a maximal torus T which splits over a finite Galois extension k'/k such that $\text{Gal}(k'/k)$ fixes a point in the apartment of $T \otimes_k k'$ in $\mathscr{B}(G, k')$. If *the valuation on k' is normalized in such a way that it extends the valuation on k,* then, for any $\ell \in \mathbf{R}$, the affinoid subgroup

$$\prod_{\beta \in p^{-1}(\alpha)} \mathrm{U}_{\beta,\ell} \times \prod_{\beta \in p^{-1}(2\alpha)} \mathrm{U}_{\beta,2\ell}$$

of the left hand side corresponds to an affinoid subgroup of the right hand side which does not depend on the ordering of the factors and is preserved by the natural action of $\mathrm{Gal}(k'|k)$; this can be checked by using calculations in [BrT72, 6.1] at the level of k'' points, for any finite extension k''/k'. By Galois descent, we obtain an affinoid subgroup $\mathrm{U}_{\alpha,\ell}$ of $\mathrm{U}_\alpha^{\mathrm{an}}$ such that

$$\mathrm{U}_{\alpha,\ell}(k) = U_\alpha(k) \cap \left(\prod_{\beta \in p^{-1}(\alpha)} \mathrm{U}_{\beta,\ell}(k') \times \prod_{\beta \in p^{-1}(2\alpha)} \mathrm{U}_{\beta,2\ell}(k') \right).$$

By [BrT84, 5.1.16 and 5.1.20], the filtrations $\{\mathrm{U}_{\alpha,\ell}(k)\}_{\ell \in \mathbf{R}}$ are induced by a valuation on the root group datum $(S(k), \{\mathrm{U}_\alpha(k)\})$.

Let us finish by mentioning why this two-step strategy is well-adapted to the case we are interested in, namely that of a semisimple group G defined over a complete, discretely valued field k with perfect residue field $\tilde{k}$: thanks to a result of R. Steinberg's [Ser94, III, 2.3], such a group is known to quasi-split over k^{sh}. Compactifications of Bruhat-Tits buildings fit in this more specific context for G and k. Indeed, the Bruhat-Tits building $\mathscr{B}(\mathrm{G}, k)$ is locally compact if and only if so is k, see the discussion of the local structure of buildings below (4.2.3). Note finally that the terminology "henselian" used in [BrT84] is a well-known algebraic generalization of "complete" (the latter "analytic" condition is the only one we consider seriously here, since we will use Berkovich geometry).

Existence of Bruhat-Tits buildings. Here is at last a general statement on existence of Bruhat-Tits buildings which will be enough for our purposes; this result was announced in [BrT72, 6.2.3 c)] and is implied by [BrT84, Th. 5.1.20].

Theorem 4.24 *Assume that k is complete, discretely valued, with perfect residue field. The root group datum of $\mathrm{G}(k)$ associated with a split maximal torus admits a valuation satisfying the conditions of Definition 4.17.*

Let us also give now an example illustrating both the statement of the theorem and the general geometric approach characterizing Bruhat-Tits theory.

Example 4.25 Let h be a Hermitian form of index 1 in three variables, say on the vector space $\mathrm{V} \simeq k^3$. We assume that h splits over a quadratic extension, say E/k, so that $\mathrm{SU}(\mathrm{V}, h)$ is isomorphic to SL_3 over E, and we denote $\mathrm{Gal}(E/k) = \{1; \sigma\}$. Then the building of $\mathrm{SU}(\mathrm{V}, h)$ can be seen as the set of fixed points for a suitable action of the Galois involution σ on the 2-dimensional Bruhat-Tits building of type $\tilde{\mathrm{A}}_2$ associated to $\mathrm{V} \otimes_k E$ as in Sect. 2.2. If k is local and if q denotes the cardinality of the residue field, then the Euclidean building $\mathscr{B}(\mathrm{SU}(\mathrm{V}, h), k)$ is a locally finite tree: indeed, it is a Euclidean building of dimension 1 because the k-rank of $\mathrm{SU}(\mathrm{V}, h)$, i.e. the dimension of maximal k-split tori, is 1. The tree is homogeneous of valency $1 + q$ when E/k is ramified, in which case the type of the group is $C\text{-}BC_1$ in Tits' classification [Tit79, p. 60, last line]. The tree is semi-homogeneous of valencies $1 + q$ and $1 + q^3$ when E/k is unramified, and then the type is $^2A_2'$ [Tit79, p. 63, line 2]. For the computation of the valencies, we refer to Sect. 4.2.3 below.

Functoriality. For our purpose (i.e. embedding of Bruhat-Tits buildings in analytic spaces and subsequent compactifications), the existence statement is not sufficient. We need a stronger result than the mere existence; in fact, we also need a good behavior of the building with respect to field extensions.

Theorem 4.26 *Whenever k is complete, discretely valued, with perfect residue field, the Bruhat-Tits building $\mathscr{B}(\mathrm{G}, K)$ depends functorially on the non-Archimedean extension K of k.*

More precisely, let us denote by $\mathrm{G} - \mathbf{Sets}$ the category whose objets are pairs $(K/k, \mathrm{X})$, where K/k is a non-Archimedean extension and X is a topological space endowed with a continuous action of $\mathrm{G}(K)$, and arrows $(K/k, \mathrm{X}) \to (K'/k, \mathrm{X}')$ are pairs (ι, f), where ι is an isometric embedding of K into K' and f is a $\mathrm{G}(K)$-equivariant and continues map from X to X'. We see the building of G as a *section* $\mathscr{B}(\mathrm{G}, -)$ of the forgetful functor

$$\mathrm{G} - \mathbf{Sets} \longrightarrow \begin{pmatrix} \text{non} - \text{Archimedean} \\ \text{extensions } K/k \end{pmatrix}.$$

Remark (Reference) It is explained in [RTW10, 1.3.4] how to deduce this from the general theory.

One word of caution is in order here. If k'/k is a Galois extension, then there is a natural action of $\mathrm{Gal}(k'/k)$ on $\mathscr{B}(\mathrm{G}, k')$ by functoriality and the smaller building $\mathscr{B}(\mathrm{G}, k)$ is contained in the Galois-fixed point set in $\mathscr{B}(\mathrm{G}, k')$. In general, this inclusion is strict, even when the group is split [Rou77, III] (see also **5.2**). However, one can show that there is equality if the extension k'/k is *tamely ramified* [**loc. cit.**] and [Pra01].

We will need to have more precise information about the behavior of apartments. As above, we assume that k is complete, discretely valued and with perfect residue field.

Definition 4.27 Let T be a maximal torus of G and let k_{T} be the minimal Galois extension of k (in some fixed algebraic closure) which splits T. We denote by $k_{\mathrm{T}}^{\mathrm{ur}}$ the maximal unramified extension of k in k_{T}.

The torus T is *well-adjusted* if the maximal split subtori of T and $\mathrm{T} \otimes_k k_{\mathrm{T}}^{\mathrm{ur}}$ are maximal split tori of G and $\mathrm{G} \otimes_k k_{\mathrm{T}}^{\mathrm{ur}}$.

Lemma 4.28 *1. Every maximal split torus S of G is contained in a well-adjusted maximal torus T.*

2. *Assume that* S *and* T *are as above, and let* K/k *be any non-Archimedean field extension which splits* T. *The embedding* $\mathscr{B}(G,k) \hookrightarrow \mathscr{B}(G,K)$ *maps* $A(S,k)$ *into* $A(T,K)$.

Proof 1. For each unramified finite Galois extension k'/k, we can find a torus $S' \subset G$ which contains S and such that $S' \otimes_k k'$ is a maximal split torus of $G \otimes_k k'$ [BrT84, Corollaire 5.1.12]. We choose a pair (k', S') such that the rank of S' is maximal, equal to the relative rank of $G \otimes_k k^{\mathrm{ur}}$; this means that $S' \otimes_k k''$ is a maximal split torus of $G \otimes_k k''$ for any unramified extension k''/k containing k'.

The centralizer of $S' \otimes_k k'$ in $G \otimes_k k'$ is a maximal torus of $G \otimes_k k'$, hence $T = Z(S')$ is a maximal torus of G. By construction, S' splits over k_T^{ur} and $S' \otimes_k k_T^{\mathrm{ur}}$ is a maximal split torus of $G \otimes_k k_T^{\mathrm{ur}}$. Since $S \subset S'$, this proves that T is well-adjusted.

2. We keep the same notation as above. The extension K/k contains k_T, hence it is enough by functoriality to check that the embedding $\mathscr{B}(G,k) \hookrightarrow \mathscr{B}(G,k_T)$ maps $A(S,k)$ into $A(T,k_T)$.

Let us consider the embeddings

$$\mathscr{B}(G,k) \hookrightarrow \mathscr{B}(G,k_T^{\mathrm{ur}}) \hookrightarrow \mathscr{B}(G,k_T).$$

The first one maps $A(S,k)$ into $A(S',k_T^{\mathrm{ur}})$ by By [BrT84, Proposition 5.1.14] and the second one maps $A(S',k_T^{\mathrm{ur}})$ into $A(T,k_T)$ by [Rou77, Théorème 2.5.6], hence their composite has the required property.   □

4.2.3 Compact Open Subgroups and Integral Structures

In what follows, we maintain the previous assumptions, in particular the group G is semisimple and k-isotropic. The building $\mathscr{B}(G,k)$ admits a strongly transitive $G(k)$-action by isometries. Moreover it is a *labelled* simplicial complex in the sense that, if d denotes the number of codimension 1 facets (called *panels*) in the closure of a given alcove, we can choose d colors and assign one of them to each panel in $\mathscr{B}(G,k)$ so that each color appears exactly once in the closure of each alcove. For some questions, it is convenient to restrict oneself to the finite index subgroup $G(k)^\bullet$ consisting of the color-preserving (or *type-preserving*) isometries in $G(k)$.

Compact open subgroups. For any facet $F \subset \mathscr{B}(G,k)$ we denote by P_F the stabilizer $\mathrm{Stab}_{G(k)}(F)$: it is a bounded subgroup of $G(k)$ and when k is local, it is a compact, open subgroup. It follows from the Bruhat-Tits fixed point theorem (2.1.3) that the conjugacy classes of maximal compact subgroups in $G(k)^\bullet$ are in one-to-one correspondence with the vertices in the closure of a given alcove. The fact that there are usually several conjugacy classes of maximal compact subgroups in $G(k)$ makes harmonic analysis more delicate than in the classical case of real Lie groups. Still, for instance thanks to the notion of a special vertex, many achievements can also be obtained in the non-Archimedean case [Mac71]. Recall that a point $x \in \mathscr{B}(G,k)$ is called *special* if for any apartment $\mathbb{A}$ containing x, the stabilizer of x in the affine Weyl group is the full vectorial part of this affine reflection group, i.e. is isomorphic to the (spherical) Weyl group of the root system R of G over k.

Integral models for some stabilizers. In what follows, we are more interested in algebraic properties of compact open subgroups obtained as facet stabilizers. The following statement is explained in [BrT84, 5.1.9].

Theorem 4.29 *For any facet $F \subset \mathcal{B}(G, k)$ there exists a smooth k°-group scheme $\mathscr{G}_F$ with generic fiber G such that $\mathscr{G}_F(k^\circ) = P_F$.*

As already mentioned, the point of view of group schemes over k° in Bruhat-Tits theory is not only an important tool to perform the descent, but it is also an important outcome of the theory. Here is an example. The "best" structure *a priori* available for a facet stabilizer is only of topological nature (and even for this, we have to assume that k is locally compact). The above models over k° provide an algebraic point of view on these groups, which allows one to define a filtration on them leading to the computation of some cohomology groups of great interest for the congruence subgroup problem, see for instance [PR84a] and [PR84b]. Filtrations are also of great importance in the representation theory of non-Archimedean Lie groups, see for instance [MP94] and [MP96].

Closed fibres and local combinatorial description of the building. We finish this brief summary of Bruhat-Tits theory by mentioning quickly two further applications of integral models for facet stabilizers.

First let us pick a facet $F \subset \mathcal{B}(G, k)$ as above and consider the associated k°-group scheme $\mathscr{G}_F$. As a scheme over k°, it has a closed fibre (so to speak obtained by reduction modulo $k^{\circ\circ}$) which we denote by $\overline{\mathscr{G}_F}$. This is a group scheme over the residue field $\tilde{k}$. It turns out that the rational points $\overline{\mathscr{G}_F}(\tilde{k})$ have a nice combinatorial structure (even though the $\tilde{k}$-group $\overline{\mathscr{G}_F}$ needn't be reductive in general); more precisely, $\overline{\mathscr{G}_F}(\tilde{k})$ has a Tits system structure (see the end of Sect. 4.1.2) with finite Weyl group. One consequence of this is that $\overline{\mathscr{G}_F}(\tilde{k})$ admits an action on a spherical building (a *spherical building* is merely a simplicial complex satisfying the axioms of Definition 2.2 with the Euclidean tiling Σ replaced by a spherical one). The nice point is that this spherical building naturally appears in the (Euclidean) Bruhat-Tits building $\mathcal{B}(G, k)$. Namely, the set of closed facets containing F is a geometric realization of the spherical building of $\overline{\mathscr{G}_F}(\tilde{k})$ [BrT84, Prop. 5.1.32]. In particular, for a complete valued field k, the building $\mathcal{B}(G, k)$ is locally finite if and only if the spherical building of $\overline{\mathscr{G}_F}(\tilde{k})$ is actually finite for each facet F, which amounts to requiring that the residue field $\tilde{k}$ be finite. Note that a metric space admits a compactification if, and only if, it is locally compact. Therefore from this combinatorial description of neighborhoods of facets, we see that *the Bruhat-Tits building $\mathcal{B}(G, k)$ admits a compactification if and only if k is a local field*.

Remark 4.30 Let us assume here that k is discretely valued. This is the context where the more classical combinatorial structure of an (affine) Tits system is relevant [Bou07, IV.2]. Let us exhibit such a structure. First, a parahoric subgroup in $G(k)$ can be defined to be the image of $(\mathscr{G}_F)^\circ(k^\circ)$ for some facet F in $\mathcal{B}(G, k)$, where

$(\mathscr{G}_F)^\circ$ denotes the identity component of $\mathscr{G}_F$ [BrT84, 5.2.8]. We also say for short that a parahoric subgroup is the connected stabilizer of a facet in the Bruhat-Tits building $\mathscr{B}(G, k)$. If G is simply connected (in the sense of algebraic groups), then the family of parahoric subgroups is the family of abstract parabolic subgroups of a Tits system with affine Weyl group [BrT84, Prop. 5.2.10]. An Iwahori subgroup corresponds to the case when F is a maximal facet. At last, if moreover k is local with residual characteristic p, then an Iwahori subgroup can be characterized as the normalizer of a maximal pro-p subgroup and an arbitrary parahoric subgroup as a subgroup containing an Iwahori subgroup.

Finally, the above integral models provide an important tool in the realization of Bruhat-Tits buildings in analytic spaces (and subsequent compactifications). Indeed, the fundamental step (see Theorem 5.5) for the whole thing consists in attaching injectively to *any* point $x \in \mathscr{B}(G, K)$ an affinoid subgroup G_x of the analytic space G^{an} attached to G, and the definition of G_x makes use of the integral models attached to vertices. But one word of caution is in order here since the connexion with integral models avoids all their subtleties! For our construction, only smooth k°-group schemes $\mathscr{G}_F$ which are *reductive* are of interest; this is not the case in general, but one can easily prove the following statement: *given a vertex* $x \in \mathscr{B}(G, k)$, *there exists a finite extension* k'/k *such that the* k'°-*group scheme* $\mathscr{G}'_x$, *attached to the point x seen as a vertex of* $\mathscr{B}(G, k')$, *is a Chevalley-Demazure group scheme over* k'°. In this situation, one can define $(G \otimes_k k')_x$ as the *generic fibre* of the formal completion of $\mathscr{G}'_x$ along its special fibre; this is a k'-affinoid subgroup of $(G \otimes_k k')^{an}$ and one invokes descent theory to produce a k-affinoid subgroup of G^{an}.

4.2.4 A Characterization of Apartments

For later use, we end this section on Bruhat-Tits theory by a useful characterization of apartments inside buildings.

Given a torus S over k, we denote by $S^1(k)$ the maximal bounded subgroup of $S(k)$. It is the subgroup of $S(k)$ defined by the equations $|\chi| = 1$, where χ runs over the character group of S.

Proposition 4.31 *Let* S *be a maximal split torus and let x be a point of* $\mathscr{B}(G, k)$. *If the residue field of k contains at least four elements, then the following conditions are equivalent:*

(i) x *belongs to the apartment* A(S, k);
(ii) x *is fixed under the action of* $S^1(k)$.

Proof Condition (i) always implies condition (ii). With our hypothesis on the cardinality of the residue field, the converse implication holds by [BrT84, Proposition 5.1.37]. □

5   Buildings and Berkovich Spaces

As above, we consider a semisimple group G over some non-Archimedean field
k. In this section, we explain how to realize the Bruhat-Tits building $\mathscr{B}(G, k)$ of
$G(k)$ in non-Archimedean analytic spaces deduced from G, and we present two
procedures that can be used to compactify Bruhat-Tits buildings in full generality;
as we pointed out before, the term "compactification" is abusive if k is not a local
field (see the discussion before Remark 4.30).

Assuming that k is locally compact, let us describe very briefly those two ways
of compactifying a building. The first is due to V. Berkovich when G is split [Ber90,
Chap. V] and it consists in two steps:

1. to define a closed embedding of the building into the analytification of the
   group (5.1);
2. to compose this closed embedding with an analytic map from the group to a
   (compact) flag variety (5.2).

By taking the closure of the image of the composed map, we obtain an
equivariant compactification which admits a Lie-theoretic description (as expected).
For instance, there is a convenient description of this $G(k)$-topological space
(convergence of sequences, boundary strata etc.) by means of invariant fans in
$(X_*(S) \otimes_{\mathbf{Z}} \mathbf{R}, W)$, where $X_*(S)$ denotes the cocharacter group of a maximal split
torus S endowed with the natural action of the Weyl group W (5.3). The finite family
of compactifications obtained in this way is indexed by $G(k)$-conjugacy classes of
parabolic subgroups.

These spaces can be recovered from a different point of view, using representa-
tion theory and the concrete compactification $\overline{\mathscr{X}}(V, k)$ of the building $\mathscr{X}(V, k)$ of
$SL(V, k)$ which was described in Sect. 3. It mimics the original strategy of I. Satake
in the case of symmetric spaces [Sat60a]: we pick a faithful linear representation of
G and, relying on analytic geometry, we embed $\mathscr{B}(G, k)$ in $\mathscr{X}(V, k)$; by taking the
closure in $\overline{\mathscr{X}}(V, k)$, we obtain our compactification.

Caution—

1. We need some functoriality assumption on the building with respect to the field:
   in a sense which was made precise after the statement of Theorem 4.26, this
   means that $\mathscr{B}(G, -)$ is functor on the category of non-Archimedean extensions
   of k.

   As explained in [RTW10, 1.3.4], these assumptions are fulfilled if k quasi-
   splits over a tamely ramified extension of k. This is in particular the case is k is
   discretely valued with perfect residue field, or if G is split.
2. There is no other restriction on the non-Archimedean field k considered in 4.1.
   From 4.2 on, we assume that k is *local*. In any case, the reader should keep in
   mind that non-local extensions of k do always appear in the study of Bruhat-Tits
   buildings from Berkovich's point of view (see Proposition 4.2).

The references for the results quoted in this section are [RTW10] and [RTW12].

5.1   Realizing Buildings Inside Berkovich Spaces

Let k be a field which is complete with respect to a non-trivial non-Archimedean absolute value. We fix a semisimple group G over k. Our first goal is to define a continuous injection of the Bruhat-Tits building $\mathscr{B}(G, k)$ in the Berkovich space G^{an} associated to the algebraic group G. Since G is affine with affine coordinate ring $\mathscr{O}(G)$, its analytification consists of all multiplicative seminorms on $\mathscr{O}(G)$ extending the absolute value on k [Tem14].

5.1.1   Non-Archimedean Extensions and Universal Points

We will have to consider infinite non-Archimedean extensions of k as in the following example.

Example 5.1 Let $\mathbf{r} = (r_1, \dots, r_n)$ be a tuple of positive real numbers such that $r_1^{i_1} \dots r_n^{i_n} \notin |k^{\times}|$ for all choices of $(i_1, \dots, i_n) \in \mathbf{Z}^n - \{0\}$. Then the k-algebra

$$
k_{\mathbf{r}} = \left\{ \sum_{I = (i_1 \dots i_n)} a_I x_1^{i_1} \dots x_n^{i_n} \in k[[x_1^{\pm 1}, \dots, x_n^{\pm 1}]] \; ; \; |a_I| r_1^{i_1} \dots r_n^{i_n} \to 0 \right.
$$

$$
\left. \text{when } |i_1| + \dots + |i_n| \to \infty \right\}
$$

is a non-Archimedean field extension of k with absolute value $|f| = \max_I \{|a_I| r_1^{i_1} \dots r_n^{i_n}\}$.

We also need to recall the notion of a *universal* point.[2] Let z be a point in G^{an}, seen as a multiplicative k-seminorm on $\mathscr{O}(G)$. For a given non-Archimedean field extension K/k, there is a natural K-seminorm $||.|| = z \otimes 1$ on $\mathscr{O}(G) \otimes_k K$, defined by

$$
||a|| = \inf \max_i |a_i(z)| \cdot |\lambda_i|
$$

where the infimum is taken over the set of all expressions $\sum_i a_i \otimes \lambda_i$ representing a, with $a_i \in \mathscr{O}(G)$ and $\lambda_i \in K$. The point z is said to be *universal* if, for any non-Archimedean field extension K/k, the above K-seminorm on $\mathscr{O}(G) \otimes_k K$ is multiplicative. One writes z_K for the corresponding point in $G^{an} \hat{\otimes}_k K$. We observe that this condition depends only on the completed residue field $\mathscr{H}(z)$ of G^{an} at z.

---

[2]This notion was introduced by Berkovich, who used the adjective *peaked* [Ber90, 5.2]. Its study was carried on by Poineau, who preferred the adjective *universal* [Poi13].

Remark 5.2 1. Obviously, points of G^{an} coming from k-rational points of G are universal.

2. Let $x \in G^{an}$ be universal. For any finite Galois extension k'/k, the canonical extension $x_{k'}$ of x to $G^{an} \otimes_k k'$ is invariant under the action of $\mathrm{Gal}(k'/k)$: indeed, the k'-norm $x \otimes 1$ on $\mathscr{O}(G) \otimes_k k'$ is Galois invariant.

3. If k is algebraically closed, Poineau proved that every point of G^{an} is universal [Poi13, Corollaire 4.10].

5.1.2 Improving Transitivity

Now let G^{an} be the Berkovich analytic space associated to the algebraic group G. Our goal is the first step mentioned in the introduction, namely the definition of a continuous injection

$$\vartheta : \mathscr{B}(G, k) \longrightarrow G^{an}.$$

We proceed as follows. For every point x in the building $\mathscr{B}(G, k)$ we construct an affinoid subgroup G_x of G^{an} such that, for any non-Archimedean extension K/k, the subgroup $G_x(K)$ of $G(K)$ is precisely the stabilizer of x in the building over K. Then we define $\vartheta(x)$ as the (multiplicative) seminorm on $\mathscr{O}(G)$ defined by taking the maximum over the compact subset G_x of G^{an}.

If the Bruhat-Tits building $\mathscr{B}(G, k)$ can be seen as non-Archimedean analogue of a Riemannian symmetric space, it is not homogeneous under $G(k)$; for example, if k is discretely valued, the building carries a polysimplicial structure which is preserved by the action of $G(k)$. There is a very simple way to remedy at this situation using field extensions, and this is where our functoriality assumption comes in.

Let us first of all recall that the notion of a special point was defined in Sect. 2, just before Definition 2.9. Its importance comes from the fact that, when G is split, the stabilizer of a special point is particularly nice (see the discussion after Theorem 5.5). As simple consequences of the definition, one should notice the following two properties: if a point $x \in \mathscr{B}(G, k)$ is special, then

– every point in the $G(k)$-orbit of x is again special;
– if moreover G is *split*, then x remains special in $\mathscr{B}(G, K)$ for any non-Archimedean field extension K/k (indeed: the local Weyl group at x over K contains the local Weyl group at x over k, and the full Weyl group of G is the same over k and over K).

We can now explain how field extensions allow to improve transitivity of the group action on the building.

Proposition 5.3 *1. Given any two points $x, y \in \mathscr{B}(G, k)$, there exists a non-Archimedean field extension K/k such that x and y, identified with points of*

$\mathscr{B}(G, K)$ *via the canonical injection* $\mathscr{B}(G, k) \hookrightarrow \mathscr{B}(G, K)$, *belong to the same orbit under* $G(K)$.

2. *For every point* $x \in \mathscr{B}(G, k)$, *there exists a non-Archimedean field extension* K/k *such that the following conditions hold:*
    (i) *The group* $G \otimes_k K$ *is split;* (ii) *The canonical injection* $\mathscr{B}(G, k) \to \mathscr{B}(G, K)$ *maps* x *to a special point.*

We give a proof of this Proposition since it is the a key result for the investigation of Bruhat-Tits buildings from Berkovich's point of view. The second assertion follows easily from the first: just pick a finite separable field extension k'/k splitting G and a special point y in $\mathscr{B}(G, k')$, then consider a non-Archimedean field extension K/k' such that x and y belong to the same $G(K)$-orbit. In order to prove the first assertion, we may and do assume that G is split. Let S denote a maximal split torus of G whose apartment $A(S, k)$ contains both x and y. As recalled in Proposition 3.17, this apartment is an affine space under $X_*(S) \otimes_{\mathbf{Z}} \mathbf{R}$, where $X_*(S)$ denotes the cocharacter space of S, and $S(k)$ acts on $A(S, k)$ by translation via a map $\nu : S(k) \to X_*(S) \otimes_{\mathbf{Z}} \mathbf{R}$. Using a basis of characters to identify $X_*(S)$ (resp. S) with $\mathbf{Z}^n$ (resp. $\mathbf{G}_m^n$), it turns out that ν is simply the map

$$k^{\times} \longrightarrow \mathbf{R}^n, (t_1, \ldots, t_n) \mapsto (-\log|t_1|, \ldots, -\log|t_n|).$$

By combining finite field extensions and transcendental extensions as described in Example 5.1, we can construct a non-Archimedean field extension K/k such that the vector $x - y \in \mathbf{R}^n$ belongs to the subgroup $\log|(K^{\times})^n|$. This implies that x and y, seen as points of $A(S, K)$, belong to the same orbit under $S(K)$, hence under $G(K)$.

Remark 5.4 If $|K^{\times}| = \mathbf{R}_{>0}$, then $G(K)$ acts transitively on $\mathscr{B}(G, K)$. However, it is more natural to work functorially than to fix arbitrarily an algebraically closed non-Archimedean extension Ω/k such that $|\Omega^{\times}| = \mathbf{R}_{>0}$.

5.1.3 Affinoid Subgroups

Let us now describe the key fact explaining the relationship between Bruhat-Tits theory and non-Archimedean analytic geometry. This result is crucial for all subsequent constructions.

Theorem 5.5 *For every point* $x \in \mathscr{B}(G, k)$ *there exists a unique k-affinoid subgroup* G_x *of* G^{an} *satisfying the following condition: for every non-Archimedean field extension* K/k, *the group* $G_x(K)$ *is the stabilizer in* $G(K)$ *of the image of* x *under the injection* $\mathscr{B}(G, k) \to \mathscr{B}(G, K)$.

The idea of the proof is the following (see [RTW10, Th. 2.1] for details). If G is split and x is a special point in the building, then the integral model $\mathscr{G}_x$ of G described in (3.2.3) is a Chevalley group scheme, and we define G_x as the generic fibre $\widehat{\mathscr{G}_{x\eta}}$ of the formal completion of $\mathscr{G}_x$ along its special fibre. This is a k-affinoid subgroup of G^{an}, and it is easy to check that it satisfies the universal property in our claim. Thanks to Proposition 5.3, we can achieve this situation after a suitable non-Archimedean extension K/k, and we apply faithfully flat descent to obtain the k-affinoid subgroup G_x [RTW10, App. A]. Let us remark that, in order to perform this descent step, it is necessary to work with an extension which is not too big (technically, the field K should be a k-affinoid algebra); since one can obtain K by combining finite extensions with the transcendental ones described in Example 5.1, this is fine.

5.1.4   Closed Embedding in the Analytic Group

The k-affinoid subgroup G_x is the Berkovich spectrum of a k-affinoid algebra A_x, i.e., G_x is the Gelfand spectrum $\mathscr{M}(A_x)$ of bounded multiplicative seminorms on A_x. This is a compact and Hausdorff topological space over which elements of A_x define non-negative real-valued functions. For any non-zero k-affinoid algebra A, one can show that its Gelfand spectrum $\mathscr{M}(A)$ contains a smallest non-empty subset, called its *Shilov boundary* and denoted $\Gamma(A)$, such that each element f of A reaches its maximum at some point in $\Gamma(A)$.

Remark 5.6   (i)   If $A = k\{T\}$ is the Tate algebra of restricted power series in one variable, then $\mathscr{M}(A)$ is Berkovich's closed unit disc and its Shilov boundary is reduced to the point o defined by the Gauss norm: for $f = \sum_{n \in \mathbf{N}} a_n T^n$, one has $|f(o)| = \max_n |a_n|$.

(ii)   Let $a \in k$ with $0 < |a| < 1$. If $A = k\{T, S\}/(ST-a)$, then $\mathscr{M}(A)$ is an annulus of modulus $|a|$ and $\Gamma(A)$ contains two points o, o': for $f = \sum_{n \in \mathbf{Z}} a_n T^n$, where $T^{-1} = a^{-1}S$, one has $|f(o)| = \max_n |a_n|$ and $|f(o')| = \max_n |a_n| \cdot |a|^n$.

(iii)   For any non-zero k-affinoid algebra A, its Shilov boundary $\Gamma(A)$ is reduced to a point if and only if the seminorm

$$A \to \mathbf{R}_{\geq 0}, \quad f \mapsto \sup_{x \in \mathscr{M}(A)} |f(x)|$$

is multiplicative.

For every point x of $\mathscr{B}(G, k)$, it turns out that the Shilov boundary of $G_x = \mathscr{M}(A_x)$ is reduced to a unique point, denoted $\vartheta(x)$. This is easily seen by combining the nice behavior of Shilov boundaries under non-Archimedean extensions, together with a natural bijection between the Shilov boundary of $\mathscr{V}_\eta$ and the set of irreducible components of $\mathscr{V} \otimes_{k^\circ} \tilde{k}$ if $\mathscr{V}$ is a normal k°-formal scheme; indeed, the smooth k°-group scheme $\mathscr{G}_x$ has a connected special fibre when it is a Chevalley group scheme. Let us also note that the affinoid subgroup G_x is completely determined by the single point $\vartheta(x)$ via

$$G_x = \{z \in G^{an} ; \ \forall f \in \mathscr{O}(G), \ |f(z)| \leq |f(\vartheta(x))|\}.$$

In this way we define the desired map

$$\vartheta : \mathscr{B}(G, k) \to G^{\mathrm{an}},$$

and we show [RTW10, Prop. 2.7] that it is injective, continuous and $G(k)$-equivariant (where $G(k)$ acts on G^{an} by conjugation). If k is a local field, ϑ induces a homeomorphism from $\mathscr{B}(G, k)$ to a closed subspace of G^{an} [RTW10, Prop. 2.11].

Finally, the map ϑ is also compatible with non-Archimedean extensions K/k, i.e., the following diagram

$$
\begin{array}{ccc}
\mathscr{B}(G, k) & \xrightarrow{\ \vartheta_K\ } & (G \otimes_k K)^{\mathrm{an}} \\[2mm]
{\scriptstyle \iota_{K/k}}\big\uparrow & & \big\downarrow {\scriptstyle p_{K/k}} \\[2mm]
\mathscr{B}(G, k) & \xrightarrow[\ \vartheta\ ]{} & G^{\mathrm{an}}
\end{array}
$$

where $\iota_{K/k}$ (resp. $p_{K/k}$) is the canonical embedding (resp. projection) is commutative. In particular, we see that this defines a *section* of $p_{K/k}$ over the image of ϑ. In fact, any point z belonging to this subset of G^{an} is universal (5.1.1) and $\vartheta_K(\iota_{K/k}(x))$ coincides with the canonical lift $\vartheta(x)_K$ of $\vartheta(x)$ to $(G \otimes_k K)^{\mathrm{an}}$ for any $x \in \mathscr{B}(G, k)$.

Moreover, if K/k is a Galois extension, then the upper arrow in the diagram is $\mathrm{Gal}(K/k)$-equivariant by [RTW10, Prop. 2.7].

5.2   Compactifying Buildings with Analytic Flag Varieties

Once the building has been realized in the analytic space G^{an}, it is easy to obtain compactifications. In order not to misuse the latter word, we assume from now one that k is *locally compact*.

5.2.1   Maps to Flag Varieties

The embedding $\vartheta : \mathscr{B}(G, k) \to G^{\mathrm{an}}$ defined in Sect. 5.1.4 can be used to compactify the Bruhat-Tits building $\mathscr{B}(G, k)$. We choose a parabolic subgroup P of G. Then the flag variety G/P is complete, and therefore the associated Berkovich space $(G/P)^{\mathrm{an}}$ is compact. Hence we can map the building to a compact space by the composition

$$\vartheta_P : \mathscr{B}(G, k) \xrightarrow{\ \vartheta\ } G^{\mathrm{an}} \longrightarrow (G/P)^{\mathrm{an}}.$$

The map ϑ_P is by construction $G(k)$-equivariant and it depends only on the $G(k)$-conjugacy class of P: we have $\vartheta_{gPg^{-1}} = g\vartheta_P g^{-1}$ for any $g \in G(k)$.

However, ϑ_P may not be injective. By the structure theory of semisimple groups, there exists a finite family of normal reductive subgroups G_i of G (each of them quasi-simple), such that the product morphism

$$\prod_i G_i \longrightarrow G$$

is a central isogeny. Then the building $\mathscr{B}(G, k)$ can be identified with the product of all $\mathscr{B}(G_i, k)$. If one of the factors G_i is contained in P, then the factor $\mathscr{B}(G_i, k)$ is squashed down to a point in the analytic flag variety $(G/P)^{an}$.

If we remove from $\mathscr{B}(G, k)$ all factors $\mathscr{B}(G_i, k)$ such that G_i is contained in P, then we obtain a building $\mathscr{B}_t(G, k)$, where t stands for the type of the parabolic subgroup P, i.e., for its $G(k)$-conjugacy class. The factor $\mathscr{B}_t(G, k)$ is mapped injectively into $(G/P)^{an}$ via ϑ_P.

Remark 5.7 If G is almost simple, then ϑ_P is injective whenever P is a proper parabolic subgroup in G; hence in this case the map ϑ_P provides an embedding of $\mathscr{B}(G, k)$ into $(G/P)^{an}$.

5.2.2 Berkovich Compactifications

Allowing compactifications of the building in which some factors are squashed down to a point, we introduce the following definition.

Definition 5.8 Let t be a $G(k)$-conjugacy class of parabolic subgroups of G. We define $\overline{\mathscr{B}}_t(G, k)$ to be the closure of the image of $\mathscr{B}(G, k)$ in $(G/P)^{an}$ under ϑ_P, where P belongs to t, and we endow this space with the induced topology. The compact space $\overline{\mathscr{B}}_t(G, k)$ is called the *Berkovich compactification of type t* of the building $\mathscr{B}(G, k)$.

Note that we obtain one compactification for each $G(k)$-conjugacy class of parabolic subgroups.

Remark 5.9 If we drop the assumption that k is locally compact, the map ϑ_P is continuous but the image of $\mathscr{B}_t(G, k)$ is not locally closed. In this case, the right way to proceed is to compactify each apartment $A_t(S, k)$ of $\mathscr{B}_t(G, k)$ by closing it in G^{an}/P^{an} and to define $\overline{\mathscr{B}}_t(G, k)$ as the union of all compactified apartments. This set is a quotient of $G(k) \times \overline{A}_t(S, k)$ and we endow it with the quotient topology [RTW10, 3.4].

5.2.3 The Boundary

Now we want to describe the boundary of the Berkovich compactifications. We fix a type t (i.e., a $G(k)$-conjugacy class) of parabolic subgroups.

Definition 5.10 Two parabolic subgroups P and Q of G are called *osculatory* if their intersection P ∩ Q is also a parabolic subgroup.

Hence P and Q are osculatory if and only if they contain a common Borel group after a suitable field extension. We can generalize this definition to semisimple groups over arbitrary base schemes. Then for every parabolic subgroup Q there is a variety $\mathrm{Osc}_t(Q)$ over k representing the functor which associates to any base scheme S the set of all parabolics of type t over S which are osculatory to Q [RTW10, Prop. 3.2].

Definition 5.11 Let Q be a parabolic subgroup. We say that Q is *t-relevant* if there is no parabolic subgroup Q' strictly containing Q such that $\mathrm{Osc}_t(Q) = \mathrm{Osc}_t(Q')$.

Let us illustrate this definition with the following example.

Example 5.12 Let G be the group SL(V), where V is a k-vector space of dimension $d + 1$. The non-trivial parabolic subgroups of G are the stabilizers of flags

$$(0 \subsetneq V_1 \subsetneq \ldots \subsetneq V_r \subsetneq V).$$

Let H be a hyperplane in V, and let P be the parabolic subgroup of SL(V) stabilizing the flag $(0 \subset H \subset V)$. We denote its type by δ. Let Q be an arbitrary parabolic subgroup, stabilizing a flag $(0 \subsetneq V_1 \subsetneq \ldots \subsetneq V_r \subsetneq V)$. Then Q and P are osculatory if and only if H contains the linear subspace V_r. This shows that all parabolic subgroups Q stabilizing flags contained in the subspace V_r give rise to the same variety $\mathrm{Osc}_\delta(Q)$. Therefore, a non-trivial parabolic is δ-relevant if and only if the corresponding flag has the form $0 \subsetneq W \subsetneq V$.

Having understood how to parametrize boundary strata, we can now give the general description of the Berkovich compactification $\overline{\mathscr{B}}_t(G, k)$. The following result is [RTW10, Theorem 4.1].

Theorem 5.13 *For every t-relevant parabolic subgroup Q, let Q_{ss} be its semisimplification (i.e., Q_{ss} is the quotient $Q/\mathscr{R}(Q)$ where $\mathscr{R}(Q)$ denotes the radical of Q). Then $\overline{\mathscr{B}}_t(G, k)$ is the disjoint union of all the buildings $\mathscr{B}_t(Q_{ss}, k)$, where Q runs over the t-relevant parabolic subgroups of G.*

The fact that the Berkovich compactifications of a given group are contained in the flag varieties of this group enables one to have natural maps between compactifications: they are the restrictions to the compactifications of (the analytic maps associated to) the natural fibrations between the flag varieties. The above combinatorics of t-relevancy is a useful tool to formulate which boundary components are shrunk when passing from a compactification to a smaller one [RTW10, Section 4.2].

Example 5.14 Let us continue the discussion in Example 5.12 by describing the stratification of $\overline{\mathscr{B}}_\delta(\mathrm{SL}(V), k)$. Any δ-relevant subgroup Q of G = SL(V) is either equal to SL(V) or equal to the stabilizer of a linear subspace $0 \subsetneq W \subsetneq V$. In the latter case Q_{ss} is isogeneous to SL(W) × SL(V/W). Now SL(W) is contained in a parabolic of type δ, hence $\mathscr{B}_\delta(Q_{ss}, k)$ coincides with $\mathscr{B}(SL(V/W), k)$. Therefore

$$\overline{\mathscr{B}}_\delta(SL(V), k) = \bigcup_{W \subsetneq V} \mathscr{B}(SL(V/W, k)),$$

where W runs over all linear subspaces $W \subsetneq V$.

Recall from 4.21 that the Euclidean building $\mathscr{B}(SL(V), k)$ can be identified with the Goldman-Iwahori space $\mathscr{X}(V, k)$ defined in 2.16. Hence $\overline{\mathscr{B}}_\delta(SL(V), k)$ is the disjoint union of all $\mathscr{X}(V/W, k)$. Therefore we can identify the seminorm compactification $\overline{\mathscr{X}}(V, k)$ from Sect. 3.2 with the Berkovich compactification of type δ.

5.3 Invariant Fans and Other Compactifications

Our next goal is to compare our approach to compactifying building with another one, developed in [Wer07] without making use of Berkovich geometry. In this work, compactified buildings are defined by a gluing procedure, similar to the one defining the Bruhat-Tits building in Theorem 4.20. In a first step, compactifications of apartments are obtained by a cone decomposition. Then these compactified apartments are glued together with the help of subgroups which turn out to be the stabilizers of points in the compactified building.

Let G be a (connected) semisimple group over k and $\mathscr{B}(G, k)$ the associated Bruhat-Tits building. We fix a maximal split torus T in G, giving rise to the cocharacter space $\Sigma_{vect} = X_*(T) \otimes \mathbf{R}$. The starting point is a faithful, geometrically irreducible representation $\rho : G \to GL(V)$ on some finite-dimensional k-vector space V.

Let $R = R(T, G) \subset X^*(T)$ be the associated root system. We fix a basis Δ of R and denote by $\lambda_0(\Delta)$ the highest weight of the representation ρ with respect to Δ. Then every other (k-rational) weight of ρ is of the form $\lambda_0(\Delta) - \sum_{\alpha \in \Delta} n_\alpha \alpha$ with coefficients $n_\alpha \geqslant 0$. We write $[\lambda_0(\Delta) - \lambda] = \{\alpha \in \Delta : n_\alpha > 0\}$. We call every such subset Y of Δ of the form $Y = [\lambda_0(\Delta) - \lambda]$ for some weight λ *admissible*.

Definition 5.15 Let $Y \subset \Delta$ be an admissible subset. We denote by C_Y^Δ the following cone in Σ_{vect}:

$$C_Y^\Delta = \left\{ x \in \Sigma_{vect} ; \begin{array}{ll} \alpha(x) = 0 & \text{for all } \alpha \in Y, \text{ and} \\ (\lambda_0(\Delta) - \lambda)(x) \geqslant 0 & \text{for all weights } \lambda \text{ such that } [\lambda_0(\Delta) - \lambda] \not\subset Y \end{array} \right\}$$

The collection of all cones C_Y^Δ, where Δ runs over all bases of the root system and Y over all admissible subsets of Δ, is a complete fan $\mathscr{F}_\rho$ in Σ_{vect}. There is a natural compactification of Σ_{vect} associated to $\mathscr{F}_\rho$, which is defined as $\overline{\Sigma}_{vect} = \bigcup_{C \in \mathscr{F}_\rho} \Sigma_{vect}/\langle C \rangle$ endowed with a topology given by tubular neighborhoods around boundary points. For details see [Wer07, Section 2] or [RTW10, Appendix B].

Fig. 1 Compactification of
an apartment: regular highest
weight

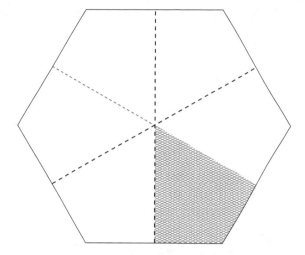

We will describe this compactification in two examples.

Example 5.16 If the highest weight of ρ is regular, then every subset Y of Δ is admissible. In this case, the fan $\mathscr{F}_\rho$ is the full Weyl fan. In the case of a root system of type A_2, the resulting compactification is shown on Fig. 1. The shaded area is a compactified Weyl chamber, whose interior contains the corresponding highest weight of ρ.

Example 5.17 Let $G = SL(V)$ be the special linear group of a $(d+1)$-dimensional k-vector space V, and let ρ be the identical representation. We look at the torus T of diagonal matrices in $SL(V)$, which gives rise to the root system $R = \{\alpha_{i,j}\}$ of type A_d described in Example 4.13. Then $\Delta = \{\alpha_{0,1}, \alpha_{1,2}, \ldots, \alpha_{d-1,d}\}$ is a basis of R and $\lambda_0(\Delta) = \varepsilon_0$ in the notation of Example 4.13. The other weights of the identical representation are $\varepsilon_1, \ldots, \varepsilon_d$. Hence the admissible subsets of Δ are precisely the sets $Y_r = \{\alpha_{0,1}, \ldots, \alpha_{r-1,r}\}$ for $r = 1, \ldots, d$, and $Y_0 = \varnothing$. Let $\eta_0, \ldots, \eta_d$ be the dual basis of $\varepsilon_0, \ldots, \varepsilon_d$. Then Σ_{vect} can be identified with $\bigoplus_{i=0}^{d} \mathbf{R}\eta_i / \mathbf{R}(\sum_i \eta_i)$, and we find

$$C_{Y_r}^\Delta = \{x = \sum_i x_i \eta_i \in \Sigma_{\text{vect}} : x_0 = \ldots = x_r \text{ and}$$

$$x_0 \geqslant x_{r+1}, x_0 \geqslant x_{r+2}, \ldots, x_0 \geqslant x_d\}/\mathbf{R}(\sum_i \eta_i)$$

The associated compactification is shown in Fig. 2. The shaded area is a compactified Weyl chamber and its codimension one face marked by an arrow contains the highest weight of ρ (with respect to this Weyl chamber).

The compactification $\overline{\Sigma}_{\text{vect}}$ induces a compactification $\overline{\Sigma}$ of the apartment $\Sigma = A(T, k)$, which is an affine space under Σ_{vect}. Note that the fan $\mathscr{F}_\rho$ and hence the

Fig. 2 Compactification of
an apartment: singular
highest weight

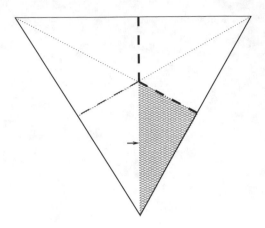

compactification $\overline{\Sigma}$ only depend on the Weyl chamber face containing the highest
weight of ρ, see [Wer07, Theorem 4.5].

Using a generalization of Bruhat-Tits theory one can define a subgroup P_x for all
$x \in \overline{\Sigma}$ such that for $x \in \Sigma$ we retrieve the groups P_x defined in section 3.2, see
[Wer07, section 3]. Note that by continuity the action of $N_G(T, k)$ on Σ extends to
an action on $\overline{\Sigma}$.

Definition 5.18 The compactification $\overline{\mathscr{B}}(G, k)_\rho$ associated to the representation ρ
is defined as the quotient of the topological space $G(k) \times \overline{\Sigma}$ by a similar equivalence
relation as in Theorem 4.20:

$$(g, x) \sim (h, y) \quad \Longleftrightarrow \quad \text{there exists } n \in N_G(T, k) \text{ such that}$$
$$y \;=\; \nu(n).x \text{ and } g^{-1}hn \in P_x.$$

The compactification of $\mathscr{B}(G, k)$ with respect to a representation with regular
highest weight coincides with the polyhedral compactification defined by Erasmus
Landvogt in [Lan96].

The connection to the compactifications defined with Berkovich spaces in
Sect. 5.2 is given by the following result, which is proved in [RTW12, Theorem
2.1].

Theorem 5.19 *Let ρ be a faithful, absolutely irreducible representation of G with
highest weight $\lambda_0(\Delta)$. Define*

$$Z = \{\alpha \in \Delta : \langle \alpha, \lambda_0(\Delta) \rangle = 0\},$$

*where $\langle \, , \, \rangle$ is a scalar product associated to the root system as in Definition 4.12.
We denote by τ the type of the standard parabolic subgroup of G associated to Z.*

Then there is a G(k)-equivariant homeomorphism

$$\overline{\mathscr{B}}(G, k)_\rho \to \overline{\mathscr{B}}_\tau(G, k)$$

restricting to the identity map on the building.

Example 5.20 In the situation of Example 5.17 we have $\lambda_0(\Delta) = \varepsilon_0$ and $Z = \{\alpha_{1,2}, \ldots, \alpha_{d-1,d}\}$. The associated standard parabolic is the stabilizer of a line. We denote its type by π. Hence the compactification of the building associated to SL(V) given by the identity representation is the one associated to type π by Theorem 5.19. This compactification was studied in [Wer01]. It is isomorphic to the seminorm compactification $\overline{\mathscr{X}}(V^\vee, k)$ of the building $\mathscr{X}(V^\vee, k)$.

5.4   Satake's Viewpoint

If G is a non-compact real Lie group with maximal compact subgroup K, Satake constructed in [Sat60b] a compactification of the Riemannian symmetric space S = G/K in the following way:

- (i) First consider the symmetric space H associated to the group PSL(n, $\mathbf{C}$) which can be identified with the space of all positive definite hermitian $n \times n$-matrices with determinant 1. Then H has a natural compactification $\overline{H}$ defined as the set of the homothety classes of all hermitian $n \times n$-matrices.
- (ii) For an arbitrary symmetric space S = G/K use a faithful representation of G to embed S into H and consider the closure of S in $\overline{H}$.

   In the setting of Bruhat-Tits buildings we can imitate this strategy in two different ways.

Functoriality of buildings—The first strategy is a generalization of functoriality results for buildings developed by Landvogt [Lan00]. Let $\rho : G \to SL(V)$ be a representation of the semisimple group G. Let S be a maximal split torus in G with normalizer N, and let A(S, k) denote the corresponding apartment in $\mathscr{B}(G, k)$. Choose a special vertex o in A(S, k). By [Lan00], there exists a maximal split torus T in SL(V) containing $\rho(S)$, and there exists a point o' in the apartment A(T, k) of T in $\mathscr{B}(SL(V), k)$ such that the following properties hold:

1. There is a unique affine map between apartments $i : A(S, k) \to A(T, k)$ such that $i(o) = o'$. Its linear part is the map on cocharacter spaces $X_*(S) \otimes_{\mathbf{Z}} \mathbf{R} \to X_*(T) \otimes_{\mathbf{Z}} \mathbf{Z}$ induced by $\rho : S \to T$.
2. The map i is such that $\rho(P_x) \subset P'_{i(x)}$ for all $x \in A(S, k)$, where P_x denotes the stabilizer of the point x with respect to the G(k)-action on $\mathscr{B}(G, k)$, and $P'_{i(x)}$ denotes the stabilizer of the point $i(x)$ with respect to the SL(V, k)-action on $\mathscr{B}(SL(V), k)$.

3. The map $\rho_* : A(S, k) \to A(T, k) \to \mathscr{B}(SL(V), k)$ defined by composing i with the natural embedding of the apartment $A(T, k)$ in the building $\mathscr{B}(SL(V), k)$ is $N(k)$-equivariant, i.e., for all $x \in A(S, k)$ and $n \in N(k)$ we have $\rho_*(nx) = \rho(n)\rho_*(x)$.

These properties imply that $\rho_* : A(S, k) \to \mathscr{B}(SL(V), k)$ can be continued to a map $\rho_* : \mathscr{B}(G, k) \to \mathscr{B}(SL(V), k)$, which is continuous and $G(k)$-equivariant. By [Lan00, 2.2.9], ρ_* is injective.

Let $\mathscr{F}$ be the fan in $X_*(T) \otimes_{\mathbf{Z}} \mathbf{R}$ associated to the identity representation, which is described in Example 5.17. It turns out that the preimage of $\mathscr{F}$ under the map $\Sigma_{\text{vect}}(S, k) \to \Sigma_{\text{vect}}(T, k)$ induced by $\rho : S \to T$ is the fan $\mathscr{F}_\rho$, see [RTW12, Lemma 5.1]. This implies that the map i can be extended to a map of compactified apartments $\overline{A}(S, k) \to \overline{A}(T, k)$. An analysis of the stabilizers of boundary points shows moreover that $\rho(P_x) \subset P'_{i(x)}$ for all $x \in \overline{A}(S, k)$, where P_x denotes the stabilizer of x in $G(k)$, and $P'_{i(x)}$ denotes the stabilizer of $i(x)$ in $SL(V, k)$ [RTW12, Lemma 5.2]. Then it follows from the definition of $\overline{\mathscr{B}}(G, k)_\rho$ in 5.18 that the embedding of buildings ρ_* may be extended to a map

$$\overline{\mathscr{B}}(G, k)_\rho \longrightarrow \overline{\mathscr{B}}(SL(V), k)_{\text{id}}.$$

It is shown in [RTW12, Theorem 5.3] that this map is a $G(k)$-equivariant homeomorphism of $\overline{\mathscr{B}}(G, k)_\rho$ onto the closure of the image of $\mathscr{B}(G, k)$ in the right hand side.

Complete flag variety—Satake's strategy of embedding the building in a fixed compactification of the building associated to $SL(V, k)$ can also be applied in the setting of Berkovich spaces. Recall from 4.21 that the building $\mathscr{B}(SL(V), k)$ can be identified with the space $\mathscr{X}(V, k)$ of (homothety classes of) non-Archimedean norms on V. In Sect. 3.2, we constructed a compactification $\overline{\mathscr{X}}(V, k)$ as the space of (homothety classes of) non-zero non-Archimedean seminorms on V and a retraction map $\tau : \mathbf{P}(V)^{\text{an}} \longrightarrow \overline{\mathscr{X}}(V, k)$.

Now let G be a (connected) semisimple k-group together with an absolutely irreducible projective representation $\rho : G \to PGL(V, k)$. Let $\text{Bor}(G)$ be the variety of all Borel groups of G. We assume for simplicity that G is quasi-split, i.e., that there exists a Borel group B defined over k; this amounts to saying that $\text{Bor}(G)(k)$ is non-empty. Then $\text{Bor}(G)$ is isomorphic to G/B. There is a natural morphism

$$\text{Bor}(G) \longrightarrow \mathbf{P}(V)$$

such that any Borel subgroup B in $G \otimes K$ for some field extension K of k is mapped to the unique K-point in $\mathbf{P}(V)$ invariant under $B \otimes_k K$, see [RTW12, Proposition 4.1]. Recall that in Sect. 5.2.1 we defined a map

$$\vartheta_\varnothing : \mathscr{B}(G, k) \to \text{Bor}(G)^{\text{an}}$$

($\varnothing$ denotes the type of Borel subgroups). Now we consider the composition

$$\mathscr{B}(G, k) \xrightarrow{\vartheta_\varnothing} \mathrm{Bor}(G)^{\mathrm{an}} \to \mathbf{P}(V)^{\mathrm{an}} \xrightarrow{\tau} \overline{\mathscr{X}}(V, k).$$

We can compactify the building $\mathscr{B}(G, k)$ by taking the closure of the image. If $\rho^\vee$ denotes the contragradient representation of ρ, then it is shown in [RTW12, 4.8 and 5.3] that in this way we obtain the compactification $\overline{\mathscr{B}}(G, k)_{\rho^\vee}$.

6 Erratum to [RTW10] and [RTW12]

Tobias Schmidt pointed out that Lemma A.10 in Appendix A to [RTW10] needed to be corrected. The problem comes from the fact that, for a finite Galois extension ℓ/k of non-Archimedean fields, the canonical map

$$\lambda : \ell \otimes_k \ell \longrightarrow \prod_{\mathrm{Gal}(\ell|k)} \ell, \quad a \otimes b \longmapsto (g(a)b)_{g \in \mathrm{Gal}(\ell|k)}$$

is not always an isometry when the left-hand side is equipped with the tensor product norm; this is the case if and only the extension is tamely ramified.

A first observation is that the algebraic isomorphism λ is an isometry with respect to spectral norms on both sides since we are working with finite dimensional k-algebras. Therefore, the question amounts to understanding when the tensor product norm $|.|_\otimes$ on $A = \ell \otimes_k \ell$ coincides with the spectral norm, which is the case if and only if $|.|_\otimes$ is power-multiplicative. Let us consider M. Temkin's *graded reduction* of $(A, |.|_\otimes)$ [Tem04], which is to say the graded ring

$$\tilde{A}_\bullet = \bigoplus_{r \in \mathbf{R}_{>0}} A_{\leqslant r}/A_{<r}$$

where $A_{\leqslant r} = \{a \in A \; ; \; |a|_\otimes \leqslant r\}$ and $A_{<r} = \{a \in A \; ; \; |a|_\otimes < r\}$. The norm $|.|_\otimes$ is power-multiplicative, hence coincides with the spectral norm, if and only if $\tilde{A}_\bullet$ is reduced. This graded ring is isomorphic to $\tilde{\ell}_\bullet \otimes_{\tilde{k}_\bullet} \tilde{\ell}_\bullet$ [Sch13, proof of Lemma 2.12] and therefore is reduced if and only if the extension of graded fields $\tilde{\ell}_\bullet/\tilde{k}_\bullet$ is separable (since ℓ/k is Galois, separability of $\tilde{\ell}/\tilde{k}$ can be checked over $\tilde{\ell}$). This is the case if and only if the field extension ℓ/k is tamely ramified [Duc13, Proposition 2.10].

By the arguments in Sect. 5.1, both the statement and the proof of Lemma 1.10 are correct if we restrict to a tamely ramified Galois extension.

Lemma A.10 was not used in [RTW10]. In the second paper [RTW12], we used it in Lemma 4.6 of [RTW12], a technical step in the proof of Proposition 4.5; therefore, both statements are proved only if the group G splits over a tamely ramified extension. Finally, the same restriction applies to Theorem 4.8 since the proof relies on Proposition 4.5.

Acknowledgements We warmly thank the organizers of the summer school "Berkovich spaces" held in Paris in July 2010. We are grateful to the referee for many comments, corrections and some relevant questions, one of which led to Proposition 5.11. Finally, we thank Tobias Schmidt for pointing out that Lemma A.10 of [RTW10] needed to be corrected.

References

[AB08]   P. Abramenko, K.S. Brown, *Buildings: Theory and Applications*. Graduate Texts in Mathematics, vol. 248 (Springer, New York, 2008)

[Bar00]   S. Barré, Immeubles de Tits triangulaires exotiques. Ann. Fac. Sci. Toulouse Math. **9**, 575–603 (2000)

[Ber90]   V.G. Berkovich, *Spectral Theory and Analytic Geometry Over Non-Archimedean Fields*. Mathematical Surveys and Monographs, vol. 33 (American Mathematical Society, Providence, RI, 1990)

[Ber95]   V.G. Berkovich, The automorphism group of the Drinfeld half-plane. C. R. Acad. Sci. Paris **321**, 1127–1132 (1995)

[BH99]   M. Bridson, A. Haefliger, *Metric Spaces of Non-positive Curvature*. Grundlehren der Mathematischen Wissenschaften, vol. 319 (Springer, New York, 1999)

[Bor91]   A. Borel, *Linear Algebraic Groups*. Graduate Texts in Mathematics, vol. 126 (Springer, New York, 1991)

[BT65]   A. Borel, J. Tits, 'Groupes réductifs. Inst. Hautes Études Sci. Publ. Math. **27**, 55–150 (1965)

[BT78]   A. Borel, J. Tits, 'Théorèmes de structure et de conjugaison pour les groupes algébriques linéaires. C. R. Acad. Sci. Paris Sér. A-B **287**(2), A55–A57 (1978)

[BLR90]   S. Bosch, W. Lütkebohmert, M. Raynaud, *Néron Models*, Ergebnisse der Mathematik und ihrer Grenzgebiete (3), vol. 21 (Springer, New York, 1990)

[Bou07]   N. Bourbaki, *Groupes et algèbres de Lie 4–6*. Éléments de mathématique. (Springer, New York, 2007)

[BrT72]   F. Bruhat, J. Tits, Groupes réductifs sur un corps local, I. Données radicielles valuées. Inst. Hautes Études Sci. Publ. Math. **41**, 5–251 (1972)

[BrT84]   F. Bruhat, J. Tits, Groupes réductifs sur un corps local, II. Schémas en groupes. Existence d'une donnée radicielle valuée. Inst. Hautes Études Sci. Publ. Math. **60**, 197–376 (1984)

[Che05]   C. Chevalley, *Classification des groupes algébriques semi-simples*. Collected Works, vol. 3 (Springer, New York, 2005)

[CGP10]   B. Conrad, O. Gabber, G. Prasad, *Pseudo-Reductive Groups*. New Mathematical Monographs, vol. 17 (Cambridge University Press, Cambridge, 2010)

[DG70]   M. Demazure, P. Gabriel, *Groupes algébriques. Tome I : Géométrie algébrique, généralités, groupes commutatifs* (Masson, 1970)

[Duc13]   A. Ducros, Toute forme modérément ramifiée d'un polydisque ouvert est triviale. Math. Z. **273**, 331–353 (2013)

[SGA3]   M. Demazure, A. Grothendieck (dir.), *Schémas en groupes, Tome III (Structure des schémas en groupes réductifs)*, nouvelle édition. Documents mathématiques, vol. 8 (SMF, Paris, 2011)

[Fur63]   H. Furstenberg, A Poisson formula for semi-simple Lie groups. Ann. Math. **77**, 335–386 (1963)

[GI63]   O. Goldman, N. Iwahori, The space of p-adic norms. Acta Math. **109**, 137–177 (1963)

[GR06]   Y. Guivarc'h, B. Rémy, Group-theoretic compactification of Bruhat-Tits buildings. Ann. Sci. École Norm. Sup. **39**, 871–920 (2006)

[KL97]   B. Kleiner, B. Leeb, Rigidity of quasi-isometries for symmetric spaces and Euclidean buildings. Inst. Hautes Études Sci. Publ. Math. **86**, 115–197 (1997)

[KMRT98] M.-A. Knus, A. Merkurjev, M. Rost, J.-P. Tignol, *The Book of Involutions*, American Mathematical Society Colloquium Publications, vol. 44, (American Mathematical Society, Providence, RI, 1998), with a preface in French by J. Tits

[Lan96] E. Landvogt, *A Compactification of the Bruhat-Tits Building*. Lecture Notes in Mathematics, vol. 1619 (Springer, New York, 1996)

[Lan00] E. Landvogt, Some functorial properties of the Bruhat-Tits building. J. Reine Angew. Math. **518**, 213–241 (2000)

[Mac71] I.G. Macdonald, *Spherical Functions on a Group of p-Adic Type* (Publications of the Ramanujan Institute, No. 2, University of Madras, 1971)

[Mas88] B. Maskit, *Kleinian Groups*. Grundlehren der Mathematischen Wissenschaften, vol. 287 (Springer, New York, 1988)

[Mau09] J. Maubon, Symmetric spaces of the non-compact type: differential geometry, in *Géométries à courbure négative ou nulle, groupes discrets et rigidités*, ed. by A. Parreau, L. Bessières, B. Rémy, Séminaires et Congrès, no. 18 (Société mathématique de France, Paris, 2009), pp. 1–38

[Moo64] C.C. Moore, Compactifications of symmetric spaces. Am. J. Math. **86**, 201–218 (1964)

[MP94] A. Moy, G. Prasad, Unrefined minimal K-types for p-adic groups. Invent. Math. **116**(1–3), 393–408 (1994)

[MP96] A. Moy, G. Prasad, Jacquet functors and unrefined minimal K-types. Comment. Math. Helv. **71**(1), 98–121 (1996)

[Par09] P.-E. Paradan, Symmetric spaces of the non-compact type: Lie groups, in *Géométries à courbure négative ou nulle, groupes discrets et rigidités*, ed. by A. Parreau, L. Bessières, B. Rémy, Séminaires et Congrès, no. 18 (Société mathématique de France, Paris, 2009), pp. 39–76

[Par00] A. Parreau, Dégénérescences de sous-groupes discrets de groupes de Lie semi-simples et actions de groupes sur des immeubles affines, PhD thesis, University Paris 11, Orsay, 2000

[PR94] A. Platonov, I. Rapinchuk, *Algebraic Groups and Number Theory*. Pure and Applied Mathematics, vol. 139 (Academic Press, Boston, 1994)

[Poi13] J. Poineau, Les espaces de Berkovich sont angéliques. Bull. Soc. Math. France **141**, 267–297 (2013)

[Pra82] G. Prasad, Elementary proof of a theorem of Bruhat-Tits-Rousseau and of a theorem of Tits. Bull. Soc. Math. France **110**, 197–202 (1982)

[Pra01] G. Prasad, Galois-fixed points in the Bruhat-Tits building of a reductive group. Bull. Soc. Math. France **129**, 169–174 (2001)

[PR84a] G. Prasad, M.S. Raghunathan, Topological central extensions of semisimple groups over local fields. Ann. Math. **119**(1), 143–201 (1984)

[PR84b] G. Prasad, M.S. Raghunathan, Topological central extensions of semisimple groups over local fields. II. Ann. Math. **119**(2), 203–268 (1984)

[Rém11] B. Rémy, Groupes algébriques pseudo-réductifs et applications, d'après J. Tits et B. Conrad-O. Gabber-G. Prasad, Séminaire Bourbaki. Vol. 2009/2010. Exposés 1012–1026. Astérisque No. 339, Exp. No. 1021, viii–ix, 259–304 (2011)

[Rou77] G. Rousseau, *Immeubles des groupes réductifs sur les corps locaux*, Thèse de doctorat, Publications Mathématiques d'Orsay, No. 221–77.68, Université Paris XI, Orsay, 1977. Disponible à l'adresse http://www.iecn.u-nancy.fr/~rousseau/Textes/

[Rou09] G. Rousseau, Euclidean buildings, in *Géométries à courbure négative ou nulle, groupes discrets et rigidités*, ed. by A. Parreau, L. Bessières, B. Rémy, Séminaires et Congrès, no. 18 (Société mathématique de France, 2009), pp. 77–116

[RTW10] B. Rémy, A. Thuillier, A. Werner, Bruhat-Tits theory from Berkovich's point of view. I: realizations and compactifications of buildings. Annales Scientifiques de l'ENS **43**, 461–554 (2010)

[RTW12] B. Rémy, A. Thuillier, A. Werner, Bruhat-Tits theory from Berkovich's point of view. II: Satake compactifications of buildings. J. Inst. Math. Jussieu **11**(2), 421–465 (2012)

[Sat60a]   I. Satake, On representations and compactifications of symmetric Riemannian spaces. Ann. Math. **71**, 77–110 (1960)

[Sat60b]   I. Satake, On compactifications of the quotient spaces for arithmetically defined discontinuous groups. Ann. Math. **72**, 555–580 (1960)

[Sch13]   T. Schmidt, Forms of an affinoid disc and ramification. Preprint arXiv:1212.3508 (2012)

[Ser77]   J.-P. Serre, *Arbres, Amalgames, SL$_2$*, (Société mathématique de France, Paris, 1977). rédigé avec la collaboration de Hyman Bass, Astérisque, No. 46

[Ser94]   J.-P. Serre, *Cohomologie Galoisienne*, 5th edn. Lecture Notes in Mathematics, vol. 5 (Springer, Berlin, 1994)

[Spr98]   T.A. Springer, *Linear Algebraic Groups*, 2nd edn. Progress in Mathematics, vol. 9 (Birkhäuser, Basel, 1998)

[Ste68]   R. Steinberg, *Lectures on Chevalley Groups* (Yale University, New Haven, CT, 1968). Notes prepared by John Faulkner and Robert Wilson

[Str11]   K. Struyve, (Non)-completeness of **R**-buildings and fixed point theorems. Groups Geom. Dyn. **5**, 177–188 (2011)

[Tem04]   M. Temkin, On local properties of non-Archimedean analytic spaces. II. Isr. J. Math. **140**, 1–27 (2004)

[Tem14]   M. Temkin, Introduction to Berkovich analytic spaces. This volume

[Tit64]   J. Tits, Algebraic and abstract simple groups. Ann. Math. **80**, 313–329 (1964)

[Tits74]   J. Tits, On buildings and their applications. Proc. Int. Congress of Mathematicians (Vancouver, B. C., 1974), Vol. 1 (1975), 209–220

[Tit79]   J. Tits, *Reductive Groups Over Local Fields*. Automorphic forms, representations and *L*-functions, in *Proc. Symp. Pure Math., vol. XXXIII, part 1*, ed. by Armand Borel, William A. Casselman (AMS, Providence, RI, 1979), pp. 29–69

[Tit86]   J. Tits, Immeubles de type affine, in *Buildings and the Geometry of Diagrams* (Como, 1984). Lecture Notes in Math., vol. 1181 (Springer, New York, 1986), pp. 159–190

[Tit92]   J. Tits, Théorie des groupes. Ann. Collège France **92**, 115–133 (1991/92)

[Tit93]   J. Tits, Théorie des groupes. Ann. Collège France **93**, 113–131 (1992/93)

[Wat79]   W.C. Waterhouse, *Introduction to Affine Group Schemes*. Graduate Texts in Mathematics, vol. 66 (Springer, New York, 1979)

[Wei60]   A. Weil, Algebras with involutions and the classical groups. J. Indian Math. Soc. (N.S.) **24**, 589–623 (1960)

[Wei74]   A. Weil, *Basic Number Theory* (third edition). Grundlehren des mathematischen Wissenschaften, vol. 144 (Springer, New York, 1974)

[Wei09]   R.M. Weiss, *The Structure of Affine Buildings*. Annals of Mathematics Studies, vol. 168 (Princeton University Press, Princeton, NJ, 2009)

[Wer01]   A. Werner, Compactification of the Bruhat-Tits building of PGL by lattices of smaller rank. Doc. Math. **6**, 315–342 (2001)

[Wer04]   A. Werner, Compactification of the Bruhat-Tits building of PGL(V) by seminorms. Math. Z. **248**, 511–526 (2004)

[Wer07]   A. Werner, Compactifications of Bruhat-Tits buildings associated to linear representations. Proc. Lond. Math. Soc. **95**, 497–518 (2007)

[Wer11]   A. Werner, A tropical view on Bruhat-Tits buildings and their compactifications. Cent. Eur. J. Math. **9**(2), 390–402 (2011)

Part III
Valuation Spaces and Dynamics

Dynamics on Berkovich Spaces in Low Dimensions

Mattias Jonsson

Contents

1   Introduction

The goal of these notes is twofold. First, I'd like to describe how Berkovich spaces enter naturally in certain instances of discrete dynamical systems. In particular, I will try to show how my own work with Charles Favre [FJ07, FJ11] on valuative dynamics relates to the dynamics of rational maps on the Berkovich projective line as initiated by Juan Rivera-Letelier in his thesis [Riv03a] and subsequently studied by him and others. In order to keep the exposition somewhat focused, I have chosen three sample problems (Theorems A, B and C below) for which I will present reasonably complete proofs.

M. Jonsson (✉)
Department of Mathematics, University of Michigan, 530 Church Street, 2076 East Hall, Ann Arbor, MI 48109-1043, USA
e-mail: mattiasj@umich.edu, Url: http://www.math.lsa.umich.edu/~mattiasj/

© Springer International Publishing Switzerland 2015   205
A. Ducros et al. (eds.), *Berkovich Spaces and Applications*, Lecture Notes
in Mathematics 2119, DOI 10.1007/978-3-319-11029-5_6

The second objective is to show some of the simplest Berkovich spaces "in action". While not necessarily representative of the general situation, they have a structure that is very rich, yet can be described in detail. In particular, they are trees, or cones over trees.

For the purposes of this introduction, the dynamical problems that we shall be interested in all arise from polynomial mappings

$$f : \mathbf{A}^n \to \mathbf{A}^n,$$

where $\mathbf{A}^n$ denotes affine n-space over a *valued field*, that is, a field K complete with respect a norm $| \cdot |$. Studying the dynamics of f means, in rather vague terms, studying the asymptotic behavior of the *iterates* of f:

$$f^m = f \circ f \circ \cdots \circ f$$

(the composition is taken m times) as $m \to \infty$. For example, one may try to identify regular as opposed to chaotic behavior. One is also interested in invariant objects such as fixed points, invariant measures, etc.

When K is the field of complex numbers, polynomial mappings can exhibit very interesting dynamics both in one and higher dimensions. We shall discuss this a little further in Sect. 1.1 below. As references we point to [CG93, Mil06] for the one-dimensional case and [Sib99] for higher dimensions.

Here we shall instead focus on the case when the norm on K is *non-Archimedean* in the sense that the strong triangle inequality $|a + b| \leq \max\{|a|, |b|\}$ holds. Interesting examples of such fields include the p-adic numbers $\mathbf{Q}_p$, the field of Laurent series $\mathbf{C}((t))$, or any field K equipped with the *trivial* norm.

One motivation for investigating the dynamics of polynomial mappings over non-Archimedean fields is simply to see to what extent the known results over the complex (or real) numbers continue to hold. However, non-Archimedean dynamics sometimes plays a role even when the original dynamical system is defined over the complex numbers. We shall see some instances of this phenomenon in these notes; other examples are provided by the work of Kiwi [Kiw06], Baker and DeMarco [BdM09], and Ghioca, Tucker and Zieve [GTZ08].

Over the complex numbers, many of the most powerful tools for studying dynamics are either topological or analytical in nature: distortion estimates, potential theory, quasiconformal mappings etc. These methods do not directly carry over to the non-Archimedean setting since K is totally disconnected.

On the other hand, a polynomial mapping f automatically induces a selfmap

$$f : \mathbf{A}^n_{\text{Berk}} \to \mathbf{A}^n_{\text{Berk}}$$

of the corresponding *Berkovich space* $\mathbf{A}^n_{\text{Berk}}$. By definition, $\mathbf{A}^n_{\text{Berk}} = \mathbf{A}^n_{\text{Berk}}(K)$ is the set of multiplicative seminorms on the coordinate ring $R \simeq K[z_1, \ldots, z_n]$ of $\mathbf{A}^n$ that extend the given norm on K. It carries a natural topology in which it locally compact and arcwise connected. It also contains a copy of $\mathbf{A}^n$: a point $x \in \mathbf{A}^n$ is identified with the seminorm $\phi \mapsto |\phi(x)|$. The action of f on $\mathbf{A}^n_{\text{Berk}}$ is given as follows. A seminorm $|\cdot|$ is mapped by f to the seminorm whose value on a polynomial $\phi \in R$ is given by $|f^*\phi|$.

The idea is now to study the dynamics on $\mathbf{A}^n_{\text{Berk}}$. At this level of generality, not very much seems to be known at the time of writing (although the time may be ripe to start looking at this). Instead, the most interesting results have appeared in situations when the structure of the space $\mathbf{A}^n_{\text{Berk}}$ is better understood, namely in sufficiently low dimensions.

We shall focus on two such situations:

(1) $f : \mathbf{A}^1 \to \mathbf{A}^1$ is a polynomial mapping of the affine line over a general valued field K;
(2) $f : \mathbf{A}^2 \to \mathbf{A}^2$ is a polynomial mapping of the affine plane over a field K equipped with the trivial norm.

In both cases we shall mainly treat the case when K is algebraically closed.

In (1), one makes essential use of the fact that the Berkovich affine line $\mathbf{A}^1_{\text{Berk}}$ is a tree.[1] This tree structure was pointed out already by Berkovich in his original work [Ber90] and is described in great detail in the book [BR10] by Baker and Rumely. It has been exploited by several authors and a very nice picture of the global dynamics on this Berkovich space has taken shape. It is beyond the scope of these notes to give an account of all the results that are known. Instead, we shall focus on one specific problem: equidistribution of preimages of points. This problem, which will be discussed in further detail in Sect. 1.1, clearly shows the advantage of working on the Berkovich space as opposed to the "classical" affine line.

As for (2), the Berkovich affine plane $\mathbf{A}^2_{\text{Berk}}$ is already quite a beast, but it is possible to get a handle on its structure. We shall be concerned not with the global dynamics of f, but the local dynamics either at a fixed point $0 = f(0) \in \mathbf{A}^2$, or at infinity. There are natural subspaces of $\mathbf{A}^2_{\text{Berk}}$ consisting of seminorms that "live" at 0 or at infinity, respectively, in a sense that can be made precise. These two spaces are cones over a tree and hence reasonably tractable.

While it is of general interest to study the dynamics in (2) for a general field K, there are surprising applications to *complex* dynamics when using $K = \mathbf{C}$ equipped with the trivial norm. We shall discuss this in Sects. 1.2 and 1.3 below.

1.1 Polynomial Dynamics in One Variable

Our first situation is that of a polynomial mapping

$$f : \mathbf{A}^1 \to \mathbf{A}^1$$

---

[1]For a precise definition of what we mean by "tree", see Sect. 2.

of degree $d > 1$ over a complete valued field K, that we here shall furthermore assume to be algebraically closed and, for simplicity, of characteristic zero.

When K is equal to the (archimedean) field $\mathbf{C}$, there is a beautiful theory describing the polynomial dynamics. The foundation of this theory was built in the 1920s by Fatou and Julia, who realized that Montel's theorem could be used to divide the phase space $\mathbf{A}^1 = \mathbf{A}^1(\mathbf{C})$ into a region where the dynamics is tame (the Fatou set) and a region where it is chaotic (the Julia set). In the 1980s and beyond, the theory was very significantly advanced, in part because of computer technology allowing people to visualize Julia sets as fractal objects, but more importantly because of the introduction of new tools, in particular quasiconformal mappings. For further information on this we refer the reader to the books [CG93, Mil06].

In between, however, a remarkable result by Hans Brolin [Bro65] appeared in the 1960s. His result seems to have gone largely unnoticed at the time, but has been of great importance for more recent developments, especially in higher dimensions. Brolin used potential theoretic methods to study the asymptotic distribution of preimages of points. To state his result, let us introduce some terminology. Given a polynomial mapping f as above, one can consider the *filled Julia set* of f, consisting of all points $x \in \mathbf{A}^1$ whose orbit is bounded. This is a compact set. Let ρ_f be *harmonic measure* on the filled Julia set, in the sense of potential theory. Now, given a point $x \in \mathbf{A}^1$ we can look at the distribution of preimages of x under f^n. There are d^n preimages of x, counted with multiplicity, and we write $f^{n*}\delta_x = \sum_{f^n y = x} \delta_y$, where the sum is taken over these preimages. Thus $d^{-n} f^{n*}\delta_x$ is a probability measure on $\mathbf{A}^1$. Brolin's theorem now states

Theorem *For all points $x \in \mathbf{A}^1$, with at most one exception, we have*

$$\lim_{n \to \infty} d^{-n} f^{n*}\delta_x \to \rho_f.$$

Furthermore, a point $x \in \mathbf{A}^1$ is exceptional iff there exists a global coordinate z on $\mathbf{A}^1$ vanishing at x such that f is given by the polynomial $z \mapsto z^d$. In this case, $d^{-n} f^{n*}\delta_x = \delta_x$ for all n.

A version of this theorem for selfmaps of $\mathbf{P}^1$ was later proved independently by Lyubich [Lyu83] and by Freire-Lopez-Mañé [FLM83]. There have also been far-reaching generalizations of Brolin's theorem to higher-dimensional complex dynamics. However, we shall stick to the one-dimensional polynomial case in this introduction.

It is now natural to ask what happens when we replace $\mathbf{C}$ by a *non-Archimedean* valued field K. We still assume that K is algebraically closed and, as above, that it is of characteristic zero. An important example is $K = \mathbf{C}_p$, the completed algebraic closure of the p-adic numbers $\mathbf{Q}_p$. However, while most of the early work focused on $\mathbf{C}_p$, and certain deep results that are true for this field do not hold for general K, we shall not assume $K = \mathbf{C}_p$ in what follows.

Early on, through work of Silverman, Benedetto, Hsia, Rivera-Letelier and others [Ben00, Ben01a, Ben02b, Hsi00, MS95, Riv03a] it became clear that there were some significant differences to the archimedean case. For example, with the most direct translations of the definitions from the complex numbers, it may well happen that the Julia set of a polynomial over a non-Archimedean field K is empty. This is in clear distinction with the complex case. Moreover, the topological structure of K is vastly different from that of $\mathbf{C}$. Indeed, K is totally disconnected and usually not even locally compact. The lack of compactness is inherited by the space of probability measures on K: there is a priori no reason for the sequence of probability measures on K to admit a convergent subsequence. This makes it unlikely that a naïve generalization of Brolin's theorem should hold.

Juan Rivera-Letelier was the first one to realize that Berkovich spaces could be effectively used to study the dynamics of rational functions over non-Archimedean fields. As we have seen above, $\mathbf{A}^1$ embeds naturally into $\mathbf{A}^1_{\text{Berk}}$ and the map f extends to a map

$$f : \mathbf{A}^1_{\text{Berk}} \to \mathbf{A}^1_{\text{Berk}} .$$

Now $\mathbf{A}^1_{\text{Berk}}$ has good topological properties. It is locally compact[2] and contractible. This is true for the Berkovich affine space $\mathbf{A}^n_{\text{Berk}}$ of any dimension. However, the structure of the Berkovich affine $\mathbf{A}^1_{\text{Berk}}$ can be understood in much greater detail, and this is quite helpful when analyzing the dynamics. Specifically, $\mathbf{A}^1_{\text{Berk}}$ has a structure of a tree and the induced map $f : \mathbf{A}^1_{\text{Berk}} \to \mathbf{A}^1_{\text{Berk}}$ preserves the tree structure, in a suitable sense.

Introducing the Berkovich space $\mathbf{A}^1_{\text{Berk}}$ is critical for even formulating many of the known results in non-Archimedean dynamics. This in particular applies to the non-Archimedean version of Brolin's theorem:

Theorem A *Let* $f : \mathbf{A}^1 \to \mathbf{A}^1$ *be a polynomial map of degree* $d > 1$ *over an algebraically closed field of characteristic zero. Then there exists a probability measure* $\rho = \rho_f$ *on* $\mathbf{A}^1_{\text{Berk}}$ *such that for all points* $x \in \mathbf{A}^1$, *with at most one exception, we have*

$$\lim_{n \to \infty} d^{-n} f^{n*} \delta_x \to \rho.$$

Furthermore, a point $x \in \mathbf{A}^1$ *is exceptional iff there exists a global coordinate* z *on* $\mathbf{A}^1$ *vanishing at* x *such that* f *is given by the polynomial* $z \mapsto z^d$. *In this case,* $d^{-n} f^{n*} \delta_x = \delta_x$ *for all* n.

In fact, we could have started with any point $x \in \mathbf{A}^1_{\text{Berk}}$ assuming we are careful with the definition of $f^{n*} \delta_x$. Notice that when $x \in \mathbf{A}^1$, the probability measures $d^{-n} f^{n*} \delta_x$ are all supported on $\mathbf{A}^1 \subseteq \mathbf{A}^1_{\text{Berk}}$, but the limit measure may very well give no mass to $\mathbf{A}^1$. It turns out that if we define the Julia set J_f of f as the support of the measure ρ_f, then J_f shares many properties of the Julia set of complex

---

[2]Its one-point compactification is the Berkovich projective line $\mathbf{P}^1_{\text{Berk}} = \mathbf{A}^1_{\text{Berk}} \cup \{\infty\}$.

polynomials. This explains why we may not see a Julia set when studying the dynamics on $\mathbf{A}^1$ itself.

Theorem A is due to Favre and Rivera-Letelier [FR10]. The proof is parallel to Brolin's original proof in that it uses *potential theory*. Namely, one can define a Laplace operator Δ on $\mathbf{A}^1_{\text{Berk}}$ and to every probability measure ρ on $\mathbf{A}^1_{\text{Berk}}$ associate a subharmonic function $\varphi = \varphi_\rho$ such that $\Delta\varphi = \rho - \rho_0$, where ρ_0 is a fixed reference measure (typically a Dirac mass at a point of $\mathbf{A}^1_{\text{Berk}} \setminus \mathbf{A}^1$). The function φ is unique up to an additive constant. One can then translate convergence of the measures in Theorem A to the more tractable statement about convergence of potentials. The Laplace operator itself can be very concretely interpreted in terms of the tree structure on $\mathbf{A}^1_{\text{Berk}}$. All of this will be explained in Sects. 2–5.

The story does not end with Theorem A. For instance, Favre and Rivera-Letelier analyze the ergodic properties of f with respect to the measure ρ_f. Okuyama has given a quantitative strengthening of the equidistribution result in Theorem A, see [Oku11b]. The measure ρ_f also describes the distribution of periodic points, see [FR10, Théorème B] as well as [Oku11a].

As already mentioned, there is also a very interesting Fatou-Julia theory. We shall discuss this a little further in Sect. 4 but the discussion will be brief due to limited space. The reader will find many more details in the book [BR10]. We also recommend the recent survey by Benedetto [Ben10].

1.2 Local Plane Polynomial Dynamics

The second and third situations that we will study both deal with polynomial mappings

$$f : \mathbf{A}^2 \to \mathbf{A}^2$$

over a valued field K. In fact, they originally arose from considerations in *complex* dynamics and give examples where non-Archimedean methods can be used to study Archimedean problems.

Thus we start out by assuming that $K = \mathbf{C}$. Polynomial mappings of $\mathbf{C}^2$ can have quite varied and very interesting dynamics; see the survey by Sibony [Sib99] for some of this. Here we will primarily consider local dynamics, so we first consider a *fixed point* $0 = f(0) \in \mathbf{A}^2$. For a detailed general discussion of local dynamics in this setting we refer to Abate's survey [Aba10].

The behavior of f at the fixed point is largely governed by the tangent map $df(0)$ and in particular on the eigenvalues λ_1, λ_2 of the latter. For example, if $|\lambda_1|, |\lambda_2| < 1$, then we have an *attracting* fixed point: there exists a small neighborhood $U \ni 0$ such that $f(\overline{U}) \subseteq U$ and $f^n \to 0$ on U. Further, when there are no *resonances* between the eigenvalues λ_1, λ_2, the dynamics can in fact be *linearized*: there exists a local biholomorphism $\phi : (\mathbf{A}^2, 0) \to (\mathbf{A}^2, 0)$ such that $f \circ \phi = \phi \circ \Lambda$, where $\Lambda(z_1, z_2) = (\lambda_1 z_1, \lambda_2 z_2)$. This in particular gives very precise information on the rate at which typical orbits converge to the origin: for a "typical" point $x \approx 0$ we have $\|f^n(x)\| \sim \max_{i=1,2} |\lambda_i|^n \|x\|$ as $n \to \infty$.

On the other hand, in the *superattracting* case, when $\lambda_1 = \lambda_2 = 0$, the action of f on the tangent space $T_0 \mathbf{C}^2$ does not provide much information about the dynamics. Let us still try to understand at what rate orbits tend to the fixed point. To this end, let

$$f = f_c + f_{c+1} + \cdots + f_d$$

be the expansion of f in homogeneous components: $f_j(\lambda z) = \lambda^j f_j(z)$ and where $f_c \not\equiv 0$. Thus $c = c(f) \geq 1$ and the number $c(f)$ in fact does not depend on the choice of coordinates. Note that for a typical point $x \approx 0$ we will have

$$\|f(x)\| \sim \|x\|^{c(f)}.$$

Therefore, one expects that the speed at which the orbit of a typical point x tends to the origin is governed by the growth of $c(f^n)$ as $n \to \infty$. This can in fact be made precise, see [FJ07], but here we shall only study the sequence $(c(f^n))_n$.

Note that this sequence is supermultiplicative: $c(f^{n+m}) \geq c(f^n)c(f^m)$. This easily implies that the limit

$$c_\infty(f) := \lim_{n \to \infty} c(f^n)^{1/n}$$

exists. Clearly $c_\infty(f^n) = c_\infty(f)^n$ for $n \geq 1$.

Example 1.1 If $f(z_1, z_2) = (z_2, z_1 z_2)$, then $c(f^n)$ is the $(n+2)$th Fibonacci number and $c_\infty(f) - \frac{1}{2}(\sqrt{5}+1)$ is the golden mean.

Our aim is to give a proof of the following result, originally proved in [FJ07].

Theorem B *The number $c_\infty = c_\infty(f)$ is a quadratic integer: there exist $a, b \in \mathbf{Z}$ such that $c_\infty^2 = ac_\infty + b$. Moreover, there exists a constant $\delta > 0$ such that*

$$\delta c_\infty^n \leq c(f^n) \leq c_\infty^n$$

for all $n \geq 1$.

Note that the right-hand inequality $c(f^n) \leq c_\infty^n$ is an immediate consequence of supermultiplicativity. It is the left-hand inequality that is nontrivial.

To prove Theorem B we study the induced dynamics

$$f : \mathbf{A}^2_{\mathrm{Berk}} \to \mathbf{A}^2_{\mathrm{Berk}}$$

of f on the Berkovich affine plane $\mathbf{A}^2_{\mathrm{Berk}}$. Now, if we consider $K = \mathbf{C}$ with its standard Archimedean norm, then it is a consequence of the Gelfand-Mazur theorem that $\mathbf{A}^2_{\mathrm{Berk}} \simeq \mathbf{A}^2$, so this may not seem like a particularly fruitful approach. If we instead, however, consider $K = \mathbf{C}$ equipped with the *trivial* norm, then the associated Berkovich affine plane $\mathbf{A}^2_{\text{Berk}}$ is a totally different creature and the induced dynamics is very interesting.

By definition, the elements of $\mathbf{A}^2_{\text{Berk}}$ are multiplicative seminorms on the coordinate ring of $\mathbf{A}^2$, that is, the polynomial ring $R \simeq K[z_1, z_2]$ in two variables over K. It turns out to be convenient to instead view these elements "additively" as *semivaluations* $v : R \to \mathbf{R} \cup \{+\infty\}$ such that $v|_{K^*} \equiv 0$. The corresponding seminorm is $|\cdot| = e^{-v}$.

Since we are interested in the local dynamics of f near a (closed) fixed point $0 \in \mathbf{A}^2$, we shall study the dynamics of f on a corresponding subspace of $\mathbf{A}^2_{\text{Berk}}$, namely the set $\hat{\mathcal{V}}_0$ of semivaluations v such that $v(\phi) > 0$ whenever ϕ vanishes at 0. In valuative terminology, these are the semivaluations $v \in \mathbf{A}^2_{\text{Berk}} \setminus \mathbf{A}^2$ whose *center* on $\mathbf{A}^2$ is the point 0. It is clear that $f(\hat{\mathcal{V}}_0) \subseteq \hat{\mathcal{V}}_0$.

Note that $\hat{\mathcal{V}}_0$ has the structure of a cone: if $v \in \hat{\mathcal{V}}_0$, then $tv \in \hat{\mathcal{V}}_0$ for $0 < t \leq \infty$. The apex of this cone is the image of the point $0 \in \mathbf{A}^2$ under the embedding $\mathbf{A}^2 \hookrightarrow \mathbf{A}^2_{\text{Berk}}$. The base of the cone can be identified with the subset $\mathcal{V}_0 \subseteq \hat{\mathcal{V}}_0$ consisting of semivaluations that are *normalized* by the condition $v(\mathfrak{m}_0) = \min_{\phi \in \mathfrak{m}_0} v(\phi) = +1$, where $\mathfrak{m}_0 \subseteq R$ denotes the maximal ideal of 0. This space $\mathcal{V}_0$ is compact and has a structure of an $\mathbf{R}$-tree. We call it the *valuative tree* at the point 0. Its structure is investigated in detail in [FJ04] and will be examined in Sect. 7.[3]

Now, $\mathcal{V}_0$ is in general not invariant by f. Instead, f induces a selfmap

$$f_\bullet : \mathcal{V}_0 \to \mathcal{V}_0$$

and a "multiplier" function $c(f, \cdot) : \mathcal{V}_0 \to \mathbf{R}_+$ such that

$$f(v) = c(f, v) f_\bullet v$$

for $v \in \mathcal{V}_0$. The number $c(f)$ above is exactly equal to $c(f, \text{ord}_0)$, where $\text{ord}_0 \in \mathcal{V}_0$ denotes the order of vanishing at $0 \in \mathbf{A}^2$. Moreover, we have

$$c(f^n) = c(f^n, \text{ord}_0) = \prod_{i=0}^{n-1} c(f, v_i), \quad \text{where } v_i = f_\bullet^i \, \text{ord}_0;$$

this equation will allow us to understand the behavior of the sequence $c(f^n)$ through the dynamics of $f_\bullet$ on $\mathcal{V}_0$.

The proof of Theorem B given in these notes is simpler than the one in [FJ07]. Here is the main idea. Suppose that there exists a valuation $v \in \mathcal{V}_0$ such that $f_\bullet v = v$, so that $f(v) = cv$, where $c = c(f, v) > 0$. Then $c(f^n, v) = c^n$ for $n \geq 1$. Suppose that v satisfies an Izumi-type bound:

$$v(\phi) \leq C \, \text{ord}_0(\phi) \quad \text{for all polynomials } \phi, \tag{1}$$

---

[3]In [FJ04, FJ07], the valuative tree is denoted by $\mathcal{V}$. We write $\mathcal{V}_0$ here in order to emphasize the choice of point $0 \in \mathbf{A}^2$.

where $C > 0$ is a constant independent of ϕ. This is true for many, but not all semivaluations $v \in \mathcal{V}_0$. The reverse inequality $v \geq \mathrm{ord}_0$ holds for all $v \in \mathcal{V}_0$ by construction. Then we have

$$C^{-1}c^n = C^{-1}c(f^n, v) \leq c(f^n) \leq c(f^n, v) \leq c^n.$$

This shows that $c_\infty(f) = c$ and that the bounds in Theorem B hold with $\delta = C^{-1}$. To see that c_∞ is a quadratic integer, we look at the value group Γ_v of v. The equality $f(v) = cv$ implies that $c\Gamma_v \subseteq \Gamma_v$. If we are lucky, then $\Gamma \simeq \mathbf{Z}^d$, where $d \in \{1, 2\}$, which implies that $c_\infty = c$ is an algebraic integer of degree one or two.

The two desired properties of v hold when the eigenvaluation v is *quasimonomial* valuation. In general, there may not exist a quasimonomial eigenvaluation, so the argument is in fact a little more involved. We refer to Sect. 8 for more details.

1.3 Plane Polynomial Dynamics at Infinity

Again consider a polynomial mapping

$$f : \mathbf{A}^2 \to \mathbf{A}^2$$

over the field $K = \mathbf{C}$ of complex numbers. In the previous subsection, we discussed the dynamics of f at a (superattracting) fixed point in $\mathbf{A}^2$. Now we shall consider the dynamics at infinity and, specifically, the rate at which orbits tend to infinity. Fix an embedding $\mathbf{A}^2 \hookrightarrow \mathbf{P}^2$. It is then reasonable to argue that the rate at which "typical" orbits tend to infinity is governed by the *degree growth sequence* $(\deg f^n)_{n \geq 1}$. Precise assertions to this end can be found in [FJ07, FJ11]. Here we shall content ourselves with the study on the degree growth sequence.

In contrast to the local case, this sequence is *sub*multiplicative: $\deg f^{n+m} \leq \deg f^n \deg f^m$, but again the limit

$$d_\infty(f) := \lim_{n \to \infty} (\deg f^n)^{1/n}$$

exists. Apart from some inequalities being reversed, the situation is very similar to the local case, so one may hope for a direct analogue of Theorem B above. However, the skew product example $f(z_1, z_2) = (z_1^2, z_1 z_2^2)$ shows that we may have $\deg f^n \sim n d_\infty^n$. What does hold true in general is

Theorem C *The number $d_\infty = d_\infty(f)$ is a quadratic integer: there exist $a, b \in \mathbf{Z}$ such that $d_\infty^2 = a d_\infty + b$. Moreover, we are in exactly one of the following two cases:*

(a) there exists $C > 0$ such that $d_\infty^n \leq \deg f^n \leq C d_\infty^n$ for all n;
(b) $\deg f^n \sim n d_\infty^n$ as $n \to \infty$.

Moreover, case (b) occurs iff f, after conjugation by a suitable polynomial automorphism of $\mathbf{C}^2$, is a skew product of the form

$$f(z_1, z_2) = (\phi(z_1), \psi(z_1)z_2^{d_\infty} + O_{z_1}(z_2^{d_\infty - 1})),$$

where $\deg \phi = d_\infty$ and $\deg \psi > 0$.

As in the local case, we approach this theorem by considering the induced dynamics

$$f : \mathbf{A}_{\mathrm{Berk}}^2 \to \mathbf{A}_{\mathrm{Berk}}^2,$$

where we consider $K = \mathbf{C}$ equipped with the trivial norm. Since we are interested in the dynamics of f at infinity, we restrict our attention to the space $\hat{\mathcal{V}}_\infty$ consisting of semivaluations $v : R \to \mathbf{R} \cup \{+\infty\}$ whose center is at infinity, that is, for which $v(\phi) < 0$ for some polynomial ϕ. This space has the structure of a pointed[4] cone. To understand its base, note that our choice of embedding $\mathbf{A}^2 \hookrightarrow \mathbf{P}^2$ determines the space $\mathcal{L}$ of affine functions on $\mathbf{A}^2$ (the polynomials of degree at most one). Define

$$\mathcal{V}_\infty := \{v \in \mathbf{A}_{\mathrm{Berk}}^2 \mid \min_{L \in \mathcal{L}} v(L) = -1\}.$$

We call $\mathcal{V}_\infty$ the *valuative tree at infinity*.[5] This subspace at first glance looks very similar to the valuative tree $\mathcal{V}_0$ at a point but there are some important differences. Notably, for a semivaluation $v \in \mathcal{V}_0$ we have $v(\phi) \geq 0$ for all polynomials ϕ. In contrast, while a semivaluations in $\mathcal{V}_\infty$ must take some negative values, it can take positive values on certain polynomials.

Assuming for simplicity that f is *proper*, we obtain a dynamical system $f : \hat{\mathcal{V}}_\infty \to \hat{\mathcal{V}}_\infty$, which we can split into an induced map $f_\bullet : \mathcal{V}_\infty \to \mathcal{V}_\infty$ and a multiplier $d(f, \cdot) : \mathcal{V}_\infty \to \mathbf{R}_+$ such that $f(v) = d(f, v) f_\bullet v$.

The basic idea in the proof of Theorem C is again to look for an eigenvaluation, that is, a semivaluation $v \in \mathcal{V}_\infty$ such that $f_\bullet v = v$. However, even if we can find a "nice" (say, quasimonomial) eigenvaluation, the proof in the local case does not automatically go through. The reason is that Izumi's inequality (1) may fail.

The remedy to this problem is to use an invariant subtree $\mathcal{V}'_\infty \subseteq \mathcal{V}_\infty$ where the Izumi bound almost always holds. In fact, the valuations $v \in \mathcal{V}'_\infty$ for which Izumi's inequality does not hold are of a very special form, and the case when we end up with a fixed point of that type corresponds exactly to the degree growth $\deg f^n \sim n d_\infty^n$. In these notes, $\mathcal{V}'_\infty$ is called the *tight tree at infinity*. I expect it to have applications beyond the situation here.

---

[4]The apex of the cone does not define an element in $\mathbf{A}_{\mathrm{Berk}}^2$.

[5]In [FJ07, FJ11], the valuative tree at infinity is denoted by $\mathcal{V}_0$, but the notation $\mathcal{V}_\infty$ seems more natural.

1.4 Philosophy and Scope

When writing these notes I was faced with the question of how much material to present, and at what level of detail to present it. Since I decided to have Theorems A, B and C as goals for the presentation, I felt it was necessary to provide enough background for the reader to go through the proofs, without too many black boxes. As it turns out, there is quite a lot of background to cover, so these notes ended up rather expansive!

All the main results that I present here can be found in the literature, However, we draw on many different sources that use different notation and terminology. In order to make the presentation coherent, I have tried to make it self-contained. Many complete proofs are included, others are sketched in reasonable detail.

While the point of these notes is to illustrate the usefulness of Berkovich spaces, we only occasionally draw on the general theory as presented in [Ber90, Ber93]. As a general rule, Berkovich spaces obtained by analytification of an algebraic variety are much simpler than the ones constructed by gluing affinoid spaces. Only at a couple of places in Sects. 3 and 4 do we rely on (somewhat) nontrivial facts from the general theory. On the other hand, these facts, mainly involving the local rings at a point on the Berkovich space, are very useful. We try to exploit them systematically. It is likely that in order to treat higher-dimensional questions, one has to avoid simple topological arguments based on the tree structure and instead use algebraic arguments involving the structure sheaf of the space in question.

At the same time, the tree structure of the spaces in question is of crucial importance. They can be viewed as the analogue of the conformal structure on Riemann surfaces. For this reason I have included a self-contained presentation of potential theory and dynamics on trees, at least to the extent that is needed for the later applications in these notes.

I have made an attempt to provide a unified point of view of dynamics on low-dimensional Berkovich spaces. One can of course try to go further and study dynamics on higher-dimensional Berkovich spaces over a field (with either trivial or nontrivial valuation). After all, there has been significant progress in higher dimensional complex dynamics over the last few years. For example, it is reasonable to hope for a version of the Briend-Duval equidistribution theorem [BD01].

Many interesting topics are not touched upon at all in these notes. For instance, we say very little about the dynamics on, or the structure of the Fatou set of a rational map and we likewise do not study the ramification locus. Important contributions to these and other issues have been made by Matt Baker, Robert Benedetto, Laura DeMarco, Xander Faber, Charles Favre, Liang-Chung Hsia, Jan Kiwi, Yûsuke Okuyama, Clayton Petsche, Juan Rivera-Letelier, Robert Rumely Lucien Szpiro, Michael Tepper, Eugenio Trucco and others.

For the relevant results we refer to the original papers [BdM09, Bak06, Bak09, BH05, BR06, Ben98, Ben00, Ben01a, Ben01b, Ben02a, Ben05a, Ben05b, Ben06, Fab09, Fab13a, Fab13b, Fab14, FKT11, FR04, FR06, FR10, Hsi00, Kiw06, Kiw14, Oku11a, Oku11b, PST09, Riv03a, Riv03b, Riv04, Riv05, Tru09]. Alternatively, many of these results can be found in the book [BR10] by Baker and Rumely or the lecture notes [Ben10] by Benedetto.

Finally, we say nothing about arithmetic aspects such as the equidistribution of points of small height [BR10, CL06, FR06, Yua08, Gub08, Fab09, YZ09a, YZ09b]. For an introduction to arithmetic dynamics, see [Sil07] and [Sil10].

1.5   Comparison to Other Surveys

Beyond research articles such as the ones mentioned above, there are several useful sources that contain a systematic treatment of material related to the topics discussed in these notes.

First, there is a significant overlap between these notes and the material in the *Thèse d'Habilitation* [Fav05] of Charles Favre. The latter thesis, which is strongly recommended reading, explains the usage of tree structures in dynamics and complex analysis. It treats Theorems A–C as well as some of my joint work with him on the singularities of plurisubharmonic functions [FJ05a, FJ05b]. However, the presentation here has a different flavor and contains more details.

The book by [BR10] by Baker and Rumely treats potential theory and dynamics on the Berkovich projective line in great detail. The main results in Sects. 3–5 are contained in this book, but the presentation in these notes is at times a little different. We also treat the case when the ground field has positive characteristic and discuss the case when it is not algebraically closed and/or trivially valued. On the other hand, [BR10] contains a great deal of material not covered here. For instance, it contains results on the structure of the Fatou and Julia sets of rational maps, and it gives a much more systematic treatment of potential theory on the Berkovich line.

The lecture notes [Ben10] by Benedetto are also recommended reading. Just as [BR10], they treat the dynamics on the Fatou and Julia sets in detail. It also contains results in "classical" non-Archimedean analysis and dynamics, not involving Berkovich spaces.

The Ph.D. thesis by Amaury Thuillier [Thu05] gives a general treatment of potential theory on Berkovich curves. It is written in a less elementary way than the treatment in, say, [BR10] but on the other hand is more amenable to generalizations to higher dimensions. Potential theory on curves is also treated in [Bak08].

The valuative tree in Sect. 7 is discussed in detail in the monograph [FJ04]. However, the exposition here is self-contained and leads more directly to the dynamical applications that we have in mind.

As already mentioned, we do not discuss arithmetic dynamics in these notes. For information on this fascinating subject we again refer to the book and lecture notes by Silverman [Sil07, Sil10].

1.6   Structure

The material is divided into three parts. In the first part, Sect. 2, we discuss trees since the spaces on which we do dynamics are either trees or cones over trees. The second part, Sects. 3–5, is devoted to the Berkovich affine and projective lines and dynamics on them. Finally, in Sects. 6–10 we study polynomial dynamics on the Berkovich affine plane over a trivially valued field.

We now describe the contents of each chapter in more detail. Each chapter ends with a section called "Notes and further references" containing further comments.

In Sect. 2 we gather some general definitions and facts about trees. Since we shall work on several spaces with a tree structure, I felt it made sense to collect the material in a separate section. See also [Fav05]. First we define what we mean by a tree, with or without a metric. Then we define a Laplace operator on a general metric tree, viewing the latter as a pro-finite tree. In our presentation, the Laplace operator is defined on the class of quasisubharmonic functions and takes values in the space of signed measures with total mass zero and whose negative part is a finite atomic measure. Finally we study maps between trees. It turns out that simply assuming that such a map is finite, open and surjective gives quite strong properties. We also prove a fixed point theorem for selfmaps of trees.

The structure of the Berkovich affine and projective lines is outlined in Sect. 3. This material is described in much more detail in [BR10]. One small way in which our presentation stands out is that we try to avoid coordinates as far as possible. We also point out some features of the local rings that turn out to be useful for analyzing the mapping properties and we make some comments about the case when the ground field is not algebraically closed and/or trivially valued.

In Sect. 4 we start considering rational maps. Since we work in arbitrary characteristic, we include a brief discussion of separable and purely inseparable maps. Then we describe how polynomial and rational maps extend to maps on the Berkovich affine and projective line, respectively. This is of course only a very special case of the analytification functor in the general theory of Berkovich spaces, but it is useful to see in detail how to do this. Again our approach differs slightly from the ones in the literature that I am aware of, in that it is coordinate free. Having extended a rational map to the Berkovich projective line, we look at the important notion of the local degree at a point.[6] We adopt an algebraic definition of the local degree and show that it can be interpreted as a local expansion factor in the hyperbolic metric. While this important result is well known, we give an algebraic proof that I believe is new. We also show that the local degree is the same as the multiplicity defined by Baker and Rumely, using the Laplacian (as was already known.) See [Fab13a, Fab13b] for more on the local degree and the ramification locus, defined as the subset where the local degree is at least two. Finally, we discuss the case when the ground field is not algebraically closed and/or is trivially valued.

---

[6]In [BR10], the local degree is called *multiplicity*.

We arrive at the dynamics on the Berkovich projective line in Sect. 5. Here we do not really try to survey the known results. While we do discuss fixed points and the Fatou and Julia sets, the exposition is very brief and the reader is encouraged to consult the book [BR10] by Baker and Rumely or the notes [Ben10] by Benedetto for much more information. Instead we focus on Theorem A in the introduction, the equidistribution theorem by Favre and Rivera-Letelier. We give a complete proof which differs in the details from the one in [FR10]. We also give some consequences of the equidistribution theorem. For example, we prove Rivera-Letelier's dichotomy that the Julia set is either a single point or else a perfect set. Finally, we discuss the case when the ground field is not algebraically closed and/or is trivially valued.

At this point, our attention turns to the Berkovich affine plane over a trivially valued field. Here it seems more natural to change from the multiplicative terminology of seminorms to the additive notion of semivaluations. We start in Sect. 6 by introducing the home and the center of a valuation. This allows us to stratify the Berkovich affine space. This stratification is very explicit in dimension one, and possible (but nontrivial) to visualize in dimension two. We also introduce the important notion of a quasimonomial valuation and discuss the Izumi-Tougeron inequality.

In Sect. 7 we come to the valuative tree at a closed point 0. It is the same object as in the monograph [FJ04] but here it is defined as a subset of the Berkovich affine plane. We give a brief, but self-contained description of its main properties with a presentation that is influenced by my joint work with Boucksom and Favre [BFJ08b, BFJ12, BFJ14] in higher dimensions. As before, our treatment is coordinate-free. A key result is that the valuative tree at 0 is homeomorphic to the inverse limit of the dual graphs over all birational morphisms above 0. Each dual graph has a natural metric, so the valuative tree is a pro-finite metric tree, and hence a metric tree in the sense of Sect. 2. In some sense, the cone over the valuative tree is an even more natural object. We define a Laplace operator on the valuative tree that takes this fact into account. The subharmonic functions turn out to be closely related to ideals in the ring of polynomials that are primary to the maximal ideal at 0. In general, the geometry of blowups of the point 0 can be well understood and we exploit this systematically.

Theorem B is proved in Sect. 8. We give a proof that is slightly different and shorter than the original one in [FJ07]. In particular, we have a significantly simpler argument for the fact that the number c_∞ is a quadratic integer. The new argument makes more systematic use of the value groups of valuations.

Next we move from a closed point in $\mathbf{A}^2$ to infinity. The valuative tree at infinity was first defined in [FJ07] and in Sect. 9 we review its main properties. Just as in the local case, the presentation is supposed to be self-contained and also more geometric than in [FJ07]. There is a dictionary between the situation at a point and at infinity. For example, a birational morphism above the closed point $0 \in \mathbf{A}^2$ corresponds to a compactification of $\mathbf{A}^2$ and indeed, the valuative tree at infinity is homeomorphic to the inverse limit of the dual graphs of all (admissible) compactifications. Unfortunately, the dictionary is not perfect, and there are many subtleties when working at infinity. For example, a polynomial in two variables tautologically defines a function on both the valuative tree at a point and at infinity. At a point, this function is always negative but at infinity, it takes on both positive and negative values. Alternatively, the subtleties can be said to stem from the fact that the geometry of compactifications of $\mathbf{A}^2$ can be much more complicated than that of blowups of a closed point.

To remedy some shortcomings of the valuative tree at infinity, we introduce a subtree, the tight tree at infinity. It is an inverse limit of dual graphs over a certain class of tight compactifications of $\mathbf{A}^2$. These have much better properties than general compactifications and should have applications to other problems. In particular, the nef cone of a tight compactification is always simplicial, whereas the nef cone in general can be quite complicated.

Finally, in Sect. 10 we come to polynomial dynamics at infinity, in particular the proof of Theorem C. We follow the strategy of the proof of Theorem B closely, but we make sure to only use tight compactifications. This causes some additional complications, but we do provide a self-contained proof, that is simpler than the one in [FJ07].

1.7 Novelties

While most of the material here is known, certain proofs and ways of presenting the results are new.

The definitions of a general tree in Sect. 2.1 and metric tree in Sect. 2.2 are new, although equivalent to the ones in [FJ04]. The class of quasisubharmonic functions on a general tree also seems new, as are the results in Sect. 2.5.6 on their singularities. The results on tree maps in Sect. 2.6 are new in this setting: they can be found in e.g. [BR10] for rational maps on the Berkovich projective line.

Our description of the Berkovich affine and projective lines is new, but only in the way that we insist on defining things in a coordinate free way whenever possible. The same applies to the extension of a polynomial or rational map from $\mathbf{A}^1$ or $\mathbf{P}^1$ to $\mathbf{A}^1_{\mathrm{Berk}}$ or $\mathbf{P}^1_{\mathrm{Berk}}$, respectively.

While Theorem 4.7, expressing the local degree as a dilatation factor in the hyperbolic metric, is due to Rivera-Letelier, the proof here is directly based on the definition of the local degree and seems to be new. The remarks in Sect. 4.11 on the non-algebraic case also seem to be new.

The structure of the Berkovich affine plane over a trivially valued field, described in Sect. 6.7 was no doubt known to experts but not described in the literature. In particular, the valuative tree at a closed point and at infinity were never explicitly identified as subsets of the Berkovich affine plane.

Our exposition of the valuative tree differs from the treatment in the book [FJ04] and instead draws on the analysis of the higher dimensional situation in [BFJ08b].

The proof of Theorem B in Sect. 8 is new and somewhat simpler than the one in [FJ07]. In particular, the fact that c_∞ is a quadratic integer is proved using value groups, whereas in [FJ07] this was done via rigidification. The same applies to Theorem C in Sect. 10.

2   Tree Structures

We shall do dynamics on certain low-dimensional Berkovich spaces, or subsets thereof. In all cases, the space/subset has the structure of a tree. Here we digress to discuss exactly what we mean by this. We also present a general version of potential theory on trees. The definitions that follow are slightly different from, but equivalent to the ones in [FJ04, BR10, Fav05], to which we refer for details. The idea is that any two points in a tree should be joined by a unique interval. This interval should look like a real interval but may or may not be equipped with a distance function.

2.1   Trees

We start by defining a general notion of a tree. All our trees will be modeled on the real line (as opposed to a general ordered group Λ).[7] In order to avoid technicalities, we shall also only consider trees that are complete in the sense that they contain all their endpoints.

Definition 2.1 An *interval structure* on a set I is a partial order $\leq$ on I under which I becomes isomorphic (as a partially ordered set) to the real interval $[0, 1]$ or to the trivial real interval $[0, 0] = \{0\}$.

Let I be a set with an interval structure. A *subinterval* of I is a subset $J \subseteq I$ that becomes a subinterval of $[0, 1]$ or $[0, 0]$ under such an isomorphism. The *opposite* interval structure on I is obtained by reversing the partial ordering.

Definition 2.2 A *tree* is a set X together with the following data. For each $x, y \in X$, there exists a subset $[x, y] \subseteq X$ containing x and y and equipped with an interval structure. Furthermore, we have:

(T1) $[x, x] = \{x\}$;

(T2)   if $x \neq y$, then $[x, y]$ and $[y, x]$ are equal as subsets of X but equipped with opposite interval structures; they have x and y as minimal elements, respectively;

(T3)   if $z \in [x, y]$ then $[x, z]$ and $[z, y]$ are subintervals of $[x, y]$ such that $[x, y] = [x, z] \cup [z, y]$ and $[x, z] \cap [z, y] = \{z\}$;

(T4)   for any $x, y, z \in X$ there exists a unique element $x \wedge_z y \in [x, y]$ such that $[z, x] \cap [y, x] = [x \wedge_z y, x]$ and $[z, y] \cap [x, y] = [x \wedge_z y, y]$;

(T5)   if $x \in X$ and $(y_\alpha)_{\alpha \in A}$ is a net in X such that the segments $[x, y_\alpha]$ increase with α, then there exists $y \in X$ such that $\bigcup_\alpha [x, y_\alpha[= [x, y[$.

---

[7]Our definition of "tree" is not the same as the one used in set theory [Jec03] but we trust that no confusion will occur. The terminology "**R**-tree" would have been natural, but has already been reserved [GH90] for slightly different objects.

In (T5) we have used the convention $[x, y[:= [x, y] \setminus \{y\}$. Recall that a *net* is a sequence indexed by a directed (possibly uncountable) set. The subsets $[x, y]$ above will be called *intervals* or *segments*.

2.1.1 Topology

A tree as above carries a natural *weak topology*. Given a point $x \in X$, define two points $y, z \in X \setminus \{x\}$ to be equivalent if $]x, y] \cap]x, z] \neq \emptyset$. An equivalence class is called a *tangent direction* at x and the set of $y \in X$ representing a tangent direction $\vec{v}$ is denoted $U(\vec{v})$. The weak topology is generated by all such sets $U(\vec{v})$. Clearly X is arcwise connected and the connected components of $X \setminus \{x\}$ are exactly the sets $U(\vec{v})$ as $\vec{v}$ ranges over tangent directions at x. A tree is in fact uniquely arc connected in the sense that if $x \neq y$ and $\gamma : [0, 1] \to X$ is an injective continuous map with $\gamma(0) = x$, $\gamma(1) = y$, then the image of γ equals $[x, y]$. Since the sets $U(\vec{v})$ are connected, any point in X admits a basis of connected open neighborhoods. We shall see shortly that X is compact in the weak topology.

If $\gamma = [x, y]$ is a nontrivial interval, then the *annulus* $A(\gamma) = A(x, y)$ is defined by $A(x, y) := U(\vec{v}_x) \cap U(\vec{v}_y)$, where $\vec{v}_x$ (resp., $\vec{v}_y$) is the tangent direction at x containing y (resp., at y containing x).

An *end* of X is a point admitting a unique tangent direction. A *branch point* is a point having at least three tangent directions.

2.1.2 Subtrees

A *subtree* of a tree X is a subset $Y \subseteq X$ such that the intersection $[x, y] \cap Y$ is either empty or a closed subinterval of $[x, y]$ for any $x, y \in X$. In particular, if $x, y \in Y$, then $[x, y] \subseteq Y$ and this interval is then equipped with the same interval structure as in X. It is easy to see that conditions (T1)–(T5) are satisfied so that Y is a tree. The intersection of any collection of subtrees of X is a subtree (if nonempty). The *convex hull* of any subset $Z \subseteq X$ is the intersection of all subtrees containing Z.

A subtree Y is a closed subset of X and the inclusion $Y \hookrightarrow X$ is an embedding. We can define a *retraction* $r : X \to Y$ as follows: for $x \in X$ and $y \in Y$ the intersection $[x, y] \cap Y$ is an interval of the form $[r(x), y]$; one checks that $r(x)$ does not depend on the choice of y. The map r is continuous and restricts to the identity on Y. A subtree of X is *finite* if it is the convex hull of a finite set.

Let $(Y_\alpha)_{\alpha \in A}$ be an increasing net of finite subtrees of X, indexed by a directed set A (i.e. $Y_\alpha \subseteq Y_\beta$ when $\alpha \leq \beta$). Assume that the net is *rich* in the sense that for any two distinct points $x_1, x_2 \in X$ there exists $\alpha \in A$ such that the retraction $r_\alpha : X \to Y_\alpha$ satisfies $r_\alpha(x_1) \neq r_\alpha(x_2)$. For example, A could be the set of *all* finite subtrees, partially ordered by inclusion. The trees (Y_α) form an inverse system via the retraction maps $r_{\alpha\beta} : Y_\beta \to Y_\alpha$ for $a \leq \beta$ defined by $r_{\alpha\beta} = r_\alpha|_{Y_\beta}$, and we can form the inverse limit $\varprojlim Y_\alpha$, consisting of points $(y_\alpha)_{\alpha \in A}$ in the product space $\prod_\alpha Y_\alpha$ such that $r_{\alpha\beta}(y_\beta) = y_\alpha$ for all $\alpha \le \beta$. This inverse limit is a compact Hausdorff space. Since X retracts to each Y_α we get a continuous map

$$r : X \to \varprojlim Y_\alpha,$$

which is injective by the assumption that A is rich. That r is surjective is a consequence of condition (T5). Let us show that the inverse of r is also continuous. This will show that r is a homeomorphism, so that X is compact. (Of course, if we knew that X was compact, the continuity of r^{-1} would be immediate.)

Fix a point $x \in X$ and a tangent direction $\vec{v}$ at x. It suffices to show that $r(U(\vec{v}))$ is open in $\varprojlim Y_\alpha$. Pick a sequence $(x_n)_{n\ge1}$ in $U(\vec{v})$ such that $[x_{n+1}, x] \subseteq [x_n, x]$ and $\bigcap_n [x_n, x[= \emptyset$. By richness there exists $\alpha_n \in A$ such that $r_{\alpha_n}(x_n) \ne r_{\alpha_n}(x)$. Let $\vec{v}_n$ be the tangent direction in X at $r_{\alpha_n}(x)$ represented by $r_{\alpha_n}(x_n)$. Then $r(U(\vec{v}_n))$ is open in $\varprojlim Y_\alpha$, and hence so is $r(U(\vec{v})) = \bigcup_n r(U(\vec{v}_n))$.

Remark 2.3 One may form the inverse limit of any inverse system of finite trees (not necessarily subtrees of a given tree). However, such an inverse limit may contain a "compactified long line" and hence not be a tree!

2.2   Metric Trees

Let I be a set with an interval structure. A *generalized metric* on I is a function $d : I \times I \to [0, +\infty]$ satisfying:

(GM1) $d(x, y) = d(y, x)$ for all x, y, and $d(x, y) = 0$ iff $x = y$;
(GM2) $d(x, y) = d(x, z) + d(z, y)$ whenever $x \le z \le y$
(GM3) $d(x, y) < \infty$ if neither x nor y is an endpoint of I.
(GM4)   if $0 < d(x, y) < \infty$, then for every $\varepsilon > 0$ there exists $z \in I$ such that $x \le z \le y$ and $0 < d(x, z) < \varepsilon$.

A *metric tree* is a tree X together with a choice of generalized metric on each interval $[x, y]$ in X such that whenever $[z, w] \subseteq [x, y]$, the inclusion $[z, w] \hookrightarrow [x, y]$ is an isometry in the obvious sense.

It is an interesting question whether or not every tree is *metrizable* in the sense that it can be equipped with a generalized metric. See Remark 2.6 below.

2.2.1   Hyperbolic Space

Let X be a metric tree containing more than one point and let $x_0 \in X$ be a point that is not an end. Define *hyperbolic space* **H** to be the set of points $x \in X$ having finite distance from x_0. This definition does not depend on the choice of x_0. Note that all points in $X \setminus \mathbf{H}$ are ends, but that some ends in X may be contained in **H**.

The generalized metric on X restricts to a *bona fide* metric on **H**. One can show that **H** is complete in this metric and that **H** is an **R**-tree in the usual sense [GH90]. In general, even if $\mathbf{H} = X$, the topology generated by the metric may be strictly stronger than the weak topology. In fact, the weak topology on X may not be metrizable. This happens, for example, when there is a point with uncountable tangent space: such a point does not admit a countable basis of open neighborhoods.

2.2.2 Limit of Finite Trees

As noted in Remark 2.3, the inverse limit of finite trees may fail to be a tree. However, this cannot happen in the setting of metric trees. A *finite metric tree* is a finite tree equipped with a generalized metric in which all distances are finite. Suppose we are given a directed set A, a finite metric tree Y_α for each $\alpha \in A$ and, for $\alpha \leq \beta$:

- an isometric embedding $\iota_{\beta\alpha} : Y_\alpha \to Y_\beta$; this means that each interval in Y_α maps isometrically onto an interval in Y_β;
- a continuous map $r_{\alpha\beta} : Y_\beta \to Y_\alpha$ such that $r_{\alpha\beta} \circ \iota_{\beta\alpha} = \mathrm{id}_{Y_\alpha}$ and such that $r_{\alpha\beta}$ maps each connected component of $Y_\beta \setminus Y_\alpha$ to a single point in Y_α.

We claim that the space

$$X := \varprojlim_{\alpha} Y_\alpha$$

is naturally a metric tree. Recall that X is the set of points $(x_\alpha)_{\alpha \in A}$ in the product space $\prod_\alpha Y_\alpha$ such that $r_{\alpha\beta}(x_\beta) = x_\alpha$ for all $\alpha \leq \beta$. It is a compact Hausdorff space. For each α we have an injective map $\iota_\alpha : Y_\alpha \to X$ mapping $x \in Y_\alpha$ to $(x_\beta)_{\beta \in A}$, where $x_\beta \in Y_\beta$ is defined as follows: $x_\beta = r_{\beta\gamma}\iota_{\gamma\alpha}(x)$, where $\gamma \in A$ dominates both α and β. Abusing notation, we view Y_α as a subset of X. For distinct points $x, y \in X$ define

$$[x, y] := \{x\} \cup \bigcup_{\alpha \in A} [x_\alpha, y_\alpha] \cup \{y\}.$$

We claim that $[x, y]$ naturally carries an interval structure as well as a generalized metric. To see this, pick α_0 such that $x_{\alpha_0} \neq y_{\alpha_0}$ and $z = (z_\alpha) \in]x_{\alpha_0}, y_{\alpha_0}[$. Then $d_\alpha(x_\alpha, z_\alpha)$ and $d_\alpha(y_\alpha, z_\alpha)$ are finite and increasing functions of α, hence converge to $\delta_x, \delta_y \in [0, +\infty]$, respectively. This gives rise to an isometry of $[x, y]$ onto the interval $[-\delta_x, \delta_y] \subseteq [-\infty, +\infty]$.

2.3   Rooted and Parametrized Trees

Sometimes there is a point in a tree that plays a special role. This leads to the following notion.

Definition 2.4 A *rooted tree* is a partially ordered set $(X, \leq)$ satisfying the following properties:

(RT1) X has a unique minimal element x_0;

(RT2)   for any $x \in X \setminus \{x_0\}$, the set $\{z \in X \mid z \leq x\}$ is isomorphic (as a partially ordered set) to the real interval $[0, 1]$;

(RT3)   any two points $x, y \in X$ admit an infimum $x \wedge y$ in X, that is, $z \leq x$ and $z \leq y$ iff $z \leq x \wedge y$;

(RT4)   any totally ordered subset of X has a least upper bound in X.

Sometimes it is natural to reverse the partial ordering so that the root is the unique *maximal* element.

Remark 2.5 In [FJ04] it was claimed that (RT3) follows from the other three axioms but this is not true. A counterexample is provided by two copies of the interval $[0, 1]$ identified along the half-open subinterval $[0, 1[$. I am grateful to Josnei Novacoski and Franz-Viktor Kuhlmann for pointing this out.

Let us compare this notion with the definition of a tree above. If $(X, \leq)$ is a rooted tree, then we can define intervals $[x, y] \subseteq X$ as follows. First, when $x \leq y \in X$, set $[x, y] := \{z \in X \mid x \leq z \leq y\}$ and $[y, x] := [x, y]$. For general $x, y \in X$ set $[x, y] := [x \wedge y, x] \cup [x \wedge y, y]$. We leave it to the reader to equip $[x, y]$ with an interval structure and to verify conditions (T1)–(T5). Conversely, given a tree X and a point $x_0 \in X$, define a partial ordering on X by declaring $x \leq y$ iff $x \in [x_0, y]$. One checks that conditions (RT1)–(RT4) are verified.

A *parametrization* of a rooted tree $(X, \leq)$ as above is a monotone function $\alpha : X \to [-\infty, +\infty]$ whose restriction to any segment $[x, y]$ with $x < y$ is a homeomorphism onto a closed subinterval of $[-\infty, +\infty]$. We also require $|\alpha(x_0)| < \infty$ unless x_0 is an endpoint of X. This induces a generalized metric on X by setting

$$d(x, y) = |\alpha(x) - \alpha(x \wedge y)| + |\alpha(y) - \alpha(x \wedge y)|$$

for distinct points $x, y \in X$. The set **H** is exactly the locus where $|\alpha| < \infty$. Conversely given a generalized metric d on a tree X, a point $x_0 \in \mathbf{H}$ and a real number $\alpha_0 \in \mathbf{R}$, we obtain an increasing parametrization α of the tree X rooted in x_0 by setting $\alpha(x) = \alpha_0 + d(x, x_0)$.

Remark 2.6 A natural question is whether or not every rooted tree admits a parametrization. In personal communication to the author, Andreas Blass has outlined an example of a rooted tree that cannot be parametrized. His construction relies on Suslin trees [Jec03], the existence of which cannot be decided from the ZFC axioms. It would be interesting to have a more explicit example.

2.4   Radon Measures on Trees

Let us review the notions of Borel and Radon measures on compact topological spaces and, more specifically, on trees.

2.4.1   Radon and Borel Measures on Compact Spaces

A reference for the material in this section is [Fol99, §7.1-2]. Let X be a compact (Hausdorff) space and $\mathcal{B}$ the associated Borel σ-algebra. A *Borel measure* on X is a function $\rho : \mathcal{B} \to [0, +\infty]$ satisfying the usual axioms. A Borel measure ρ is *regular* if for every Borel set $E \subseteq X$ and every $\varepsilon > 0$ there exists a compact set F and an open set U such that $F \subseteq E \subseteq U$ and $\rho(U \setminus F) < \varepsilon$.

A *Radon measure* on X is a positive linear functional on the vector space $C^0(X)$ of continuous functions on X. By the Riesz representation theorem, Radon measures can be identified with regular Borel measures.

If X has the property that every open set of X is σ-compact, that is, a countable union of compact sets, then every Borel measure on X is Radon. However, many Berkovich spaces do not have this property. For example, the Berkovich projective line over any non-Archimedean field K is a tree, but if the residue field of K is uncountable, then the complement of any Type 2 point (see Sect. 3.3.4) is an open set that is not σ-compact.

We write $\mathcal{M}^+(X)$ for the set of positive Radon measures on X and endow it with the topology of weak (or vague) convergence. By the Banach-Alaoglu Theorem, the subspace $\mathcal{M}_1^+(X)$ of Radon probability measure is compact.

A *finite atomic measure* on X is a Radon measure of the form $\rho = \sum_{i=1}^{N} c_i \delta_{x_i}$, where $c_i > 0$. A *signed* Radon measure is a real-valued linear functional on $C^0(X; \mathbf{R})$. The only signed measures that we shall consider will be of the form $\rho - \rho_0$, where ρ is a Radon measure and ρ_0 a finite atomic measure.

2.4.2   Measures on Finite Trees

Let X be a *finite* tree. It is then easy to see that every connected open set is of the form $\bigcap_{i=1}^{n} U(\vec{v}_i)$, where $\vec{v}_1, \ldots, \vec{v}_n$ are tangent directions in X such that $U(\vec{v}_i) \cap U(\vec{v}_j) \neq \emptyset$ but $U(\vec{v}_i) \not\subseteq U(\vec{v}_j)$ for $i \neq j$. Each such set is a countable union of compact subsets, so it follows from the above that every Borel measure is in fact a Radon measure.

2.4.3 Radon Measures on General Trees

Now let X be an arbitrary tree in the sense of Definition 2.2. It was claimed in [FJ04] and [BR10] that in this case, too, every Borel measure is Radon, but there is a gap in the proofs.

Example 2.7 Let Y be a set with the following property: there exists a probability measure μ on the maximal σ-algebra (that contains all subsets of Y) that gives zero mass to any finite set. The existence of such a set, whose cardinality is said to be a *real-valued measurable cardinal* is a well known problem in set theory [Fre93]: suffice it to say that its existence or nonexistence cannot be decided from the ZFC axioms. Now equip Y with the discrete topology and let X be the cone over Y, that is $X = Y \times [0, 1]/ \sim$, where $(y, 0) \sim (y', 0)$ for all $y, y' \in Y$. Let $\phi : Y \to X$ be the continuous map defined by $\phi(y) = (y, 1)$. Then $\rho := \phi_* \mu$ is a Borel measure on X which is not Radon. Indeed, the open set $U := X \setminus \{0\}$ has measure 1, but any compact subset of U is contained in a *finite* union of intervals $\{y\} \times]0, 1]$ and thus has measure zero.

Fortunately, this does not really lead to any problems. The message to take away is that on a general tree, one should systematically use Radon measures, and this is indeed what we shall do here.

2.4.4 Coherent Systems of Measures

The description of a general tree X as a pro-finite tree is well adapted to describe Radon measures on X. Namely, let $(Y_\alpha)_{\alpha \in A}$ be a rich net of finite subtrees of X, in the sense of Sect. 2.1.2. The homeomorphism $X \xrightarrow{\sim} \varprojlim Y_\alpha$ then induces a homeomorphism $\mathcal{M}_1^+(X) \xrightarrow{\sim} \varprojlim \mathcal{M}_1^+(Y_\alpha)$. Concretely, the right hand side consists of collections $(\rho_\alpha)_{\alpha \in A}$ of Radon measures on each Y_α satisfying $(r_{\alpha\beta})_* \rho_\beta = \rho_\alpha$ for $\alpha \leq \beta$. Such a collection of measures is called a *coherent system of measures* in [BR10]. The homeomorphism above assigns to a Radon probability measure ρ on X the collection $(\rho_\alpha)_{\alpha \in A}$ defined by $\rho_\alpha := (r_\alpha)_* \rho$.

2.5 Potential Theory

Next we outline how to do potential theory on a metric tree. The presentation is adapted to our needs but basically follows [BR10], especially §1.4 and §2.5. The Laplacian on a tree is a combination of the usual real Laplacian with the combinatorially defined Laplacian on a simplicial tree.

2.5.1 Quasisubharmonic Functions on Finite Metric Trees

Let X be a *finite* metric tree. The Laplacian Δ on X is naturally defined on the class $\mathrm{BDV}(X) \subseteq C^0(X)$ of functions with bounded differential variation, see [BR10, §3.5], but we shall restrict our attention to the subclass $\mathrm{QSH}(X) \subseteq \mathrm{BDV}(X)$ of *quasisubharmonic* functions.

Let $\rho_0 = \sum_{i=1}^{N} c_i \delta_{x_i}$ be a finite atomic measure on X. Define the class $\mathrm{SH}(X, \rho_0)$ of ρ_0-*subharmonic* functions as the set of continuous functions φ that are convex on any segment disjoint from the support of ρ_0 and such that, for any $x \in X$:

$$\rho_0\{x\} + \sum_{\vec{v}} D_{\vec{v}}\varphi \geq 0,$$

where the sum is over all tangent directions $\vec{v}$ at x. Here $D_{\vec{v}}\varphi$ denotes the directional derivative of φ in the direction $\vec{v}$ (outward from x): this derivative is well defined by the convexity of φ. We leave it to the reader to verify that

$$D_{\vec{v}}\varphi \leq 0 \quad \text{whenever} \quad \rho_0(U(\vec{v})) = 0 \tag{2}$$

for any $\varphi \in \mathrm{SH}(X, \rho_0)$; this inequality is quite useful.

Define $\mathrm{QSH}(X)$ as the union of $\mathrm{SH}(X, \rho_0)$ over all finite atomic measures ρ_0. Note that if ρ_0, ρ_0' are two finite atomic measures with $\rho_0' \geq \rho_0$, then $\mathrm{SH}(X, \rho_0) \subseteq \mathrm{SH}(X, \rho_0')$. We also write $\mathrm{SH}(X, x_0) := \mathrm{SH}(X, \delta_{x_0})$ and refer to its elements as x_0-*subharmonic*.

Let $Y \subseteq X$ be a subtree of X containing the support of ρ_0. We have an injection $\iota : Y \hookrightarrow X$ and a retraction $r : X \to Y$. It follows easily from (2) that

$$\iota^* \mathrm{SH}(X, \rho_0) \subseteq \mathrm{SH}(Y, \rho_0) \quad \text{and} \quad r^* \mathrm{SH}(Y, \rho_0) \subseteq \mathrm{SH}(X, \rho_0).$$

Moreover, $\varphi \leq r^* \iota^* \varphi$ for any $\varphi \in \mathrm{SH}(X, \rho_0)$.

2.5.2 Laplacian

For $\varphi \in \mathrm{QSH}(X)$, we let $\Delta\varphi$ be the signed (Borel) measure on X defined as follows: if $\vec{v}_1, \ldots, \vec{v}_n$ are tangent directions in X such that $U(\vec{v}_i) \cap U(\vec{v}_j) \neq \emptyset$ but $U(\vec{v}_i) \not\subseteq U(\vec{v}_j)$ for $i \neq j$, then

$$\Delta\varphi\left(\bigcap_{i=1}^{n} U(\vec{v}_i)\right) = \sum_{i=1}^{n} D_{\vec{v}_i}\varphi.$$

This equation defines $\Delta\varphi$ uniquely as every open set in X is a countable disjoint union of open sets of the form $\bigcap U(\vec{v}_i)$. The mass of $\Delta\varphi$ at a point $x \in X$ is given by $\sum_{\vec{v} \in T_x} D_{\vec{v}} \varphi$ and the restriction of $\Delta \varphi$ to any open segment $I \subseteq X$ containing no branch point is equal to the usual real Laplacian of $\varphi|_I$.

The Laplace operator is essentially injective. Indeed, suppose $\varphi_1, \varphi_2 \in \mathrm{QSH}(X)$ and $\Delta \varphi_1 = \Delta \varphi_2$. We may assume $\varphi_1, \varphi_2 \in \mathrm{SH}(X, \rho_0)$ for a common positive measure ρ_0. If $\varphi = \varphi_1 - \varphi_2$, then φ is affine on any closed interval whose interior is disjoint from the support of ρ_0. Moreover, at any point $x \in X$ we have $\sum_{\vec{v} \in T_x} D_{\vec{v}} \varphi = 0$. These two conditions easily imply that φ is constant. (Too see this, first check that φ is locally constant at any end of X.)

If $\varphi \in \mathrm{SH}(X, \rho_0)$, then $\rho_0 + \Delta \varphi$ is a positive Borel measure on X of the same mass as ρ_0. In particular, when ρ_0 is a probability measure, we obtain a map

$$\mathrm{SH}(X, \rho_0) \ni \varphi \mapsto \rho_0 + \Delta \varphi \in \mathcal{M}_1^+(X), \tag{3}$$

where $\mathcal{M}_1^+(X)$ denotes the set of probability measures on X. We claim that this map is surjective. To see this, first note that the function $\varphi_{y,z}$ given by

$$\varphi_{y,z}(x) = -d(z, x \wedge_z y), \tag{4}$$

with $x \wedge_z y \in X$ as in (T4), belongs to $\mathrm{SH}(X, z)$ and satisfies $\Delta \varphi = \delta_y - \delta_z$. For a general probability measure ρ and finite atomic probability measure ρ_0, the function

$$\varphi(x) = \iint \varphi_{y,z}(x) \, d\rho(y) d\rho_0(z) \tag{5}$$

belongs to $\mathrm{SH}(X, \rho_0)$ and satisfies $\Delta \varphi = \rho - \rho_0$.

Let $Y \subseteq X$ be a subtree containing the support of ρ_0 and denote the Laplacians on X and Y by Δ_X and Δ_Y, respectively. Then, with notation as above,

$$\Delta_Y(\iota^* \varphi) = r_*(\Delta_X \varphi) \quad \text{for } \varphi \in \mathrm{SH}(X, \rho_0) \tag{6}$$

$$\Delta_X(r^* \varphi) = \iota_*(\Delta_Y \varphi) \quad \text{for } \varphi \in \mathrm{SH}(Y, \rho_0), \tag{7}$$

where $\iota : Y \hookrightarrow X$ and $r : X \to Y$ are the inclusion and retraction, respectively.

2.5.3 Equicontinuity

The spaces $\mathrm{SH}(X, \rho_0)$ have very nice compactness properties deriving from the fact that if ρ_0 is a probability measure then

$$|D_{\vec{v}} \varphi| \leq 1 \quad \text{for all tangent directions } \vec{v} \text{ and all } \varphi \in \mathrm{SH}(X, \rho_0). \tag{8}$$

Indeed, using the fact that a function in $\mathrm{QSH}(X)$ is determined, up to an additive constant, by its Laplacian (8) follows from (4) when ρ_0 and $\rho_0 + \Delta \varphi$ are Dirac masses, and from (5) in general.

As a consequence of (8), the functions in $SH(X, \rho_0)$ are uniformly Lipschitz continuous and in particular equicontinuous. This shows that pointwise convergence in $SH(X, \rho_0)$ implies uniform convergence.

The space $SH(X, \rho_0)$ is easily seen to be closed in the C^0-topology, so we obtain several compactness assertions from the Arzela-Ascoli theorem. For example, the set $SH^0(X, \rho_0)$ of $\varphi \in SH(X, \rho_0)$ for which $\max \varphi = 0$ is compact.

Finally, we have an exact sequence of topological vector spaces

$$0 \to \mathbf{R} \to SH(X, \rho_0) \to \mathcal{M}_1^+(X) \to 0; \tag{9}$$

here $\mathcal{M}_1^+(X)$ is equipped with the weak topology on measures. Indeed, the construction in (4)–(5) gives rise to a continuous bijection between $\mathcal{M}_1^+(X)$ and $SH(X, \rho_0)/\mathbf{R} \simeq SH^0(X, \rho_0)$. By compactness, the inverse is also continuous.

2.5.4 Quasisubharmonic Functions on General Metric Trees

Now let X be a general metric tree and ρ_0 a finite atomic measure supported on the associated hyperbolic space $\mathbf{H} \subseteq X$.

Let A be the set of finite metric subtrees of X that contain the support of ρ_0. This is a directed set, partially ordered by inclusion. For $\alpha \in A$, denote the associated metric tree by Y_α. The net $(Y_\alpha)_{\alpha \in A}$ is rich in the sense of Sect. 2.1.2, so the retractions $r_\alpha : X \to Y_\alpha$ induce a homeomorphism $r : X \overset{\sim}{\to} \varprojlim Y_\alpha$.

Define $SH(X, \rho_0)$ to be the set of functions $\varphi : X \to [-\infty, 0]$ such that $\varphi|_{Y_\alpha} \in SH(Y_\alpha, \rho_0)$ for all $\alpha \in A$ and such that $\varphi = \lim r_\alpha^* \varphi$. Notice that in this case $r_\alpha^* \varphi$ in fact decreases to φ. Since $r_\alpha^* \varphi$ is continuous for all α, this implies that ψ is upper semicontinuous.

We define the topology on $SH(X, \rho_0)$ in terms of pointwise convergence on $\mathbf{H}$. Thus a net φ_i converges to φ in $SH(X, \rho_0)$ iff $\varphi_i|_{Y_\alpha}$ converges to $\varphi|_{Y_\alpha}$ for all α. Note, however, that the convergence $\varphi_i \to \varphi$ is not required to hold on all of X.

Since, for all α, $SH(Y_\alpha, \rho_0)$ is compact in the topology of pointwise convergence on Y_α, it follows that $SH(X, \rho_0)$ is also compact. The space $SH(X, \rho_0)$ has many nice properties beyond compactness. For example, if $(\varphi_i)_i$ is a decreasing net in $SH(X, \rho_0)$, and $\varphi := \lim \varphi_i$, then either $\varphi_i \equiv -\infty$ on X or $\varphi \in SH(X, \rho_0)$. Further, if $(\varphi_i)_i$ is a family in $SH(X, \rho_0)$ with $\sup_i \max_X \varphi_i < \infty$, then the upper semicontinuous regularization of $\varphi := \sup_i \varphi_i$ belongs to $SH(X, \rho_0)$.

As before, we define $QSH(X)$, the space of *quasisubharmonic functions*, to be the union of $SH(X, \rho_0)$ over all finite atomic measures ρ_0 supported on $\mathbf{H}$.

2.5.5 Laplacian

Let X, ρ_0 and A be as above. Recall that a Radon probability measure ρ on X is given by a coherent system $(\rho_\alpha)_{\alpha \in A}$ of (Radon) probability measures on Y_α.

For $\varphi \in SH(X, \rho_0)$ we define $\rho_0 + \Delta\varphi \in \mathcal{M}_1^+(X)$ to be the unique Radon probability measure such that

$$(r_\alpha)_*(\rho_0 + \Delta\varphi) = \rho_0 + \Delta_{Y_\alpha}(\varphi|_{Y_\alpha})$$

for all $\alpha \in A$. This makes sense in view of (6).

The construction in (4)–(5) remains valid and the sequence (9) of topological vector spaces is exact. For future reference we record that if $(\varphi_i)_i$ is a net in $SH^0(X, \rho_0)$, then $\varphi_i \to 0$ (pointwise on **H**) iff $\Delta\varphi_i \to 0$ in $\mathcal{M}_1^+(X)$.

2.5.6 Singularities of Quasisubharmonic Functions

Any quasisubharmonic function on a metric tree X is bounded from above on all of X and Lipschitz continuous on hyperbolic space **H**, but can take the value $-\infty$ at infinity. For example, if $x_0 \in \mathbf{H}$ and $y \in X \setminus \mathbf{H}$, then the function $\varphi(x) = -d_\mathbf{H}(x_0, x \wedge_{x_0} y)$ is x_0-subharmonic and $\varphi(y) = -\infty$. Note that $\Delta\varphi = \delta_y - \delta_{x_0}$. The following result allows us to estimate a quasisubharmonic function from below in terms of the mass of its Laplacian at infinity. It will be used in the proof of the equidistribution result in Sect. 5.7.

Proposition 2.8 *Let ρ_0 be a finite atomic probability measure on* **H** *and let $\varphi \in$ SH(X, ρ_0). Pick $x_0 \in \mathbf{H}$ and any number $\lambda > \sup_{y \in X \setminus \mathbf{H}} \Delta\varphi\{y\}$. Then there exists a constant $C = C(x_0, \rho_0, \varphi, \lambda) > 0$ such that*

$$\varphi(x) \geq \varphi(x_0) - C - \lambda d_\mathbf{H}(x, x_0)$$

for all $x \in \mathbf{H}$.

We shall use the following estimates, which are of independent interest.

Lemma 2.9 *Let ρ_0 be a finite atomic probability measure on* **H** *and let $x_0 \in \mathbf{H}$. Pick $\varphi \in$ SH(X, ρ_0) and set $\rho = \rho_0 + \Delta\varphi$. Then*

$$\varphi(x) - \varphi(x_0) \geq -\int_{x_0}^x \rho\{z \geq y\} d\alpha(y) \geq -d_\mathbf{H}(x, x_0) \cdot \rho\{z \geq x\},$$

where $\leq$ is the partial ordering on X rooted in x_0.

Proof of Lemma 2.9 It follows from (5) that

$$\varphi(x) - \varphi(x_0) = -\int_{x_0}^x (\Delta\varphi)\{z \geq y\} d\alpha(y)$$

$$\geq -\int_{x_0}^x \rho\{z \geq y\} d\alpha(y) \geq -\int_{x_0}^x \rho\{z \geq x\} d\alpha(y) = -d_\mathbf{H}(x, x_0) \cdot \rho\{z \geq x\},$$

where we have used that $\rho \geq \Delta\varphi$ and $x \geq y$. □

Proof of Proposition 2.8 Let $\leq$ denote the partial ordering rooted in x_0 and set

$$Y_\lambda := \{y \in X \mid (\rho_0 + \Delta\varphi)\{z \geq y\} \geq \lambda\}.$$

Recall that $\rho_0 + \Delta\varphi$ is a probability measure. Thus $Y_\lambda = \emptyset$ if $\lambda > 1$. If $\lambda \leq 1$, then Y_λ is a finite subtree of X containing x_0 and having at most $1/\lambda$ ends. The assumption that $\lambda > \sup_{y \in X \setminus \mathbf{H}} \Delta\varphi\{y\}$ implies that Y_λ is in fact contained in $\mathbf{H}$. In particular, the number $C := \sup_{y \in Y_\lambda} d_{\mathbf{H}}(x_0, y)$ is finite.

It now follows from Lemma 2.9 that

$$\varphi(x) - \varphi(x_0) \geq -\int_{x_0}^x (\rho_0 + \Delta\varphi)\{z \geq y\} d\alpha(y) \geq -C - \lambda d_{\mathbf{H}}(x, x_0),$$

completing the proof. $\qquad\square$

2.5.7 Regularization

In complex analysis, it is often useful to approximate a quasisubharmonic function by a decreasing sequence of smooth quasisubharmonic functions. In higher dimensions, regularization results of this type play a fundamental role in pluripotential theory, as developed by Bedford and Taylor [BT82, BT87]. They are also crucial to the approach to non-Archimedean pluripotential theory in [BFJ08b, BFJ12, BFJ14].

Let us say that a function $\varphi \in \mathrm{SH}(X, \rho_0)$ is *regular* if it is piecewise affine in the sense that $\Delta\varphi = \rho - \rho_0$, where ρ is a finite atomic measure supported on $\mathbf{H}$.

Theorem 2.10 *For any $\varphi \in \mathrm{SH}(X, \rho_0)$ there exists a decreasing sequence of regular functions $(\psi_n)_{n=1}^\infty$ in $\mathrm{SH}(X, \rho_0)$ such that φ_n converges pointwise to φ on X.*

Proof Let $Y_0 \subset X$ be a finite tree containing the support of ρ_0 and pick a point $x_0 \in Y_0$. Set $\rho = \rho_0 + \Delta\varphi$.

First assume that ρ is supported on a finite subtree contained in $\mathbf{H}$. We may assume $Y_0 \subseteq Y$. For each $n \geq 1$, write $Y \setminus \{x_0\}$ as a finite disjoint union of half-open segments $\gamma_i =]x_i, y_i]$, $i \in I_n$, called segments of order n, in such a way that each segment of order n has length at most 2^{-n} and is the disjoint union of two segments of order $n + 1$. Define finite atomic measures ρ_n by

$$\rho_n = \rho\{x_0\}\delta_{x_0} + \sum_{i \in I_n} \rho(\gamma_i)\delta_{y_i}$$

and define $\varphi_n \in \mathrm{SH}(X, x_0)$ by $\Delta\varphi_n = \rho_n - \rho_0$, $\varphi_n(x_0) = \varphi(x_0)$. From (4) and (5) it follows that φ_n decreases to φ pointwise on X, as $n \to \infty$. Since $\varphi = r_Y^*\varphi$ is continuous, the convergence is in fact uniform by Dini's Theorem.

Now consider a general $\varphi \in \mathrm{SH}(X, \rho_0)$. For $n \geq 1$, define $Y'_n \subseteq X$ by

$$Y'_n := \{y \in X \mid \rho\{z \geq y\} \geq 2^{-n} \quad \text{and} \quad d_{\mathbf{H}}(x_0, y) \leq 2^n\},$$

where $\leq$ denotes the partial ordering rooted in x_0. Then Y'_n is a finite subtree of X and $Y'_n \subseteq Y'_{n+1}$ for $n \geq 1$. Let Y_n be the convex hull of the union of Y'_n and Y_0 and set $\psi_n = r^*_{Y_n} \varphi_n$. Since $Y_n \subseteq Y_{n+1}$, we have $\varphi \leq \psi_{n+1} \leq \psi_n$ for all n. We claim that $\psi_n(x)$ converges to $\varphi(x)$ as $n \to \infty$ for every $x \in X$. Write $x_n := r_{Y_n}(x)$ so that $\psi_n(x) = \varphi(x_n)$. The points x_n converge to a point $y \in [x_0, x]$ and $\lim_n \psi_n(x) = \varphi(y)$. If $y = x$, then we are done. But if $y \neq x$, then by construction of Y'_n, the measure ρ puts no mass on the interval $]y, x]$, so it follows from (4) and (5) that $\varphi(x) = \varphi(y)$.

Hence ψ_n decreases to φ pointwise on X as $n \to \infty$. By the first part of the proof, we can find a regular $\varphi_n \in \mathrm{SH}(X, \rho_0)$ such that $\psi_n \leq \varphi_n \leq \psi_n + 2^{-n}$ on X. Then φ_n decreases to φ pointwise on X, as desired.   □

Remark 2.11 A different kind of regularization is used in [FR06, §4.6]. Fix a point $x_0 \in \mathbf{H}$ and for each $n \geq 1$ let $X_n \subseteq X$ be the (a priori not finite) subtree defined by $X_n = \{x \in X \mid d_{\mathbf{H}}(x_0, x) \leq n^{-1}\}$. Let $\varphi_n \in \mathrm{SH}(X, \rho_0)$ be defined by $\rho_0 + \Delta \varphi_n = (r_n)_*(\rho_0 + \Delta \varphi)$ and $\varphi_n(x_0) = \varphi(x_0)$, where $r_n : X \to X_n$ is the retraction. Then φ_n is bounded and φ_n decreases to φ as $n \to \infty$.

2.6   Tree Maps

Let X and X' be trees in the sense of Sect. 2.2. We say that a continuous map $f : X \to X'$ is a *tree map* if it is open, surjective and finite in the sense that there exists a number d such that every point in X' has at most d preimages in X. The smallest such number d is the *topological degree* of f.

Proposition 2.12 *Let $f : X \to X'$ be a tree map of topological degree d.*

(i) *if $U \subseteq X$ is a connected open set, then so is $f(U)$ and $\partial f(U) \subseteq f(\partial U)$;*

(ii) *if $U' \subseteq X'$ is a connected open set and U is a connected component of $f^{-1}(U')$, then $f(U) = U'$ and $f(\partial U) = \partial U'$; as a consequence, $f^{-1}(U')$ has at most d connected components;*

(iii) *if $U \subseteq X$ is a connected open set and $U' = f(U)$, then U is a connected component of $f^{-1}(U')$ iff $f(\partial U) \subseteq \partial U'$.*

The statement is valid for finite surjective open continuous maps $f : X \to X'$ between compact Hausdorff spaces, under the assumption that every point of X admits a basis of connected open neighborhoods. We omit the elementary proof; see Lemma 9.11, Lemma 9.12 and Proposition 9.15 in [BR10] for details.

Corollary 2.13 *Consider a point $x \in X$ and set $x' := f(x) \in X'$. Then there exists a connected open neighborhood V of x with the following properties:*

(i) *if $\vec{v}$ is a tangent direction at x, then there exists a tangent direction $\vec{v}'$ at x' such that $f(V \cap U(\vec{v})) \subseteq U(\vec{v}')$; furthermore, either $f(U(\vec{v})) = U(\vec{v}')$ or $f(U(\vec{v})) = X'$;*

(ii) *if $\vec{v}'$ is a tangent direction at x' then there exists a tangent direction $\vec{v}$ at x such that $f(V \cap U(\vec{v})) \subseteq U(\vec{v}')$.*

Definition 2.14 *The tangent map of f at x is the map that associates $\vec{v}'$ to $\vec{v}$.*

The tangent map is surjective and every tangent direction has at most d preimages. Since the ends of X are characterized by the tangent space being a singleton, it follows that f maps ends to ends.

Proof of Corollary 2.13 Pick V small enough so that it contains no preimage of x' besides x. Note that (ii) follows from (i) and the fact that $f(V)$ is an open neighborhood of x'.

To prove (i), note that $V \cap U(\vec{v})$ is connected for every $\vec{v}$. Hence $f(V \cap U(\vec{v}))$ is connected and does not contain x', so it must be contained in $U(\vec{v}')$ for some $\vec{v}'$. Moreover, the fact that f is open implies $\partial f(U(\vec{v})) \subseteq f(\partial U(\vec{v})) = \{x'\}$. Thus either $f(U(\vec{v})) = X'$ or $f(U(\vec{v}))$ is a connected open set with boundary $\{x'\}$. In the latter case, we must have $f(U(\vec{v})) = U(\vec{v}')$.   □

2.6.1   Images and Preimages of Segments

The following result makes the role of the tangent map more precise.

Corollary 2.15 *Let $f : X \to X'$ be a tree map as above. Then:*

(i) *if $\vec{v}$ is a tangent direction at a point $x \in X$, then there exists a point $y \in U(\vec{v})$ such that f is a homeomorphism of the interval $[x, y] \subseteq X$ onto the interval $[f(x), f(y)] \subseteq X'$; furthermore, f maps the annulus $A(x, y)$ onto the annulus $A(f(x), f(y))$;*

(ii) *if $\vec{v}'$ is a tangent direction at a point $x' \in X'$, then there exists $y' \in U(\vec{v}')$ such that if $\gamma' := [x', y']$ then $f^{-1}\gamma' = \bigcup_i \gamma_i$, where the $\gamma_i = [x_i, y_i]$ are closed intervals in X with pairwise disjoint interiors and f maps γ_i homeomorphically onto γ' for all i; furthermore we have $f(A(x_i, y_i)) = A(x', y')$ for all i and $f^{-1}(A(x', y')) = \bigcup_i A(x_i, y_i)$.*

Proof We first prove (ii). Set $U' = U(\vec{v}')$ and let U be a connected component of $f^{-1}(U')$. By Proposition 2.12 (ii), the boundary of U consists of finitely many preimages $x_1, \ldots x_m$ of x'. (The same preimage of x' can lie on the boundary of several connected components U.) Since U is connected, there exists, for $1 \leq i \leq m$, a unique tangent direction $\vec{v}_i$ at x_i such that $U \subseteq U(\vec{v}_i)$.

Pick any point $z' \in U'$. Also pick points $z_1, \ldots, z_m$ in U such that the segments $[x_i, z_i]$ are pairwise disjoint. Then $f(]x_i, z_i]) \cap]x', z'] \neq \emptyset$ for all i, so we can find

$y' \in]x', z']$ and $y_i \in]x_i, z_i]$ arbitrarily close to x_i such that $f(y_i) = y'$ for all i. In particular, we may assume that the annulus $A_i := A(x_i, y_i)$ contains no preimage of z'. By construction it contains no preimage of x' either. Proposition 2.12 (i) first shows that $\partial f(A_i) \subseteq \{x', y'\}$, so $f(A_i) = A' := A(x', y')$ for all i. Proposition 2.12 (iii) then implies that A_i is a connected component of $f^{-1}(A')$. Hence $f^{-1}(A') \cap U = \bigcup_i A_i$.

Write $\gamma_i = [x_i, y_i]$ and $\gamma' = [x', y']$. Pick any $\xi \in]x_i, y_i[$ and set $\xi' := f(\xi)$. On the one hand, $f(A(\xi, y_i)) \subseteq f(A_i) = A'$. On the other hand, $\partial f(A(\xi, y_i)) \subseteq \{\xi', y'\}$ so we must have $f(A(\xi, y_i)) = A(\xi', y')$ and $\xi' \in \gamma'$. We conclude that $f(\gamma_i) = \gamma'$ and that $f : \gamma_i \to \gamma'$ is injective, hence a homeomorphism.

The same argument gives $f(A(x_i, \xi)) = A(x', \xi)$. Consider any tangent direction $\vec{w}$ at ξ such that $U(\vec{w}) \subseteq A_i$. As above we have $f(U(\vec{w})) \subseteq A'$ and $\partial f(U(\vec{w})) \subseteq \{\xi'\}$, which implies $f(U(\vec{w})) = U(\vec{w}')$ for some tangent direction $\vec{w}$ at ξ' for which $U(\vec{w}) \subseteq A'$. We conclude that $f^{-1}(\gamma') \cap A_i \subseteq \gamma_i$.

This completes the proof of (ii), and (i) is an easy consequence. □

Using compactness, we easily deduce the following result from Corollary 2.15. See the proof of Theorem 9.35 in [BR10].

Corollary 2.16 *Let $f : X \to X'$ be a tree map as above. Then:*

(i) *any closed interval γ in X can be written as a finite union of closed intervals γ_i with pairwise disjoint interiors, such that $\gamma'_i := f(\gamma_i) \subseteq X'$ is an interval and $f : \gamma_i \to \gamma'_i$ is a homeomorphism for all i; furthermore, f maps the annulus $A(\gamma_i)$ onto the annulus $A(\gamma'_i)$;*

(ii) *any closed interval γ' in X' can be written as a union of finitely many intervals γ'_i with pairwise disjoint interiors, such that, for all i, $f^{-1}(\gamma'_i)$ is a finite union of closed intervals γ_{ij} with pairwise disjoint interiors, such that $f : \gamma_{ij} \to \gamma'_i$ is a homeomorphism for each j; furthermore, f maps the annulus $A(\gamma_{ij})$ onto the annulus $A(\gamma'_i)$; and $A(\gamma_{ij})$ is a connected component of $f^{-1}(A(\gamma'_i))$.*

2.6.2 Fixed Point Theorem

It is an elementary fact that any continuous selfmap of a *finite* tree admits a fixed point. This can be generalized to arbitrary trees. Versions of the following fixed point theorem can be found in [FJ04, Riv04, BR10].

Proposition 2.17 *Any tree map $f : X \to X$ admits a fixed point $x = f(x) \in X$. Moreover, we can assume that one of the following two conditions hold:*

(i) *x is not an end of X;*

(ii) *x is an end of X and x is an attracting fixed point: there exists an open neighborhood $U \subseteq X$ of x such that $f(U) \subseteq U$ and $\bigcap_{n \geq 0} f^n(U) = \{x\}$.*

In the proof we will need the following easy consequence of Corollary 2.16 (i).

Lemma 2.18 *Suppose there are points* $x, y \in X$, $x \neq y$, *with* $r(f(x))) = x$ *and* $r(f(y)) = y$, *where* r *denotes the retraction of* X *onto the segment* $[x, y]$. *Then* f *has a fixed point on* $[x, y]$.

Proof of Proposition 2.17 We may suppose that f does not have any fixed point that is not an end of X, or else we are in case (i). Pick any non-end $x_0 \in X$ and pick a finite subtree X_0 that contains x_0, all preimages of x_0, but does not contain any ends of X. Let A be the set of finite subtrees of X that contain X_0 but does not contain any end of X. For $\alpha \in A$, let Y_α be the corresponding subtree. Then $(Y_\alpha)_{\alpha \in A}$ is a rich net of subtrees in the sense of Sect. 2.1.2, so $X \xrightarrow{\sim} \varprojlim Y_\alpha$.

For each α, define $f_\alpha : Y_\alpha \to Y_\alpha$ by $f_\alpha = f \circ r_\alpha$. This is a continuous selfmap of a finite tree so the set F_α of its fixed points is a nonempty compact set. We will show that $r_\alpha(F_\beta) = F_\alpha$ when $\beta \geq \alpha$. This will imply that there exists $x \in X$ such that $r_\alpha(f(r_\alpha(x)) = r_\alpha(x)$ for all α. By assumption, x is an end in X. Pick a sequence $(x_n)_{n=0}^\infty$ of points in X such that $x_{n+1} \in]x_n, x[$ and $x_n \to x$ as $n \to \infty$. Applying what precedes to the subtrees $Y_{\alpha_n} = X_0 \cup [x_0, x_n]$ we easily conclude that x is an attracting fixed point.

It remains to show that $r_\alpha(F_\beta) = F_\alpha$ when $\beta \geq \alpha$. First pick $x_\beta \in F_\beta$. We will show that $x_\alpha := r_\alpha(x_\beta) \in F_\alpha$. This is clear if $x_\beta \in Y_\alpha$ since $r_\alpha = r_{\alpha\beta} \circ r_\beta$, so suppose $x_\beta \notin Y_\alpha$. By assumption, $f(x_\alpha) \neq x_\alpha$ and $f(x_\beta) \neq x_\beta$. Let $\vec{v}$ be the tangent direction at x_α represented by x_β. Then $U(\vec{v}) \cap Y_\alpha = \emptyset$ so $x_0 \notin f(U(\vec{v}))$ and hence $f(U(\vec{v})) = U(\vec{v}')$ for some tangent direction $\vec{v}'$ at $f(x_\alpha)$. Note that $f(x_\beta) \in U(\vec{v}')$. If $f(x_\alpha) \notin U(\vec{v})$, then Lemma 2.18 applied to $x = x_\alpha$, $y = x_\beta$ gives a fixed point for f in $[x_\alpha, x_\beta] \subseteq Y_\beta$, a contradiction. Hence $f(x_\alpha) \notin U(\vec{v})$, so that $r_\alpha(f(x_\alpha)) = x_\alpha$, that is, $x_\alpha \in F_\alpha$.

Conversely, pick $x_\alpha \in F_\alpha$. By assumption, $f(x_\alpha) \neq x_\alpha$. Let $\vec{v}$ be the tangent direction at x_α defined by $U(\vec{v})$. Then $U(\vec{v}) \cap Y_\alpha = \emptyset$ so $f(\overline{U(\vec{v})}) \subseteq U(\vec{v})$. Now $U(\vec{v}) \cap Y_\beta$ is a finite nonempty subtree of X that is invariant under f_β. Hence f_β admits a fixed point x_β in this subtree. Then $x_\beta \in Y_\beta$ and $r_\alpha(x_\beta) = x_\alpha$. □

2.7 Notes and Further References

Our definition of "tree" differs from the one in set theory, see [Jec03]. It is also not equivalent to the notion of "**R**-tree" that has been around for quite some time (see [GH90]) and found striking applications. An **R**-tree is a metric space and usually considered with its metric topology. On the other hand, the notion of the weak topology on an **R**-tree seems to have been rediscovered several times, sometimes under different names (see [CLM07]).

Our definitions of trees and metric trees are new but equivalent[8] to the ones given in [FJ04], where rooted trees are defined first and general (non-rooted) trees are

---

[8]Except for the missing condition (RT3), see Remark 2.5.

defined as equivalence classes of rooted trees. The presentation here seems more natural. Following Baker and Rumely [BR10] we have emphasized viewing a tree as a pro-finite tree, that is, an inverse limit of finite trees.

Potential theory on simplicial graphs is a quite old subject but the possibility of doing potential theory on general metric trees seems to have been discovered independently by Favre and myself [FJ04], Baker and Rumely [BR10] and Thuillier [Thu05]; see also [Fav05]. Our approach here follows [BR10] quite closely in how the Laplacian is extended from finite to general trees. The class of quasisubharmonic functions is modeled on its complex counterpart, where its compactness properties makes this class very useful in complex dynamics and geometry. It is sufficiently large for our purposes and technically easier to handle than the class of functions of bounded differential variations studied in [BR10].

Note that the interpretation of "potential theory" used here is quite narrow; for further results and questions we refer to [BR10, Thu05]. It is also worth mentioning that while potential theory on the Berkovich projective line can be done in a purely tree theoretic way, this approach has its limitations. In other situations, and especially in higher dimensions, it seems advantageous to take a more geometric approach. This point of view is used already in [Thu05] and is hinted at in our exposition of the valuative tree in Sects. 7 and 9. We should remark that Thuillier in [Thu05] does potential theory on general Berkovich curves. These are not always trees in our sense as they can contain loops.

Most of the results on tree maps in Sect. 2.6 are well known and can be found in [BR10] in the context of the Berkovich projective line. I felt it would be useful to isolate some properties that are purely topological and only depend on the map between trees being continuous, open and finite. In fact, these properties turn out to be quite plentiful.

As noted in the text, versions of the fixed point result in Proposition 2.17 can be found in the work of Favre and myself [FJ07] and of Rivera-Letelier [Riv04]. The proof here is new.

3 The Berkovich Affine and Projective Lines

Let us briefly describe the Berkovich affine and projective lines. A comprehensive reference for this material is the recent book by Baker and Rumely [BR10]. See also Berkovich's original work [Ber90]. One minor difference to the presentation in [BR10] is that we emphasize working in a coordinate free way.

3.1 Non-Archimedean Fields

We start by recalling some facts about non-Archimedean fields. A comprehensive reference for this material is [BGR84].

3.1.1 Seminorms and Semivaluations

Let R be a integral domain. A *multiplicative, non-Archimedean seminorm* on R is a function $|\cdot| : R \to \mathbf{R}_+$ satisfying $|0| = 0$, $|1| = 1$, $|ab| = |a||b|$ and $|a + b| \leq \max\{|a|, |b|\}$. If $|a| > 0$ for all nonzero a, then $|\cdot|$ is a *norm*. In any case, the set $\mathfrak{p} \subseteq R$ consisting of elements of norm zero is a prime ideal and $|\cdot|$ descends to a norm on the quotient ring $R/\mathfrak{p}$ and in turn extends to a norm on the fraction field of the latter.

Sometimes it is more convenient to work additively and consider the associated *semi-valuation*[9] $v : R \to \mathbf{R} \cup \{+\infty\}$ defined by $v = -\log|\cdot|$. It satisfies the axioms $v(0) = +\infty$, $v(1) = 0$, $v(ab) = v(a) + v(b)$ and $v(a + b) \geq \min\{v(a), v(b)\}$. The prime ideal $\mathfrak{p}$ above is now given by $\mathfrak{p} = \{v = +\infty\}$ and v extends uniquely to a real-valued valuation on the fraction field of $R/\mathfrak{p}$.

Any seminorm on a field K is a norm. A *non-Archimedean field* is a field K equipped with a non-Archimedean, multiplicative norm $|\cdot| = |\cdot|_K$ such that K is complete in the induced metric. In general, we allow the norm on K be trivial: see Example 3.1. As a topological space, K is totally disconnected. We write $|K^*| = \{|a| \mid a \in K \setminus \{0\}\} \subseteq \mathbf{R}_+^*$ for the (multiplicative) *value group* of K.

3.1.2 Discs

A *closed disc* in K is a set of the form $D(a, r) = \{b \in K \mid |a - b| \leq r\}$. This disc is *degenerate* if $r = 0$, *rational* if $r \in |K^*|$ and *irrational* otherwise. Similarly, $D^-(a, r) := \{b \in K \mid |a - b| < r\}$, $r > 0$, is an *open disc*.

The terminology is natural but slightly misleading since nondegenerate discs are both open and closed in K. Further, if $0 < r \notin |K^*|$, then $D^-(a, r) = D(a, r)$. Note that any point in a disc in K can serve as a center and that when two discs intersect, one must contain the other. As a consequence, any two closed discs admit a unique smallest closed disc containing them both.

3.1.3 The Residue Field

The *valuation ring* of K is the ring $\mathfrak{o}_K := \{|\cdot| \leq 1\}$. It is a local ring with maximal ideal $\mathfrak{m}_K := \{|\cdot| < 1\}$. The *residue field* of K is $\tilde{K} := \mathfrak{o}_K/\mathfrak{m}_K$. We can identify $\mathfrak{o}_K$ and $\mathfrak{m}_K$ with the closed and open unit discs in K, respectively. The *residue characteristic* of K is the characteristic of $\tilde{K}$. Note that if $\tilde{K}$ has characteristic zero, then so does K.

---

[9]Unfortunately, the terminology is not uniform across the literature. In [BGR84, Ber90] 'valuation' is used to denoted multiplicative norms. In [FJ04], 'valuation' instead of 'semi-valuation' is used even when the prime ideal $\{v = +\infty\}$ is nontrivial.

Example 3.1 We can equip any field K with the *trivial* norm in which $|a| = 1$ whenever $a \neq 0$. Then $\mathfrak{o}_K = K$, $\mathfrak{m}_K = 0$ and $\tilde{K} = K$.

Example 3.2 The field $K = \mathbf{Q}_p$ of p-adic numbers is the completion of $\mathbf{Q}$ with respect to the p-adic norm. Its valuation ring $\mathfrak{o}_K$ is the ring of p-adic integers $\mathbf{Z}_p$ and the residue field $\tilde{K}$ is the finite field $\mathbf{F}_p$. In particular, $\mathbf{Q}_p$ has characteristic zero and residue characteristic $p > 0$.

Example 3.3 The algebraic closure of $\mathbf{Q}_p$ is not complete. Luckily, the completed algebraic closure $\mathbf{C}_p$ of $\mathbf{Q}_p$ is both algebraically closed and complete. Its residue field is $\overline{\mathbf{F}_p}$, the algebraic closure of $\mathbf{F}_p$. Again, $\mathbf{C}_p$ has characteristic zero and residue characteristic $p > 0$.

Example 3.4 Consider the field $\mathbf{C}$ of complex numbers (or any algebraically closed field of characteristic zero) *equipped with the trivial norm*. Let $K = \mathbf{C}((u))$ be the field of Laurent series with coefficients in $\mathbf{C}$. The norm $|\cdot|$ on K is given by $\log |\sum_{n\in\mathbf{Z}} a_n u^n| = -\min\{n \mid a_n \neq 0\}$. Then $\mathfrak{o}_K = \mathbf{C}[[u]]$, $\mathfrak{m}_K = u\mathfrak{o}_K$ and $\tilde{K} = \mathbf{C}$. We see that K is complete and of residue characteristic zero. However, it is not algebraically closed.

Example 3.5 Let $K = \mathbf{C}((u))$ be the field of Laurent series. By the Newton-Puiseux theorem, the algebraic closure K^a of K is the field of *Puiseux series*

$$a = \sum_{\beta\in B} a_\beta u^\beta, \tag{10}$$

where the sum is over a (countable) subset $B \subseteq \mathbf{Q}$ for which there exists $m, N \in \mathbf{N}$ (depending on a) such that $m + NB \subseteq \mathbf{N}$. This field is not complete; its completion $\widehat{K^a}$ is algebraically closed as well as complete. It has residue characteristic zero.

Example 3.6 A giant extension of $\mathbf{C}((u))$ is given by the field K consisting of series of the form (10), where B ranges over well-ordered subsets of $\mathbf{R}$. In this case, $|K^*| = \mathbf{R}^*$.

3.2 The Berkovich Affine Line

Write $R \simeq K[z]$ for the ring of polynomials in one variable with coefficients in K. The *affine line* $\mathbf{A}^1$ over K is the set of maximal ideals in R. Any choice of coordinate z (i.e. $R = K[z]$) defines an isomorphism $\mathbf{A}^1 \xrightarrow{\sim} K$. A (closed or open) disc in $\mathbf{A}^1$ is a disc in K under this isomorphism. This makes sense since any automorphism $z \mapsto az + b$ of K maps discs to discs. We can also talk about rational and irrational discs. However, the radius of a disc in $\mathbf{A}^1$ is not well defined.

Definition 3.7 The *Berkovich affine line* $\mathbf{A}^1_{\text{Berk}} = \mathbf{A}^1_{\text{Berk}}(K)$ is the set of multiplicative seminorms $|\cdot| : R \to \mathbf{R}_+$ whose restriction to the ground field $K \subseteq R$ is equal to the given norm $|\cdot|_K$.

Such a seminorm is necessarily non-Archimedean. Elements of $\mathbf{A}^1_{\text{Berk}}$ are usually denoted x and the associated seminorm on R by $|\cdot|_x$. The topology on $\mathbf{A}^1_{\text{Berk}}$ is the weakest topology in which all evaluation maps $x \mapsto |\phi|_x$, $\phi \in R$, are continuous. There is a natural partial ordering on $\mathbf{A}^1_{\text{Berk}}$: $x \leq y$ iff $|\phi|_x \leq |\phi|_y$ for all $\phi \in R$.

3.3 Classification of Points

One very nice feature of the Berkovich affine line is that we can completely and precisely classify its elements. The situation is typically much more complicated in higher dimensions. Following Berkovich [Ber90] we shall describe four types of points in $\mathbf{A}^1_{\text{Berk}}$, then show that this list is in fact complete.

For simplicity we shall from now on and until Sect. 3.9 assume that K is *algebraically closed* and that the valuation on K is *nontrivial*. The situation when one or both of these conditions is not satisfied is discussed briefly in Sect. 3.9. See also Sect. 6.6 for a different presentation of the trivially valued case.

3.3.1 Seminorms from Points

Any closed point $x \in \mathbf{A}^1$ defines a seminorm $|\cdot|_x$ on R through

$$|\phi|_x := |\phi(x)|.$$

This gives rise to an embedding $\mathbf{A}^1 \hookrightarrow \mathbf{A}^1_{\text{Berk}}$. The images of this map will be called *classical points*.[10]

Remark 3.8 If we define $\mathbf{A}^1_{\text{Berk}}$ as above when $K = \mathbf{C}$, then it follows from the Gel'fand-Mazur Theorem that all points are classical, that is, the map $\mathbf{A}^1 \to \mathbf{A}^1_{\text{Berk}}$ is surjective. The non-Archimedean case is vastly different.

3.3.2 Seminorms from Discs

Next, let $D \subseteq \mathbf{A}^1$ be a closed disc and define a seminorm $|\cdot|_D$ on R by

$$|\phi|_D := \max_{x \in D} |\phi(x)|.$$

---

[10]They are sometimes called *rigid points* as they are the points that show up rigid analytic geometry [BGR84].

It follows from Gauss' Lemma that this indeed defines a multiplicative seminorm on R. In fact, the maximum above is attained for a "generic" $x \in D$. We denote the corresponding element of $\mathbf{A}_{\mathrm{Berk}}^1$ by x_D. In the degenerate case $D = \{x\}$, $x \in \mathbf{A}^1$, this reduces to the previous construction: $x_D = x$.

3.3.3 Seminorms from Nested Collections of Discs

It is clear from the construction that if D, D' are closed discs in $\mathbf{A}^1$, then

$$|\phi|_D \leq |\phi|_{D'} \text{ for all } \phi \in R \text{ iff } D \subseteq D'. \tag{11}$$

Definition 3.9 A collection $\mathcal{E}$ of closed discs in $\mathbf{A}^1$ is *nested* if the following conditions are satisfied:

(a) if $D, D' \in \mathcal{E}$ then $D \subseteq D'$ or $D' \subseteq D$;
(b) if D and D' are closed discs in $\mathbf{A}^1$ with $D' \in \mathcal{E}$ and $D' \subseteq D$, then $D \in \mathcal{E}$;
(c) if $(D_n)_{n \geq 1}$ is a decreasing sequence of discs in $\mathcal{E}$ whose intersection is a disc D in $\mathbf{A}^1$, then $D \in \mathcal{E}$.

In view of (11) we can associate a seminorm $x_{\mathcal{E}} \in \mathbf{A}_{\mathrm{Berk}}^1$ to a nested collection $\mathcal{E}$ of discs by

$$x_{\mathcal{E}} = \inf_{D \in \mathcal{E}} x_D;$$

indeed, the limit of an decreasing sequence of seminorms is a seminorm. When the intersection $\bigcap_{D \in \mathcal{E}} D$ is nonempty, it is a closed disc $D(\mathcal{E})$ (possibly of radius 0). In this case $x_{\mathcal{E}}$ is the seminorm associated to the disc $D(\mathcal{E})$. In general, however, the intersection above may be empty (the existence of a nested collection of discs with nonempty intersection is equivalent to the field K not being *spherically complete*).

The set of nested collections of discs is partially ordered by inclusion and we have $x_{\mathcal{E}} \leq x_{\mathcal{E}'}$ iff $\mathcal{E}' \subseteq \mathcal{E}$.

3.3.4 Classification

Berkovich proved that all seminorms in $\mathbf{A}_{\mathrm{Berk}}^1$ arise from the construction above.

Theorem 3.10 *For any $x \in \mathbf{A}_{\mathrm{Berk}}^1$ there exists a unique nested collection $\mathcal{E}$ of discs in $\mathbf{A}^1$ such that $x = x_{\mathcal{E}}$. Moreover, the map $\mathcal{E} \to x_{\mathcal{E}}$ is an order-preserving isomorphism.*

Sketch of proof The strategy is clear enough: given $x \in \mathbf{A}_{\mathrm{Berk}}^1$ define $\mathcal{E}(x)$ as the collection of discs D such that $x_D \geq x$. However, it requires a little work to show that the maps $\mathcal{E} \mapsto x_{\mathcal{E}}$ and $x \mapsto \mathcal{E}(x)$ are order-preserving and inverse one to another. Here we have to use the assumptions that K is algebraically closed and that the norm on K is nontrivial. The first assumption implies that x is uniquely determined by its values on *linear* polynomials in R. The second assumption is necessary to ensure surjectivity of $\mathcal{E} \mapsto x_{\mathcal{E}}$: if the norm on K is trivial, then there are too few discs in $\mathbf{A}^1$. See the proof of [BR10, Theorem 1.2] for details. $\square$

3.3.5 Tree Structure

Using the classification theorem above, we can already see that the Berkovich affine line is naturally a tree. Namely, let $\mathfrak{E}$ denote the set of nested collections of discs in $\mathbf{A}^1$. We also consider the empty collection as an element of $\mathfrak{E}$. It is then straightforward to verify that $\mathfrak{E}$, partially ordered by inclusion, is a rooted tree in the sense of Sect. 2.3. As a consequence, the set $\mathbf{A}^1_{\mathrm{Berk}} \cup \{\infty\}$ is a rooted metric tree. Here ∞ corresponds to the empty collection of discs in $\mathbf{A}^1$ and can be viewed as the function $|\cdot|_\infty : R \to [0, +\infty]$ given by $|\phi| = \infty$ for any nonconstant polynomial $\phi \in R$ and $|\cdot|_\infty = |\cdot|_K$ on K. Then $\mathbf{A}^1_{\mathrm{Berk}} \cup \{\infty\}$ is a rooted tree with the partial ordering $x \leq x'$ iff $|\cdot|_x \geq |\cdot|_{x'}$ on R. See Fig. 1.

3.3.6 Types of Points

Using the identification with nested collections of discs, Berkovich classifies the points in $\mathbf{A}^1_{\mathrm{Berk}}$ as follows:

- a point of *Type 1* is a classical point, that is, a point in the image of the embedding $\mathbf{A}^1 \hookrightarrow \mathbf{A}^1_{\mathrm{Berk}}$;
- a point of *Type 2* is of the form x_D where D is a rational disc in $\mathbf{A}^1$;
- a point of *Type 3* is of the form x_D where D is an irrational disc in $\mathbf{A}^1$;
- a point of *Type 4* is of the form $x_{\mathcal{E}}$, where $\mathcal{E}$ is a nested collection of discs with empty intersection.

Note that Type 3 points exist iff $|K| \subsetneq \mathbf{R}_+$, while Type 4 points exist iff K is not spherically complete.

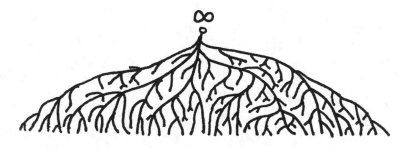

Fig. 1 The Berkovich affine line

3.3.7 Action by Automorphisms

Any automorphism $A \in \mathrm{Aut}(\mathbf{A}^1)$ arises from a K-algebra automorphism A^* of R, hence extends to an automorphism of $\mathbf{A}^1_{\mathrm{Berk}}$ by setting

$$|\phi|_{A(x)} := |A^*\phi|_x$$

for any polynomial $\phi \in R$. Note that A is order-preserving. If $\mathcal{E}$ is a nested collection of discs in $\mathbf{A}^1$, then so is $A(\mathcal{E})$ and $A(x_{\mathcal{E}}) = x_{A(\mathcal{E})}$. It follows that A preserves the type of a point in $\mathbf{A}^1_{\mathrm{Berk}}$.

Clearly $\mathrm{Aut}(\mathbf{A}^1)$ acts transitively on $\mathbf{A}^1$, hence on the Type 1 points in $\mathbf{A}^1_{\mathrm{Berk}}$. It also acts transitively on the rational discs in $\mathbf{A}^1$, hence the Type 2 points. In general, it will not act transitively on the set of Type 3 or Type 4 points, see Sect. 3.3.8.

3.3.8 Coordinates, Radii and the Gauss Norm

The description of $\mathbf{A}^1_{\mathrm{Berk}}$ above was coordinate independent. Now fix a coordinate $z : \mathbf{A}^1 \xrightarrow{\sim} K$. Using z, every disc $D \subseteq \mathbf{A}^1$ becomes a disc in K, hence has a well-defined *radius* $r_z(D)$. If D is a closed disc of radius $r = r_z(D)$ centered at point in $\mathbf{A}^1$ with coordinate $a \in K$, then

$$|z - b|_D = \max\{|a - b|, r\}. \tag{12}$$

We can also define the radius $r_z(\mathcal{E}) := \inf_{D \in \mathcal{E}} r_z(D)$ of a nested collection of discs. The completeness of K implies that if $r_z(\mathcal{E}) = 0$, then $\bigcap_{D \in \mathcal{E}} D$ is a point in $\mathbf{A}^1$.

The *Gauss norm* is the norm in $\mathbf{A}^1_{\mathrm{Berk}}$ defined by the unit disc in K. We emphasize that the Gauss norm depends on a choice of coordinate z. In fact, any Type 2 point is the Gauss norm in some coordinate.

The radius $r_z(D)$ of a disc depends on z. However, if we have two closed discs $D \subseteq D'$ in $\mathbf{A}^1$, then the ratio $r_z(D')/r_z(D)$ does *not* depend on z. Indeed, any other coordinate w is of the form $w = az + b$, with $a \in K^*$, $b \in K$ and so $r_w(D) = |a|r_z(D)$, $r_w(D') = |a|r_z(D')$. We think of the quantity $\log \frac{r_z(D')}{r_z(D)}$ as the modulus of the annulus $D' \setminus D$. It will play an important role in what follows.

In the same spirit, the class $[r_z(x)]$ of $r_z(x)$ in $\mathbf{R}_+^*/|K^*|$ does not depend on the choice of coordinate z. This implies that if $|K| \neq \mathbf{R}_+$, then $\mathrm{Aut}(\mathbf{A}^1)$ does not act transitively on Type 3 points. Indeed, if $|K| \neq \mathbf{R}_+$, then given any Type 3 point x we can find another Type 3 point $y \in [\infty, x]$ such that $[r_z(x)] \neq [r_z(y)]$. Then $A(x) \neq y$ for any $A \in \mathrm{Aut}(\mathbf{A}^1)$. The same argument shows that if K admits Type 4 points of any given radius, then A does not always act transitively on Type 4 points. For $K = \mathbf{C}_p$, there does indeed exist Type 4 points of any given radius, see [Rob00, p.143].

3.4 The Berkovich Projective Line

We can view the projective line $\mathbf{P}^1$ over K as the set of proper valuation rings A of F/K, where $F \simeq K(z)$ is the field of rational functions in one variable with coefficients in K. In other words, $A \subsetneq F$ is a subring containing K such that for every nonzero $\phi \in F$ we have $\phi \in A$ or $\phi^{-1} \in A$. Since $A \neq F$, there exists $z \in F \setminus A$ such that $F = K(z)$ and $z^{-1} \in A$. The other elements of $\mathbf{P}^1$ are then the localizations of the ring $R := K[z]$ at its maximal ideals. This gives rise to a decomposition $\mathbf{P}^1 = \mathbf{A}^1 \cup \{\infty\}$ in which A becomes the point $\infty \in \mathbf{P}^1$.

Given such a decomposition we define a *closed disc* in $\mathbf{P}^1$ to be a closed disc in $\mathbf{A}^1$, the singleton $\{\infty\}$, or the complement of an open disc in $\mathbf{A}^1$. Open discs are defined in the same way. A disc is *rational* if it comes from a rational disc in $\mathbf{A}^1$. These notions do not depend on the choice of point $\infty \in \mathbf{P}^1$.

Definition 3.11 The *Berkovich projective line* $\mathbf{P}^1_{\text{Berk}}$ over K is the set of functions $|\cdot| : F \to [0, +\infty]$ extending the norm on $K \subseteq F$ and satisfying $|\phi + \psi| \leq \max\{|\phi|, |\psi|\}$ for all $\phi, \psi \in F$, and $|\phi\psi| = |\phi||\psi|$ unless $|\phi| = 0, |\psi| = +\infty$ or $|\psi| = 0, |\phi| = +\infty$.

To understand this, pick a rational function $z \in F$ such that $F = K(z)$. Then $R := K[z]$ is the coordinate ring of $\mathbf{A}^1 := \mathbf{P}^1 \setminus \{z = \infty\}$. There are two cases. Either $|z| = \infty$, in which case $|\phi| = \infty$ for all nonconstant polynomials $\phi \in R$, or $|\cdot|$ is a seminorm on R, hence an element of $\mathbf{A}^1_{\text{Berk}}$. Conversely, any element $x \in \mathbf{A}^1_{\text{Berk}}$ defines an element of $\mathbf{P}^1_{\text{Berk}}$ in the sense above. Indeed, every nonzero $\phi \in F$ is of the form $\phi = \phi_1/\phi_2$ with $\phi_1, \phi_2 \in R$ having no common factor. Then we can set $|\phi|_x := |\phi_1|_x/|\phi_2|_x$; this is well defined by the assumption on ϕ_1 and ϕ_2. Similarly, the function which is identically ∞ on all nonconstant polynomials defines a unique element of $\mathbf{P}^1_{\text{Berk}}$: each $\phi \in F$ defines a rational function on $\mathbf{P}^1$ and $|\phi| := |\phi(\infty)| \in [0, +\infty]$. This leads to a decomposition

$$\mathbf{P}^1_{\text{Berk}} = \mathbf{A}^1_{\text{Berk}} \cup \{\infty\},$$

corresponding to the decomposition $\mathbf{P}^1 = \mathbf{A}^1 \cup \{\infty\}$.

We equip $\mathbf{P}^1_{\text{Berk}}$ with the topology of pointwise convergence. By Tychonoff, $\mathbf{P}^1_{\text{Berk}}$ is a compact Hausdorff space and, as a consequence, $\mathbf{A}^1_{\text{Berk}}$ is locally compact. The injection $\mathbf{A}^1 \hookrightarrow \mathbf{A}^1_{\text{Berk}}$ extends to an injection $\mathbf{P}^1 \hookrightarrow \mathbf{P}^1_{\text{Berk}}$ by associating the function $\infty \in \mathbf{P}^1_{\text{Berk}}$ to the point $\infty \in \mathbf{P}^1$.

Any automorphism $A \in \text{Aut}(\mathbf{P}^1)$ is given by an element $A^* \in \text{Aut}(F/K)$. hence extends to an automorphism of $\mathbf{P}^1_{\text{Berk}}$ by setting

$$|\phi|_{A(x)} := |A^*\phi|_x$$

for any rational function $\phi \in F$. As in the case of $\mathbf{A}^1_{\text{Berk}}$, the type of a point is preserved. Further, $\text{Aut}(\mathbf{P}^1)$ acts transitively on the set of Type 1 and Type 2 points, but not on the Type 3 or Type 4 points in general, see Sect. 3.3.8.

3.5 Tree Structure

We now show that $\mathbf{P}^1_{\mathrm{Berk}}$ admits natural structures as a tree and a metric tree. See Sect. 2 for the relevant definitions.

Consider a decomposition $\mathbf{P}^1 = \mathbf{A}^1 \cup \{\infty\}$ and the corresponding decomposition $\mathbf{P}^1_{\mathrm{Berk}} = \mathbf{A}^1_{\mathrm{Berk}} \cup \{\infty\}$. The elements of $\mathbf{P}^1_{\mathrm{Berk}}$ define functions on the polynomial ring R with values in $[0, +\infty]$. This gives rise to a partial ordering on $\mathbf{P}^1_{\mathrm{Berk}}$: $x \leq x'$ iff and only if $|\phi|_x \geq |\phi|_{x'}$ for all polynomials ϕ. As already observed in Sect. 3.3.5, $\mathbf{P}^1_{\mathrm{Berk}}$ then becomes a rooted tree in the sense of Sect. 2.3, with ∞ as its root. The partial ordering on $\mathbf{P}^1_{\mathrm{Berk}}$ depends on a choice of point $\infty \in \mathbf{P}^1$, but the associated (nonrooted) tree structure does not.

The ends of $\mathbf{P}^1_{\mathrm{Berk}}$ are the points of Type 1 and 4, whereas the branch points are the Type 2 points. See Fig. 2.

Given a coordinate $z : \mathbf{A}^1 \xrightarrow{\sim} K$ we can parametrize $\mathbf{P}^1_{\mathrm{Berk}}$ rooted in ∞ using radii of discs. Instead of doing so literally, we define an decreasing parametrization $\alpha_z : \mathbf{P}^1_{\mathrm{Berk}} \to [-\infty, +\infty]$ using

$$\alpha_z(x_{\mathcal{E}}) := \log r_z(\mathcal{E}). \tag{13}$$

One checks that this is a parametrization in the sense of Sect. 2.3. The induced metric tree structure on $\mathbf{P}^1_{\mathrm{Berk}}$ does *not* depend on the choice of coordinate z and any

Fig. 2 The Berkovich projective line

automorphism of $\mathbf{P}^1$ induces an isometry of $\mathbf{P}^1_{\text{Berk}}$ in this generalized metric. This is one reason for using the logarithm in (13). Another reason has to do with potential theory, see Sect. 3.6. Note that $\alpha_z(\infty) = \infty$ and $\alpha_z(x) = -\infty$ iff x is of Type 1.

The associated *hyperbolic space* in the sense of Sect. 2 is given by

$$\mathbf{H} := \mathbf{P}^1_{\text{Berk}} \setminus \mathbf{P}^1.$$

The generalized metric on $\mathbf{P}^1_{\text{Berk}}$ above induces a complete metric on $\mathbf{H}$ (in the usual sense). Any automorphism of $\mathbf{P}^1$ induces an isometry of $\mathbf{H}$.

3.6   Topology and Tree Structure

The topology on $\mathbf{P}^1_{\text{Berk}}$ defined above agrees with the weak topology associated to the tree structure. To see this, note that $\mathbf{P}^1_{\text{Berk}}$ is compact in both topologies. It therefore suffices to show that if $\vec{v}$ is a tree tangent direction $\vec{v}$ at a point $x \in \mathbf{P}^1_{\text{Berk}}$, then the set $U(\vec{v})$ is open in the Berkovich topology. We may assume that x is of Type 2 or 3. In a suitable coordinate z, $x = x_{D(0,r)}$ and $\vec{v}$ is represented by the point x_0. Then $U(\vec{v}) = \{y \in \mathbf{P}^1_{\text{Berk}} \mid |z|_x < r\}$, which is open in the Berkovich topology.

A *generalized open Berkovich disc* is a connected component of $\mathbf{P}^1_{\text{Berk}} \setminus \{x\}$ for some $x \in \mathbf{P}^1_{\text{Berk}}$. When x is of Type 2 or 3 we call it an *open Berkovich disc* and when x of Type 2 a *strict open Berkovich disc*. A (strict) *simple domain* is a finite intersection of (strict) open Berkovich discs. The collection of all (strict) simple domains is a basis for the topology on $\mathbf{P}^1_{\text{Berk}}$.

3.7   Potential Theory

As $\mathbf{P}^1_{\text{Berk}}$ is a metric tree we can do potential theory on it, following Sect. 2.5. See also [BR10] for a comprehensive treatment, and the thesis of Thuillier [Thu05] for potential theory on general Berkovich analytic curves.

We shall not repeat the material in Sect. 2.5 here, but given a finite atomic probability measure ρ_0 on X with support on $\mathbf{H}$, we have a space $\text{SH}(\mathbf{P}^1_{\text{Berk}}, \rho_0)$ of ρ_0-subharmonic functions, as well as a homeomorphism

$$\rho_0 + \Delta : \text{SH}(\mathbf{P}^1_{\text{Berk}}, \rho_0)/\mathbf{R} \xrightarrow{\sim} \mathcal{M}^+_1(\mathbf{P}^1_{\text{Berk}}).$$

Over the complex numbers, the analogue of $\text{SH}(\mathbf{P}^1_{\text{Berk}}, \rho_0)$ is the space $\text{SH}(\mathbf{P}^1, \omega)$ of ω-subharmonic functions on $\mathbf{P}^1$, where ω is a Kähler form.

Lemma 3.12 *If $\phi \in F \setminus \{0\}$ is a rational function, then the function $\log |\phi| : \mathbf{H} \to \mathbf{R}$ is Lipschitz continuous with Lipschitz constant $\deg(\phi)$.*

Proof Pick any coordinate z on $\mathbf{P}^1$ and write $\phi = \phi_1/\phi_2$, with ϕ_1, ϕ_2 polynomials. The functions $\log|\phi_1|$ and $\log|\phi_2|$ are decreasing in the partial ordering rooted at ∞ and $\log|\phi| = \log|\phi_1| - \log|\phi_2|$. Hence we may assume that ϕ is a polynomial. Using that K is algebraically closed we further reduce to the case $\phi = z - b$, where $b \in K$. But then the result follows from (12). □

Remark 3.13 The function $\log|\phi|$ belongs to the space $\mathrm{BDV}(\mathbf{P}^1_{\mathrm{Berk}})$ of functions of bounded differential variation and $\Delta \log|\phi|$ is the divisor of ϕ, viewed as a signed, finite atomic measure on $\mathbf{P}^1 \subseteq \mathbf{P}^1_{\mathrm{Berk}}$; see [BR10, Lemma 9.1]. Lemma 3.12 then also follows from a version of (8) for functions in $\mathrm{BDV}(\mathbf{P}^1_{\mathrm{Berk}})$. These considerations also show that the generalized metric on $\mathbf{P}^1_{\mathrm{Berk}}$ is the correct one from the point of potential theory.

3.8 Structure Sheaf and Numerical Invariants

Above, we have defined the Berkovich projective line as a topological space, but it also an analytic space in the sense of Berkovich and carries a structure sheaf $\mathcal{O}$. The local rings $\mathcal{O}_x$ are useful for defining and studying the local degree of a rational map. They also allow us to recover Berkovich's classification via certain numerical invariants.

3.8.1 Structure Sheaf

A holomorphic function on an open set $U \subseteq \mathbf{P}^1_{\mathrm{Berk}}$ is a locally uniform limit of rational functions without poles in U. To make sense of this, we first need to say where the holomorphic functions take their values: the value at a point $x \in \mathbf{P}^1_{\mathrm{Berk}}$ is in a non-Archimedean field $\mathcal{H}(x)$.

To define $\mathcal{H}(x)$, assume $x \in \mathbf{A}^1_{\mathrm{Berk}}$. The kernel of the seminorm $|\cdot|_x$ is a prime ideal in R and $|\cdot|_x$ defines a norm on the fraction field of $R/\ker(|\cdot|_x)$; the field $\mathcal{H}(x)$ is its completion.

When x is of Type 1, $\mathcal{H}(x) \simeq K$. If instead x is of Type 3, pick a coordinate $z \in R$ such that $r := |z|_x \notin |K|$. Then $\mathcal{H}(x)$ is isomorphic to the set of series $\sum_{-\infty}^{\infty} a_j z^j$ with $a_j \in K$ and $|a_j|r^j \to 0$ as $j \to \pm\infty$. For x of Type 2 or 4, I am not aware of a similar explicit description of $\mathcal{H}(x)$.

The pole set of a rational function $\phi \in F$ can be viewed as a set of Type 1 points in $\mathbf{P}^1_{\mathrm{Berk}}$. If x is not a pole of ϕ, then $\phi(x) \in \mathcal{H}(x)$ is well defined. The definition of a holomorphic function on an open subset $U \subseteq \mathbf{P}^1_{\mathrm{Berk}}$ now makes sense and gives rise to the structure sheaf $\mathcal{O}$.

3.8.2 Local Rings and Residue Fields

The ring $\mathcal{O}_x$ for $x \in \mathbf{P}^1_{\text{Berk}}$ is the ring of germs of holomorphic functions at x. Denote by $\mathfrak{m}_x$ the maximal ideal of $\mathcal{O}_x$ and by $\kappa(x) := \mathcal{O}_x/\mathfrak{m}_x$ the residue field. Note that the seminorm $|\cdot|_x$ on $\mathcal{O}_x$ induces a norm on $\kappa(x)$. The field $\mathcal{H}(x)$ above is the completion of $\kappa(x)$ with respect to the residue norm and is therefore called the completed residue field.

When x is of Type 1, $\mathcal{O}_x$ is isomorphic to the ring of power series $\sum_0^\infty a_j z^j$ such that $\limsup |a_j|^{1/j} < \infty$, and $\kappa(x) = \mathcal{H}(x) = K$.

If x is not of Type 1, then $\mathfrak{m}_x = 0$ and $\mathcal{O}_x = \kappa(x)$ is a field. This field is usually difficult to describe explicitly. However, when x is of Type 3 it has a description analogous to the one of its completion $\mathcal{H}(x)$ given above. Namely, pick a coordinate $z \in R$ such that $r := |z|_x \notin |K|$. Then $\mathcal{O}_x$ is isomorphic to the set of series $\sum_{-\infty}^\infty a_j z^j$ with $a_j \in K$ for which there exists $r' < r < r''$ such that $|a_j|(r'')^j, |a_{-j}|(r')^{-j} \to 0$ as $j \to +\infty$.

3.8.3 Numerical Invariants

While the local rings $\mathcal{O}_x$ and the completed residue fields $\mathcal{H}(x)$ are not always easy to describe explicitly, certain numerical invariants of them are more tractable and allow us to recover Berkovich's classification.

First, x is of Type 1 iff the seminorm $|\cdot|_x$ has nontrivial kernel. Now suppose the kernel is trivial. Then $\mathcal{O}_x$ is a field and contains $F \simeq K(z)$ as a subfield. Both these fields are dense in $\mathcal{H}(x)$ with respect to the norm $|\cdot|_x$. In this situation we have two basic invariants.

First, the (additive) *value group* is defined by

$$\Gamma_x := \log |\mathcal{H}(x)^*|_x = \log |\mathcal{O}_x^*|_x = \log |F^*|_x.$$

This is an additive subgroup of $\mathbf{R}$ containing $\Gamma_K := \log |K^*|$. The *rational rank* rat. rk x of x is the dimension of the $\mathbf{Q}$-vector space $(\Gamma_x/\Gamma_K) \otimes_{\mathbf{Z}} \mathbf{Q}$.

Second, the three fields $\mathcal{H}(x)$, $\mathcal{O}_x$ and F have the same residue field with respect to the norm $|\cdot|_x$. We denote this field by $\widetilde{\mathcal{H}(x)}$; it contains the residue field $\tilde{K}$ of K as a subfield. The *transcendence degree* tr. deg x of x is the transcendence degree of the field extension $\widetilde{\mathcal{H}(x)}/\tilde{K}$.

One shows as in [BR10, Proposition 2.3] that

- if x is of Type 2, then tr. deg $x = 1$ and rat. rk $x = 0$; more precisely $\Gamma_x = \Gamma_K$ and $\widetilde{\mathcal{H}(x)} \simeq \tilde{K}(z)$;
- if x is of Type 3, then tr. deg $x = 0$ and rat. rk $x = 1$; more precisely, $\Gamma_x = \Gamma_K \oplus \mathbf{Z}\alpha$, where $\alpha \in \Gamma_x \setminus \Gamma_K$, and $\widetilde{\mathcal{H}(x)} \simeq \tilde{K}$;
- if x is of Type 4, then tr. deg $x = 0$ and rat. rk $x = 0$; more precisely, $\Gamma_x = \Gamma_K$ and $\widetilde{\mathcal{H}(x)} \simeq \tilde{K}$;

3.8.4 Quasicompleteness of the Residue Field

Berkovich proved in [Ber93, 2.3.3] that the residue field $\kappa(x)$ is *quasicomplete* in the sense that the induced norm $|\cdot|_x$ on $\kappa(x)$ extends uniquely to any algebraic extension of $\kappa(x)$. This fact is true for any point of a "good" Berkovich space. It will be exploited (only) in Sect. 4.8.2.

3.8.5 Weak Stability of the Residue Field

If x is of Type 2 or 3, then the residue field $\kappa(x) = \mathcal{O}_x$ is *weakly stable*. By definition [BGR84, 3.5.2/1] this means that any finite extension $L/\kappa(x)$ is *weakly Cartesian*, that is, there exists a linear homeomorphism $L \xrightarrow{\sim} \kappa(x)^n$, where $n = [L : \kappa(x)]$, see [BGR84, 2.3.2/4]. Here the norm on L is the unique extension of the norm on the quasicomplete field $\kappa(x)$. The homeomorphism above is not necessarily an isometry.

The only consequence of weak stability that we shall use is that if $L/\kappa(x)$ is a finite extension, then $[L : \kappa(x)] = [\hat{L} : \mathcal{H}(x)]$, where $\hat{L}$ denotes the completion of L, see [BGR84, 2.3.3/6]. This, in turn, will be used (only) in Sect. 4.8.2.

Let us sketch a proof that $\kappa(x) = \mathcal{O}_x$ is weakly stable when x is of Type 2 or 3. Using the remark at the end of [BGR84, 3.5.2] it suffices to show that the field extension $\mathcal{H}(x)/\mathcal{O}_x$ is separable. This is automatic if the ground field K has characteristic zero, so suppose K has characteristic $p > 0$. Pick a coordinate $z \in R$ such that x is associated to a disc centered at $0 \in K$. It is then not hard to see that $\mathcal{O}_x^{1/p} = \mathcal{O}_x[z^{1/p}]$ and it suffices to show that $z^{1/p} \notin \mathcal{H}(x)$. If x is of Type 3, then this follows from the fact that $\frac{1}{p} \log r = \log |z^{1/p}|_x \notin \Gamma_K + \mathbf{Z} \log r = \Gamma_x$. If instead x is of Type 2, then we may assume that x is the Gauss point with respect to the coordinate z. Then $\widetilde{\mathcal{H}(x)} \simeq \tilde{K}(z) \not\ni z^{1/p}$ and hence $z^{1/p} \notin \mathcal{H}(x)$.

3.8.6 Stability of the Completed Residue Field

When x is a Type 2 or Type 3 point, the completed residue field $\mathcal{H}(x)$ is *stable* field in the sense of [BGR84, 3.6.1/1]. This means that any finite extension $L/\mathcal{H}(x)$ admits a basis $e_1, \ldots, e_m$ such that $|\sum_i a_i e_i| = \max_i |a_i||e_i|$ for $a_i \in K$. Here the norm on L is the unique extension of the norm on the complete field $\mathcal{H}(x)$. The stability of $\mathcal{H}(x)$ is proved in [Tem10, 6.3.6] (the case of a Type 2 point also follows from [BGR84, 5.3.2/1]).

Let x be of Type 2 or 3. The stability of $\mathcal{H}(x)$ implies that for any finite extension $L/\mathcal{H}(x)$ we have $[L : \mathcal{H}(x)] = [\Gamma_L : \Gamma_x] \cdot [\tilde{L} : \widetilde{\mathcal{H}(x)}]$, where Γ_L and $\tilde{L}$ are the value group and residue field of L, see [BGR84, 3.6.2/4].

3.8.7 Tangent Space and Reduction Map

Fix $x \in \mathbf{P}^1_{\mathrm{Berk}}$. Using the tree structure, we define as in Sect. 2.1.1 the tangent space T_x of $\mathbf{P}^1_{\mathrm{Berk}}$ at x as well as a tautological "reduction" map from $\mathbf{P}^1_{\mathrm{Berk}} \setminus \{x\}$ onto T_x. Let us interpret this procedure algebraically in the case when x is a Type 2 point.

The tangent space T_x at a Type 2 point x is the set of valuation rings $A \subsetneq \tilde{H}(x)$ containing $\tilde{K}$. Fix a coordinate z such that x becomes the Gauss point. Then $\tilde{H}(x) \xrightarrow{\sim} \tilde{K}(z)$ and $T_x \simeq \mathbf{P}^1(\tilde{K})$. Let us define the reduction map r_x of $\mathbf{P}^1_{\mathrm{Berk}} \setminus \{x\}$ onto $T_x \simeq \mathbf{P}^1(\tilde{K})$. Pick a point $y \in \mathbf{P}^1_{\mathrm{Berk}} \setminus \{x\}$. If $|z|_y > 1$, then we declare $r_x(y) = \infty$. If $|z|_y \leq 1$, then, since $y \neq x$, there exists $a \in o_K$ such that $|z - a|_y < 1$. The element a is not uniquely defined, but its class $\tilde{a} \in \tilde{K}$ is and we set $r_x(y) = \tilde{a}$. One can check that this definition does not depend on the choice of coordinate z and gives the same result as the tree-theoretic construction.

The reduction map can be naturally understood in the context of formal models, but we shall not discuss this here.

3.9 Other Ground Fields

Recall that from Sect. 3.4 onwards, we assumed that the field K was algebraically closed and nontrivially valued. These assumptions were used in the proof of Theorem 3.10. Let us briefly discuss what happens when they are removed.

As before, $\mathbf{A}^1_{\mathrm{Berk}}(K)$ is the set of multiplicative seminorms on $R \simeq K[z]$ extending the norm on K and $\mathbf{P}^1_{\mathrm{Berk}}(K) \simeq \mathbf{A}^1_{\mathrm{Berk}}(K) \cup \{\infty\}$. We can equip $\mathbf{A}^1_{\mathrm{Berk}}(K)$ and $\mathbf{P}^1_{\mathrm{Berk}}(K)$ with a partial ordering defined by $x \leq x'$ iff $|\phi(x)| \geq |\phi(x')|$ for all polynomials $\phi \in R$.

3.9.1 Non-Algebraically Closed Fields

First assume that K is nontrivially valued but not algebraically closed. Our discussion follows [Ber90, §4.2]; see also [Ked11b, §2.2], [Ked10, §5.1] and [Ked11a, §6.1].

Denote by K^a the algebraic closure of K and by $\widehat{K^a}$ its completion. Since K is complete, the norm on K has a unique extension to $\widehat{K^a}$.

The Galois group $G := \mathrm{Gal}(K^a/K)$ acts on the field $\widehat{K^a}$ and induces an action on $\mathbf{A}^1_{\mathrm{Berk}}(\widehat{K^a})$, which in turn extends to $\mathbf{P}^1_{\mathrm{Berk}}(\widehat{K^a}) = \mathbf{A}^1_{\mathrm{Berk}}(\widehat{K^a}) \cup \{\infty\}$ using $g(\infty) = \infty$ for all $g \in G$. It is a general fact that $\mathbf{P}^1_{\mathrm{Berk}}(K)$ is isomorphic to the quotient $\mathbf{P}^1_{\mathrm{Berk}}(\widehat{K^a})/G$. The quotient map $\pi : \mathbf{P}^1_{\mathrm{Berk}}(\widehat{K^a}) \to \mathbf{P}^1_{\mathrm{Berk}}(K)$ is continuous, open and surjective.

It is easy to see that g maps any segment $[x, \infty]$ homeomorphically onto the segment $[g(x), \infty]$. This implies that $\mathbf{P}^1_{\mathrm{Berk}}(K)$ is a tree in the sense of Sect. 2.1. In fact, the rooted tree structure on $\mathbf{P}^1_{\mathrm{Berk}}(K)$ is defined by the partial ordering above.

If $g \in G$ and $x \in \mathbf{P}^1_{\mathrm{Berk}}(\widehat{K^a})$, then x and $g(x)$ have the same type. This leads to a classification of points in $\mathbf{P}^1_{\mathrm{Berk}}(K)$ into Types 1-4. Note that since $\widehat{K^a} \neq K^a$ in general, there may exist Type 1 points $x \neq \infty$ such that $|\phi(x)| > 0$ for all polynomials $\phi \in R = K[z]$.

We can equip the Berkovich projective line $\mathbf{P}^1_{\mathrm{Berk}}(K)$ with a generalized metric. In fact, there are two natural ways of doing this. Fix a coordinate $z \in R$. Let $\hat{\alpha}_z : \mathbf{P}^1_{\mathrm{Berk}}(\widehat{K^a}) \to [-\infty, +\infty]$ be the parametrization defined in Sect. 3.5. It satisfies $\hat{\alpha}_z \circ g = \hat{\alpha}_z$ for all $g \in G$ and hence induces a parametrization $\hat{\alpha}_z : \mathbf{P}^1_{\mathrm{Berk}}(K) \to [-\infty, +\infty]$. The associated generalized metric on $\mathbf{P}^1_{\mathrm{Berk}}(K)$ does not depend on the choice of coordinate z and has the nice feature that the associated hyperbolic space consists exactly of points of Types 2-4.

However, for potential theoretic considerations, it is better to use a slightly different metric. For this, first define the *multiplicity*[11] $m(x) \in \mathbf{Z}_+ \cup \{\infty\}$ of a point $x \in \mathbf{P}^1_{\mathrm{Berk}}(K)$ as the number of preimages of x in $\mathbf{P}^1_{\mathrm{Berk}}(\widehat{K^a})$. The multiplicity of a Type 2 or Type 3 point is finite and if $x \leq y$, then $m(x)$ divides $m(y)$. Note that $m(0) = 1$ so all points on the interval $[\infty, 0]$ have multiplicity 1. We now define a decreasing parametrization $\alpha_z : \mathbf{P}^1_{\mathrm{Berk}}(K) \to [-\infty, +\infty]$ as follows. Given $x \in \mathbf{P}^1_{\mathrm{Berk}}(K)$, set $x_0 := x \wedge 0$ and

$$\alpha_z(x) = \alpha_z(x_0) - \int_{x_0}^x \frac{1}{m(y)} \, d\hat{\alpha}_z(y) \tag{14}$$

Again, the associated generalized metric on $\mathbf{P}^1_{\mathrm{Berk}}(K)$ does not depend on the choice of coordinate z. The hyperbolic space $\mathbf{H}$ now contains all points of Types 2-4 but may also contain some points of Type 1.

One nice feature of the generalized metric induced by α_z is that if ρ_0 is a finite positive measure on $\mathbf{P}^1_{\mathrm{Berk}}(K)$ supported on points of finite multiplicity and if $\varphi \in \mathrm{SH}(\mathbf{P}^1_{\mathrm{Berk}}(K), \rho_0)$, then $\pi^*\varphi \in \mathrm{QSH}(\mathbf{P}^1_{\mathrm{Berk}}(\widehat{K^a}))$ and

$$\Delta\varphi = \pi_* \Delta(\pi^*\varphi).$$

Furthermore, for any rational function $\phi \in F$, the measure $\Delta \log |\phi|$ on $\mathbf{P}^1_{\mathrm{Berk}}(K)$ can be identified with the divisor of ϕ, see Remark 3.13.

3.9.2 Trivially Valued Fields

Finally we discuss the case when K is trivially valued, adapting the methods above. A different approach is presented in Sect. 6.6.

First assume K is algebraically closed. Then a multiplicative seminorm on R is determined by its values on linear polynomials. Given a coordinate $z \in R$ it is easy

---

[11]This differs from the "algebraic degree" used by Trucco, see [Tru09, Definition 5.1].

to see that any point $x \in \mathbf{A}^1_{\text{Berk}}$ is of one of the following three types:

- we have $|z - a|_x = 1$ for all $a \in K$; this point x is the *Gauss point*;
- there exists a unique $a \in K$ such that $|z - a|_x < 1$;
- there exists $r > 1$ such that $|z - a|_x = r$ for all $a \in K$.

Thus we can view $\mathbf{A}^1_{\text{Berk}}$ as the quotient $K \times [0, \infty[\, / \sim$, where $(a, r) \sim (b, s)$ iff $r = s$ and $|a - b| \leq r$. Note that if $r \geq 1$, then $(a, r) \sim (b, r)$ for all r, whereas if $0 \leq r < 1$, then $(a, r) \simeq (b, r)$ iff $a = b$.

We see that the Berkovich projective line $\mathbf{P}^1_{\text{Berk}} = \mathbf{A}^1_{\text{Berk}} \cup \{\infty\}$ is a tree naturally rooted at ∞ with the Gauss point as its only branch point. See Fig. 3. The hyperbolic metric is induced by the parametrization $\alpha_z(a, r) = \log r$. In fact, this parametrization does not depend on the choice of coordinate $z \in R$.

If we instead choose the Gauss point as the root of the tree, then we can view the topological space underlying $\mathbf{P}^1_{\text{Berk}}$ as the cone over $\mathbf{P}^1$, that is, as the quotient $\mathbf{P}^1 \times [0, \infty]$, where $(a, s) \sim (b, t)$ if $s = t = 0$. The Gauss point is the apex of the cone and its distance to (a, t) is t in the hyperbolic metric. See Fig. 4.

Just as in the nontrivially valued case, the generalized metric on $\mathbf{P}^1_{\text{Berk}}$ is the correct one in the sense that Remark 3.13 holds also in this case.

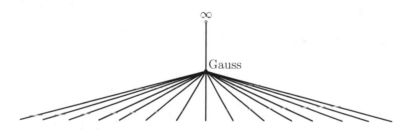

Fig. 3 The Berkovich affine line over a trivially valued field

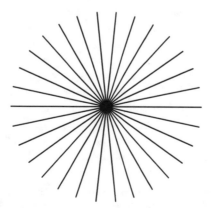

Fig. 4 The Berkovich projective line over a trivially valued field

Following the terminology of Sect. 3.3.4, a point of the form (a, t) is of Type 1 and Type 2 iff $t = 0$ and $t = \infty$, respectively. All other points are of Type 3; there are no Type 4 points.

We can also describe the structure sheaf $\mathcal{O}$. When x is the Gauss point, the local ring $\mathcal{O}_x$ is the field F of rational functions and $\mathcal{H}(x) = \mathcal{O}_x = F$ is equipped with the trivial norm. Further, $\Gamma_x = \Gamma_K = 0$, so rat. rk $x = 0$ and tr. deg $x = 1$.

Now assume $x \in \mathbf{P}^1_{\mathrm{Berk}}$ is not the Gauss point and pick a coordinate $z \in F$ such that $|z|_x < 1$. If x is of Type 3, that is, $0 < |z|_x < 1$, then $\mathcal{O}_x = K((z))$ is the field of formal Laurent series and $\mathcal{H}(x) = \mathcal{O}_x$ is equipped with the norm $|\sum_j a_j z^j|_x = r^{\max\{j \,|\, a_j \neq 0\}}$. Further, $\Gamma_x = \mathbf{Z} \log r$, so rat. rk $x = 1$, tr. deg $x = 0$.

If instead $|z|_x = 0$ so that x is of Type 1, then we have $\mathcal{O}_x = K[[z]]$, whereas $\mathcal{H}(x) \simeq K$ is equipped with the trivial norm.

Finally, when K is not algebraically closed, we view $\mathbf{P}^1_{\mathrm{Berk}}(K)$ as a quotient of $\mathbf{P}^1_{\mathrm{Berk}}(K^a)$, where K^a is the algebraic closure of K (note that K^a is already complete in this case). We can still view the Berkovich projective line as the quotient $\mathbf{P}^1(K) \times [0, \infty]/\sim$, with $\mathbf{P}^1(K)$ the set of closed (but not necessarily K-rational) points of the projective line over K and where $(a, 0) \sim (b, 0)$ for all a, b. The multiplicity (i.e. the number of preimages in $\mathbf{P}^1_{\mathrm{Berk}}(K^a)$) of the Gauss point is 1 and the multiplicity of any point (a, t) is equal to the degree $[K(a) : K]$ if $t > 0$, where $K(a)$ is the residue field of a. We define a parametrization of $\mathbf{P}^1_{\mathrm{Berk}}(K)$ using (14). Then the result in Remark 3.13 remains valid.

3.10 Notes and Further References

The construction of the Berkovich affine and projective lines is, of course, due to Berkovich and everything in this section is, at least implicitly, contained in his book [Ber90].

For general facts on Berkovich spaces we refer to the original works [Ber90, Ber93] or to some of the recent surveys, e.g. the ones by Conrad [Con08] and Temkin [Tem14]. However, the affine and projective lines are very special cases of Berkovich spaces and in fact we need very little of the general theory in order to understand them. I can offer a personal testimony to this fact as I was doing dynamics on Berkovich spaces before I even knew what a Berkovich space was!

Having said that, it is sometimes advantageous to use some general notions, and in particular the structure sheaf, which will be used to define the local degree of a rational map in Sect. 4.6. Further, the stability of the residue field at Type 2 and Type 3 points is quite useful. In higher dimensions, simple arguments using the tree structure are probably less useful than in dimension 1.

The Berkovich affine and projective lines are studied in great detail in the book [BR10] by Baker and Rumely, to which we refer for more details. However, our presentation here is slightly different and adapted to our purposes. In particular, we insist on trying to work in a coordinate free way whenever possible. For example, the Berkovich unit disc and its associated Gauss norm play an important role in

most descriptions of the Berkovich projective line, but they are only defined once
we have chosen a coordinate; without coordinates all Type 2 points are equivalent.
When studying the dynamics of rational maps, there is usually no canonical choice
of coordinate and hence no natural Gauss point (the one exception being maps of
simple reduction, see Sect. 5.5).

One thing that we do not talk about at all are formal models. They constitute a
powerful geometric tool for studying Berkovich spaces, see [Ber99, Ber04] but we
do not need them here. However, the corresponding notion for trivially valued fields
is used systematically in Sects. 6–10.

4   Action by Polynomial and Rational Maps

We now study how a polynomial or a rational map acts on the Berkovich affine and
projective lines, respectively. Much of the material in this chapter can be found with
details in the literature. However, as a general rule our presentation is self-contained,
the exception being when we draw more heavily on the general theory of Berkovich
spaces or non-Archimedean geometry. As before, we strive to work in a coordinate
free way whenever possible.

Recall that over the complex numbers, the projective line $\mathbf{P}^1$ is topologically a
sphere. Globally a rational map $f : \mathbf{P}^1 \to \mathbf{P}^1$ is a branched covering. Locally it is of
the form $z \mapsto z^m$, where $m \geq 1$ is the local degree of f at the point. In fact, $m = 1$
outside the ramification locus of f, which is a finite set.

The non-Archimedean case is superficially very different but in fact exhibits
many of the same properties when correctly translated. The projective line is a
tree and a rational map is a tree map in the sense of Sect. 2.6. Furthermore,
there is a natural notion of local degree that we shall explore in some detail.
The ramification locus can be quite large and has been studied in detail by
Faber [Fab13a, Fab13b, Fab14]. Finally, it is possible to give local normal forms,
at least at points of Types 1-3.

4.1   Setup

As before, K is a non-Archimedean field. We assume that the norm on K
is non-trivial and that K is algebraically closed but of arbitrary characteristic.
See Sect. 4.11 for extensions.

Recall the notation $R \simeq K[z]$ for the polynomial ring in one variable with
coefficients in K, and $F \simeq K(z)$ for its fraction field.

4.2   Polynomial and Rational Maps

We start by recalling some general algebraic facts about polynomial and rational maps. The material in Sects. 4.2.3–4.2.5 is interesting mainly when the ground field K has positive characteristic. General references for that part are [Lan02, VII.7] and [Har77, IV.2].

4.2.1   Polynomial Maps

A nonconstant polynomial map $f : \mathbf{A}^1 \to \mathbf{A}^1$ of the affine line over K is given by an injective K-algebra homomorphism $f^* : R \to R$. The *degree* $\deg f$ of f is the length of R as a module over f^*R. Given coordinates $z, w \in R$ on $\mathbf{A}^1$, f^*w is a polynomial in z of degree $\deg f$.

4.2.2   Rational Maps

A nonconstant regular map $f : \mathbf{P}^1 \to \mathbf{P}^1$ of the projective line over K is is defined by an injective homomorphism $f^* : F \to F$ of fields over K, where $F \simeq K(z)$ is the fraction field of R. The degree of f is the degree of the field extension F/f^*F. Given coordinates $z, w \in F$ on $\mathbf{P}^1$, f^*w is a rational function of z of degree $d :=$ $\deg f$, that is, $f^*w = \phi/\psi$, where $\phi, \psi \in K[z]$ are polynomials without common factor and $\max\{\deg\phi, \deg\psi\} = d$. Thus we refer to f as a rational map, even though it is of course regular.

Any polynomial map $f : \mathbf{A}^1 \to \mathbf{A}^1$ extends to a rational map $f : \mathbf{P}^1 \to \mathbf{P}^1$ satisfying $f(\infty) = \infty$. In fact, polynomial maps can be identified with rational maps $f : \mathbf{P}^1 \to \mathbf{P}^1$ admitting a totally invariant point $\infty = f^{-1}(\infty)$.

4.2.3   Separable Maps

We say that a rational map f is *separable* if the field extension F/f^*F is separable, see [Lan02, VII.4]. This is always the case if K has characteristic zero.

If f is separable, of degree d, then, by the Riemann-Hurwitz Theorem [Har77, IV.2/4] the *ramification divisor* R_f on $\mathbf{P}^1$ is well defined and of degree $2d - 2$. In particular, all but finitely many points of $\mathbf{P}^1$ have exactly d preimages under f, so f has *topological degree* d.

4.2.4   Purely Inseparable Maps

We say that a rational map f is *purely inseparable* if the field extension F/f^*F is purely inseparable. Assuming $\deg f > 1$, this can only happen when K has characteristic $p > 0$ and means that for every $\phi \in F^*$ there exists $n \geq 0$ such that $\phi^{p^n} \in f^*F$, see [Lan02, VII.7]. Any purely inseparable map $f : \mathbf{P}^1 \to \mathbf{P}^1$ is bijective. We shall see in Sect. 5.3 that if f is purely inseparable of degree $d > 1$, then $d = p^n$ for some $n \geq 1$ and there exists a coordinate $z \in F$ on $\mathbf{P}^1$ such that $f^*z = z^d$.

4.2.5 Decomposition

In general, any algebraic field extension can be decomposed into a separable extension followed by a purely inseparable extension, see [Lan02, VII.7]. As a consequence, any rational map f can be factored as $f = g \circ h$, where g is separable and h is purely inseparable. The topological degree of f is equal to the degree of g or, equivalently, the separable degree of the field extension F/f^*F, see [Lan02, VII.4].

4.2.6 Totally Ramified Points

We say that a rational map $f : \mathbf{P}^1 \to \mathbf{P}^1$ is *totally ramified* at a point $x \in \mathbf{P}^1$ if $f^{-1}(f(x)) = \{x\}$.

Proposition 4.1 *Let* $f : \mathbf{P}^1 \to \mathbf{P}^1$ *be a rational map of degree* $d > 1$.

(i) *If* f *is purely inseparable, then* f *is totally ramified at every point* $x \in \mathbf{P}^1$.
(ii) *If* f *is not purely inseparable, then there are at most two points at which* f *is totally ramified.*

Proof If f is purely inseparable, then $f : \mathbf{P}^1 \to \mathbf{P}^1$ is bijective and hence totally ramified at every point.

Now suppose f is not purely inseparable. Then $f = g \circ h$, where h is purely inseparable and g is separable, of degree $\deg g > 1$. If f is totally ramified at x, then so is g, so we may assume f is separable. In this case, a direct calculation shows that the ramification divisor has order $d - 1$ at x. The result follows since the ramification divisor has degree $2(d - 1)$. $\qquad\square$

4.3 Action on the Berkovich Space

Recall that the affine and projective line $\mathbf{A}^1$ and $\mathbf{P}^1$ embed in the corresponding Berkovich spaces $\mathbf{A}^1_{\mathrm{Berk}}$ and $\mathbf{P}^1_{\mathrm{Berk}}$, respectively.

4.3.1 Polynomial Maps

Any nonconstant polynomial map $f : \mathbf{A}^1 \to \mathbf{A}^1$ extends to

$$f : \mathbf{A}^1_{\mathrm{Berk}} \to \mathbf{A}^1_{\mathrm{Berk}}$$

as follows. If $x \in \mathbf{A}^1_{\mathrm{Berk}}$, then $x' = f(x)$ is the multiplicative seminorm $|\cdot|_{x'}$ on R defined by

$$|\phi|_{x'} := |f^*\phi|_x.$$

It is clear that $f : \mathbf{A}^1_{\mathrm{Berk}} \to \mathbf{A}^1_{\mathrm{Berk}}$ is continuous, as the topology on $\mathbf{A}^1_{\mathrm{Berk}}$ was defined in terms of pointwise convergence. Further, f is order-preserving in the partial ordering on $\mathbf{A}^1_{\mathrm{Berk}}$ given by $x \leq x'$ iff $|\phi|_x \leq |\phi|_{x'}$ for all polynomials ϕ.

4.3.2   Rational Maps

Similarly, we can extend any nonconstant rational map $f : \mathbf{P}^1 \to \mathbf{P}^1$ to a map

$$f : \mathbf{P}^1_{\mathrm{Berk}} \to \mathbf{P}^1_{\mathrm{Berk}}.$$

Recall that we defined $\mathbf{P}^1_{\mathrm{Berk}}$ as the set of generalized seminorms $|\cdot| : F \to [0, +\infty]$. If $x \in \mathbf{P}^1_{\mathrm{Berk}}$, then the value of the seminorm $|\cdot|_{f(x)}$ on a rational function $\phi \in F$ is given by

$$|\phi|_{f(x)} := |f^*\phi|_x.$$

On the Berkovich projective line $\mathbf{P}^1_{\mathrm{Berk}}$ there is no canonical partial ordering, so in general it does not make sense to expect f to be order preserving. The one exception to this is when there exist points $x, x' \in \mathbf{P}^1_{\mathrm{Berk}}$ such that $f^{-1}(x') = \{x\}$. In this case one can show that $f : \mathbf{P}^1_{\mathrm{Berk}} \to \mathbf{P}^1_{\mathrm{Berk}}$ becomes order preserving when the source and target spaces are equipped with the partial orderings rooted in x and x'. If x and x' are both of Type 2, we can find coordinates on the source and target in which x and x' are both equal to the Gauss point, in which case one says that f has *good reduction*, see Sect. 5.5.

4.4   Preservation of Type

There are many ways of analyzing the mapping properties of a rational map $f : \mathbf{P}^1_{\mathrm{Berk}} \to \mathbf{P}^1_{\mathrm{Berk}}$. First we show the type of a point is invariant under f. For this, we use the numerical classification in Sect. 3.8.3.

Lemma 4.2 *The map $f : \mathbf{P}^1_{\mathrm{Berk}} \to \mathbf{P}^1_{\mathrm{Berk}}$ sends a point of Type 1-4 to a point of the same type.*

Proof We follow the proof of [BR10, Proposition 2.15]. Fix $x \in \mathbf{P}^1_{\mathrm{Berk}}$ and write $x' = f(x)$.

If $|\cdot|_{x'}$ has nontrivial kernel, then clearly so does $|\cdot|_x$ and it is not hard to prove the converse, using that K is algebraically closed.

Now suppose $|\cdot|_x$ and $|\cdot|_{x'}$ have trivial kernels. In this case, the value group $\Gamma_{x'}$ is a subgroup of Γ_x of finite index. As a consequence, x and x' have the same rational rank. Similarly, $\widetilde{\mathcal{H}(x)}/\widetilde{\mathcal{H}(x')}$ is a finite field extension, so x and x' have the same transcendence degree. In view of the numerical classification, x and x' must have the same type.

$\square$

4.5   Topological Properties

Next we explore the basic topological properties of a rational map.

Proposition 4.3 *The map* $f : \mathbf{P}^1_{\mathrm{Berk}} \to \mathbf{P}^1_{\mathrm{Berk}}$ *is continuous, finite, open and surjective. Any point in* $\mathbf{P}^1_{\mathrm{Berk}}$ *has at least one and at most d preimages, where* $d = \deg f$.

We shall see shortly that any point has *exactly d* preimages, counted with multiplicity. However, note that for a purely inseparable map, this multiplicity is equal to $\deg f$ at every point.

Proof All the properties follow quite easily from more general results in [Ber90, Ber93], but we recall the proof from [FR10, p. 126].

Continuity of f is clear from the definition, as is the fact that a point of Type 1 has at least one and at most d preimages. A point in $\mathbf{H} = \mathbf{P}^1_{\mathrm{Berk}} \setminus \mathbf{P}^1$ defines a norm on F, hence also on the subfield $f^* F$. The field extension $F/f^* F$ has degree d, so by [ZS75] a valuation on $f^* F$ has at least one and at most d extensions to F. This means that a point in $\mathbf{H}$ also has at least one and at most d preimages.

In particular, f is finite and surjective. By general results about morphisms of Berkovich spaces, this implies that f is open, see [Ber90, 3.2.4]. $\square$

Since $\mathbf{P}^1_{\mathrm{Berk}}$ is a tree, Proposition 4.3 shows that all the results of Sect. 2.6 apply and give rather strong information on the topological properties of f.

One should note, however, that these purely topological results seem very hard to replicated for Berkovich spaces of higher dimensions. The situation over the complex numbers is similar, where the one-dimensional and higher-dimensional analyses are quite different.

4.6   Local Degree

It is reasonable to expect that any point in $\mathbf{P}^1_{\mathrm{Berk}}$ should have exactly $d = \deg f$ preimages under f counted with multiplicity. This is indeed true, the only problem being to define this multiplicity. There are several (equivalent) definitions in the literature. Here we shall give the one spelled out by Favre and Rivera-Letelier [FR10], but also used by Thuillier [Thu05]. It is the direct translation of the corresponding notion in algebraic geometry.

Fix a point $x \in \mathbf{P}_{\text{Berk}}^1$ and write $x' = f(x)$. Let $\mathfrak{m}_x$ be the maximal ideal in the local ring $\mathcal{O}_x$ and $\kappa(x) := \mathcal{O}_x / \mathfrak{m}_x$ the residue field. Using f we can $\mathcal{O}_x$ as an $\mathcal{O}_{x'}$-module and $\mathcal{O}_x / \mathfrak{m}_{x'} \mathcal{O}_x$ as a $\kappa(x')$-vector space.

Definition 4.4 The local degree of f at x is $\deg_x f = \dim_{\kappa(x')}(\mathcal{O}_x / \mathfrak{m}_{x'} \mathcal{O}_x)$.

Alternatively, since f is finite, it follows [Ber90, 3.1.6] that $\mathcal{O}_x$ is a *finite* $\mathcal{O}_{x'}$-module. The local degree $\deg_x f$ is therefore also equal to the rank of the module $\mathcal{O}_x$ viewed as $\mathcal{O}_{x'}$-module, see [Mat89, Theorem 2.3]. From this remark it follows that if $f, g : \mathbf{P}^1 \to \mathbf{P}^1$ are nonconstant rational maps, then

$$\deg_x(f \circ g) = \deg_x g \cdot \deg_{g(x)} f$$

for any $x \in \mathbf{P}_{\text{Berk}}^1$.

The definition above of the local degree works also over the complex numbers. A difficulty in the non-Archimedean setting is that the local rings $\mathcal{O}_x$ are not as concrete as in the complex case, where they are isomorphic to the ring of convergent power series.

The following result shows that local degree behaves as one would expect from the complex case. See [FR10, Proposition-Definition 2.1].

Proposition 4.5 *For every simple domain V and every connected component U of $f^{-1}(V)$, the integer*

$$\sum_{f(y)=x, y \in U} \deg_y f \tag{15}$$

is independent of the point $x \in V$.

Recall that a simple domain is a finite intersection of open Berkovich discs; see Sect. 3.6. The integer in (15) should be interpreted as the degree of the map from U to V. If we put $U = V = \mathbf{P}_{\text{Berk}}^1$, then this degree is d.

We refer to [FR10, p.126] for a proof. The idea is to view $f : U \to V$ as a map between Berkovich analytic curves. In fact, this is one of the few places in these notes where we draw more heavily on the general theory of Berkovich spaces.

We would like to give a more concrete interpretation of the local degree. First, at a Type 1 point, it can be read off from a local expansion of f:

Proposition 4.6 *Let $x \in \mathbf{P}_{\text{Berk}}^1$ be a Type 1 point and pick coordinates z, w on $\mathbf{P}^1$ such that $x = f(x) = 0$. Then $\mathcal{O}_x \simeq K\{z\}$, $\mathcal{O}_{f(x)} = K\{w\}$ and we have*

$$f^* w = a z^k (1 + h(z)), \tag{16}$$

where $a \neq 0$, $k = \deg_x(f)$ and $h(0) = 0$.

Proof The only thing that needs to be checked is that $k = \deg_x(f)$. We may assume $a = 1$. First suppose char $K = 0$. Then we can find $\phi(z) \in K\{z\}$ such that $1 +$

$h(z) = (1 + \phi(z))^k$ in $K\{z\}$. It is now clear that $\mathcal{O}_x \sim K\{z\}$ is a free module over $f^*\mathcal{O}_{f(x)}$ of rank k, with basis given by $(z(1 + \phi(z)))^j$, $0 \le j \le k - 1$, so $\deg_x(f) = k$. A similar argument can be used the case when K has characteristic $p > 0$; we refer to [FR10, p.126] for the proof. □

We shall later see how the local degree at a Type 2 or Type 3 points also appears in a suitable local expansion of f.

The following crucial result allows us to interpret the local degree quite concretely as a local expansion factor in the hyperbolic metric.

Theorem 4.7 *Let* $f : \mathbf{P}^1_{\text{Berk}} \to \mathbf{P}^1_{\text{Berk}}$ *be as above.*

(i) *If* x *is a point of Type 1 or 4 and* $\gamma = [x, y]$ *is a sufficiently small segment, then* f *maps* γ *homeomorphically onto* $f(\gamma)$ *and expands the hyperbolic metric on* γ *by a factor* $\deg_x(f)$.

(ii) *If* x *is a point of Type 3 and* γ *is a sufficiently small segment containing* x *in its interior, then* f *maps* γ *homeomorphically onto* $f(\gamma)$ *and expands the hyperbolic metric on* γ *by a factor* $\deg_x(f)$.

(iii) *If* x *is a point of Type 2, then for every tangent direction* $\vec{v}$ *at* x *there exists an integer* $m_{\vec{v}}(f)$ *such that the following holds:*

(a) *for any sufficiently small segment* $\gamma = [x, y]$ *representing* $\vec{v}$, f *maps* γ *homeomorphically onto* $f(\gamma)$ *and expands the hyperbolic metric on* γ *by a factor* $m_{\vec{v}}(f)$;

(b) *if* $\vec{v}$ *is any tangent direction at* x *and* $\vec{v}_1, \dots, \vec{v}_m$ *are the preimages of* $\vec{v}$ *under the tangent map, then* $\sum_i m_{\vec{v}_i}(f) = \deg_x(f)$.

Theorem 4.7 is due to Rivera-Letelier [Riv05, Proposition 3.1] (see also [BR10, Theorem 9.26]). However, in these references, different (but equivalent) definitions of local degree were used. In Sect. 4.8 below we will indicate a direct proof of Theorem 4.7 using the above definition of the local degree.

Since the local degree is bounded by the algebraic degree, we obtain as an immediate consequence

Corollary 4.8 *If* $f : \mathbf{P}^1_{\text{Berk}} \to \mathbf{P}^1_{\text{Berk}}$ *is as above, then*

$$d_{\mathbf{H}}(f(x), f(y)) \le \deg f \cdot d_{\mathbf{H}}(x, y)$$

for all $x, y \in \mathbf{H}$.

Using Theorem 4.7 we can also make Corollary 2.16 more precise:

Corollary 4.9 *Let* $\gamma \subseteq \mathbf{P}^1_{\text{Berk}}$ *be a segment such that the local degree is constant on the interior of* γ. *Then* f *maps* γ *homeomorphically onto* $\gamma' := f(\gamma)$.

Proof By Corollary 2.16 the first assertion is a local statement: it suffices to prove that if x belongs to the interior of γ then the tangent map of f is injective on the set of tangent directions at x defined by γ. But if this were not the case, the local degree at x would be too high in view of assertion (iii) (b) in Theorem 4.7. □

Remark 4.10 Using similar arguments, Rivera-Letelier was able to improve Proposition 2.12 and describe $f(U)$ for a simple domain U. For example, he described when the image of an open disc is an open disc as opposed to all of $\mathbf{P}^1_{\mathrm{Berk}}$ and similarly described the image of an annulus. See Theorems 9.42 and 9.46 in [BR10] and also the original papers [Riv03a, Riv03b].

4.7 Ramification Locus

Recall that, over the complex numbers, a rational map has local degree 1 except at finitely many points. In the non-Archimedean setting, the situation is more subtle.

Definition 4.11 The *ramification locus* R_f of f is the set of $x \in \mathbf{P}^1_{\mathrm{Berk}}$ such that $\deg_x(f) > 1$. We say that f is *tame*[12] if R_f is contained in the convex hull of a finite subset of $\mathbf{P}^1$.

Lemma 4.12 *If K has residue characteristic zero, then f is tame. More precisely, R_f is a finite union of segments in $\mathbf{P}^1_{\mathrm{Berk}}$ and is contained in the convex hull of the critical set of $f : \mathbf{P}^1 \to \mathbf{P}^1$. As a consequence, the local degree is one at all Type 4 points.*

We will not prove this lemma here. Instead we refer to the papers [Fab13a, Fab13b] by X. Faber for a detailed analysis of the ramification locus, including the case of positive residue characteristic. The main reason why the zero residue characteristic case is easier stems from the following version of Rolle's Theorem (see e.g. [BR10, Proposition A.20]): if char $\tilde{K} = 0$ and $D \subseteq \mathbf{P}^1$ is an open disc such that $f(D) \neq \mathbf{P}^1$ and f is not injective on D, then f has a critical point in D.

See Sect. 4.10 below for some examples of ramification loci.

4.8 Proof of Theorem 4.7

While several proofs of Theorem 4.7 exist in the literature, I am not aware of any that directly uses Definition 4.4 of the local degree. Instead, they use different definitions, which in view of Proposition 4.5 are equivalent to the one we use. Our proof of Theorem 4.7 uses some basic non-Archimedean analysis in the spirit of [BGR84].

4.8.1 Type 1 Points

First suppose $x \in \mathbf{P}^1$ is a classical point. As in the proof of Proposition 4.6, we find coordinates z and w on $\mathbf{P}^1$ vanishing at x and x', respectively, such that $f^*w =$

---

[12]The terminology "tame" follows Trucco [Tru09].

$az^k(1 + h(z))$, where $a \neq 0$, $k = \deg_x(f) \geq 1$ and $h(0) = 0$. In fact, we may assume $a = 1$. Pick $r_0 > 0$ so small that $|h(z)|_{D(0,r)} < 1$ for $r \leq r_0$. It then follows easily that $f(x_{D(0,r)}) = x_{D(0,r^k)}$ for $0 \leq r \leq r_0$. Thus f maps the segment $[x_0, x_{D(0,r_0)}]$ homeomorphically onto the segment $[x_0, x_{D(0,r_0^k)}]$ and the hyperbolic metric is expanded by a factor k.

4.8.2   Completion

Suppose x is of Type 2 or 3. Then the seminorm $|\cdot|_x$ is a norm, $\mathcal{O}_x$ is a field having $\mathcal{O}_{x'}$ as a subfield and $\deg_x(f)$ is the degree $[\mathcal{O}_x : \mathcal{O}_{x'}]$ of the field extension $\mathcal{O}_x/\mathcal{O}_{x'}$. Recall that $\mathcal{H}(x)$ is the completion of $\mathcal{O}_x$.

In general, the degree of a field extension can change when passing to the completion. However, we have

Proposition 4.13 *For any point $x \in \mathbf{P}^1_{\mathrm{Berk}}$ of Type 2 or 3 we have*

$$\deg_x(f) = [\mathcal{O}_x : \mathcal{O}_{x'}] = [\mathcal{H}(x) : \mathcal{H}(x')] = [\Gamma_x : \Gamma_{x'}] \cdot [\widetilde{\mathcal{H}(x)} : \widetilde{\mathcal{H}(x')}], \qquad (17)$$

where Γ and $\tilde{H}$ denotes the value groups and residue fields of the norms under consideration.

Proof Recall from Sect. 3.8.4 that the field $\mathcal{O}_{x'}$ is quasicomplete in the sense that the norm $|\cdot|_{x'}$ on $\mathcal{O}_{x'}$ extends uniquely to any algebraic extension. In particular, the norm $|\cdot|_x$ is the unique extension of this norm to $\mathcal{O}_x$. Also recall from Sect. 3.8.5 that the field $\mathcal{O}_{x'}$ is weakly stable. Thus $\mathcal{O}_x$ is weakly Cartesian over $\mathcal{O}_{x'}$, which by [BGR84, 2.3.3/6] implies the second equality in (17).

Finally recall from Sect. 3.8.6 that the field $\mathcal{H}(x')$ is stable. The third equality in (17) then follows from [BGR84, 3.6.2/4].   □

4.8.3   Approximation

In order to understand the local degree of a rational map, it is useful to simplify the map in a way similar to (16). Suppose x and $x' = f(x)$ are Type 2 or Type 3 points. In suitable coordinates on the source and target, we can write $x = x_{D(0,r)}$ and $x' = x_{D(0,r')}$, where $0 < r, r' \leq 1$. If x and x' are Type 2 points, we can further assume $r = r' = 1$.

Write $f^*w = f(z)$ for some rational function $f(z) \in F \simeq K(z)$. Suppose we can find a decomposition in F of the form

$$f(z) = g(z)(1 + h(z)), \quad \text{where } |h(z)|_x < 1.$$

The rational function $g(z) \in F$ induces a rational map $g : \mathbf{P}^1 \to \mathbf{P}^1$, which extends to $g : \mathbf{P}^1_{\mathrm{Berk}} \to \mathbf{P}^1_{\mathrm{Berk}}$.

Lemma 4.14 *There exists $\delta > 0$ such that $g(y) = f(y)$ and $\deg_y(g) = \deg_y(f)$ for all $y \in \mathbf{H}$ with $d_\mathbf{H}(y, x) \leq \delta$.*

Proof We may assume that $h(z) \not\equiv 0$, or else there is nothing to prove. Thus we have $|h(z)|_x > 0$. Pick $0 < \varepsilon < 1$ such that $|h(z)|_x \leq \varepsilon^3$, set

$$\delta = (1 - \varepsilon) \min \left\{ \frac{|h(z)|_x}{\deg h(z)}, \frac{r'}{2 \deg f} \right\}$$

and assume $d_\mathbf{H}(y, x) \leq \delta$. We claim that

$$|f^*\phi - g^*\phi|_y \leq \varepsilon |f^*\phi|_y \quad \text{for all } \phi \in F. \tag{18}$$

Granting (18), we get $|g^*\phi|_y = |f^*\phi|_y$ for all ϕ and hence $g(y) = f(y) =: y'$. Furthermore, f and g give rise to isometric embeddings $f^*, g^* : \mathcal{H}(y') \to \mathcal{H}(y)$. By Proposition 4.13, the degrees of the two induced field extensions $\mathcal{H}(y)/\mathcal{H}(y')$ are equal to $\deg_y f$ and $\deg_y g$, respectively. By continuity, the inequality (18) extends to all $\phi \in \mathcal{H}(y')$. It then follows from [Tem10, 6.3.3] that $\deg_y f = \deg_y g$.

We also remark that (18) implies

$$f^*\Gamma_{y'} = g^*\Gamma_{y'} \quad \text{and} \quad f^*\widetilde{\mathcal{H}(y')} = g^*\widetilde{\mathcal{H}(y')}. \tag{19}$$

Thus f and g give the same embeddings of $\Gamma_{y'}$ and $\widetilde{\mathcal{H}(y')}$ into Γ_y and $\widetilde{\mathcal{H}(y)}$, respectively. When y, and hence y' is of Type 2 or 3, the field $\mathcal{H}(y')$ is stable, and so (17) gives another proof of the equality $\deg_y f = \deg_y g$.

It remains to prove (18). A simple calculation shows that if (18) holds for $\phi, \psi \in F$, then it also holds for $\phi\psi$, $1/\phi$ and $a\phi$ for any $a \in K$. Since K is algebraically closed, it thus suffices to prove (18) for $\phi = w - b$, where $b \in K$.

Using Lemma 3.12 and the fact that $f(x) = x_{D(0,r')}$, we get

$$|f(z) - b|_y \geq |f(z) - b|_x - \delta \deg f = |w - b|_{f(x)} - \delta \deg f \geq$$
$$\geq r' - \delta \deg f \geq \varepsilon(r' + \delta \deg f) = \varepsilon(|f(z)|_x + \delta \deg f) \geq \varepsilon|f(z)|_y.$$

Now Lemma 3.12 and the choice of δ imply $|h(z)|_y \leq \varepsilon^2 < 1$. As a consequence, $|g(z)|_y = |f(z)|_y$. We conclude that

$$|f^*(w - b) - g^*(w - b)|_y = |h(z)|_y |g(z)|_y \leq \varepsilon^2 |f(z)|_y \leq \varepsilon|f(z) - b|_y$$
$$= \varepsilon|f^*(w - b)|_y,$$

establishing (18) and completing the proof of Lemma 4.14.   □

4.8.4 Type 3 Points

Now consider a point x of Type 3. In suitable coordinates z, w we may assume that x and $x' = f(x)$ are associated to irrational closed discs $D(0, r)$ and $D(0, r')$, respectively. In these coordinates, f is locally approximately monomial at x; there exist $\theta \in K^*$ and $k \in \mathbf{Z} \setminus \{0\}$ such that $f^* w = \theta z^k (1 + h(z))$, where $h(z) \in K(z)$ satisfies $|h(z)|_x < 1$. Replacing w by $(\theta^{-1} w)^{\pm 1}$ we may assume $\theta = 1$ and $k > 0$. In particular, $r' = r^k$.

Let $g : \mathbf{P}^1 \to \mathbf{P}^1$ be defined by $g^* w = z^k$. We claim that $\deg_x(g) = k$. Indeed, the field $\mathcal{H}(x)$ (resp. $\mathcal{H}(x')$) can be concretely described as the set of formal series $\sum_{-\infty}^{\infty} a_j z^j$ (resp. $\sum_{-\infty}^{\infty} b_j w^j$) with $|a_j| r^j \to 0$ as $|j| \to \infty$ (resp. $|b_j| r^{kj} \to 0$ as $|j| \to \infty$). Then $1, z, \ldots, z^{k-1}$ form a basis for $\mathcal{H}(x)/\mathcal{H}(x')$. We can also see that $\deg_x(g) = k$ from (17) using that $\widetilde{\mathcal{H}(x)} = \widetilde{\mathcal{H}(x')} = \tilde{K}$, $\Gamma_x = \Gamma_K + \mathbf{Z} \log r$ and $\Gamma_{x'} = \Gamma_K + k\mathbf{Z} \log r$.

Lemma 4.14 gives $\deg_x(f) = \deg_x(g)$. Moreover, we must have $f(x_{D(0,s)}) = x_{D(0,s^k)}$ for $s \approx r$, so f expands the hyperbolic metric by a factor $k = \deg_x(f)$. Thus we have established all statements in Theorem 4.7 for Type 3 points.

4.8.5 Type 2 Points

Now suppose x and hence $x' = f(x)$ is of Type 2. Then $\Gamma_x = \Gamma_{x'} = \Gamma_K$. We may assume x and x' both equal the Gauss point in suitable coordinates z and w. The algebraic tangent spaces $T_x, T_{x'} \simeq \mathbf{P}^1(\tilde{K})$ defined in Sect. 3.8.7 have $\widetilde{\mathcal{H}(x)} \simeq \tilde{K}(z)$ and $\widetilde{\mathcal{H}(x')} \simeq \tilde{K}(w)$ as function fields. Now f induces a map $f^* : \widetilde{\mathcal{H}(x')} \to \widetilde{\mathcal{H}(x)}$ and hence a map $T_x \to T_{x'}$. By (17), the latter has degree $\deg_x(f)$.

As opposed to the Type 3 case, we cannot necessarily approximate f by a monomial map. However, after applying a coordinate change of the form $z \mapsto (\theta z)^{\pm 1}$, we can find $g(z) \in F = K(z)$ of the form

$$g(z) = z^m \frac{\prod_{i=1}^{l-m}(z - a_i)}{\prod_{j=1}^{k}(z - b_j)}, \tag{20}$$

with $m \geq 0$, $|a_i| = |b_j| = 1$, $a_i \neq b_j$ and $a_i b_j \neq 0$ for all i, j, such that

$$f^* w = g(z)(1 + h(z)),$$

in F, where $|h(z)|_x < 1 = |g(z)|_x$.

On the one hand, $g(z)$ induces a map $g : \mathbf{P}^1(K) \to \mathbf{P}^1(K)$ and hence also a map $g : \mathbf{P}^1_{\text{Berk}} \to \mathbf{P}^1_{\text{Berk}}$. We clearly have $g(x) = x'$ and Lemma 4.14 gives $\deg_x(g) = \deg_x(f)$. On the other hand, $g(z)$ also induces a map $g : \mathbf{P}^1(\tilde{K}) \to \mathbf{P}^1(\tilde{K})$, which can be identified with the common tangent map $T_x \to T_{x'}$ of f and g. Both these maps g have degree $\max\{l, k\}$, so in accordance with (17), we see that $\deg_x(f) = [\widetilde{\mathcal{H}(x)} : \widetilde{\mathcal{H}(x')}]$.

To prove the remaining statements in Theorem 4.7 (iii), define $m_{\vec{v}}(f)$ as the local degree of the algebraic tangent map $T_x \rightarrow T'_x$ at the tangent direction $\vec{v}$. Statement (a) in Theorem 4.7 (iii) is then clear, so it suffices to show (b). We may assume that $\vec{v}$ and its image $\vec{v}'$ are both represented by x_0. Then $m(\vec{v})$ is the integer m in (20). We see from (20) and from Lemma 4.14 that $f(x_{D(0,r)}) = x_{D(0,r^m)}$ when $0 \ll 1 - r < 1$. Thus (b) holds.

4.8.6 Type 4 Points

Finally suppose x is a Type 4 point. By Corollary 2.15 we can find $y \in \mathbf{P}^1_{\text{Berk}}$ such that f is a homeomorphism of the segment $\gamma = [x, y]$ onto $f(\gamma)$. We first claim that by moving y closer to x, f will expand the hyperbolic metric on γ by a fixed integer constant $m \geq 1$.

Let $\vec{w}$ be the tangent direction at y represented by x. By moving y closer to x, if necessary, we may assume that x is the unique preimage of x' in $U(\vec{w})$.

Consider a point $\xi \in]x, y[$. If ξ is of Type 3, then we know that f locally expands the hyperbolic metric along γ by a factor $m(\xi)$. Now suppose ξ is a Type 2 point and let $\vec{v}_+$ and $\vec{v}_-$ be the tangent directions at ξ represented by x and y, respectively. Then f locally expands the hyperbolic metric along $\vec{v}_\pm$ by factors $m(\vec{v}_\pm)$. Suppose that $m(\vec{v}_+) < m(\vec{v}_-)$. Then there must exist a tangent direction $\vec{v}$ at ξ different from $\vec{v}_+$ but having the same image as $\vec{v}_+$ under the tangent map. By Corollary 2.13 this implies that $x' \in f(U(\vec{v})) \subseteq f(U(\vec{w}) \setminus \{x\})$, a contradiction. Hence $m(\vec{v}_+) \geq m(\vec{v}_-)$. Since $m(\vec{v}_+)$ is bounded from above by $d = \deg f$, we may assume that $m(\vec{v}_+) = m(\vec{v}_-)$ at all Type 2 points on γ. This shows that f expands the hyperbolic metric on γ by a constant factor m.

To see that $m = \deg_x(f)$, note that the above argument shows that $\deg_\xi(f) = m$ for all $\xi \in \gamma \setminus \{x\}$. Moreover, if $\vec{w}'$ is the tangent direction at $f(y)$ represented by $f(x)$, then the above reasoning shows that $U(\vec{w})$ is a connected component of $f^{-1}(U(\vec{w}'))$ and that ξ is the unique preimage of $f(\xi)$ in $U(\vec{w})$ for any $\xi \in \gamma$. It then follows from Proposition 4.5 that $\deg_x f = m$.

4.9 Laplacian and Pullbacks

Using the local degree we can pull back Radon measures on $\mathbf{P}^1_{\text{Berk}}$ by f. This we do by first defining a push-forward operator on continuous functions:

$$f_* H(x) = \sum_{f(y)=x} \deg_y(f) H(y)$$

for any $H \in C^0(\mathbf{P}^1_{\text{Berk}})$. It follows from Proposition 4.5 that $f_* H$ is continuous and it is clear that $\|f_* H\|_\infty \leq d \|H\|_\infty$, where $d = \deg f$. We then define the pull-back of Radon measures by duality:

$$\langle f^*\rho, H \rangle = \langle \rho, f_*H \rangle.$$

The pull-back operator is continuous in the weak topology of measures. If ρ is a probability measure, then so is $d^{-1}f^*\rho$. Note that the pull-back of a Dirac mass becomes

$$f^*\delta_x = \sum_{f(y)=x} \deg_y(f)\delta_y.$$

Recall from Sect. 2.5 that given a positive Radon measure ρ on $\mathbf{P}^1_{\text{Berk}}$ and a finite atomic measure ρ_0 supported on $\mathbf{H}$ of the same mass as ρ, we can write $\rho = \rho_0 + \Delta\varphi$ for a unique function $\varphi \in \text{SH}^0(\mathbf{P}^1_{\text{Berk}}, \rho_0)$. A key property is

Proposition 4.15 *If* $\varphi \in \text{SH}^0(\mathbf{P}^1_{\text{Berk}}, \rho_0)$, *then* $f^*\varphi \in \text{SH}^0(\mathbf{P}^1_{\text{Berk}}, f^*\rho_0)$ *and*

$$\Delta(f^*\varphi) = f^*(\Delta\varphi). \tag{21}$$

This formula, which will be crucial for the proof of the equidistribution in the next section, confirms that the generalized metric $d_{\mathbf{H}}$ on the tree $\mathbf{P}^1_{\text{Berk}}$ is the correct one. See also Remark 3.13.

Proof By approximating φ by its retractions $\varphi \circ r_X$, where X ranges over finite subtrees of $\mathbf{H}$ containing the support of ρ_0 we may assume that $\rho := \rho_0 + \Delta\varphi$ is supported on such a finite subtree X. This means that φ is locally constant outside X. By further approximation we reduce to the case when ρ is a finite atomic measure supported on Type 2 points of X.

Let $Y = f^{-1}(X)$. Using Corollary 2.15 and Theorem 4.7 we can write X (resp. Y) as a finite union γ_i (resp. γ_{ij}) of intervals with mutually disjoint interiors such that f maps γ_{ij} homeomorphically onto γ_i and the local degree is constant, equal to d_{ij} on the interior of γ_{ij}. We may also assume that the interior of each γ_i (resp. γ_{ij}) is disjoint from the support of ρ and ρ_0 (resp. $f^*\rho$ and $f^*\rho_0$). Since f expands the hyperbolic metric on each γ_{ij} with a constant factor d_{ij}, it follows that $\Delta(f^*\varphi) = 0$ on the interior of γ_{ij}.

In particular, $\Delta(f^*\varphi)$ is a finite atomic measure. Let us compute its mass at a point x. If $\vec{v}$ is a tangent direction at x and $\vec{v}' = Df(\vec{v})$ its image under the tangent map, then it follows from Theorem 4.7 (iii) that

$$D_{\vec{v}}(f^*\varphi) = m_{\vec{v}}(f)D_{\vec{v}'}(\varphi) \tag{22}$$

and hence

$$\Delta(f^*\varphi)\{x\} = \sum_{\vec{v}} D_{\vec{v}}(f^*\varphi) = \sum_{\vec{v}} m_{\vec{v}}(f)D_{\vec{v}'}(\varphi) = \sum_{\vec{v}'} D_{\vec{v}'}\varphi \sum_{Df(\vec{v})=\vec{v}'} m_{\vec{v}}(f)$$

$$= \deg_x(f) \sum_{\vec{v}} D_{\vec{v}}(\varphi) = \deg_x(f)(\Delta\varphi)\{f(x)\} = f^*(\Delta\varphi)\{x\},$$

which completes the proof.   □

4.10 Examples

To illustrate the ideas above, let us study three concrete examples of rational maps. Fix a coordinate $z \in F$ on $\mathbf{P}^1$. Following standard practice we write $f(z)$ for the rational function f^*z.

Example 4.16 Consider the polynomial map defined by

$$f(z) = a(z^3 - 3z^2)$$

where $a \in K$. Here K has residue characteristic zero. The critical points of f : $\mathbf{P}^1 \rightarrow \mathbf{P}^1$ are $z = 0$, $z = 2$ and $z = \infty$, where the local degree is 2, 2 and 3, respectively. On $\mathbf{P}^1_{\text{Berk}}$, the local degree is 3 on the interval $[x_G, \infty]$, where x_G is the Gauss norm. The local degree is 2 on the intervals $[0, x_G[$ and $[2, x_G[$ and it is 1 everywhere else. See Fig. 5.

Example 4.17 Next consider the polynomial map defined by

$$f(z) = z^p$$

for a prime p. Here the ground field K has characteristic zero. If the residue characteristic is different from p, then f is tamely ramified and the ramification

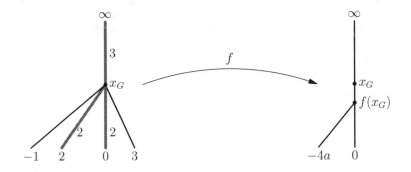

Fig. 5 The ramification locus of the map $f(z) = a(z^3 - 3z^2)$ in Example 4.16 when $|a| < 1$. Here x_G is the Gauss point. The preimage of the interval $[0, f(x_G)]$ is $[0, x_G]$ (with multiplicity 2) and $[3, x_G]$. The preimage of the interval $[-4a, f(x_G)]$ is $[2, x_G]$ (with multiplicity 2) and $[-1, x_G]$. The preimage of the interval $[\infty, f(x_G)]$ is $[\infty, x_G]$ (with multiplicity 3)

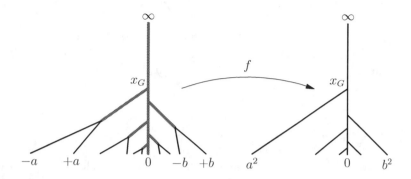

Fig. 6 The ramification locus of the map $f(z) = z^2$ in residual characteristic 2. A point in $\mathbf{A}^1_{\mathrm{Berk}}$ corresponding to a disc $D(a, r)$ belongs to the ramification locus iff $r \geq 2|a|$. The point x_G is the Gauss point

locus is the segment $[0, \infty]$. On the other hand, if the residue characteristic is p, then f is not tamely ramified. A point in $\mathbf{A}^1_{\mathrm{Berk}}$ corresponding to a disc $D(a, r)$ belongs to the ramification locus iff $r \geq p^{-1}|a|$. The ramification locus is therefore quite large and can be visualized as an "inverted Christmas tree", as illustrated in Fig. 6. It is the set of points in $\mathbf{P}^1_{\mathrm{Berk}}$ having hyperbolic distance at most $\log p$ to the segment $[0, \infty]$. See [BR10, Example 9.30] for more details.

Example 4.18 Again consider the polynomial map defined by

$$f(z) = z^p$$

for a prime p, but now assume that K has characteristic $p > 0$. Then f is purely inseparable and usually called the *Frobenius map*. We will see in Sect. 5.3 that every purely inseparable map of degree > 1 is an iterate of the Frobenius map in some coordinate z.

The mapping properties of f on the Berkovich projective line are easy to describe. Since f is a bijection, the local degree is equal to p at *all* points of $\mathbf{P}^1_{\mathrm{Berk}}$. Hence the ramification locus is equal to $\mathbf{P}^1_{\mathrm{Berk}}$. The Gauss point x_G in the coordinate z is a fixed point: $f(x_G) = x_G$. If $x \in \mathbf{P}^1_{\mathrm{Berk}}$, then f maps the segment $[x_G, x]$ homeomorphically onto the segment $[x_G, f(x)]$ and expands the hyperbolic metric by a constant factor p.

For many more interesting examples, see [BR10, §10.10].

4.11   Other Ground Fields

Above we worked with the assumption that our non-Archimedean field K was algebraically closed and nontrivially valued. Let us briefly discuss what happens when one or both of these assumption is dropped.

4.11.1   Non-Algebraically Closed Fields

First suppose K is nontrivially valued but not algebraically closed. Most of the results above remain true in this more general setting and can be proved by passing to the completed algebraic closure $\widehat{K^a}$ as in Sect. 3.9. Let us outline how to do this.

The definitions and results in Sect. 4.3 go through unchanged. Note that f induces a map $\hat{f} : \mathbf{P}^1_{\mathrm{Berk}}(\widehat{K^a}) \to \mathbf{P}^1_{\mathrm{Berk}}(\widehat{K^a})$ that is equivariant under the action of the Galois group $G = \mathrm{Gal}(K^a/K)$. Thus $f \circ \pi = \pi \circ \hat{f}$, where $\pi : \mathbf{P}^1_{\mathrm{Berk}}(\widehat{K^a}) \to \mathbf{P}^1_{\mathrm{Berk}}(K)$ is the projection. The fact that $\hat{f}$ preserves the type of a point (Lemma 4.2) implies that f does so as well. Proposition 4.3 remains valid and implies that f is a tree map in the sense of Sect. 2.6.

We define the local degree of f as in Sect. 4.6. Proposition 4.5 remains valid. The local degrees of f and $\hat{f}$ are related as follows. Pick a point $\hat{x} \in \mathbf{P}^1_{\mathrm{Berk}}(\widehat{K^a})$ and set $x = \pi(\hat{x})$, $\hat{x}' := f(\hat{x})$ and $x' := \pi(\hat{x}') = f(x)$. The stabilizer $G_{\hat{x}} := \{\sigma \in G \mid \sigma(\hat{x}) = \hat{x}\}$ is a subgroup of G and we have $G_{\hat{x}} \subseteq G_{\hat{x}'}$. The index of $G_{\hat{x}}$ in $G_{\hat{x}'}$ only depends on the projection $x = \pi(\hat{x})$ and we set

$$\delta_x(f) := [G_{\hat{x}'} : G_{\hat{x}}];$$

this is an integer bounded by the (topological) degree of f. We have $m(x) = \delta_x(f) m(f(x))$ for any $x \in \mathbf{P}^1_{\mathrm{Berk}}(K)$, where $m(x)$ is the multiplicity of x, i.e. the number of preimages of x under π. Now

$$\deg_x(f) = \delta_x(f) \deg_{\hat{x}}(\hat{f}).$$

Using this relation (and doing some work), one reduces the assertions in Theorem 4.7 to the corresponding statements for f. Thus the local degree can still be interpreted as a local expansion factor for the hyperbolic metric on $\mathbf{P}^1_{\mathrm{Berk}}(K)$, when this metric is defined as in Sect. 3.9. In particular, Corollaries 4.8 and 4.9 remain valid. Finally, the pullback of measures is defined using the local degree as in Sect. 4.9 and formulas (21)–(22) continue to hold.

4.11.2   Trivially Valued Fields

Finally, let us consider the case when K is trivially valued. First assume K is algebraically closed. The Berkovich projective line $\mathbf{P}^1_{\mathrm{Berk}}$ is discussed in Sect. 3.9.2

(see also Sect. 6.6 below). In particular, the Berkovich projective line is a cone over the usual projective line. In other words, $\mathbf{P}^1_{\mathrm{Berk}} \simeq \mathbf{P}^1 \times [0, \infty] / \sim$, where $(x, 0) \sim (y, 0)$ for any $x, y \in \mathbf{P}^1$. This common point $(x, 0)$ is the Gauss point in any coordinate. See Fig. 10. The generalized metric on $\mathbf{P}^1_{\mathrm{Berk}}$ is induced by the parametrization $\alpha : \mathbf{P}^1_{\mathrm{Berk}} \to [0, +\infty]$ given by $\alpha(x, t) = t$.

Any rational map $f : \mathbf{P}^1 \to \mathbf{P}^1$ of degree $d \geq 1$ induces a selfmap of $\mathbf{P}^1_{\mathrm{Berk}}$ that fixes the Gauss point. The local degree is d at the Gauss point. At any point (x, t) with $t > 0$, the local degree is equal to the local degree of $f : \mathbf{P}^1 \to \mathbf{P}^1$ at x. Moreover, $f(x, t) = (f(x), t \deg_x(f))$, so f expands the hyperbolic metric by a factor equal to the local degree, in accordance with Theorem 4.7.

Finally, the case when K is trivially valued but not algebraically closed can be treated by passing to the algebraic closure K^a (which is of course already complete under the trivial norm).

4.12 Notes and Further References

A rational map on the Berkovich projective line is a special case of a finite morphism between Berkovich curves, so various results from [Ber90, Ber93] apply. Nevertheless, it is instructive to see the mapping properties in more detail, in particular the interaction with the tree structure.

The fact that the Berkovich projective line can be understood from many different points of view means that there are several ways of defining the action of a rational map. In his thesis and early work, Rivera-Letelier viewed the action as an extension from $\mathbf{P}^1$ to the hyperbolic space $\mathbf{H}$, whose points he identified with nested collections of closed discs as in Sect. 3.3.4. The definition in [BR10, §2.3] uses homogeneous coordinates through a "Proj" construction of the Berkovich projective line whereas [FR10] simply used the (coordinate-dependent) decomposition $\mathbf{P}^1_{\mathrm{Berk}} = \mathbf{A}^1_{\mathrm{Berk}} \cup \{\infty\}$. Our definition here seems to be new, but it is of course not very different from what is already in the literature. As in Sect. 3, it is guided by the principle of trying to work without coordinates whenever possible.

There are some important techniques that we have not touched upon, in particular those that take place on the classical (as opposed to Berkovich) affine and projective lines. For example, we never employ Newton polygons even though these can be useful see [BR10, §A.10] or [Ben10, §3.2].

The definition of the local degree is taken from [FR10] but appears already in [Thu05] and is the natural one in the general context of finite maps between Berkovich spaces. In the early work of Rivera-Letelier, a different definition was used, modeled on Theorem 4.7. The definition of the local degree (called multiplicity there) in [BR10] uses potential theory and is designed to make (21) hold.

As noted by Favre and Rivera-Letelier, Proposition 4.5 implies that all these different definitions coincide. Having said that, I felt it was useful to have a proof of Theorem 4.7 that is directly based on the algebraic definition of the local degree. The proof presented here seems to be new although many of the ingredients are not.

The structure of the ramification locus in the case of positive residue characteristic is very interesting. We refer to [Fab13a, Fab13b, Fab14] for details.

5 Dynamics of Rational Maps in One Variable

Now that we have defined the action of a rational map on the Berkovich projective line, we would like to study the dynamical system obtained by iterating the map. While it took people some time to realize that considering the dynamics on $\mathbf{P}^1_{\text{Berk}}$ (as opposed to $\mathbf{P}^1$) could be useful, it has become abundantly clear that this is the right thing to do for many questions.

It is beyond the scope of these notes to give an overview of all the known results in this setting. Instead, in order to illustrate some of the main ideas, we shall focus on an equidistribution theorem due to Favre and Rivera-Letelier [FR10], as well as some of its consequences. For these results we shall, on the other hand, give more or less self-contained proofs.

For results not covered here—notably on the structure of Fatou and Julia sets— we recommend the book [BR10] by Baker and Rumely and the survey [Ben10] by Benedetto.

5.1 Setup

We work over a fixed non-Archimedean field K, of any characteristic. For simplicity we shall assume that K is algebraically closed and nontrivially valued. The general case is discussed in Sect. 5.10.

Fix a rational map $f : \mathbf{P}^1 \to \mathbf{P}^1$ of degree $d > 1$. Our approach will be largely coordinate free, but in any case, note that since we are to study the dynamics of f, we must choose the same coordinates on the source and target. Given a coordinate z, f^*z is a rational function in z of degree d.

5.2 Periodic Points

When analyzing a dynamical system, one of the first things to look at are periodic points. We say that $x \in \mathbf{P}^1_{\text{Berk}}$ is a *fixed point* if $f(x) = x$ and a *periodic point* if $f^n(x) = x$ for some $n \geq 1$.

5.2.1 Classical Periodic Points

First suppose $x = f^n(x) \in \mathbf{P}^1$ is a classical periodic point and pick a coordinate z on $\mathbf{P}^1$ vanishing at x. Then

$$f^{*n}z = \lambda z + O(z^2)$$

where $\lambda \in K$ is the *multiplier* of the periodic point. We say that x is *attracting* if $|\lambda| < 1$, *neutral* if $|\lambda| = 1$ and *repelling* if $|\lambda| > 1$. The terminology is more or less self-explanatory. For example, if x is attracting, then there exists a small disc $D \subseteq \mathbf{P}^1$ containing x such that $f^n(D) \subseteq D$ and $f^{nm}(y) \to x$ as $m \to \infty$ for every $y \in D$.

The *multiplicity* of a periodic point $x = f^n(x)$ is the order of vanishing at x of the rational function $f^{n*}z - z$ for any coordinate $z \in F$ vanishing at x. It is easy to see that f has $d + 1$ fixed points counted with multiplicity. Any periodic point of multiplicity at least two must have multiplier $\lambda = 1$.

Proposition 5.1 *Let $f : \mathbf{P}^1 \to \mathbf{P}^1$ be a rational map of degree $d > 1$.*

(i) There exist infinitely many distinct classical periodic points.

(ii) There exists at least one classical nonrepelling fixed point.

(iii) Any nonrepelling classical fixed point admits a basis of open neighborhoods $U \subseteq \mathbf{P}^1_{\mathrm{Berk}}$ that are invariant, i.e. $f(U) \subseteq U$.

Statement (i) when $K = \mathbf{C}$ goes back at least to Julia. A much more precise result was proved by by I. N. Baker [Bak64]. Statements (ii) and (iii) are due to Benedetto [Ben98] who adapted an argument used by Julia.

Sketch of proof To prove (i) we follow [Bca91, pp.102–103] and [Sil07, Corollary 4.7]. We claim that the following holds for all but at most $d + 2$ primes q: any classical point x with $f(x) = x$ has the same multiplicity as a fixed point of f and as a fixed point of f^q. This will show that f^q has $d^q - d > 1$ fixed points (counted with multiplicity) that are not fixed points of f. In particular, f has infinitely many distinct classical periodic points.

To prove the claim, consider a fixed point $x \in \mathbf{P}^1$ and pick a coordinate $z \in F$ vanishing at x. We can write $f^*z = az + bz^{r+1} + O(z^{r+2})$, where $a, b \in K^*$ and $r > 0$. One proves by induction that

$$f^{n*}z = a^n z + b_n z^{r+1} + O(z^{r+2}),$$

where $b_n = a^{n-1}b(1 + a^r + \cdots + a^{(n-1)r})$. If $a \neq 1$, then for all but at most one prime q we have $a^q \neq 1$ and hence x is a fixed point of multiplicity one for both f and f^q. If instead $a = 1$, then $b_q = qb$, so if q is different from the characteristic of K, then x is a fixed point of multiplicity r for both f and f^q.

Next we prove (ii), following [Ben10, §1.3]. Any fixed point of f of multiplicity at least two is nonrepelling, so we may assume that f has exactly $d + 1$ fixed points $(x_i)_{i=1}^{d+1}$. Let $(\lambda_i)_{i=1}^{d+1}$ be the corresponding multipliers. Hence $\lambda_i \neq 1$ for all i. it follows from the Residue Theorem (see [Ben10, Theorem 1.6]) that

$$\sum_{i=1}^{d+1} \frac{1}{1 - \lambda_i} = 1.$$

If $|\lambda_i| > 1$ for all i, then the left hand side would have norm < 1, a contradiction. Hence $|\lambda_i| \leq 1$ for some i and then x_i is a nonrepelling fixed point.

Finally we prove (iii). Pick a coordinate $z \in F$ vanishing at x and write $f^* z = \lambda z + O(z^2)$, with $|\lambda| \leq 1$. For $0 < r \ll 1$ we have $f(x_{D(0,r)}) = x_{D(0,r')}$, where $r' = |\lambda| r \leq r$. Let $U_r := U(\vec{v}_r)$, where $\vec{v}_r$ is the tangent direction at $x_{D(0,r)}$ determined by x. The sets U_r form a basis of open neighborhoods of x and it follows from Corollary 2.13 (ii) that $f(U_r) \subseteq U_r$ for r small enough.   □

5.2.2 Nonclassical Periodic Points

We say that a fixed point $x = f(x) \in \mathbf{H}$ is *repelling* if $\deg_x(f) > 1$ and *neutral* otherwise (points in $\mathbf{H}$ cannot be attracting). This is justified by the interpretation of the local degree as an expansion factor in the hyperbolic metric, see Theorem 4.7.

The following result is due to Rivera-Letelier [Riv03b, Lemme 5.4].

Proposition 5.2 *Any repelling fixed point $x \in \mathbf{H}$ must be of Type 2.*

Sketch of proof We can rule out that x is of Type 3 using value groups. Indeed, by (17) the local degree of f at a Type 3 point is equal to index of the value group $\Gamma_{f(x)}$ as a subgroup of Γ_x, so if $f(x) = x$, then the local degree is one.

I am not aware of an argument of the same style to rule out repelling points of Type 4. Instead, Rivera-Letelier argues by contradiction. Using Newton polygons he shows that any neighborhood of a repelling fixed point of Type 4 would contain a classical fixed point. Since there are only finitely many classical fixed points, this gives a contradiction. See the original paper by Rivera-Letelier or [BR10, Lemma 10.80].   □

5.2.3 Construction of Fixed Points

Beyond Proposition 5.1 there are at least two other methods for producing fixed points.

First, one can use Newton polygons to produce classical fixed points. This was alluded to in the proof of Proposition 5.2 above. We shall not describe this further here but instead refer the reader to [Ben10, §3.2] and [BR10, §A.10].

Second, one can use topology. Since f can be viewed as a tree map, Proposition 2.17 applies and provides a fixed point in $\mathbf{P}^1_{\text{Berk}}$. This argument can be refined, using that f expands the hyperbolic metric, to produce either attracting or repelling fixed points. See [BR10, §10.7].

5.3 Purely Inseparable Maps

Suppose f is purely inseparable of degree $d > 1$. In particular, char $K = p > 0$. We claim that there exists a coordinate $z \in F$ and $n \geq 1$ such that $f^*z = z^{p^n}$. A rational map f such that $f^*z = z^p$ is usually called the *Frobenius map*, see [Har77, 2.4.1–2.4.2].

To prove the claim, we use the fact that f admits exactly $d + 1$ classical fixed points. Indeed, the multiplier of each fixed point is zero. Pick a coordinate $z \in F$ such that $z = 0$ and $z = \infty$ are fixed points of f. Since f is purely inseparable there exists $n \geq 0$ such that $z^{p^n} \in f^*F$. Choose n minimal with this property. Since $\deg f > 1$ we must have $n \geq 1$. On the other hand, the minimality of n shows that $z^{p^n} = f^*w$ for some coordinate $w \in F$. The fact that $z = 0$ and $z = \infty$ are fixed points imply that $z = aw$ for some $a \in K^*$, so $f^*z = az^{p^n}$. After multiplying z by a suitable power of a, we get $a = 1$, proving the claim.

5.4 The Exceptional Set

A classical point $x \in \mathbf{P}^1$ is called *exceptional* for f if its total backward orbit $\bigcup_{n \geq 0} f^{-n}(x)$ is finite. The *exceptional set* of f is the set of exceptional points and denoted E_f. Since f is surjective, it is clear that $E_{f^n} = E_f$ for any $n \geq 1$. We emphasize that E_f by definition consists of classical points only.

Lemma 5.3 *Let* $f : \mathbf{P}^1 \to \mathbf{P}^1$ *be a rational map of degree* $d > 1$.

(i) *If* f *is not purely inseparable, then there are at most two exceptional points. Moreover:*

  (a) *if there are two exceptional points, then* $f(z) = z^{\pm d}$ *in a suitable coordinate* z *on* $\mathbf{P}^1$ *and* $E_f = \{0, \infty\}$;

  (b) *if there is exactly one exceptional point, then* f *is a polynomial in a suitable coordinate and* $E_f = \{\infty\}$.

(ii) *If* f *is purely inseparable, then the exceptional set is countably infinite and consists of all periodic points of* f.

Case (ii) only occurs when char $K = p > 0$ and f is an iterate of the Frobenius map: $f^*z = z^d$ for d a power of p in some coordinate $z \in F$, see Sect. 5.3.

Proof For $x \in E_f$ set $F_x := \bigcup_{n \geq 0} f^{-n}(x)$. Then F_x is a finite set with $f^{-1}(F_x) \subseteq F_x \subseteq E_f$. Since f is surjective, $f^{-1}(F_x) = F_x = f(F_x)$. Hence each point in F_x must be totally ramified in the sense that $f^{-1}(f(x)) = \{x\}$.

If f is purely inseparable, then every point in $\mathbf{P}^1$ is totally ramified, so F_x is finite iff x is periodic.

If f is not purely inseparable, then it follows from Proposition 4.1 (i) that E_f has at most two elements. The remaining statements are easily verified. $\qquad\square$

5.5 Maps of Simple Reduction

By definition, the exceptional set consists of classical points only. The following result by Rivera-Letelier [Riv03b] characterizes totally invariant points in hyperbolic space.

Proposition 5.4 *If $x_0 \in \mathbf{H}$ is a totally invariant point, $f^{-1}(x_0) = x_0$, then x_0 is a Type 2 point.*

Definition 5.5 A rational map $f : \mathbf{P}^1 \to \mathbf{P}^1$ has *simple reduction* if there exists a Type 2 point that is totally invariant for f.

Remark 5.6 Suppose f has simple reduction and pick a coordinate z in which the totally invariant Type 2 point becomes the Gauss point. Then we can write $f^* z = \phi/\psi$, where $\phi, \psi \in \mathfrak{o}_K[z]$ and where the rational function $\tilde{\phi}/\tilde{\psi} \in \tilde{K}(z)$ has degree $d = \deg f$. Such a map is usually said to have *good reduction* [MS95]. Some authors refer to simple reduction as *potentially good reduction*. One could argue that dynamically speaking, maps of good or simple reduction are not the most interesting ones, but they do play an important role. For more on this, see [Ben05b, Bak09, PST09].

Proof of Proposition 5.4 A totally invariant point in $\mathbf{H}$ is repelling so the result follows from Proposition 5.2. Nevertheless, we give an alternative proof.
   Define a function $G : \mathbf{P}^1_{\text{Berk}} \times \mathbf{P}^1_{\text{Berk}} \to [-\infty, 0]$ by[13]

$$G(x, y) = -d_{\mathbf{H}}(x_0, x \wedge_{x_0} y).$$

It is characterized by the following properties: $G(y, x) = G(x, y)$, $G(x_0, y) = 0$ and $\Delta G(\cdot, y) = \delta_y - \delta_{x_0}$.
   Pick any point $y \in \mathbf{P}^1_{\text{Berk}}$. Let $(y_i)^m_{i=1}$ be the preimages of y under f and $d_i = \deg_{y_i}(f)$ the corresponding local degrees. We claim that

$$G(f(x), y) = \sum_{i=1}^{m} d_i \, G(x, y_i) \tag{23}$$

for any $x \in \mathbf{P}^1_{\text{Berk}}$. To see this, note that since $f^* \delta_{x_0} = d \delta_{x_0}$ it follows from Proposition 4.15 that both sides of (23) are $d \delta_{x_0}$-subharmonic as a function of x, with Laplacian $f^*(\delta_y - \delta_{x_0}) = \sum_i d_i (\delta_{y_i} - \delta_{x_0})$. Now, the Laplacian determines a quasisubharmonic function up to a constant, so since both sides of (23) vanish when $x = x_0$ they must be equal for all x, proving the claim.
   Now pick x and y as distinct classical fixed points of f. Such points exist after replacing f by an iterate, see Proposition 5.1. We may assume $y_1 = y$.

---

[13]In [Bak09, BR10], the function $-G$ is called the normalized Arakelov-Green's function with respect to the Dirac mass at x_0.

Then (23) gives

$$(d_1 - 1)G(x, y) + \sum_{i \geq 2} d_i G(x, y_i) = 0 \tag{24}$$

Since $G \leq 0$, we must have $G(x, y_i) = 0$ for $i \geq 2$ and $(d_1 - 1)G(x, y) = 0$.

First assume x_0 is of Type 4. Then x_0 is an end in the tree $\mathbf{P}^1_{\text{Berk}}$, so since $x \neq x_0$ and $y_i \neq x_0$ for all i, we have $x \wedge_{x_0} y_i \neq x_0$ and hence $G(x, y_i) < 0$. This contradicts (24).

Now assume x_0 is of Type 3. Then there are exactly two tangent directions at x_0 in the tree $\mathbf{P}^1_{\text{Berk}}$. Replacing f by an iterate, we may assume that these are invariant under the tangent map. We may assume that the classical fixed points $x, y \in \mathbf{P}^1$ above represent the same tangent direction, so that $x \wedge_{x_0} y \neq x_0$. Since x_0 is totally invariant, it follows from Corollary 2.13 (i) that all the preimages y_i of y also represent this tangent vector at x_0. Thus $G(x, y_i) < 0$ for all i which again contradicts (24). $\qquad\square$

Remark 5.7 The proof in [Bak09] also uses the function G above and analyzes the lifting of f as a homogeneous polynomial map of $K \times K$.

5.6 Fatou and Julia Sets

In the early part of the 20th century, Fatou and Julia developed a general theory of iteration of rational maps on the Riemann sphere. Based upon some of those results, we make the following definition.

Definition 5.8 The *Julia set* $\mathcal{J} = \mathcal{J}_f$ is the set of points $x \in \mathbf{P}^1_{\text{Berk}}$ such that for every open neighborhood U of x we have $\bigcup_{n \geq 0} f^n(U) \supseteq \mathbf{P}^1_{\text{Berk}} \setminus E_f$. The *Fatou set* is the complement of the Julia set.

Remark 5.9 Over the complex numbers, one usually defines the Fatou set as the largest open subset of the Riemann sphere where the sequence of iterates is locally equicontinuous. One then shows that the Julia set is characterized by the conditions in the definition above. Very recently, a non-Archimedean version of this was found by Favre, Kiwi and Trucco, see [FKT11, Theorem 5.4]. Namely, a point $x \in \mathbf{P}^1_{\text{Berk}}$ belongs to the Fatou set of f iff the family $\{f^n\}_{n \geq 1}$ is normal in a neighborhood of x in a suitable sense. We refer to [FKT11, §5] for the definition of normality, but point out that the situation is more subtle in the non-Archimedean case than over the complex numbers.

Theorem 5.10 Let $f : \mathbf{P}^1 \to \mathbf{P}^1$ be any rational map of degree $d > 1$.

(i) The Fatou set $\mathcal{F}$ and Julia set $\mathcal{J}$ are totally invariant: $\mathcal{F} = f(\mathcal{F}) = f^{-1}(\mathcal{F})$ and $\mathcal{J} = f(\mathcal{J}) = f^{-1}(\mathcal{J})$.

(ii) We have $\mathcal{F}_f = \mathcal{F}_{f^n}$ and $\mathcal{J}_f = \mathcal{J}_{f^n}$ for all $n \geq 1$.

(iii) *The Fatou set is open and dense in* $\mathbf{P}^1_{\mathrm{Berk}}$. *It contains any nonrepelling classical periodic point and in particular any exceptional point.*

(iv) *The Julia set is nonempty, compact and has empty interior. Further:*

    (a) *if f has simple reduction, then* $\mathcal{J}$ *consists of a single Type 2 point;*

    (b) *if f does not have simple reduction, then* $\mathcal{J}$ *is a perfect set, that is, it has no isolated points.*

Proof It is clear that $\mathcal{F}$ is open. Since $f : \mathbf{P}^1_{\mathrm{Berk}} \to \mathbf{P}^1_{\mathrm{Berk}}$ is an open continuous map, it follows that $\mathcal{F}$ is totally invariant. Hence $\mathcal{J}$ is compact and totally invariant. The fact that $\mathcal{F}_{f^n} = \mathcal{F}_f$, and hence $\mathcal{J}_{f^n} = \mathcal{J}_f$, follow from the total invariance of $E_f = E_{f^n}$.

It follows from Proposition 5.1 that any nonrepelling classical periodic point is in the Fatou set. Since such points exist, the Fatou set is nonempty. This also implies that the Julia set has nonempty interior. Indeed, if U were an open set contained in the Julia set, then the set $U' := \bigcup_{n \geq 1} f^n(U)$ would be contained in the Julia set for all $n \geq 1$. Since the Fatou set is open and nonempty, it is not contained in E_f, hence $\mathbf{P}^1_{\mathrm{Berk}} \setminus U' \not\subseteq E_f$, so that $U \subseteq \mathcal{F}$, a contradiction.

The fact that the Julia set is nonempty and that properties (a) and (b) hold is nontrivial and will be proved in Sect. 5.8 as a consequence of the equidistribution theorem below. See Propositions 5.14 and 5.16. $\square$

Much more is known about the Fatou and Julia set than what is presented here. For example, as an analogue of the classical result by Fatou and Julia, Rivera-Letelier proved that $\mathcal{J}$ is the closure of the repelling periodic points of f.

For a polynomial map, the Julia set is also the boundary of the filled Julia set, that is, the set of points whose orbits are bounded in the sense that they are disjoint from a fixed open neighborhood of infinity. See [BR10, Theorem 10.91].

Finally, a great deal is known about the dynamics on the Fatou set. We shall not study this here. Instead we refer to [BR10, Ben10].

5.7 Equidistribution Theorem

The following result that describes the distribution of preimages of points under iteration was proved by Favre and Rivera-Letelier [FR04, FR10]. The corresponding result over the complex numbers is due to Brolin [Bro65] for polynomials and to Lyubich [Lyu83] and Freire-Lopez-Mañé [FLM83] for rational functions.

Theorem 5.11 *Let* $f : \mathbf{P}^1 \to \mathbf{P}^1$ *be a rational map of degree* $d > 1$. *Then there exists a unique Radon probability measure* ρ_f *on* $\mathbf{P}^1_{\mathrm{Berk}}$ *with the following property: if* ρ *is a Radon probability measure on* $\mathbf{P}^1_{\mathrm{Berk}}$, *then*

$$\frac{1}{d^n} f^{n*}\rho \to \rho_f \quad as\ n \to \infty,$$

in the weak sense of measures, iff $\rho(E_f) = 0$. The measure ρ_f puts no mass on any classical point; in particular $\rho_f(E_f) = 0$. It is totally invariant in the sense that $f^*\rho_f = d\rho_f$.

Recall that we have assumed that the ground field K is algebraically closed and nontrivially valued. See Sect. 5.10 for the general case.

As a consequence of Theorem 5.11, we obtain a more general version of Theorem A from the introduction, namely

Corollary 5.12 *With f as above, we have*

$$\frac{1}{d^n} \sum_{f^n(y)=x} \deg_y(f^n)\delta_y \to \rho_f \quad as\ n \to \infty,$$

for any non-exceptional point $x \in \mathbf{P}^1_{\text{Berk}} \setminus E_f$.

Following [BR10] we call ρ_f the *canonical measure* of f. It is clear that $\rho_f = \rho_{f^n}$ for $n \geq 1$. The proof of Theorem 5.11 will be given in Sect. 5.9.

Remark 5.13 Okuyama [Oku11b] has proved a quantitative strengthening of Corollary 5.12. The canonical measure is also expected to describe the distribution of repelling periodic points. This does not seem to be established full generality, but is known in many cases [Oku11a].

5.8 Consequences of the Equidistribution Theorem

In this section we collect some result that follow from Theorem 5.11.

Proposition 5.14 *The support of the measure ρ_f is exactly the Julia set $\mathcal{J} = \mathcal{J}_f$. In particular, $\mathcal{J}$ is nonempty.*

Proof First note that the support of ρ_f is totally invariant. This follows formally from the total invariance of ρ_f. Further, the support of ρ_f cannot be contained in the exceptional set E_f since $\rho_f(E_f) = 0$.

Consider a point $x \in \mathbf{P}^1_{\text{Berk}}$. If x is not in the support of ρ_f, let $U = \mathbf{P}^1_{\text{Berk}} \setminus \operatorname{supp} \rho_f$. Then $f^n(U) = U$ for all n. In particular, $\bigcup_{n\geq 0} f^n(U)$ is disjoint from $\operatorname{supp} \rho_f$. Since $\operatorname{supp} \rho_f \not\subseteq E_f$, x must belong to the Fatou set.

Conversely, if $x \in \operatorname{supp} \rho_f$ and U is any open neighborhood of x, then $\rho_f(U) > 0$. For any $y \in \mathbf{P}^1_{\text{Berk}} \setminus E_f$, Corollary 5.12 implies that $f^{-n}(y) \cap U \neq \emptyset$ for $n \gg 0$. We conclude that $\bigcup_{n\geq 0} f^n(U) \supseteq \mathbf{P}^1_{\text{Berk}} \setminus E_f$, so x belongs to the Julia set. □

We will not study the equilibrium measure ρ_f in detail, but the following result is not hard to deduce from what we already know.

Proposition 5.15 *The following conditions are equivalent.*

(i) ρ_f puts mass at some point in $\mathbf{P}^1_{\text{Berk}}$;
(ii) ρ_f is a Dirac mass at a Type 2 point;

(iii) f has simple reduction;
(iv) f^n has simple reduction for all $n \geq 1$;
(v) f^n has simple reduction for some $n \geq 1$.

Proof If f has simple reduction then, by definition, there exists a totally invariant Type 2 point x_0. We then have $d^{-n} f^{n*} \delta_{x_0} = \delta_{x_0}$ so Corollary 5.12 implies $\rho_f = \delta_{x_0}$. Conversely, if $\rho_f = \delta_{x_0}$ for some Type 2 point x_0, then $f^* \rho_f = d \rho_f$ implies that x_0 is totally invariant, so that f has simple reduction. Thus (ii) and (iii) are equivalent. Since $\rho_f = \rho_{f^n}$, this implies that (ii)–(v) are equivalent.

Clearly (ii) implies (i). We complete the proof by proving that (i) implies (v). Thus suppose $\rho_f \{x_0\} > 0$ for some $x_0 \in \mathbf{P}^1_{\text{Berk}}$. Since ρ_f does not put mass on classical points we have $x_0 \in \mathbf{H}$. The total invariance of ρ_f implies

$$0 < \rho_f \{x_0\} = \frac{1}{d} (f^* \rho_f) \{x_0\} = \frac{1}{d} \deg_{x_0}(f) \rho_f \{f(x_0)\} \leq \rho_f \{f(x_0)\},$$

with equality iff $\deg_{x_0}(f) = d$. Write $x_n = f^n(x_0)$ for $n \geq 0$. Now the total mass of ρ_f is finite, so after replacing x_0 by x_m for some $m \geq 0$ we may assume that $x_n = x_0$ and $\deg_{x_j}(f) = d$ for $0 \leq j < n$ and some $n \geq 1$. This implies that x_0 is totally invariant under f^n. By Proposition 5.4, x_0 is then a Type 2 point and f^n has simple reduction.   □

With the following result we complete the proof of Theorem 5.10.

Proposition 5.16 *Let $f : \mathbf{P}^1 \to \mathbf{P}^1$ be a rational map of degree $d > 1$ and let $\mathcal{J} = \mathcal{J}_f$ be the Julia set of f.*

(i) If f has simple reduction, then $\mathcal{J}$ consists of a single Type 2 point.
(ii) If f does not have simple reduction, then $\mathcal{J}$ is a perfect set.

Proof Statement (i) is a direct consequence of Proposition 5.15. Now suppose f does not have simple reduction. Pick any point $x \in \mathcal{J}$ and an open neighborhood U of x. It suffices to prove that there exists a point $y \in U$ with $y \neq x$ and $f^n(y) = x$ for some $n \geq 1$. After replacing f by an iterate we may assume that x is either fixed or not periodic. Set $m := \deg_x(f)$ if $f(x) = x$ and $m := 0$ otherwise. Note that $m < d$ as x is not totally invariant.

Since $x \notin E_f$, Corollary 5.12 shows that the measure $d^{-n} f^{n*} \delta_x$ converges weakly to ρ_f. Write $f^{n*} \delta_x = m^n \delta_x + \rho'_n$, where

$$\rho'_n = \sum_{y \neq x, f^n(y) = x} \deg_y(f^n) \delta_y.$$

We have $x \in \mathcal{J} = \operatorname{supp} \rho_f$ so $\rho_f(U) > 0$ and hence $\liminf_{n \to \infty} (d^{-n} f^{n*} \delta_x)(U) > 0$. Since $m < d$ it follows that $\rho'_n(U) > 0$ for $n \gg 0$. Thus there exist points $y \in U$ with $y \neq x$ and $f^n(y) = x$.   □

5.9 Proof of the Equidistribution Theorem

To prove the equidistribution theorem we follow the approach of Favre and Rivera-Letelier [FR10], who in turn adapted potential-theoretic techniques from complex dynamics developed by Fornæss-Sibony and others. Using the tree Laplacian defined in Sect. 2.5 we can study convergence of measures in terms of convergence of quasisubharmonic functions, a problem for which there are good techniques. If anything, the analysis is easier in the nonarchimedean case. Our proof does differ from the one in [FR10] in that it avoids studying the dynamics on the Fatou set.

5.9.1 Construction of the Canonical Measure

Fix a point $x_0 \in \mathbf{H}$. Since $d^{-1} f^* \delta_{x_0}$ is a probability measure, we have

$$d^{-1} f^* \delta_{x_0} = \delta_{x_0} + \Delta u \qquad (25)$$

for an x_0-subharmonic function u. In fact, (4) gives an explicit expression for u and shows that u is continuous, since $f^{-1}(x_0) \subseteq \mathbf{H}$.

Iterating (25) and using (21) leads to

$$d^{-n} f^{n*} \delta_{x_0} = \delta_{x_0} + \Delta u_n, \qquad (26)$$

where $u_n = \sum_{j=0}^{n-1} d^{-j} u \circ f^j$. It is clear that the sequence u_n converges uniformly to a continuous x_0-subharmonic function u_∞. We set

$$\rho_f := \delta_{x_0} + \Delta u_\infty.$$

Since u_∞ is bounded, it follows from (5) that ρ_f does not put mass on any classical point. In particular, $\rho_f(E_f) = 0$, since E_f is at most countable.

5.9.2 Auxiliary Results

Before starting the proof of equidistribution, let us record a few results that we need.

Lemma 5.17 *If $x_0, x \in \mathbf{H}$, then $d_{\mathbf{H}}(f^n(x), x_0) = O(d^n)$ as $n \to \infty$.*

Proof We know that f expands the hyperbolic metric by a factor at most d, see Corollary 4.8. Using the triangle inequality and the assumption $d \geq 2$, this yields

$$d_{\mathbf{H}}(f^n(x), x) \leq \sum_{j=0}^{n-1} d_{\mathbf{H}}(f^{j+1}(x), f^j(x)) \leq \sum_{j=0}^{n-1} d^j d_{\mathbf{H}}(f(x), x) \leq d^n d_{\mathbf{H}}(f(x), x),$$

so that

$$d_{\mathbf{H}}(f^n(x), x_0) \leq d_{\mathbf{H}}(f^n(x), f^n(x_0)) + d_{\mathbf{H}}(f^n(x_0), x_0)$$
$$\leq d^n(d_{\mathbf{H}}(x, x_0) + d_{\mathbf{H}}(f(x_0), x_0)),$$

completing the proof. $\square$

Lemma 5.18 *Suppose that f is not purely inseparable. If ρ is a Radon probability measure on $\mathbf{P}^1_{\text{Berk}}$ such that $\rho(E_f) = 0$ and we set $\rho_n := d^{-n} f^{n*} \rho$, then $\sup_{y \in \mathbf{P}^1} \rho_n\{y\} \to 0$ as $n \to \infty$.*

Note that the supremum is taken over classical points only. Also note that the lemma always applies if the ground field is of characteristic zero. However, the lemma is false for purely inseparable maps.

Proof We have $\rho_n\{y\} = d^{-n} \deg_y(f^n) \rho\{f^n(y)\}$, so it suffices to show that

$$\sup_{y \in \mathbf{P}^1 \setminus E_f} \deg_y(f^n) = o(d^n). \tag{27}$$

For $y \in \mathbf{P}^1$ and $n \geq 0$, write $y_n = f^n(y)$. If $\deg_{y_n}(f) = d$ for $n = 0, 1, 2$, then Proposition 4.1 (i) implies $y \in E_f$. Thus $\deg_y(f^3) \leq d^3 - 1$ and hence $\deg_y(f^n) \leq d^2(d^3 - 1)^{n/3}$ for $y \in \mathbf{P}^1 \setminus E_f$, completing the proof. $\square$

5.9.3   Proof of the Equidistribution Theorem

Let ρ be a Radon probability measure on $\mathbf{P}^1_{\text{Berk}}$ and set $\rho_n = d^{-n} f^{n*} \rho$. If $\rho(E_f) > 0$, then $\rho_n(E_f) = \rho(E_f) > 0$ for all n. Any accumulation point of $\{\rho_n\}$ must also put mass on E_f, so $\rho_n \not\to \rho_f$ as $n \to \infty$.

Conversely, assume $\rho(E_f) = 0$ and let us show that $\rho_n \to \rho_f$ as $n \to \infty$. Let $\varphi \in \text{SH}(\mathbf{P}^1_{\text{Berk}}, x_0)$ be a solution to the equation $\rho = \delta_{x_0} + \Delta\varphi$. Applying $d^{-n} f^{n*}$ to both sides of this equation and using (21), we get

$$\rho_n = d^{-n} f^{n*} \delta_{x_0} + \Delta\varphi_n = \delta_{x_0} + \Delta(u_n + \varphi_n),$$

where $\varphi_n = d^{-n} \varphi \circ f^n$. Here $\delta_{x_0} + \Delta u_n$ tends to ρ_f by construction. We must show that $\delta_{x_0} + \Delta(u_n + \varphi_n)$ also tends to ρ_f. By Sect. 2.5.4, this amounts to showing that φ_n tends to zero pointwise on $\mathbf{H}$. Since φ is bounded from above, we always have $\limsup_n \varphi_n \leq 0$. Hence it remains to show that

$$\liminf_{n \to \infty} \varphi_n(x) \geq 0 \quad \text{for any } x \in \mathbf{H}. \tag{28}$$

To prove (28) we first consider the case when f is not purely inseparable. Set $\varepsilon_m = \sup_{y \in \mathbf{P}^1} \rho_m\{y\}$ for $m \geq 0$. Then $\varepsilon_m \to 0$ as $m \to \infty$ by Lemma 5.18. Using

Lemma 5.17 and Proposition 2.8 we get, for $m, n \geq 0$

$$\varphi_{n+m}(x) = d^{-n}\varphi_m(f^n(x))$$
$$\geq d^{-n}\varphi_m(x_0) - d^{-n}(C_m + \varepsilon_m d_{\mathbf{H}}(f^n(x), x_0))$$
$$\geq -D\varepsilon_m - C_m d^{-n}$$

for some constant D independent of m and n and some constant C_m independent of n. Letting first $n \to \infty$ and then $m \to \infty$ yields $\liminf_n \varphi_n(x) \geq 0$, completing the proof.

Now assume f is purely inseparable. In particular, K has characteristic $p > 0$, f has degree $d = p^m$ for some $m \geq 1$ and there exists a coordinate $z \in F$ such that f becomes an iterate of the Frobenius map: $f^*z = z^d$.

In this case, we cannot use Lemma 5.18 since (27) is evidently false: the local degree is d everywhere on $\mathbf{P}^1_{\mathrm{Berk}}$. On the other hand, the dynamics is simple to describe, see Example 4.18. The Gauss point x_0 in the coordinate z is (totally) invariant. Hence $\rho_f = \delta_{x_0}$. The exceptional set E_f is countably infinite and consists of all classical periodic points. Consider the partial ordering on $\mathbf{P}^1_{\mathrm{Berk}}$ rooted in x_0. Then f is order preserving and $d_{\mathbf{H}}(f^n(x), x_0) = d^n d_{\mathbf{H}}(x, x_0)$ for any $x \in \mathbf{P}^1_{\mathrm{Berk}}$.

As above, write $\rho = \delta_{x_0} + \Delta\varphi$, with $\varphi \in \mathrm{SH}(\mathbf{P}^1_{\mathrm{Berk}}, x_0)$. Pick any point $x \in \mathbf{H}$. It suffices to prove that (28) holds, where $\varphi_n = d^{-n}\varphi(f^n(x))$. Using Lemma 2.9 and the fact that $d_{\mathbf{H}}(f^n(x), x_0) = d^n d_{\mathbf{H}}(x, x_0)$ it suffices to show that

$$\lim_{n\to\infty} \rho(Y_n) = 0, \quad \text{where } Y_n := \{y \geq f^n(x)\}. \tag{29}$$

Note that for $m, n \geq 1$, either $Y_{m+n} \subseteq Y_n$ or Y_n, Y_{n+m} are disjoint. If $\rho(Y_n) \not\to 0$, there must exist a subsequence $(n_j)_j$ such that $Y_{n_{j+1}} \subseteq Y_{n_j}$ for all j and $\rho(Y_{n_j}) \not\to 0$. Since $d_{\mathbf{H}}(f^n(x), x_0) \to \infty$ we must have $\bigcap_j Y_{n_j} = \{y_0\}$ for a classical point $y_0 \in \mathbf{P}^1$. Thus $\rho\{y_0\} > 0$. On the other hand, we claim that y_0 is periodic, hence exceptional, contradicting $\rho(E_f) = 0$.

To prove the claim, pick $m_1 \geq 1$ minimal such that $Y_{n_1+m_1} = f^{m_1}(Y_{n_1}) \subseteq Y_{n_1}$ and set $Z_r = Y_{n_1+rm_1} = f^{rm_1}(Y_{n_1})$ for $r \geq 0$. Then Z_r forms a decreasing sequence of compact sets whose intersection consists of a single classical point y, which moreover is periodic: $f^{m_1}(y) = y$. On the other hand, for $m \geq 1$ we have $Y_{n_1+m} \subseteq Y_{n_1}$ iff m_1 divides m. Thus we can write $n_j = n_1 + r_j m_1$ with $r_j \to \infty$. This implies that $\{y_0\} = \bigcap_j Y_{n_j} \subseteq \bigcap_r Z_r = \{y\}$ so that $y_0 = y$ is periodic.

The proof of Theorem 5.11 is now complete.

5.10 Other Ground Fields

Above we worked with the assumption that our non-Archimedean field K was algebraically closed and nontrivially valued. Let us briefly discuss what happens for other fields, focusing on the equidistribution theorem and its consequences.

5.10.1 Non-Algebraically Closed Fields

Suppose K is of arbitrary characteristic and nontrivially valued but not algebraically closed. The Berkovich projective line $\mathbf{P}_{\mathrm{Berk}}^1(K)$ and the action by a rational map were outlined in Sects. 3.9.1 and 4.11.1, respectively. Let K^a be the algebraic closure of K and $\widehat{K^a}$ its completion. Denote by $\pi : \mathbf{P}_{\mathrm{Berk}}^1(\widehat{K^a}) \to \mathbf{P}_{\mathrm{Berk}}^1(K)$ the natural projection. Write $\hat{f} : \mathbf{P}_{\mathrm{Berk}}^1(\widehat{K^a}) \to \mathbf{P}_{\mathrm{Berk}}^1(\widehat{K^a})$ for the induced map. Define $E_{\hat{f}}$ as the exceptional set for $\hat{f}$ and set $E_f = \pi(E_{\hat{f}})$. Then $f^{-1}(E_f) = E_f$ and E_f has at most two elements, except if K has characteristic p and f is purely inseparable, in which case E_f is countable.

We will deduce the equidistribution result in Theorem 5.11 for f from the corresponding theorem for $\hat{f}$. Let $\rho_{\hat{f}}$ be the measure on $\mathbf{P}_{\mathrm{Berk}}^1(\widehat{K^a})$ given by Theorem 5.11 and set $\rho_f = \pi_*(\rho_{\hat{f}})$. Since $E_{\hat{f}} = \pi^{-1}(E_f)$, the measure ρ_f puts no mass on E_f.

Let ρ be a Radon probability measure on $\mathbf{P}_{\mathrm{Berk}}^1(K)$. If $\rho(E_f) > 0$, then any limit point of $d^{-n} f^{n*} \rho$ puts mass on E_f, hence $d^{-n} f^{n*} \rho \not\to \rho_f$. Now assume $\rho(E_f) = 0$. Write x_0 and $\hat{x}_0$ for the Gauss point on $\mathbf{P}_{\mathrm{Berk}}^1(K)$ and $\mathbf{P}_{\mathrm{Berk}}^1(\widehat{K^a})$, respectively, in some coordinate on K. We have $\rho = \delta_{x_0} + \Delta\varphi$ for some $\varphi \in \mathrm{SH}(\mathbf{P}_{\mathrm{Berk}}^1(K), x_0)$. The generalized metric on $\mathbf{P}_{\mathrm{Berk}}^1(K)$ was defined in such a way that $\pi^*\varphi \in \mathrm{SH}(\mathbf{P}_{\mathrm{Berk}}^1(\widehat{K^a}), x_0)$. Set $\hat{\rho} := \delta_{\hat{x}_0} + \Delta(\pi^*\varphi)$. Then $\hat{\rho}$ is a Radon probability measure on $\mathbf{P}_{\mathrm{Berk}}^1(\widehat{K^a})$ such that $\pi_*\hat{\rho} = \rho$. Since $E_{\hat{f}}$ is countable, $\pi(E_{\hat{f}}) = E_f$ and $\rho(E_f) = 0$ we must have $\hat{\rho}(E_{\hat{f}}) = 0$. Theorem 5.11 therefore gives $d^{-n} \hat{f}^{n*} \hat{\rho} \to \rho_{\hat{f}}$ and hence $d^{-n} f^{n*} \rho \to \rho_f$ as $n \to \infty$.

5.10.2 Trivially Valued Fields

Finally let us consider the case when K is equipped with the trivial valuation. Then the Berkovich projective line is a cone over $\mathbf{P}^1(K)$, see Sect. 3.9.2. The equidistribution theorem can be proved essentially as above, but the proof is in fact much easier. The measure ρ_f is a Dirac mass at the Gauss point and the exceptional set consists of at most two points, except if f is purely inseparable, The details are left as an exercise to the reader.

5.11   Notes and Further References

The equidistribution theorem is due to Favre and Rivera-Letelier. Our proof basically follows [FR10] but avoids studying the dynamics on the Fatou set and instead uses the hyperbolic metric more systematically through Proposition 2.8 and Lemmas 5.17 and 5.18. In any case, both the proof here and the one in [FR10] are modeled on arguments from complex dynamics. The remarks in Sect. 5.10 about general ground fields seem to be new.

The measure ρ_f is conjectured to describe the distribution of repelling periodic points, see [FR10, Question 1, p.119]. This is known in certain cases but not in general. In characteristic zero, Favre and Rivera-Letelier proved that the classical periodic points (a priori not repelling) are distributed according to ρ_f, see [FR10, Théorème B] as well as [Oku11a].

Again motivated by results over the complex numbers, Favre and Rivera also go beyond equidistribution and study the ergodic properties of ρ_f.

Needless to say, I have not even scratched the surface when describing the dynamics of rational maps. I decided to focus on the equidistribution theorem since its proof uses potential theoretic techniques related to some of the analysis in later sections.

One of the many omissions is the Fatou-Julia theory, in particular the classification of Fatou components, existence and properties of wandering components etc. See [BR10, §10] and [Ben10, §§6–7] for this.

Finally, we have said nothing at all about arithmetic aspects of dynamical systems. For this, see e.g. the book [Sil07] and lecture notes [Sil10] by Silverman.

6   The Berkovich Affine Plane Over a Trivially Valued Field

In the remainder of the paper we will consider polynomial dynamics on the Berkovich affine plane over a trivially valued field, at a fixed point and at infinity. Here we digress and discuss the general structure of the Berkovich affine space $\mathbf{A}^n_{\mathrm{Berk}}$ in the case of a trivially valued field. While we are primarily interested in the case $n = 2$, many of the notions and results are valid in any dimension.

6.1   Setup

Let K be any field equipped with the trivial norm. (In Sects. 6.10–6.11 we shall make further restriction on K.) Let $R \simeq K[z_1, \ldots, z_n]$ denote the polynomial ring in n variables with coefficients in K. Thus R is the coordinate ring of the affine n-space $\mathbf{A}^n$ over K. We shall view $\mathbf{A}^n$ as a scheme equipped with the Zariski topology. Points of $\mathbf{A}^n$ are thus prime ideals of R and closed points are maximal ideals.

6.2 The Berkovich Affine Space and Analytification

We start by introducing the basic object that we shall study.

Definition 6.1 The Berkovich affine space $\mathbf{A}^n_{\mathrm{Berk}}$ of dimension n is the set of multiplicative seminorms on the polynomial ring R whose restriction to K is the trivial norm.

This definition is a special case of the *analytification* of a variety (or scheme) over K. Let $Y \subseteq \mathbf{A}^n$ be an irreducible subvariety defined by a prime ideal $I_Y \subseteq R$ and having coordinate ring $K[Y] = R/I_Y$. Then the analytification Y_{Berk} of Y is the set of multiplicative seminorms on $K[Y]$ restricting to the trivial norm on K.[14] We equip Y_{Berk} with the topology of pointwise convergence. The map $R \to R/I_Y$ induces a continuous injection $Y_{\mathrm{Berk}} \hookrightarrow \mathbf{A}^n_{\mathrm{Berk}}$.

As before, points in $\mathbf{A}^n_{\mathrm{Berk}}$ will be denoted x and the associated seminorm by $|\cdot|_x$. It is customary to write $|\phi(x)| := |\phi|_x$ for a polynomial $\phi \in R$. Let $\mathfrak{p}_x \subset R$ be the kernel of the seminorm $|\cdot|_x$. The completed residue field $\mathcal{H}(x)$ is the completion of the ring $R/\mathfrak{p}_x$ with respect to the norm induced by $|\cdot|_x$. The structure sheaf $\mathcal{O}$ on $\mathbf{A}^n_{\mathrm{Berk}}$ can now be defined in the same way as in Sect. 3.8.1, following [Ber90, §1.5.3], but we will not directly us it.

Closely related to $\mathbf{A}^n_{\mathrm{Berk}}$ is the *Berkovich unit polydisc* $\mathbf{D}^n_{\mathrm{Berk}}$. This is defined[15] in [Ber90, §1.5.2] as the spectrum of the Tate algebra over K. Since K is trivially valued, the Tate algebra is the polynomial ring R and $\mathbf{D}^n_{\mathrm{Berk}}$ is the set of multiplicative seminorms on R bounded by the trivial norm, that is, the set of points $x \in \mathbf{A}^n_{\mathrm{Berk}}$ such that $|\phi(x)| \leq 1$ for all polynomials $\phi \in R$.

6.3 Home and Center

To a seminorm $x \in \mathbf{A}^n_{\mathrm{Berk}}$ we can associate two basic geometric objects. First, the kernel $\mathfrak{p}_x$ of $|\cdot|_x$ defines a point in $\mathbf{A}^n$ that we call the *home* of x. Note that the home of x is equal to $\mathbf{A}^n$ iff $|\cdot|_x$ is a norm on R. We obtain a continuous *home map*

$$\mathbf{A}^n_{\mathrm{Berk}} \to \mathbf{A}^n.$$

---

[14]The analytification of a general variety or scheme over K is defined by gluing the analytifications of open affine subsets, see [Ber90, §3.5].

[15]The unit polydisc is denoted by $E(0, 1)$ in [Ber90, §1.5.2].

Recall that $\mathbf{A}^n$ is viewed as a scheme with the Zariski topology.

Second, we define the *center* of x on $\mathbf{A}^n$ as follows. If there exists a polynomial $\phi \in R$ such that $|\phi(x)| > 1$, then we say that x has *center at infinity*. Otherwise x belongs to the Berkovich unit polydisc $\mathbf{D}^n_{\mathrm{Berk}}$, in which case we define the center of x to be the point of $\mathbf{A}^n$ defined by the prime ideal $\{\phi \in R \mid |\phi(x)| < 1\}$. Thus we obtain a *center map*[16]

$$\mathbf{D}^n_{\mathrm{Berk}} \to \mathbf{A}^n$$

which has the curious property of being *anticontinuous* in the sense that preimages of open/closed sets are closed/open.

The only seminorm in $\mathbf{A}^n_{\mathrm{Berk}}$ whose center is all of $\mathbf{A}^n$ is the trivial norm on R. More generally, if $Y \subseteq \mathbf{A}^n$ is any irreducible subvariety, there is a unique seminorm in $\mathbf{A}^n_{\mathrm{Berk}}$ whose home and center are both equal to Y, namely the image of the trivial norm on $K[Y]$ under the embedding $Y_{\mathrm{Berk}} \hookrightarrow \mathbf{A}^n_{\mathrm{Berk}}$, see also (31) below. This gives rise to an embedding

$$\mathbf{A}^n \hookrightarrow \mathbf{A}^n_{\mathrm{Berk}}$$

and shows that the home and center maps are both surjective.

The home of a seminorm always contains the center, provided the latter is not at infinity. By letting the home and center vary over pairs of points of $\mathbf{A}^n$ we obtain various partitions of the Berkovich affine space, see Sect. 6.5.

It will occasionally be convenient to identify irreducible subvarieties of $\mathbf{A}^n$ with their generic points. Therefore, we shall sometimes think of the center and home of a seminorm as irreducible subvarieties (rather than points) of $\mathbf{A}^n$.

There is a natural action of $\mathbf{R}^*_+$ on $\mathbf{A}^n_{\mathrm{Berk}}$ which to a real number $t > 0$ and a seminorm $|\cdot|$ associates the seminorm $|\cdot|^t$. The fixed points under this action are precisely the images under the embedding $\mathbf{A}^n \hookrightarrow \mathbf{A}^n_{\mathrm{Berk}}$ above.

6.4   Semivaluations

In what follows, it will be convenient to work additively rather than multiplicatively. Thus we identify a seminorm $|\cdot| \in \mathbf{A}^n_{\mathrm{Berk}}$ with the corresponding *semivaluation*

$$v = -\log|\cdot|. \tag{30}$$

---

[16]The center map is called the *reduction map* in [Ber90, §2.4]. We use the valuative terminology center as in [Vaq00, §6] since it will be convenient to view the elements of $\mathbf{A}^n_{\mathrm{Berk}}$ as semivaluations rather than seminorms.

The home of v is now given by the prime ideal $(v = +\infty)$ of R. We say that v is a *valuation* if the home is all of $\mathbf{A}^n$. If $v(\phi) < 0$ for some polynomial $\phi \in R$, then v has center at infinity; otherwise v belongs to the $\mathbf{D}^n_{\text{Berk}}$ and its center is defined by the prime ideal $\{v > 0\}$. The action of $\mathbf{R}^*_+$ on $\mathbf{A}^n_{\text{Berk}}$ is now given by multiplication: $(t, v) \mapsto tv$. The image of an irreducible subvariety $Y \subseteq \mathbf{A}^n$ under the embedding $\mathbf{A}^n \hookrightarrow \mathbf{A}^n_{\text{Berk}}$ is the semivaluation triv_Y, defined by

$$\text{triv}_Y(\phi) = \begin{cases} +\infty & \text{if } \phi \in I_Y \\ 0 & \text{if } \phi \notin I_Y, \end{cases} \tag{31}$$

where I_Y is the ideal of Y. Note that $\text{triv}_{\mathbf{A}^n}$ is the trivial valuation on R.

For $v \in \mathbf{D}^n_{\text{Berk}}$ we write

$$v(\mathfrak{a}) := \min_{p \in \mathfrak{a}} v(\phi)$$

for any ideal $\mathfrak{a} \subseteq R$; here it suffices to take the minimum over any set of generators of $\mathfrak{a}$.

6.5   Stratification

Let $Y \subseteq \mathbf{A}^n$ be an irreducible subvariety. To Y we can associate two natural elements of $\mathbf{A}^n_{\text{Berk}}$: the semivaluation triv_Y above and the valuation ord_Y [17] defined by

$$\text{ord}_Y(\phi) = \max\{k \geq 0 \mid \phi \in I_Y^k\}.$$

As we explain next, Y also determines several natural subsets of $\mathbf{A}^n_{\text{Berk}}$.

6.5.1   Stratification by Home

Define

$$\mathcal{W}_{\supseteq Y}, \quad \mathcal{W}_{\subseteq Y} \quad \text{and} \quad \mathcal{W}_Y$$

as the set of semivaluations in $\mathbf{A}^n_{\text{Berk}}$ whose home in $\mathbf{A}^n$ contains Y, is contained in Y and is equal to Y, respectively. Note that $\mathcal{W}_{\subseteq Y}$ is closed by the continuity of the home map. We can identify $\mathcal{W}_{\subseteq Y}$ with the analytification Y_{Berk} of the affine

---

[17]This is a divisorial valuation given by the order of vanishing along the exceptional divisor of the blowup of Y, see Sect. 6.10.

variety Y as defined in Sect. 6.2. In particular, $\mathrm{triv}_Y \in \mathcal{W}_{\subseteq Y}$ corresponds to the trivial valuation on $K[Y]$.

The set $\mathcal{W}_{\supseteq Y}$ is open, since it is the complement in $\mathbf{A}^n_{\mathrm{Berk}}$ of the union of all $\mathcal{W}_{\subseteq Z}$, where Z ranges over irreducible subvarieties of $\mathbf{A}^n$ not containing Y. The set $\mathcal{W}_Y$, on the other hand, is neither open nor closed unless Y is a point or all of $\mathbf{A}^n$. It can be identified with the set of *valuations* on the coordinate ring $K[Y]$.

6.5.2 Valuations Centered at Infinity

We define $\hat{\mathcal{V}}_\infty$ to be the open subset of $\mathbf{A}^n_{\mathrm{Berk}}$ consisting of semivaluations having center at infinity. Note that $\hat{\mathcal{V}}_\infty$ is the complement of $\mathbf{D}^n_{\mathrm{Berk}}$ in $\mathbf{A}^n_{\mathrm{Berk}}$:

$$\mathbf{A}^n_{\mathrm{Berk}} = \mathbf{D}^n_{\mathrm{Berk}} \cup \hat{\mathcal{V}}_\infty \quad \text{and} \quad \mathbf{D}^n_{\mathrm{Berk}} \cap \hat{\mathcal{V}}_\infty = \emptyset.$$

The space $\hat{\mathcal{V}}_\infty$ is useful for the study of polynomial mappings of $\mathbf{A}^n$ at infinity and will be explored in Sect. 9 in the two-dimensional case. Notice that the action of $\mathbf{R}^*_+$ on $\hat{\mathcal{V}}_\infty$ is fixed point free. We denote the quotient by $\mathcal{V}_\infty$:

$$\mathcal{V}_\infty := \hat{\mathcal{V}}_\infty / \mathbf{R}^*_+.$$

If we write $R = K[z_1, \ldots, z_n]$, then we can identify $\mathcal{V}_\infty$ with the set of semivaluations for which $\min_{1 \le i \le n}\{v(z_i)\} = -1$. However, this identification depends on the choice of coordinates, or at least on the embedding of $\mathbf{A}^n \hookrightarrow \mathbf{P}^n$.

6.5.3 Stratification by Center

We can classify the semivaluations in the Berkovich unit polydisc $\mathbf{D}^n_{\mathrm{Berk}}$ according to their centers. Given an irreducible subvariety $Y \subseteq \mathbf{A}^n$ we define

$$\hat{\mathcal{V}}_{\supseteq Y}, \quad \hat{\mathcal{V}}_{\subseteq Y} \quad \text{and} \quad \hat{\mathcal{V}}_Y$$

as the set of semivaluations in $\mathbf{D}^n_{\mathrm{Berk}}$ whose center contains Y, is contained in Y and is equal to Y, respectively. By anticontinuity of the center map, $\hat{\mathcal{V}}_{\subseteq Y}$ is open and, consequently, $\hat{\mathcal{V}}_{\supseteq Y}$ closed in $\mathbf{D}^n_{\mathrm{Berk}}$. Note that $v \in \hat{\mathcal{V}}_{\subseteq Y}$ iff $v(I_Y) > 0$. As before, $\hat{\mathcal{V}}_Y$ is neither open nor closed unless Y is a closed point or all of $\mathbf{A}^n$.

Note that $\mathcal{W}_{\subseteq Y} \cap \mathbf{D}^n_{\mathrm{Berk}} \subseteq \hat{\mathcal{V}}_{\subseteq Y}$. The difference $\hat{\mathcal{V}}_{\subseteq Y} \setminus \mathcal{W}_{\subseteq Y}$ is the open subset of $\mathbf{D}^n_{\mathrm{Berk}}$ consisting of semivaluations v satisfying $0 < v(I_Y) < \infty$. If we define

$$\mathcal{V}_Y := \{v \in \mathbf{D}^n_{\mathrm{Berk}} \mid v(I_Y) = 1\}, \tag{32}$$

then $\mathcal{V}_Y$ is a closed subset of $\mathbf{D}^n_{\mathrm{Berk}}$ (hence also of $\mathbf{A}^n_{\mathrm{Berk}}$) and the map $v \mapsto v/v(I_Y)$ induces a homeomorphism

$$(\hat{\mathcal{V}}_{\subseteq Y} \setminus \mathcal{W}_{\subseteq Y})/\mathbf{R}^*_+ \xrightarrow{\sim} \mathcal{V}_Y.$$

Remark 6.2 In the terminology of Thuillier [Thu07], $\hat{\mathcal{V}}_{\subseteq Y}$ is the Berkovich space associated to the completion of $\mathbf{A}^n$ along the closed subscheme Y. Similarly, the open subset $\hat{\mathcal{V}}_{\subseteq Y} \setminus \mathcal{W}_{\subseteq Y}$ is the generic fiber of this formal subscheme. This terminology differs slightly from that of Berkovich [Ber94] who refers to $\hat{\mathcal{V}}_{\subseteq Y}$ as the generic fiber, see [Thu07, p.383].

6.5.4 Extremal Cases

Let us describe the subsets of $\mathbf{A}^n_{\mathrm{Berk}}$ introduced above in the case when the subvariety Y has maximal or minimal dimension. First, it is clear that

$$\mathcal{W}_{\subseteq \mathbf{A}^n} = \mathbf{A}^n_{\mathrm{Berk}} \quad \text{and} \quad \hat{\mathcal{V}}_{\subseteq \mathbf{A}^n} = \mathbf{D}^n_{\mathrm{Berk}}.$$

Furthermore,

$$\hat{\mathcal{V}}_{\supseteq \mathbf{A}^n} = \hat{\mathcal{V}}_{\mathbf{A}^n} = \mathcal{W}_{\supseteq \mathbf{A}^n} = \mathcal{W}_{\mathbf{A}^n} = \{\mathrm{triv}_{\mathbf{A}^n}\},$$

the trivial valuation on R. Since $I_{\mathbf{A}^n} = 0$, we also have

$$\mathcal{V}_{\mathbf{A}^n} = \emptyset.$$

At the other extreme, for a closed point $\xi \in \mathbf{A}^n$, we have

$$\mathcal{W}_{\subseteq \xi} = \mathcal{W}_\xi = \{\mathrm{triv}_\xi\}.$$

The space $\mathcal{V}_\xi$ is a singleton when $n = 1$ (see Sect. 6.6) but has a rich structure when $n > 1$. We shall describe in dimension two in Sect. 7, in which case it is a tree in the sense of Sect. 2.1. See [BFJ08b] for the higher-dimensional case.

6.5.5 Passing to the Completion

A semivaluation $v \in \mathbf{D}^n_{\mathrm{Berk}}$ whose center is equal to an irreducible subvariety Y extends uniquely to a semivaluation on the local ring $\mathcal{O}_{\mathbf{A}^n, Y}$ such that $v(\mathfrak{m}_Y) > 0$, where $\mathfrak{m}_Y$ is the maximal ideal. By $\mathfrak{m}_Y$-adic continuity, v further extends uniquely as a semivaluation on the completion and by Cohen's structure theorem, the latter is isomorphic to the power series ring $\kappa(Y)[\![z_1, \ldots z_r]\!]$, where r is the codimension of Y. Therefore we can view $\hat{\mathcal{V}}_Y$ as the set of semivaluations v on $\kappa(Y)[\![z_1, \ldots z_r]\!]$

whose restriction to $\kappa(Y)$ is trivial and such that $v(\mathfrak{m}_Y) > 0$. In particular, for a closed point ξ, we can view $\hat{\mathcal{V}}_\xi$ (resp., $\mathcal{V}_\xi$) as the set of semivaluations v on $\kappa(\xi)[\![z_1, \ldots z_n]\!]$ whose restriction to $\kappa(\xi)$ is trivial and such that $v(\mathfrak{m}_\xi) > 0$ (resp., $v(\mathfrak{m}_\xi) = 1$). This shows that when K is algebraically closed, the set $\mathcal{V}_\xi$ above is isomorphic to the space considered in [BFJ08b]. This space was first introduced in dimension $n = 2$ in [FJ04] where it was called the valuative tree. We shall study it from a slightly different point of view in Sect. 7. Note that it may happen that a valuation $v \in \hat{\mathcal{V}}_\xi$ has home ξ but that the extension of v to $\hat{\mathcal{O}}_{\mathbf{A}^n, \xi}$ is a semivaluation for which the ideal $\{v = \infty\} \subseteq \hat{\mathcal{O}}_{\mathbf{A}^n, \xi}$ is nontrivial.

6.6 The Affine Line

Using the definitions above, let us describe the Berkovich affine line $\mathbf{A}^1_{\mathrm{Berk}}$ over a trivially valued field K.

An irreducible subvariety of $\mathbf{A}^1$ is either $\mathbf{A}^1$ itself or a closed point. As we noted in Sect. 6.5.4

$$\hat{\mathcal{V}}_{\subseteq \mathbf{A}^1} = \mathbf{D}_{\mathrm{Berk}}, \quad \mathcal{W}_{\subseteq \mathbf{A}^1} = \mathbf{A}^1_{\mathrm{Berk}}, \quad \hat{\mathcal{V}}_{\supseteq \mathbf{A}^1} = \hat{\mathcal{V}}_{\mathbf{A}^1} = \mathcal{W}_{\supseteq \mathbf{A}^1} = \mathcal{W}_{\mathbf{A}^1} = \{\mathrm{triv}_{\mathbf{A}^1}\}$$

whereas $\mathcal{V}_{\mathbf{A}^1}$ is empty.

Now suppose the center of $v \in \mathbf{A}^1_{\mathrm{Berk}}$ is a closed point $\xi \in \mathbf{A}^1$. If the home of v is also equal to ξ, then $v = \mathrm{triv}_\xi$. Now suppose the home of v is $\mathbf{A}^1$, so that $0 < v(I_\xi) < \infty$. After scaling we may assume $v(I_\xi) = 1$ so that $v \in \mathcal{V}_\xi$. Since $R \simeq K[z]$ is a PID is follows easily that $v = \mathrm{ord}_\xi$. This shows that

$$\mathcal{W}_{\subseteq \xi} = \mathcal{W}_\xi = \{\mathrm{triv}_\xi\} \quad \text{and} \quad \mathcal{V}_\xi = \{\mathrm{ord}_\xi\},$$

Similarly, if $v \in \mathbf{A}^1_{\mathrm{Berk}}$ has center at infinity, then, after scaling, we may assume that $v(z) = -1$, where $z \in R$ is a coordinate. It is then clear that $v = \mathrm{ord}_\infty$, where ord_∞ is the valuation on R defined by $\mathrm{ord}_\infty(\phi) = -\deg \phi$. Thus we have

$$\mathcal{V}_\infty = \{\mathrm{ord}_\infty\}.$$

Note that any polynomial $\phi \in R$ can be viewed as a rational function on $\mathbf{P}^1 = \mathbf{A}^1 \cup \{\infty\}$ and $\mathrm{ord}_\infty(\phi) \leq 0$ is the order of vanishing of ϕ at ∞.

We leave it as an exercise to the reader to compare the terminology above with the one in Sect. 3.9.2. See Fig. 7 for a picture of the Berkovich affine line over a trivially valued field.

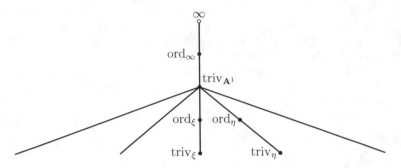

Fig. 7 The Berkovich affine line over a trivially valued field. The trivial valuation $\mathrm{triv}_{\mathbf{A}^1}$ is the only point with center $\mathbf{A}^1$. The point triv_ξ for $\xi \in \mathbf{A}^1$ has home ξ. All the points on the open segment $]\,\mathrm{triv}_{\mathbf{A}^1}, \mathrm{triv}_\xi[$ have home $\mathbf{A}^1$ and center ξ and are proportional to the valuation ord_ξ. The point ∞ does not belong to $\mathbf{A}^1_{\mathrm{Berk}}$. The points on the open segment $]\,\mathrm{triv}_{\mathbf{A}^1}, \infty[$ have home $\mathbf{A}^1$, center at infinity and are proportional to the valuation ord_∞.

6.7 The Affine Plane

In dimension $n = 2$, the Berkovich affine space is significantly more complicated than in dimension one, but can still—with some effort—be visualized.

An irreducible subvariety of $\mathbf{A}^2$ is either all of $\mathbf{A}^2$, a curve, or a closed point. As we have seen,

$$\hat{\mathcal{V}}_{\subseteq \mathbf{A}^2} = \mathbf{D}^2_{\mathrm{Berk}}, \quad \mathcal{W}_{\subseteq \mathbf{A}^2} = \mathbf{A}^2_{\mathrm{Berk}}, \quad \hat{\mathcal{V}}_{\supseteq \mathbf{A}^2} = \hat{\mathcal{V}}_{\mathbf{A}^2} = \mathcal{W}_{\supseteq \mathbf{A}^2} = \mathcal{W}_{\mathbf{A}^2} = \{\mathrm{triv}_{\mathbf{A}^2}\}$$

whereas $\mathcal{V}_{\mathbf{A}^2}$ is empty.

Now let ξ be a closed point. As before, $\mathcal{W}_{\subseteq \xi} = \mathcal{W}_\xi = \{\mathrm{triv}_\xi\}$, where triv_ξ is the image of ξ under the embedding $\mathbf{A}^2 \hookrightarrow \mathbf{A}^2_{\mathrm{Berk}}$. The set $\hat{\mathcal{V}}_{\subseteq \xi} = \hat{\mathcal{V}}_\xi$ is open and $\hat{\mathcal{V}}_\xi \setminus \{\mathrm{triv}_\xi\} = \hat{\mathcal{V}}_\xi \setminus \mathcal{W}_\xi$ is naturally a punctured cone with base $\mathcal{V}_\xi$. The latter will be called *the valuative tree* (at the point ξ) and is studied in detail in Sect. 7. Suffice it here to say that it is a tree in the sense of Sect. 2.1. The whole space $\hat{\mathcal{V}}_\xi$ is a cone over the valuative tree with its apex at triv_ξ. The boundary of $\hat{\mathcal{V}}_\xi$ consists of all semivaluations whose center strictly contains ξ, so it is the union of $\mathrm{triv}_{\mathbf{A}^2}$ and $\hat{\mathcal{V}}_C$, where C ranges over curves containing C. As we shall see, the boundary therefore has the structure of a tree naturally rooted in $\mathrm{triv}_{\mathbf{A}^2}$. See Fig. 8. If ξ and η are two different closed points, then the open sets $\hat{\mathcal{V}}_\xi$ and $\hat{\mathcal{V}}_\eta$ are disjoint.

Next consider a curve $C \subseteq \mathbf{A}^2$. By definition, the set $\mathcal{W}_{\subseteq C}$ consists all semivaluations whose home is contained in C. This means that $\mathcal{W}_{\subseteq C}$ is the image of the analytification C_{Berk} of C under the embedding $C_{\mathrm{Berk}} \hookrightarrow \mathbf{A}^2_{\mathrm{Berk}}$. As such, it looks quite similar to the Berkovich affine line $\mathbf{A}^1_{\mathrm{Berk}}$, see [Ber90, §1.4.2]. More precisely, the semivaluation triv_C is the unique semivaluation in $\mathcal{W}_{\subseteq C}$ having center C. All other semivaluations in $\mathcal{W}_{\subseteq C}$ have center at a closed point $\xi \in C$. The only such semivaluation having home ξ is triv_ξ; the other semivaluations in $\mathcal{W}_{\subseteq C} \cap \hat{\mathcal{V}}_\xi$ have

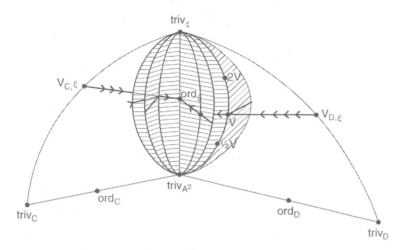

Fig. 8 The Berkovich affine plane over a trivially valued field. The picture shows the closure of the set $\hat{\mathcal{V}}_\xi$ of semivaluations having center at a closed point $\xi \in \mathbf{A}^2$. Here C, D are irreducible curves containing ξ. The semivaluation $\mathrm{triv}_\xi \in \hat{\mathcal{V}}_\xi$ has home ξ. All semivaluations in $\hat{\mathcal{V}}_\xi \setminus \{\mathrm{triv}_\xi\}$ are proportional to a semivaluation v in the valuative tree $\mathcal{V}_\xi$ at ξ. We have $tv \to \mathrm{triv}_\xi$ as $t \to \infty$. As $t \to 0+$, tv converges to the semivaluation triv_Y, where Y is the home of v. The semivaluations $v_{C,\xi}$ and $v_{D,\xi}$ belong to $\mathcal{V}_\xi$ and have home C and D, respectively. The boundary of $\hat{\mathcal{V}}_\xi$ is a tree consisting of all segments $[\mathrm{triv}_{\mathbf{A}^2}, \mathrm{triv}_C]$ for all irreducible affine curves C containing both ξ. Note that the segment $[\mathrm{triv}_C, \mathrm{triv}_\xi]$ in the closure of $\hat{\mathcal{V}}_\xi$ is also a segment in the analytification $C_{\mathrm{Berk}} \subseteq \mathbf{A}^2_{\mathrm{Berk}}$ of C, see Fig. 10

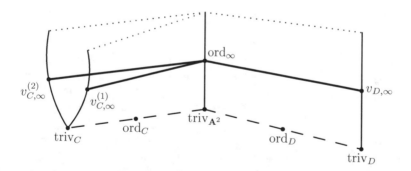

Fig. 9 The Berkovich affine plane over a trivially valued field. The picture shows (part of) the closure of the set $\hat{\mathcal{V}}_\infty$ of semivaluations having center at infinity. Here C and D are affine curves having two and one places at infinity, respectively. The set $\hat{\mathcal{V}}_\infty$ is a cone whose base is $\mathcal{V}_\infty$, the valuative tree at infinity. Fixing an embedding $\mathbf{A}^2 \hookrightarrow \mathbf{P}^2$ allows us to identify $\mathcal{V}_\infty$ with a subset of $\hat{\mathcal{V}}_\infty$ and the valuation ord_∞ is the order of vanishing along the line at infinity in $\mathbf{P}^2$. The semivaluations $v_{D,\infty}$ and $v_{C,\infty}^{(i)}$, $i = 1, 2$ have home D and C, respectively; the segments $[\mathrm{ord}_\infty, v_{D,\infty}]$ and $[\mathrm{ord}_\infty, v_{C,\infty}^{(i)}]$, $i = 1, 2$ belong to $\mathcal{V}_\infty$. The segments $[\mathrm{triv}_{\mathbf{A}^2}, \mathrm{triv}_C]$ and $[\mathrm{triv}_{\mathbf{A}^2}, \mathrm{triv}_D]$ at the bottom of the picture belong to the boundary of $\hat{\mathcal{V}}_\infty$: the full boundary is a tree consisting of all such segments and whose only branch point is $\mathrm{triv}_{\mathbf{A}^2}$. The dotted segments in the top of the picture do not belong to the Berkovich affine plane

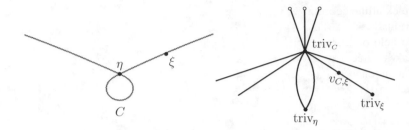

Fig. 10 The analytification C_{Berk} of an affine curve C over a trivially valued field. The semivaluation triv_C is the only semivaluation in C_{Berk} having center C and home C. To each closed point $\xi \in C$ is associated a unique semivaluation $\mathrm{triv}_\xi \in C_{Berk}$ with center and home ξ. The set of elements of C_{Berk} with home C and center at a given closed point ξ is a disjoint union of open intervals, one for each local branch of C at ξ. Similarly, the set of elements of C_{Berk} with home C and center at infinity is a disjoint union of open intervals, one for each branch of C at infinity. The left side of the picture shows a nodal cubic curve C and the right side shows its analytification C_{Berk}. Note that for a smooth point ξ on C, the segment $[\mathrm{triv}_C, \mathrm{triv}_\xi]$ in C_{Berk} also lies in the closure of the cone $\hat{\mathcal{V}}_\xi$, see Fig. 8

home C and center ξ. We can normalize them by $v(I_\xi) = 1$. If ξ is a nonsingular point on C, then there is a unique normalized semivaluation $v_{C,\xi} \in \mathbf{A}_{Berk}^2$ having home C and center ξ. When ξ is a singular point on C, the set of such semivaluations is instead in bijection with the set of local branches[18] of C at ξ. We see that $\mathcal{W}_{\subseteq C}$ looks like $\mathbf{A}_{Berk}^1$ except that there may be several intervals joining triv_C and triv_ξ: one for each local branch of C at ξ. See Fig. 10.

Now look at the closed set $\hat{\mathcal{V}}_{\supseteq C}$ of semivaluations whose center contains C. It consists of all semivaluations $t \, \mathrm{ord}_C$ for $0 \le t \le \infty$. Here $t = \infty$ and $t = 0$ correspond to triv_C and $\mathrm{triv}_{\mathbf{A}^2}$, respectively. As a consequence, for any closed point ξ, $\partial \hat{\mathcal{V}}_\xi$ has the structure of a tree, much like the Berkovich affine line $\mathbf{A}_{Berk}^1$.

The set $\hat{\mathcal{V}}_{\subseteq C}$ is open and its boundary consists of semivaluations whose center strictly contains C. In other words, the boundary is the singleton $\{\mathrm{triv}_{\mathbf{A}^2}\}$. For two curves C, D, the intersection $\hat{\mathcal{V}}_{\subseteq C} \cap \hat{\mathcal{V}}_{\subseteq D}$ is the union of sets $\hat{\mathcal{V}}_\xi$ over all closed points $\xi \in C \cap D$.

The set $\mathcal{V}_C \simeq (\hat{\mathcal{V}}_{\subseteq C} \setminus \mathcal{W}_{\subseteq C})/\mathbf{R}_+^*$ looks quite similar to the valuative tree at a closed point. To see this, note that the valuation ord_C is the only semivaluation in $\mathcal{V}_C$ whose center is equal to C. All other semivaluations in $\mathcal{V}$ have center at a closed point $\xi \in C$. For each semivaluation $v \in \mathcal{V}_\xi$ whose home is not equal to C, there exists a unique $t = t(\xi, C) > 0$ such that $tv \in \mathcal{V}_C$; indeed, $t = v(I_C)$. Therefore, $\mathcal{V}_C$ can be obtained by taking the disjoint union of the trees $\mathcal{V}_\xi$ over all $\xi \in C$ and identifying the semivaluations having home C with the point ord_C. If C is nonsingular, then $\mathcal{V}_C$ will be a tree naturally rooted in ord_C.

We claim that if C is a line, then $\mathcal{V}_C$ can be identified with the Berkovich unit disc over the field of Laurent series in one variable with coefficients in K. To see

---

[18] A local branch is a preimage of a point of C under the normalization map.

this, pick affine coordinates (z_1, z_2) such that $C = \{z_1 = 0\}$. Then $\mathcal{V}_C$ is the set of semivaluations $v : K[z_1, z_2] \to \mathbf{R}_+ \cup \{\infty\}$ such that $v(z_1) = 1$. Let $L = K((z_1))$ be the field of Laurent series, equipped with the valuation v_L that is trivial on K and takes value 1 on z_1. Then the Berkovich unit disc $\mathbf{D}_{\text{Berk}}$ over L is the set of semivaluations $L[z_2] \to \mathbf{R}_+ \cup \{\infty\}$ extending v_L. Every element of $\mathbf{D}_{\text{Berk}}$ defines an element of $\mathcal{V}_C$ by restriction. Conversely, pick $v \in \mathcal{V}_C$. If $v = \text{ord}_C$, then v extends uniquely to an element of $\mathbf{D}_{\text{Berk}}$, namely the Gauss point. If $v \neq \text{ord}_C$, then the center of v is a closed point $\xi \in C$ and v extends uniquely to the fraction field of the completion $\hat{\mathcal{O}}_\xi$. This fraction field contains $L[z_2]$.

The open subset $\hat{\mathcal{V}}_\infty = \mathbf{A}^2_{\text{Berk}} \setminus \mathbf{D}^2_{\text{Berk}}$ of semivaluations centered at infinity is a punctured cone over a base $\mathcal{V}_\infty$. The latter space is called *the valuative tree at infinity* and will be studied in detail in Sect. 9. Superficially, its structure is quite similar to the valuative tree at a closed point ξ. In particular it is a tree in the sense of Sect. 2.1. The boundary of $\hat{\mathcal{V}}_\infty$ is the union of $\hat{\mathcal{V}}_{\supseteq C}$ over *all* affine curves C, that is, the set of semivaluations in $\mathbf{D}^2_{\text{Berk}}$ whose center is *not* a closed point. Thus the boundary has a structure of a tree rooted in $\text{triv}_{\mathbf{A}^2}$. See Fig. 9. We emphasize that there is no point triv_∞ in $\hat{\mathcal{V}}_\infty$.

To summarize the discussion, $\mathbf{A}^2_{\text{Berk}}$ contains a closed subset Σ with empty interior consisting of semivaluations having center of dimension one or two. This set is a naturally a tree, which can be viewed as the cone over the collection of all irreducible affine curves. The complement of Σ is an open dense subset whose connected components are $\hat{\mathcal{V}}_\infty$, and $\hat{\mathcal{V}}_\xi$, where ξ ranges over closed points of $\mathbf{A}^2$. The set $\hat{\mathcal{V}}_\infty$ is a punctured cone over a tree $\mathcal{V}_\infty$ and its boundary is all of Σ. For a closed point ξ, $\hat{\mathcal{V}}_\xi$ is a cone over a tree $\mathcal{V}_\xi$ and its boundary is a subtree of Σ, namely the cone over the collection of all irreducible affine curves containing ξ.

6.8 Valuations

A semivaluation v on $R \simeq K[z_1, \ldots, z_n]$ is a *valuation* if the corresponding seminorm is a norm, that is, if $v(\phi) < \infty$ for all nonzero polynomials $\phi \in R$. A valuation v extends to the fraction field $F \simeq K(z_1, \ldots, z_n)$ of R by setting $v(\phi_1/\phi_2) = v(\phi_1) - v(\phi_2)$.

Let X be a variety over K whose function field is equal to F. The *center* of a valuation v on X, if it exists, is the unique (not necessarily closed) point $\xi \in X$ defined by the properties that $v \geq 0$ on the local ring $\mathcal{O}_{X,\xi}$ and $\{v > 0\} \cap \mathcal{O}_{X,\xi} = \mathfrak{m}_{X,\xi}$. By the valuative criterion of properness, the center always exists and is unique when X is proper over K.

Following [JM12] we write Val_X for the set of valuations of F that admit a center on X. As usual, this set is endowed with the topology of pointwise convergence. Note that Val_X is a subset of $\mathbf{A}^n_{\text{Berk}}$ that can in fact be shown to be dense. One nice feature of Val_X is that any proper birational morphism $X' \to X$ induces an isomorphism $\mathrm{Val}_{X'} \xrightarrow{\sim} \mathrm{Val}_X$. (In the same situation, the analytification X'_{Berk} maps onto X_{Berk}, but this map is not injective.)

We can view the Berkovich unit polydisc $\mathbf{D}^n_{\mathrm{Berk}}$ as the disjoint union of Val_Y, where Y ranges over irreducible subvarieties of X.

6.9 Numerical Invariants

To a valuation $v \in \mathbf{A}^n_{\mathrm{Berk}}$ we can associate several invariants. First, the *value group* of v is defined by $\Gamma_v := \{v(\phi) \mid \phi \in F \setminus \{0\}\}$. The *rational rank* rat. rk v of v is the dimension of the $\mathbf{Q}$-vector space $\Gamma_v \otimes_{\mathbf{Z}} \mathbf{Q}$.

Second, the valuation ring $R_v = \{\phi \in F \mid v(\phi) \geq 0\}$ of v is a local ring with maximal ideal $\mathfrak{m}_v = \{v(\phi) > 0\}$. The *residue field* $\kappa(v) = R_v/\mathfrak{m}_v$ contains K as a subfield and the *transcendence degree* of v is the transcendence degree of the field extension $\kappa(v)/K$.

In our setting, the fundamental *Abhyankar inequality* states that

$$\mathrm{rat.\,rk}\, v + \mathrm{tr.\,deg}\, v \leq n. \tag{33}$$

The valuations for which equality holds are of particular importance. At least in characteristic zero, they admit a nice geometric description that we discuss next.

6.10 Quasimonomial and Divisorial Valuations

Let X be a smooth variety over K with function field F. We shall assume in this section that the field K has characteristic zero or that X has dimension at most two. This allows us to freely use resolutions of singularities.

Let $\xi \in X$ be a point (not necessarily closed) with residue field $\kappa(\xi)$. Let $(\zeta_1, \ldots, \zeta_r)$ be a system of algebraic coordinates at ξ (i.e. a regular system of parameters of $\mathcal{O}_{X,\xi}$). We say that a valuation $v \in \mathrm{Val}_X$ is *monomial* in coordinates $(\zeta_1, \ldots, \zeta_r)$ with weights $t_1, \ldots, t_r \geq 0$ if the following holds: if we write $\phi \in \hat{\mathcal{O}}_{X,\xi}$ as $\phi = \sum_{\beta \in \mathbf{Z}^m_{\geq 0}} c_\beta \zeta^\beta$ with each $c_\beta \in \hat{\mathcal{O}}_{X,\xi}$ either zero or a unit, then

$$v(\phi) = \min\{\langle t, \beta \rangle \mid c_\beta \neq 0\},$$

where $\langle t, \beta \rangle = t_1 \beta_1 + \cdots + t_r \beta_r$. After replacing ξ by the (generic point of the) intersection of all divisors $\{\zeta_i = 0\}$ we may in fact assume that $t_i > 0$ for all i.

We say that a valuation $v \in \mathrm{Val}_X$ is *quasimonomial* (on X) if it is monomial in some birational model of X. More precisely, we require that there exists a proper birational morphism $\pi : X' \to X$, with X' smooth, such that v is monomial in some algebraic coordinates at some point $\xi \in X'$. As explained in [JM12], in this case we can assume that the divisors $\{\zeta_i = 0\}$ are irreducible components of a reduced, effective simple normal crossings divisor D on X' that contains the exceptional locus of π. (In the two-dimensional situation that we shall be primarily interested in, arranging this is quite elementary.)

It is a fact that a valuation $v \in \mathrm{Val}_X$ is quasimonomial iff equality holds in Abhyankar's inequality (33). For this reason, quasimonomial valuations are sometimes called Abhyankar valuations. See [ELS03, Proposition 2.8].

Furthermore, we can arrange the situation so that the weights t_i are all strictly positive and linearly independent over $\mathbf{Q}$: see [JM12, Proposition 3.7]. In this case the residue field of v is isomorphic to the residue field of ξ, and hence $\mathrm{tr.\,deg}\, v = \dim(\bar{\xi}) = n - r$. Furthermore, the value group of v is equal to

$$\Gamma_v = \sum_{i=1}^{r} \mathbf{Z} t_i, \tag{34}$$

so $\mathrm{rat.\,rk}\, v = r$.

A very important special case of quasimonomial valuations are given by *divisorial valuations*. Numerically, they are characterized by $\mathrm{rat.\,rk} = 1$, $\mathrm{tr.\,deg} = n - 1$. Geometrically, they are described as follows: there exists a birational morphism $X' \to X$, a prime divisor $D \subseteq X'$ and a constant $t > 0$ such that $t^{-1} v(\phi)$ is the order of vanishing along D for all $\phi \in F$.

6.11 The Izumi-Tougeron Inequality

Keep the same assumptions on K and X as in Sect. 6.10. Consider a valuation $v \in \mathrm{Val}_X$ and let ξ be its center on X. Thus ξ is a (not necessarily closed) point of X. By definition, v is nonnegative on the local ring $\mathcal{O}_{X,\xi}$ and strictly positive on the maximal ideal $\mathfrak{m}_{X,\xi}$. Let ord_ξ be the order of vanishing at ξ. It follows from the valuation axioms that

$$v \geq c\, \mathrm{ord}_\xi, \tag{35}$$

on $\mathcal{O}_{X,\xi}$, where $c = v(\mathfrak{m}_{X,\xi}) > 0$.

It will be of great importance to us that if $v \in \mathrm{Val}_X$ is quasimonomial then the reverse inequality holds in (35). Namely, there exists a constant $C = C(v) > 0$ such that

$$c\, \mathrm{ord}_\xi \leq v \leq C\, \mathrm{ord}_\xi \tag{36}$$

on $\mathcal{O}_{X,\xi}$. This inequality is often referred to as Izumi's inequality (see [Izu85, Ree89, HS01, ELS03]) but in the smooth case we are considering it goes back at least to Tougeron [Tou72, p.178]. More precisely, Tougeron proved this inequality for divisorial valuations, but that easily implies the general case.

As in Sect. 4.8.2, a valuation $v \in \mathrm{Val}_X$ having center ξ on X extends uniquely to a semivaluation on $\hat{\mathcal{O}}_{X,\xi}$. The Izumi-Tougeron inequality (36) implies that if v is quasimonomial, then this extension is in fact a valuation. In general, however, the extension may not be a valuation, so the Izumi-Tougeron inequality certainly does not hold for *all* valuations in Val_X having center ξ on X. For a concrete example, let $X = \mathbf{A}^2$, let ξ be the origin in coordinates (z, w) and let $v(\phi)$ be defined as the order of vanishing at $u = 0$ of $\phi(u, \sum_{i=1}^{\infty} \frac{u^i}{i!})$. Then $v(\phi) < \infty$ for all nonzero polynomials ϕ, whereas $v(w - \sum_{i=1}^{\infty} \frac{u^i}{i!}) = 0$.

6.12   Notes and Further References

It is a interesting feature of Berkovich's theory that one can work with trivially valued fields: this is definitely not possible in rigid geometry (see e.g. [Con08] for a general discussion of rigid geometry and various other aspects of non-Archimedean geometry).

In fact, Berkovich spaces over trivially valued fields have by now seen several interesting and unexpected applications. In these notes we focus on dynamics, but one can also study use Berkovich spaces to study the singularities of plurisubharmonic functions [FJ05a, BFJ08b] and various asymptotic singularities in algebraic geometry, such as multiplier ideals [FJ05b, JM12]. In other directions, Thuillier [Thu07] exploited Berkovich spaces to give a new proof of a theorem by Stepanov in birational geometry, and Berkovich [Ber09] has used them in the context of mixed Hodge structures.

The Berkovich affine space of course also comes with a structure sheaf $\mathcal{O}$. We shall not need use it in what follows but it is surely a useful tool for a more systematic study of polynomial mappings on the $\mathbf{A}_{\mathrm{Berk}}^{\mathrm{n}}$.

The spaces $\hat{\mathcal{V}}_{\xi}$, $\mathcal{V}_{\xi}$ and $\mathcal{V}_{\infty}$ were introduced (in the case of K algebraically closed of characteristic zero) and studied in [FJ04, FJ07, BFJ08b] but not explicitly identified as subset of the Berkovich affine plane. The structure of the Berkovich affine space does not seem to have been written down in detail before, but see [YZ09b].

The terminology "home" is not standard. Berkovich [Ber90, §1.2.5] uses this construction but does not give it a name. The name "center" comes from valuation theory, see [Vaq00, §6] whereas non-Archimedean geometry tends to use the term "reduction". Our distinction between (additive) valuations and (multiplicative) norms is not always made in the literature. Furthermore, in [FJ04, BFJ08b], the term 'valuation' instead of 'semi-valuation' is used even when the prime ideal $\{v = +\infty\}$ is nontrivial.

The space Val_X was introduced in [JM12] for the study of asymptotic invariants of graded sequences of ideals. In *loc. cit.* it is proved that Val_X is an inverse limit of cone complexes, in the same spirit as Sect. 7.5.4 below.

7 The Valuative Tree at a Point

Having given an overview of the Berkovich affine plane over a trivially valued field, we now study the set of semivaluations centered at a closed point. As indicated in Sect. 6.7, this is a cone over a space that we call the valuative tree.

The valuative tree is treated in detail in the monograph [FJ04]. However, the self-contained presentation here has a different focus. In particular, we emphasize aspects that generalize to higher dimension. See [BFJ08b] for some of these generalizations.

7.1 Setup

Let K be field equipped with the trivial norm. For now we assume that K is algebraically closed but of arbitrary characteristic. (See Sect. 7.11 for a more general case). In applications to complex dynamics we would of course pick $K = \mathbf{C}$, but we emphasize that the norm is then *not* the Archimedean one. As in Sect. 6 we work additively rather than multiplicatively and consider K equipped with the trivial valuation, whose value on nonzero elements is zero and whose value on 0 is $+\infty$.

Let R and F be the coordinate ring and function field of $\mathbf{A}^2$. Fix a closed point $0 \in \mathbf{A}^2$ and write $\mathfrak{m}_0 \subseteq R$ for the corresponding maximal ideal. If (z_1, z_2) are global coordinates on $\mathbf{A}^2$ vanishing at 0, then $R = K[z_1, z_2]$, $F = K(z_1, z_2)$ and $\mathfrak{m}_0 = (z_1, z_2)$. We say that an ideal $\mathfrak{a} \subseteq R$ is $\mathfrak{m}_0$-*primary* or simply *primary* if it contains some power of $\mathfrak{m}_0$.

Recall that the Berkovich affine plane $\mathbf{A}^2_{\text{Berk}}$ is the set of semivaluations on R that restrict to the trivial valuation on K. Similarly, the Berkovich unit bidisc $\mathbf{D}^2_{\text{Berk}}$ is the set of semivaluations $v \in \mathbf{A}^2_{\text{Berk}}$ that are nonnegative on R. If $\mathfrak{a} \subseteq R$ is an ideal and $v \in \mathbf{D}^2_{\text{Berk}}$, then we write $v(\mathfrak{a}) = \min\{v(\phi) \mid \phi \in \mathfrak{a}\}$. In particular, $v(\mathfrak{m}_0) = \min\{v(z_1), v(z_2)\}$.

7.2 The Valuative Tree

Let us recall some definitions from Sects. 6.5.3 and 6.7. Let $\hat{\mathcal{V}}_0 \subseteq \mathbf{D}^2_{\text{Berk}}$ be the subset of semivaluations whose center on $\mathbf{A}^2$ is equal to the closed point $0 \in \mathbf{A}^2$. In other words, $\hat{\mathcal{V}}_0$ is the set of semivaluations $v : R \to [0, +\infty]$ such that $v|_{K^*} \equiv 0$ and $v(\mathfrak{m}_0) > 0$.

There are now two cases. Either $v(\mathfrak{m}_0) = +\infty$, in which case $v = \text{triv}_0 \in \mathbf{A}^2_{\text{Berk}}$ is the trivial valuation associated to the point $0 \in \mathbf{A}^2$, or $0 < v(\mathfrak{m}_0) < \infty$. Define $\hat{\mathcal{V}}_0^*$ as the set of semivaluations of the latter type. This set is naturally a pointed cone and admits the following set as a "section".

Definition 7.1 The *valuative tree* $\mathcal{V}_0$ at the point $0 \in \mathbf{A}^2$ is the set of semivaluations $v : R \to [0, +\infty]$ satisfying $v(\mathfrak{m}_0) = 1$.

To repeat, we have

$$\hat{\mathcal{V}}_0 = \{\mathrm{triv}_0\} \cup \hat{\mathcal{V}}_0^* \quad \text{and} \quad \hat{\mathcal{V}}_0^* = \mathbf{R}_+^* \mathcal{V}_0.$$

We equip $\mathcal{V}_0$ and $\hat{\mathcal{V}}_0$ with the subspace topology from $\mathbf{A}_{\mathrm{Berk}}^2$, that is, the weakest topology for which all evaluation maps $v \mapsto v(\phi)$ are continuous, where ϕ ranges over polynomials in R. It follows easily from Tychonoff's theorem that $\mathcal{V}_0$ is a compact Hausdorff space.

Equivalently, we could demand that $v \mapsto v(\mathfrak{a})$ be continuous for any primary ideal $\mathfrak{a} \subseteq R$. For many purposes it is indeed quite natural to evaluate semivaluations in $\hat{\mathcal{V}}_0^*$ on primary ideals rather than polynomials. For example, we have $v(\mathfrak{a} + \mathfrak{b}) = \min\{v(\mathfrak{a}), v(\mathfrak{b})\}$ for any primary ideals $\mathfrak{a}, \mathfrak{b}$, whereas we only have $v(\phi + \psi) \geq \min\{v(\phi), v(\psi)\}$ for polynomials ϕ, ψ.

An important element of $\mathcal{V}_0$ is the valuation ord_0 defined by

$$\mathrm{ord}_0(\phi) = \max\{k \geq 0 \mid \phi \in \mathfrak{m}_0^k\}.$$

Note that $v(\phi) \geq \mathrm{ord}_0(\phi)$ for all $v \in \mathcal{V}_0$ and all $\phi \in R$.

Any semivaluation $v \in \mathbf{A}_{\mathrm{Berk}}^2$ extends as a function $v : F \to [-\infty, +\infty]$, where F is the fraction field of R, by setting $v(\phi_1/\phi_2) = v(\phi_1) - v(\phi_2)$; this is well defined since $\{v = +\infty\} \subseteq R$ is a prime ideal.

Our goal for now is to justify the name "valuative tree" by showing that $\mathcal{V}_0$ can be equipped with a natural tree structure, rooted at ord_0. This structure can be obtained from many different points of view, as explained in [FJ04]. Here we focus on a geometric approach that is partially generalizable to higher dimensions (see [BFJ08b]).

7.3 Blowups and Log Resolutions

We will consider birational morphisms

$$\pi : X_\pi \to \mathbf{A}^2,$$

with X_π smooth, that are isomorphisms above $\mathbf{A}^2 \setminus \{0\}$. Such a morphism is necessarily a finite composition of point blowups; somewhat sloppily we will refer to it simply as a *blowup*. The set $\mathfrak{B}_0$ of blowups is a partially ordered set: we say $\pi \leq \pi'$ if the induced birational map $X_{\pi'} \to X_\pi$ is a morphism (and hence itself a composition of point blowups). In fact, $\mathfrak{B}_0$ is a directed system: any two blowups can be dominated by a third.

7.3.1 Exceptional Primes

An irreducible component $E \subseteq \pi^{-1}(0)$ is called an *exceptional prime* (divisor) of π. There are as many exceptional primes as the number of points blown up. We often identify an exceptional prime of π with its strict transform to any blowup $\pi' \in \mathfrak{B}_0$ dominating π. In this way we can identify an exceptional prime E (of some blowup π) with the corresponding divisorial valuation ord_E.

If π_0 is the simple blowup of the origin, then there is a unique exceptional prime E_0 of π_0 whose associated divisorial valuation is $\mathrm{ord}_{E_0} = \mathrm{ord}_0$. Since any blowup $\pi \in \mathfrak{B}_0$ factors through π_0, E_0 is an exceptional prime of any π.

7.3.2 Free and Satellite Points

The following terminology is convenient and commonly used in the literature. Consider a closed point $\xi \in \pi^{-1}(0)$ for some blowup $\pi \in \mathfrak{B}_0$. We say that ξ is a *free* point if it belongs to a unique exceptional prime; otherwise it is the intersection point of two distinct exceptional primes and is called a *satellite* point.

7.3.3 Exceptional Divisors

A divisor on X_π is *exceptional* if its support is contained in $\pi^{-1}(0)$. We write $\mathrm{Div}(\pi)$ for the abelian group of exceptional divisors on X_π. If $E_i, i \in I$, are the exceptional primes of π, then $\mathrm{Div}(\pi) \simeq \bigoplus_{i \in I} \mathbf{Z} E_i$.

If π, π' are blowups and $\pi' = \pi \circ \mu \geq \pi$, then there are natural maps

$$\mu^* : \mathrm{Div}(\pi) \to \mathrm{Div}(\pi') \quad \text{and} \quad \mu_* : \mathrm{Div}(\pi') \to \mathrm{Div}(\pi)$$

satisfying the projection formula $\mu_* \mu^* = \mathrm{id}$. In many circumstances it is natural to identify an exceptional divisor $Z \in \mathrm{Div}(\pi)$ with its pullback $\mu^* Z \in \mathrm{Div}(\pi')$.

7.3.4 Intersection Form

We denote by $(Z \cdot W)$ the intersection number between exceptional divisors $Z, W \in \mathrm{Div}(\pi)$. If $\pi' = \pi \circ \mu$, then $(\mu^* Z \cdot W') = (Z \cdot \mu_* W')$ and hence $(\mu^* Z \cdot \mu^* W) = (Z \cdot W)$ for $Z, W \in \mathrm{Div}(\pi), Z' \in \mathrm{Div}(\pi')$.

Proposition 7.2 *The intersection form on* $\mathrm{Div}(\pi)$ *is negative definite and unimodular.*

Proof We argue by induction on the number of blowups in π. If $\pi = \pi_0$ is the simple blowup of $0 \in \mathbf{A}^2$, then $\mathrm{Div}(\pi) = \mathbf{Z} E_0$ and $(E_0 \cdot E_0) = -1$. For the inductive step, suppose $\pi' = \pi \circ \mu$, where μ is the simple blowup of a closed point on $\pi^{-1}(0)$, resulting in an exceptional prime E. Then we have an orthogonal decomposition $\mathrm{Div}(\pi') = \mu^* \mathrm{Div}(\pi) \oplus \mathbf{Z}E$. The result follows since $(E \cdot E) = -1$.

Alternatively, we may view $\mathbf{A}^2$ as embedded in $\mathbf{P}^2$ and X_π accordingly embedded in a smooth compact surface $\bar{X}_\pi$. The proposition can then be obtained as a consequence of the Hodge Index Theorem [Har77, p.364] and Poincaré Duality applied to the smooth rational surface $\bar{X}_\pi$.   □

7.3.5 Positivity

It follows from Proposition 7.2 that for any $i \in I$ there exists a unique divisor $\check{E}_i \in \mathrm{Div}(\pi)$ such that $(\check{E}_i \cdot E_i) = 1$ and $(\check{E}_i \cdot E_j) = 0$ for $j \neq i$.

An exceptional divisor $Z \in \mathrm{Div}(\pi)$ is *relatively nef*[19] if $(Z \cdot E_i) \geq 0$ for all exceptional primes E_i. We see that the set of relatively nef divisors is a free semigroup generated by the $\check{E}_i$, $i \in I$. Similarly, the set of *effective* divisors is a free semigroup generated by the E_i, $i \in I$.

Using the negativity of the intersection form and some elementary linear algebra, one shows that the divisors $\check{E}_i$ have strictly negative coefficients in the basis $(E_j)_{j \in I}$. Hence any relatively nef divisor is antieffective.[20]

We encourage the reader to explicitly construct the divisors $\check{E}_i$ using the procedure in the proof of Proposition 7.2. Doing this, one sees directly that $\check{E}_i$ is antieffective. See also Sect. 7.4.7.

7.3.6 Invariants of Exceptional Primes

To any exceptional prime E (or the associated divisorial valuation $\mathrm{ord}_E \in \hat{\mathcal{V}}_0^*$) we can associate two basic numerical invariants α_E and A_E. We shall not directly use them in this paper, but they seem quite fundamental and their cousins at infinity (see Sect. 9.3.3) will be of great importance.

To define α_E, pick a blowup $\pi \in \mathfrak{B}_0$ for which E is an exceptional prime. Above we defined the divisor $\check{E} = \check{E}_\pi \in \mathrm{Div}(\pi)$ by duality: $(\check{E}_\pi \cdot E) = 1$ and $(\check{E}_\pi \cdot F) = 0$ for all exceptional primes $F \neq E$ of π. Note that if $\pi' \in \mathfrak{B}_0$ dominates π, then the divisor $\check{E}_{\pi'} \in \mathrm{Div}(\pi')$ is the pullback of $\check{E}_\pi$ under the morphism $X_{\pi'} \to X_\pi$. In particular, the self-intersection number

$$\alpha_E := \alpha(\mathrm{ord}_E) := (\check{E} \cdot \check{E})$$

is an integer independent of the choice of π. Since $\check{E}$ is antieffective, $\alpha_E \leq -1$.

---

[19]The acronym "nef" is due to M. Reid who meant it to stand for "numerically eventually free" although many authors refer to it as "numerically effective".

[20]A higher-dimensional version of this result is known as the "Negativity Lemma" in birational geometry: see [KM98, Lemma 3.39] and also [BdFF10, Proposition 2.11].

The second invariant is the *log discrepancy* A_E.[21] This is an important invariant in higher dimensional birational geometry, see [Kol97]. Here we shall use a definition adapted to our purposes. Let ω be a nonvanishing regular 2-form on $\mathbf{A}^2$. If $\pi \in \mathfrak{B}_0$ is a blowup, then $\pi^*\omega$ is a regular 2-form on X_π. For any exceptional prime E of π with associated divisorial valuation $\mathrm{ord}_E \in \hat{\mathcal{V}}_0^*$, we define

$$A_E := A(\mathrm{ord}_E) := 1 + \mathrm{ord}_E(\pi^*\omega). \tag{37}$$

Note that $\mathrm{ord}_E(\pi^*\omega)$ is simply the order of vanishing along E of the Jacobian determinant of π. The log discrepancy A_E is a positive integer whose value does not depend on the choice of π or ω. A direct calculation shows that $A(\mathrm{ord}_0) = 2$.

7.3.7 Ideals and Log Resolutions

A *log resolution* of a primary ideal $\mathfrak{a} \subseteq R$ is a blowup $\pi \in \mathfrak{B}_0$ such that the ideal sheaf $\mathfrak{a} \cdot \mathcal{O}_{X_\pi}$ on X_π is locally principal:

$$\mathfrak{a} \cdot \mathcal{O}_{X_\pi} = \mathcal{O}_{X_\pi}(Z) \tag{38}$$

for some exceptional divisor $Z = Z_\pi(\mathfrak{a}) \in \mathrm{Div}(\pi)$. This means that the pullback of the ideal $\mathfrak{a}$ to X_π is locally generated by a single monomial in coordinates defining the exceptional primes. It is an important basic fact that any primary ideal $\mathfrak{a} \subseteq R$ admits a log resolution.

If π is a log resolution of $\mathfrak{a}$ and $\pi' = \pi \circ \mu \geq \pi$, then π' is also a log resolution of $\mathfrak{a}$ and $Z_{\pi'}(\mathfrak{a}) = \mu^* Z_\pi(\mathfrak{a})$.

Example 7.3 The ideal $\mathfrak{a} = (z_2^2 - z_1^3, z_1^2 z_2)$ admits a log resolution that is a composition of four point blowups. Each time we blow up the base locus of the strict transform of $\mathfrak{a}$. The first blowup is at the origin. In the terminology of Sect. 7.3.2, the second and fourth blowups occur at free points whereas the third blowup is at a satellite point. See Fig. 11.

7.3.8 Ideals and Positivity

The line bundle $\mathcal{O}_{X_\pi}(Z)$ on X_π in (38) is *relatively base point free*, that is, it admits a nonvanishing section at any point of $\pi^{-1}(0)$. Conversely, if $Z \in \mathrm{Div}(\pi)$ is an exceptional divisor such that $\mathcal{O}_{X_\pi}(Z)$ is relatively base point free, then $Z = Z_\pi(\mathfrak{a})$ for $\mathfrak{a} = \pi_* \mathcal{O}_{X_\pi}(Z)$.

If a line bundle $\mathcal{O}_{X_\pi}(Z)$ is relatively base point free, then its restriction to any exceptional prime E is also base point free, implying $(Z \cdot E) = \deg(\mathcal{O}_{X_\pi}(Z)|_E) \geq$

---

[21]The log discrepancy is called *thinness* in [FJ04, FJ05a, FJ05b, FJ07].

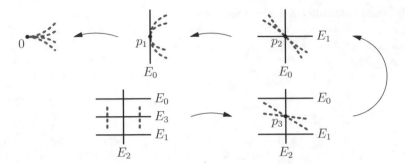

Fig. 11 A log resolution of the primary ideal $\mathfrak{a} = (z_2^2 - z_1^3, z_1^2 z_2)$. The dotted curves show the strict transforms of curves of the form $C_a = \{z_2^2 - z_1^3 = a z_1^2 z_2\}$ for two different values of $a \in K^*$. The first blowup is the blowup of the origin; then we successively blow up the intersection of the exceptional divisor with the strict transform of the curves C_a. In the terminology of Sect. 7.3.2, the second and fourth blowups occur at free points whereas the third blowup is at a satellite point

0, so that Z is relatively nef. It is an important fact that the converse implication also holds:

Proposition 7.4 *If $Z \in \mathrm{Div}(\pi)$ is relatively nef, then the line bundle $\mathcal{O}_{X_\pi}(Z)$ is relatively base point free.*

Since $0 \in \mathbf{A}^2$ is a trivial example of a rational singularity, Proposition 7.4 is merely a special case of a result by Lipman, see [Lip69, Proposition 12.1 (ii)]. The proof in *loc. cit.* uses sheaf cohomology as well as the Zariski-Grothendieck theorem on formal functions, techniques that will not be exploited elsewhere in the paper. Here we outline a more elementary proof, taking advantage of $0 \in \mathbf{A}^2$ being a smooth point and working over an algebraically closed ground field.

Sketch of proof of Proposition 7.4 By the structure of the semigroup of relatively nef divisors, we may assume $Z = \check{E}$ for an exceptional prime E of π. Pick two distinct free points ξ_1, ξ_2 on E and formal curves $\tilde{C}_i$ at ξ_i, $i = 1, 2$, intersecting E transversely. Then $C_i := \pi(\tilde{C}_i)$, $i = 1, 2$ are formal curves at $0 \in \mathbf{A}^2$ satisfying $\pi^* C_i = \tilde{C}_i + G_i$, where $G_i \in \mathrm{Div}(\pi)$ is an exceptional divisor. Now $(\pi^* C_i \cdot F) = 0$ for every exceptional prime F of π, so $(G_i \cdot F) = -(\tilde{C}_i \cdot F) = -\delta_{EF} = (-\check{E} \cdot F)$. Since the intersection pairing on $\mathrm{Div}(\pi)$ is nondegenerate, this implies $G_i = -\check{E}$, that is, $\pi^* C_i = \tilde{C}_i - \check{E}$ for $i = 1, 2$.

Pick $\phi_i \in \hat{\mathcal{O}}_{\mathbf{A}^2,0}$ defining C_i. Then the ideal $\hat{\mathfrak{a}}$ generated by ϕ_1 and ϕ_2 is primary so the ideal $\mathfrak{a} := \hat{\mathfrak{a}} \cap \mathcal{O}_{\mathbf{A}^2,0}$ is also primary and satisfies $\mathfrak{a} \cdot \hat{\mathcal{O}}_{\mathbf{A}^2,0} = \hat{\mathfrak{a}}$. Since $\mathrm{ord}_F(\mathfrak{a}) = \mathrm{ord}_F(\phi_i) = -\mathrm{ord}_F(\check{E})$, $i = 1, 2$, for any exceptional prime F and the (formal) curves $\tilde{C}_i$ are disjoint, it follows that $\mathfrak{a} \cdot \mathcal{O}_{X_\pi} = \mathcal{O}_{X_\pi}(\check{E})$ as desired. □

7.4 Dual Graphs and Fans

To a blowup $\pi \in \mathfrak{B}_0$ we can associate two basic combinatorial objects, equipped with additional structure.

7.4.1 Dual Graph

First we have the classical notion of the *dual graph* $\Delta(\pi)$. This is an abstract simplicial complex of dimension one. Its vertices correspond to exceptional primes of π and its edges to proper intersections between exceptional primes. In the literature one often labels each vertex with the self-intersection number of the corresponding exceptional prime. We shall not do so here since this number is not an invariant of the corresponding divisorial valuation but depends also on the blowup π. From the point of view of these notes, it is more natural to use invariants such as the ones in Sect. 7.3.6.

The dual graph $\Delta(\pi)$ is connected and simply connected. This can be seen using the decomposition of π as a composition of point blowups, see Sect. 7.4.3. Alternatively, the connectedness of $\Delta(\pi)$ follows from Zariski's Main Theorem [Har77, p.280] and the simple connectedness can be deduced from sheaf cohomology considerations, see [Art66, Corollary 7].

See Fig. 12 for an example of a dual graph.

7.4.2 Dual Fan

While the dual graph $\Delta(\pi)$ is a natural object, the *dual fan* $\hat{\Delta}(\pi)$ is arguably more canonical. To describe it, we use basic notation and terminology from toric varieties, see [KKMS73, Ful93, Oda88].[22] Set

$$N(\pi) := \mathrm{Hom}(\mathrm{Div}(\pi), \mathbf{Z}).$$

If we label the exceptional primes E_i, $i \in I$, then we can write $N(\pi) = \bigoplus_{i \in I} \mathbf{Z} e_i \simeq \mathbf{Z}^I$ with e_i satisfying $\langle e_i, E_j \rangle = \delta_{ij}$. Note that if we identify $N(\pi)$

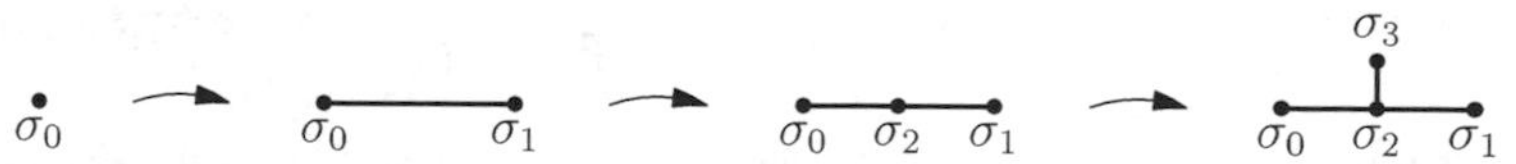

Fig. 12 The dual graphs of the blowups leading up to the log resolution of the primary ideal $\mathfrak{a} = (z_2^2 - z_1^3, z_1^2 z_2)$ described in Example 7.3 and depicted in Fig. 11. Here σ_i is the vertex corresponding to E_i

---

[22]We shall not, however, actually consider the toric variety defined by the fan $\hat{\Delta}(\pi)$.

with $\text{Div}(\pi)$ using the unimodularity of the intersection product (Proposition 7.2), then e_i corresponds to the divisor $\check{E}_i$ in Sect. 7.3.5.

Set $N_{\mathbf{R}}(\pi) := N(\pi) \otimes_{\mathbf{Z}} \mathbf{R} \simeq \mathbf{R}^I$. The one-dimensional cones in $\hat{\Delta}(\pi)$ are then of the form $\hat{\sigma}_i := \mathbf{R}_+ e_i$, $i \in I$, and the two-dimensional cones are of the form $\hat{\sigma}_{ij} := \mathbf{R}_+ e_i + \mathbf{R}_+ e_j$, where $i, j \in I$ are such that E_i and E_j intersect properly. Somewhat abusively, we will write $\hat{\Delta}(\pi)$ both for the fan and for its support (which is a subset of $N_{\mathbf{R}}(\pi)$).

Note that the dual fan $\hat{\Delta}(\pi)$ is naturally a cone over the dual graph $\Delta(\pi)$. In Sect. 7.4.6 we shall see how to embed the dual graph inside the dual fan.

A point $t \in \hat{\Delta}(\pi)$ is *irrational* if $t = t_1 e_1 + t_2 e_2$ with $t_i > 0$ and $t_1/t_2 \notin \mathbf{Q}$; otherwise t is *rational*. Note that the rational points are always dense in $\hat{\Delta}(\pi)$. The irrational points are also dense except if $\pi = \pi_0$, the simple blowup of $0 \in \mathbf{A}^2$.

7.4.3 Free and Satellite Blowups

Using the factorization of birational surface maps into simple point blowups, we can understand the structure of the dual graph and fan of a blowup $\pi \in \mathfrak{B}_0$.

First, when $\pi = \pi_0$ is a single blowup of the origin, there is a unique exceptional prime E_0, so $\hat{\Delta}(\pi_0)$ consists of a single, one-dimensional cone $\hat{\sigma}_0 = \mathbf{R}_+ e_0$ and $\Delta(\pi) = \{\sigma_0\}$ is a singleton.

Now suppose π' is obtained from π by blowing up a closed point $\xi \in \pi^{-1}(0)$. Let E_i, $i \in I$ be the exceptional primes of π. Write $I = \{1, 2, \ldots, n-1\}$, where $n \geq 2$. If $E_n \subseteq X_{\pi'}$ is the preimage of ξ, then the exceptional primes of π' are E_i, $i \in I'$, where $I' = \{1, 2, \ldots, n\}$. Recall that we are identifying an exceptional prime of π with its strict transform in $X_{\pi'}$.

To see what happens in detail, first suppose ξ is a free point, belonging to a unique exceptional prime of π, say E_1. In this case, the dual graph $\Delta(\pi')$ is obtained from $\Delta(\pi)$ by connecting a new vertex σ_n to σ_1. See Fig. 13.

If instead ξ is a satellite point, belonging to two distinct exceptional primes of π, say E_1 and E_2, then we obtain $\Delta(\pi')$ from $\Delta(\pi)$ by subdividing the edge σ_{12} into two edges σ_{1n} and σ_{2n}. Again see Fig. 13.

Fig. 13 Behavior of the dual graph under a single blowup. The left part of the picture illustrates the blowup of a free point on E_1, creating a new vertex σ_n connected to the vertex σ_1. The right part of the picture illustrates the blowup of the satellite point $E_1 \cap E_2$, creating a new vertex σ_n and subdividing the segment σ_{12} into two segments σ_{1n} and σ_{2n}

7.4.4 Integral Affine Structure

We define the *integral affine structure* on $\hat{\Delta}(\pi)$ to be the lattice

$$\text{Aff}(\pi) := \text{Hom}(N(\pi), \mathbf{Z}) \simeq \mathbf{Z}^I$$

and refer to its elements as integral affine functions. By definition, $\text{Aff}(\pi)$ can be identified with the group $\text{Div}(\pi)$ of exceptional divisors on X_π.

7.4.5 Projections and Embeddings

Consider blowups $\pi, \pi' \in \mathfrak{B}_0$ with $\pi \leq \pi'$, say $\pi' = \pi \circ \mu$, with $\mu : X_{\pi'} \to X_\pi$ a birational morphism. Then μ gives rise to an injective homomorphism $\mu^* : \text{Div}(\pi) \to \text{Div}(\pi')$ and we let

$$r_{\pi\pi'} : N(\pi') \to N(\pi)$$

denote its transpose. It is clear that $r_{\pi\pi'} \circ r_{\pi'\pi''} = r_{\pi\pi''}$ when $\pi \leq \pi' \leq \pi''$.

Lemma 7.5 *Suppose $\pi, \pi' \in \mathfrak{B}_0$ and $\pi \leq \pi'$. Then:*

(i) $r_{\pi\pi'}(\hat{\Delta}(\pi')) = \hat{\Delta}(\pi)$;
(ii) any irrational point in $\hat{\Delta}(\pi)$ has a unique preimage in $\hat{\Delta}(\pi')$;
(iii) if $\hat{\sigma}'$ is a 2-dimensional cone in $\hat{\Delta}(\pi)$ then either $r_{\pi\pi'}(\hat{\sigma}')$ is a one-dimensional cone in $\hat{\Delta}(\pi)$, or $r_{\pi\pi'}(\hat{\sigma}')$ is a 2-dimensional cone contained in a 2-dimensional cone $\hat{\sigma}$ of $\hat{\Delta}(\pi)$. In the latter case, the restriction of $r_{\pi\pi'}$ to $\hat{\sigma}'$ is unimodular in the sense that $r^*_{\pi\pi'} \text{Aff}(\pi)|_{\hat{\sigma}'} = \text{Aff}(\pi')|_{\hat{\sigma}'}$.

We use the following notation. If e_i is a basis element of $N(\pi)$ associated to an exceptional prime E_i, then e_i' denotes the basis element of $N(\pi')$ associated to the strict transform of E_i.

Proof It suffices to treat the case when $\pi' = \pi \circ \mu$, where μ is a single blowup of a closed point $\xi \in \pi^{-1}(0)$. As in Sect. 7.4.3 we let E_i, $i \in I$ be the exceptional primes of π. Write $I = \{1, 2, \ldots, n-1\}$, where $n \geq 2$. If $E_n \subseteq X_{\pi'}$ is the preimage of ξ, then the exceptional primes of π' are E_i, $i \in I'$, where $I' = \{1, 2, \ldots, n\}$.

First suppose $\xi \in E_1$ is a free point. Then $r_{\pi\pi'}(e_i') = e_i$ for $1 \leq i < n$ and $r_{\pi\pi'}(e_n') = e_1$. Conditions (i)–(iii) are immediately verified: $r_{\pi\pi'}$ maps the cone $\hat{\sigma}'_{1n}$ onto $\hat{\sigma}_1$ and maps all other cones $\hat{\sigma}'_{ij}$ onto the corresponding cones $\hat{\sigma}_{ij}$, preserving the integral affine structure.

Now suppose $\xi \in E_1 \cap E_2$ is a satellite point. The linear map $r_{\pi\pi'}$ is then determined by $r_{\pi\pi'}(e_i') = e_i$ for $1 \leq i < n$ and $r_{\pi\pi'}(e_n') = e_1 + e_2$. We see that the cones $\hat{\sigma}'_{1n}$ and $\hat{\sigma}'_{2n}$ in $\hat{\Delta}(\pi')$ map onto the subcones $\mathbf{R}_+ e_1 + \mathbf{R}_+(e_1 + e_2)$ and $\mathbf{R}_+ e_2 + \mathbf{R}_+(e_1 + e_2)$, respectively, of the cone $\hat{\sigma}_{12}$ in $\hat{\Delta}(\pi)$. Any other cone $\hat{\sigma}'_{ij}$

of $\hat{\Delta}(\pi')$ is mapped onto the corresponding cone $\hat{\sigma}_{ij}$ of $\hat{\Delta}(\pi)$, preserving the integral affine structure. Conditions (i)–(iii) follow. $\square$

Using Lemma 7.5 we can show that $r_{\pi\pi'}$ admits a natural one-side inverse.

Lemma 7.6 *Let $\pi, \pi' \in \mathfrak{B}_0$ be as above. Then there exists a unique continuous, homogeneous map $\iota_{\pi'\pi} : \hat{\Delta}(\pi) \to \hat{\Delta}(\pi')$ such that:*

(i) $r_{\pi\pi'} \circ \iota_{\pi'\pi} = \mathrm{id}$ on $\hat{\Delta}(\pi)$;
(ii) $\iota_{\pi'\pi}(e_i) = e'_i$ for all i.

Further, a two-dimensional cone $\hat{\sigma}'$ in $\hat{\Delta}(\pi')$ is contained in the image of $\iota_{\pi'\pi}$ iff $r_{\pi\pi'}(\hat{\sigma}')$ is two-dimensional.

It follows easily from the uniqueness statement that $\iota_{\pi''\pi} = \iota_{\pi''\pi'} \circ \iota_{\pi'\pi}$ when $\pi \le \pi' \le \pi''$. We emphasize that $\iota_{\pi'\pi}$ is only *piecewise* linear and not the restriction to $\hat{\Delta}(\pi)$ of a linear map $N_{\mathbf{R}}(\pi) \to N_{\mathbf{R}}(\pi')$.

Proof of Lemma 7.6 Uniqueness is clear: when $\pi = \pi_0$ is the simple blowup of $0 \in \mathbf{A}^2$, $\iota_{\pi'\pi}$ is determined by (ii) and when $\pi \ne \pi_0$, the irrational points are dense in $\hat{\Delta}(\pi)$ and uniqueness is a consequence of Lemma 7.5 (ii).

As for existence, it suffices to treat the case when $\pi' = \pi \circ \mu$, where μ is a simple blowup of a closed point $\xi \in \pi^{-1}(0)$.

When $\xi \in E_1$ is a free point, $\iota_{\pi'\pi}$ maps e_i to e'_i for $1 \le i < n$ and maps any cone $\hat{\sigma}_{ij}$ in $\hat{\Delta}(\pi)$ onto the corresponding cone $\hat{\sigma}'_{ij}$ in $\hat{\Delta}(\pi')$ linearly via $\iota_{\pi'\pi}(t_i e_i + t_j e_j) = (t_i e'_i + t_j e'_j)$.

If instead $\xi \in E_1 \cap E_2$ is a satellite point, then $\iota_{\pi'\pi}(e_i) = e'_i$ for $1 \le i < n$. Further, $\iota_{\pi'\pi}$ is piecewise linear on the cone $\hat{\sigma}_{12}$:

$$\iota_{\pi'\pi}(t_1 e_1 + t_2 e_2) = \begin{cases} (t_1 - t_2)e'_1 + t_2 e'_n & \text{if } t_1 \ge t_2 \\ (t_2 - t_1)e'_2 + t_1 e'_n & \text{if } t_1 \le t_2 \end{cases} \tag{39}$$

and maps any other two-dimensional cone $\hat{\sigma}_{ij}$ onto $\hat{\sigma}'_{ij}$ linearly via $\iota_{\pi'\pi}(t_i e_i + t_j e_j) = (t_i e'_i + t_j e'_j)$. $\square$

7.4.6 Embedding the Dual Graph in the Dual Fan

We have noted that $\hat{\Delta}(\pi)$ can be viewed as a cone over $\Delta(\pi)$. Now we embed $\Delta(\pi)$ in $\hat{\Delta}(\pi) \subseteq N_{\mathbf{R}}$, in a way that remembers the maximal ideal $\mathfrak{m}_0$. For $i \in I$ define an integer $b_i \ge 1$ by

$$b_i := \mathrm{ord}_{E_i}(\mathfrak{m}_0),$$

where ord_{E_i} is the divisorial valuation given by order of vanishing along E_i. There exists a unique function $\varphi_0 \in \mathrm{Aff}(\pi)$ such that $\varphi_0(e_i) = b_i$. It is the integral affine function corresponding to the exceptional divisor $-Z_0 \in \mathrm{Div}(\pi)$, where $Z_0 =$

$-\sum_{i \in I} b_i E_i$. Note that π is a log resolution of the maximal ideal $\mathfrak{m}_0$ and that $\mathfrak{m}_0 \cdot \mathcal{O}_{X_\pi} = \mathcal{O}_{X_\pi}(Z_0)$.

We now define $\Delta(\pi)$ as the subset of $\hat{\Delta}(\pi)$ given by $\varphi_0 = 1$. In other words, the vertices of $\Delta(\pi)$ are of the form

$$\sigma_i := \hat{\sigma}_i \cap \Delta(\pi) = b_i^{-1} e_i$$

and the edges of the form

$$\sigma_{ij} := \hat{\sigma}_{ij} \cap \Delta(\pi) = \{t_i e_i + t_j e_j \mid t_i, t_j \geq 0, b_i t_i + b_j t_j = 1\}.$$

If $\pi, \pi' \in \mathfrak{B}_0$ and $\pi' \geq \pi$, then $r_{\pi\pi'}(\Delta(\pi')) = \Delta(\pi)$ and $\iota_{\pi'\pi}(\Delta(\pi)) \subseteq \Delta(\pi')$.

7.4.7 Auxiliary Calculations

For further reference let us record a few calculations involving the numerical invariants A, α and b above.

If $\pi_0 \in \mathfrak{B}_0$ is the simple blowup of the origin, then

$$A_{E_0} = 2, \quad b_{E_0} = 1, \quad \check{E}_0 = -E_0 \quad \text{and} \quad \alpha_{E_0} = -1.$$

Now suppose $\pi' = \pi \circ \mu$, where μ is the simple blowup of a closed point ξ and let us check how the numerical invariants behave. We use the notation of Sect. 7.4.3. In the case of a free blowup we have

$$A_{E_n} = A_{E_1} + 1, \quad b_{E_n} = b_{E_1} \quad \text{and} \quad \check{E}_n = \check{E}_1 - E_n, \tag{40}$$

where, in the right hand side, we identify the divisor $\check{E}_1 \in \mathrm{Div}(\pi)$ with its pullback in $\mathrm{Div}(\pi')$. Since $(E_n \cdot E_n) = -1$ we derive as a consequence,

$$\alpha_{E_n} := (\check{E}_n \cdot \check{E}_n) = (\check{E}_1 \cdot \check{E}_1) - 1 = \alpha_{E_1} - 1. \tag{41}$$

In the case of a satellite blowup,

$$A_{E_n} = A_{E_1} + A_{E_2}, \quad b_{E_n} = b_{E_1} + b_{E_2} \quad \text{and} \quad \check{E}_n = \check{E}_1 + \check{E}_2 - E_n. \tag{42}$$

Using $(E_n \cdot E_n) = -1$ this implies

$$\alpha_{E_n} := \alpha_{E_1} + \alpha_{E_2} + 2(\check{E}_1 \cdot \check{E}_2) - 1. \tag{43}$$

We also claim that if E_i, E_j are exceptional primes that intersect properly in some X_π, then

$$((b_i \check{E}_j - b_j \check{E}_i) \cdot (b_i \check{E}_j - b_j \check{E}_i)) = -b_i b_j. \tag{44}$$

Note that both sides of (44) are independent of the blowup $\pi \in \mathfrak{B}_0$ but we have to assume that E_i and E_j intersect properly in *some* blowup.

To prove (44), we proceed inductively. It suffices to consider the case when E_i is obtained by blowing up a closed point $\xi \in E_j$. When ξ is free, we have $b_i = b_j$, $\check{E}_i = \check{E}_j - E_i$ and (44) reduces to the fact that $(E_i \cdot E_i) = -1$. When instead $\xi \in E_j \cap E_k$ is a satellite point, we have $((b_i \check{E}_k - b_k \check{E}_i) \cdot (b_i \check{E}_k - b_k \check{E}_i)) = -b_i b_k$ by induction. Furthermore, $b_i = b_j + b_k$, $\check{E}_i = \check{E}_j + \check{E}_k - E_i$; we obtain (44) from these equations and from simple algebra.

In the dual graph depicted in Fig. 12 we have $b_0 = b_1 = 1$, $b_2 = b_3 = 2$, $\alpha_0 = -1$, $\alpha_1 = -2$, $\alpha_2 = -6$, $\alpha_3 = -7$, $A_0 = 2$, $A_1 = 3$, $A_2 = 5$ and $A_3 = 6$.

7.4.8 Extension of the Numerical Invariants

We extend the numerical invariants A and α in Sect. 6.9 to functions on the dual fan

$$A_\pi : \hat{\Delta}(\pi) \to \mathbf{R}_+ \quad \text{and} \quad \alpha_\pi : \hat{\Delta}(\pi) \to \mathbf{R}_-$$

as follows. First we set $A_\pi(e_i) = A_{E_i}$ and extend A_π uniquely as an (integral) linear function on $\hat{\Delta}(\pi)$. Thus we set $A_\pi(t_i e_i) = t_i A_\pi(e_i)$ and

$$A_\pi(t_i e_i + t_j e_j) = t_i A_\pi(e_i) + t_j A_\pi(e_j). \tag{45}$$

In particular, A_π is integral affine on each simplex in the dual graph $\Delta(\pi)$.

Second, we set $\alpha_\pi(e_i) = \alpha_{E_i} = (\check{E}_i \cdot \check{E}_i)$ and extend α_π as a homogeneous function of order *two* on $\hat{\Delta}(\pi)$ which is affine on each simplex in the dual graph $\Delta(\pi)$. In other words, we set $\alpha_\pi(t_i e_i) = t_i^2 \alpha_\pi(e_i)$ for any $i \in I$ and

$$\alpha_\pi(t_i e_i + t_j e_j) = (b_i t_i + b_j t_j)^2 \left(\frac{b_i t_l}{b_i t_i + b_j t_j} \alpha_\pi(\sigma_i) + \frac{b_j t_j}{b_i t_i + b_j t_j} \alpha_\pi(\sigma_j) \right) \tag{46}$$

$$= (b_i t_i + b_j t_j) \left(\frac{t_i}{b_i} \alpha_\pi(e_i) + \frac{t_j}{b_j} \alpha_\pi(e_j) \right)$$

whenever E_i and E_j intersect properly.

Let us check that

$$A_{\pi'} \circ \iota_{\pi'\pi} = A_\pi \quad \text{and} \quad \alpha_{\pi'} \circ \iota_{\pi'\pi} = \alpha_\pi$$

on $\hat{\Delta}(\pi)$ whenever $\pi' \geq \pi$. It suffices to do this when $\pi' = \pi \circ \mu$ and μ is the blowup of X_π at a closed point ξ. Further, the only case that requires verification is when $\xi \in E_1 \cap E_2$ is a satellite point, in which case it suffices to prove $A_\pi(e_1 + e_2) = A_{\pi'}(e_n')$ and $\alpha_\pi(e_1 + e_2) = \alpha_{\pi'}(e_n')$. The first of these formulas follows from (42) and (45) whereas the second results from (43), (44) and (46). The details are left to the reader.

In the dual graph depicted in Fig. 12 we have $A_\pi(\sigma_0) = 2$, $A_\pi(\sigma_1) = 3$, $A_\pi(\sigma_2) = 5/2$, $A_\pi(\sigma_3) = 3$, $\alpha_\pi(\sigma_0) = -1$, $\alpha_\pi(\sigma_1) = -2$, $\alpha_\pi(\sigma_2) = -3/2$, and $\alpha_\pi(\sigma_3) = -7/4$.

7.4.9 Multiplicity of Edges in the Dual Graph

We define the *multiplicity* $m(\sigma)$ of an edge σ in a dual graph $\Delta(\pi)$ as follows. Let $\sigma = \sigma_{ij}$ have endpoints $v_i = b_i^{-1} e_i$ and $v_j = b_j^{-1} e_j$. We set

$$m(\sigma_{ij}) := \gcd(b_i, b_j). \tag{47}$$

Let us see what happens when π' is obtained from π by blowing up a closed point $\xi \in \pi^{-1}(0)$. We use the notation above. See also Fig. 13.

If $\xi \in E_1$ is a free point, then we have seen in (40) that $b_n = b_1$ and hence

$$m(\sigma_{1n}) = b_1. \tag{48}$$

If instead $\xi \in E_1 \cap E_2$ is a satellite point, then (42) gives $b_n = b_1 + b_2$ and hence

$$m(\sigma_{1n}) = m(\sigma_{2n}) = m(\sigma_{12}). \tag{49}$$

This shows that the multiplicity does not change when subdividing a segment.

In the dual graph depicted in Fig. 12 we have $m_{02} = m_{12} = 1$ and $m_{23} = 2$.

7.4.10 Metric on the Dual Graph

Having embedded $\Delta(\pi)$ inside $\hat{\Delta}(\pi)$, the integral affine structure $\mathrm{Aff}(\pi)$ gives rise to an abelian group of functions on $\Delta(\pi)$ by restriction. Following [KKMS73, p.95], this further induces a volume form on each simplex in $\Delta(\pi)$. In our case, this simply means a metric on each edge σ_{ij}. The length of σ_{ij} is the largest positive number l_{ij} such that $\varphi(\sigma_i) - \varphi(\sigma_j)$ is an integer multiple of l_{ij} for all $\varphi \in \mathrm{Aff}(\pi)$. From this description it follows that $l_{ij} = \mathrm{lcm}(b_i, b_j)^{-1}$.

However, it turns out that the "correct" metric for doing potential theory is the one for which

$$d_\pi(\sigma_i, \sigma_j) = \frac{1}{b_i b_j} = \frac{1}{m_{ij}} \cdot \frac{1}{\mathrm{lcm}(b_i, b_j)}, \tag{50}$$

where $m_{ij} = \gcd(b_i, b_j)$ is the multiplicity of the edge σ_{ij} as in Sect. 7.4.9.

We have seen that the dual graph is connected and simply connected. It follows that $\Delta(\pi)$ is a metric tree. The above results imply that if $\pi, \pi' \in \mathfrak{B}_0$ and $\pi' \geq \pi$, then $\iota_{\pi'\pi} : \Delta(\pi) \hookrightarrow \Delta(\pi')$ is an isometric embedding.

Let us see more concretely what happens when π' is obtained from π by blowing up a closed point $\xi \in \pi^{-1}(0)$. We use the notation above.

If $\xi \in E_1$ is a free point, then $b_n = b_1$ and the dual graph $\Delta(\pi')$ is obtained from $\Delta(\pi)$ by connecting a new vertex σ_n to σ_1 using an edge of length b_1^{-2}. See Fig. 13.

If instead $\xi \in E_1 \cap E_2$ is a satellite point, then $b_n = b_1 + b_2$ and we obtain $\Delta(\pi')$ from $\Delta(\pi)$ by subdividing the edge σ_{12}, which is of length $\frac{1}{b_1 b_2}$ into two edges σ_{1n} and σ_{2n}, of lengths $\frac{1}{b_1(b_1+b_2)}$ and $\frac{1}{b_2(b_1+b_2)}$, respectively. Note that these lengths add up to $\frac{1}{b_1 b_2}$. Again see Fig. 13.

In the dual graph depicted in Fig. 12 we have $d(\sigma_0, \sigma_2) = d(\sigma_1, \sigma_2) = 1/2$ and $d(\sigma_2, \sigma_3) = 1/4$.

7.4.11   Rooted Tree Structure

The dual graph $\Delta(\pi)$ is a tree in the sense of Sect. 2.1. We turn it into a rooted tree by declaring the root to be the vertex σ_0 corresponding to the strict transform of E_0, the exceptional prime of π_0, the simple blowup of 0.

When restricted to the dual graph, the functions α_π and A_π on the dual fan $\hat{\Delta}(\pi)$ described in Sect. 7.4.8 define parametrizations

$$\alpha_\pi : \Delta(\pi) \rightarrow]-\infty, -1] \quad \text{and} \quad A_\pi : \Delta(\pi) \rightarrow [2, \infty[\tag{51}$$

satisfying $A_{\pi'} \circ \iota_{\pi'\pi} = A_\pi$ and $\alpha_{\pi'} \circ \iota_{\pi'\pi} = \alpha_\pi$ whenever $\pi' \geq \pi$.

We claim that α_π induces the metric on the dual graph given by (50). For this, it suffices to show that $|\alpha_\pi(\sigma_i) - \alpha_\pi(\sigma_j)| = \frac{1}{b_i b_j}$ when E_i, E_j are exceptional primes intersecting properly. In fact, it suffices to verify this when E_i is obtained by blowing up a free point on E_j. But then $b_i = b_j$ and it follows from (41) that

$$\alpha_\pi(\sigma_i) - \alpha_\pi(\sigma_j) = b_i^{-2}(\alpha_{E_i} - \alpha_{E_j}) = -b_i^{-2} = -d(\sigma_i, \sigma_j).$$

In a similar way we see that the parametrization A_π of $\Delta(\pi)$ induces by the log discrepancy gives rise to the metric induced by the integral affine structure as in Sect. 7.4.10. In other words, if E_i, E_j are exceptional primes of X_π intersecting properly, then

$$A(\sigma_j) - A(\sigma_i) = -m_{ij}(\alpha(\sigma_j) - \alpha(\sigma_i)), \tag{52}$$

where $m_{ij} = \gcd(b_i, b_j)$ is the multiplicity of the edge σ_{ij}.

7.5 Valuations and Dual Graphs

Now we shall show how to embed the dual graph into the valuative tree.

7.5.1 Center

It follows from the valuative criterion of properness that any semivaluation $v \in \hat{\mathcal{V}}_0^*$ admits a *center* on X_π, for any blowup $\pi \in \mathfrak{B}_0$. The center is the unique (not necessarily closed) point $\xi = c_\pi(v) \in X_\pi$ such that $v \geq 0$ on the local ring $\mathcal{O}_{X_\pi,\xi}$ and such that $\{v > 0\} \cap \mathcal{O}_{X_\pi,\xi}$ equals the maximal ideal $\mathfrak{m}_{X_\pi,\xi}$. If $\pi' \geq \pi$, then the map $X_{\pi'} \to X_\pi$ sends $c_{\pi'}(v)$ to $c_\pi(v)$.

7.5.2 Evaluation

Consider a semivaluation $v \in \hat{\mathcal{V}}_0^*$ and a blowup $\pi \in \mathfrak{B}_0$. We can evaluate v on exceptional divisors $Z \in \mathrm{Div}(\pi)$. Concretely, if $Z = \sum_{i \in I} r_i E_i$, $\xi = c_\pi(v)$ is the center of v on X_π and E_j, $j \in J$ are the exceptional primes containing ξ, then $v(Z) = \sum_{j \in J} r_j v(\zeta_j)$, where $\zeta_j \in \mathcal{O}_{X_\pi,\xi}$ and $E_j = \{\zeta_j = 0\}$.

This gives rise to an *evaluation map*

$$\mathrm{ev}_\pi : \hat{\mathcal{V}}_0^* \to N_{\mathbf{R}}(\pi) \tag{53}$$

that is continuous, more or less by definition. The image of ev_π is contained in the dual fan $\hat{\Delta}(\pi)$. Furthermore, the embedding of the dual graph $\Delta(\pi)$ in the dual fan $\hat{\Delta}(\pi)$ was exactly designed so that $\mathrm{cv}_\pi(\mathcal{V}_0) \subseteq \Delta(\pi)$. In fact, we will see shortly that these inclusions are equalities.

It follows immediately from the definitions that

$$r_{\pi\pi'} \circ \mathrm{ev}_{\pi'} = \mathrm{ev}_\pi \tag{54}$$

when $\pi' \geq \pi$.

Notice that if the center of $v \in \hat{\mathcal{V}}_0^*$ on X_π is the generic point of $\bigcap_{i \in J} E_i$, then $\mathrm{ev}_\pi(v)$ lies in the relative interior of the cone $\sum_{i \in J} \mathbf{R}_+ e_i$.

7.5.3 Embedding and Quasimonomial Valuations

Next we construct a one-sided inverse to the evaluation map in (53).

Lemma 7.7 *Let* $\pi \in \mathfrak{B}_0$ *be a blowup. Then there exists a unique continuous map* $\mathrm{emb}_\pi : \hat{\Delta}^*(\pi) \to \hat{\mathcal{V}}_0^*$ *such that:*

(i) $\mathrm{ev}_\pi \circ \mathrm{emb}_\pi = \mathrm{id}$ *on* $\hat{\Delta}^*(\pi)$;
(ii) for $t \in \hat{\Delta}^*(\pi)$, *the center of* $\mathrm{emb}_\pi(t)$ *is the generic point of the intersection of all exceptional primes* E_i *of* π *such that* $\langle t, E_i \rangle > 0$.

Furthermore, condition (ii) is superfluous except in the case when $\pi = \pi_0$ *is a simple blowup of* $0 \in \mathbf{A}^2$ *in which case the dual graph* $\Delta(\pi)$ *is a singleton.*

As a consequence of (i), $\mathrm{emb}_\pi : \hat{\Delta}^*(\pi) \to \hat{\mathcal{V}}_0^*$ is injective and $\mathrm{ev}_\pi : \hat{\mathcal{V}}_0^* \to \hat{\Delta}^*(\pi)$ surjective.

Corollary 7.8 *If* $\pi, \pi' \in \mathfrak{B}_0$ *and* $\pi' \geq \pi$, *then* $\mathrm{emb}_{\pi'} \circ \iota_{\pi'\pi} = \mathrm{emb}_\pi$.

As in Sect. 6.10 we say that a valuation $v \in \hat{\mathcal{V}}_0^*$ is *quasimonomial* if it lies in the image of emb_π for some blowup $\pi \in \mathfrak{B}_0$. By Corollary 7.8, v then lies in the image of $\mathrm{emb}_{\pi'}$ for all $\pi' \geq \pi$.

Proof of Corollary 7.8 We may assume $\pi' \neq \pi$ so that π' is not the simple blowup of $0 \in \mathbf{A}^2$. The map $\mathrm{emb}_\pi' := \mathrm{emb}_{\pi'} \circ \iota_{\pi'\pi} : \Delta(\pi) \to \mathcal{V}_0$ is continuous and satisfies

$$\mathrm{ev}_\pi \circ \mathrm{emb}_\pi' = r_{\pi\pi'} \circ \mathrm{ev}_{\pi'} \circ \mathrm{emb}_{\pi'} \circ \iota_{\pi'\pi} = r_{\pi\pi'} \circ \iota_{\pi'\pi} = \mathrm{id}.$$

By Lemma 7.7 this implies $\mathrm{emb}_\pi' = \mathrm{emb}_\pi$.   □

Proof of Lemma 7.7 We first prove existence. Consider a point $t = \sum_{i \in I} t_i e_i \in \hat{\Delta}^*(\pi)$ and let $J \subseteq I$ be the set of indices i such that $t_i > 0$. Let ξ be the generic point of $\bigcap_{i \in J} E_i$ and write $E_i = (\zeta_i = 0)$ in local algebraic coordinates ζ_i, $i \in J$ at ξ. Then we let $\mathrm{emb}_\pi(t)$ be the monomial valuation with weights t_i on ζ_i as in Sect. 6.10. More concretely, after relabeling we may assume that either $J = \{1\}$ is a singleton and $\mathrm{emb}_\pi(t) = t_1 \mathrm{ord}_{E_1}$ is a divisorial valuation, or $J = \{1, 2\}$ in which case v_t is defined on $R \subseteq \hat{\mathcal{O}}_{X_\pi, \xi} \simeq K[\![\zeta_1, \zeta_2]\!]$ by

$$\mathrm{emb}_\pi(t)\Big(\sum_{\beta_1, \beta_2 \geq 0} c_{\beta_1 \beta_2} \zeta_1^{\beta_1} \zeta_2^{\beta_2} \Big) = \min\{t_1 \beta_1 + t_2 \beta_2 \mid c_\beta \neq 0\}. \tag{55}$$

It is clear that emb_π is continuous and that $\mathrm{ev}_\pi \circ \mathrm{emb}_\pi = \mathrm{id}$.

The uniqueness statement is clear when $\pi = \pi_0$ since the only valuation whose center on X_π is the generic point of the exceptional divisor E_0 is proportional to $\mathrm{ord}_{E_0} = \mathrm{ord}_0$.

Now suppose $\pi \neq \pi_0$ and that $\mathrm{emb}_\pi' : \hat{\Delta}^*(\pi) \to \hat{\mathcal{V}}_0^*$ is another continuous map satisfying $\mathrm{ev}_\pi \circ \iota_\pi = \mathrm{id}$. It suffices to show that $\mathrm{emb}_\pi'(t) = \mathrm{emb}_\pi(t)$ for any irrational $t \in \hat{\Delta}^*(\pi)$. But if t is irrational, the value of $\mathrm{emb}_\pi'(t)$ on a monomial $\zeta_1^{\beta_1} \zeta_2^{\beta_2}$ is $t_1 \beta_1 + t_2 \beta_2$. In particular, the values on distinct monomials are distinct, so it follows that the value of $\mathrm{emb}_\pi'(t)$ on a formal power series is given as in (55). Hence $\mathrm{emb}_\pi'(t) = \mathrm{emb}_\pi(t)$, which completes the proof.

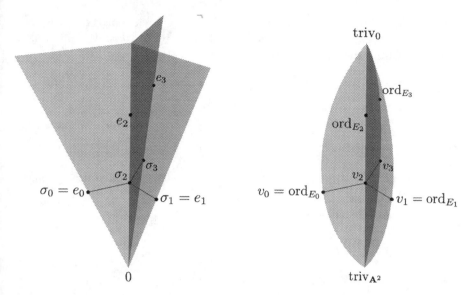

Fig. 14 The dual fan of a blowup. The picture on the left illustrates the dual fan $\hat{\Delta}(\pi)$, where π is the log resolution illustrated in Fig. 11. The picture on the left illustrates the closure of the embedding of the dual fan inside the Berkovich affine plane. The line segments illustrate the dual graph $\Delta(\pi)$ and its embedding inside the valuative tree $\mathcal{V}_0$

In particular the divisorial valuation in $\mathcal{V}_0$ associated to the exceptional prime E_i is given by

$$v_i := b_i^{-1} \operatorname{ord}_{E_i} \quad \text{where} \quad b_i := \operatorname{ord}_{E_i}(\mathfrak{m}_0) \in \mathbf{N} \qquad \square$$

The embedding $\operatorname{emb}_\pi : \hat{\Delta}^*(\pi) \hookrightarrow \hat{\mathcal{V}}_0^* \subseteq \mathbf{A}^2_{\mathrm{Berk}}$ extends to the full cone fan $\hat{\Delta}(\pi)$ and maps the apex $0 \in \hat{\Delta}(\pi)$ to the trivial valuation $\operatorname{triv}_{\mathbf{A}^2}$ on R. The boundary of $\operatorname{emb}_\pi : \hat{\Delta}^*(\pi)$ inside $\mathbf{A}^2_{\mathrm{Berk}}$ consists of $\operatorname{triv}_{\mathbf{A}^2}$ and the semivaluation triv_0. Thus $\operatorname{emb}_\pi(\hat{\Delta}(\pi))$ looks like a "double cone". See Fig. 14.

7.5.4 Structure Theorem

Because of (54), the evaluation maps ev_π induce a continuous map

$$\operatorname{ev} : \mathcal{V}_0 \to \varprojlim_\pi \Delta(\pi), \tag{56}$$

where the right hand side is equipped with the inverse limit topology. Similarly, the embeddings emb_π define an embedding

$$\operatorname{emb} : \varinjlim_\pi \Delta(\pi) \to \mathcal{V}_0, \tag{57}$$

where the direct limit is defined using the maps $\iota_{\pi'\pi}$ and is equipped with the direct limit topology. The direct limit is naturally a dense subset of the inverse limit and under this identification we have ev $\circ$ emb $=$ id.

Theorem 7.9 *The map* ev $: \mathcal{V}_0 \to \varprojlim \Delta(\pi)$ *is a homeomorphism.*

By homogeneity, we also obtain a homeomorphism ev $: \hat{\mathcal{V}}_0^* \to \varprojlim \hat{\Delta}^*(\pi)$.

Proof Since r is continuous and both sides of (56) are compact, it suffices to show that r is bijective. The image of r contains the dense subset $\varinjlim \Delta(\pi)$ so surjectivity is clear.

To prove injectivity, pick $v, w \in \mathcal{V}_0$ with $v \neq w$. Then there exists a primary ideal $\mathfrak{a} \subseteq R$ such that $v(\mathfrak{a}) \neq w(\mathfrak{a})$. Let $\pi \in \mathfrak{B}_0$ be a log resolution of $\mathfrak{a}$ and write $\mathfrak{a} \cdot \mathcal{O}_{X_\pi} = \mathcal{O}_{X_\pi}(Z)$, where $Z \in \mathrm{Div}(\pi)$. Then

$$\langle \mathrm{ev}_\pi(v), Z \rangle = -v(\mathfrak{a}) \neq -w(\mathfrak{a}) = \mathrm{ev}_\pi(Z) = \langle \mathrm{ev}_\pi(w), Z \rangle,$$

so that $\mathrm{ev}_\pi(v) \neq \mathrm{ev}_\pi(w)$ and hence $\mathrm{ev}(v) \neq \mathrm{ev}(w)$. $\qquad\square$

7.5.5 Integral Affine Structure

We set

$$\mathrm{Aff}(\hat{\mathcal{V}}_0^*) = \varinjlim_\pi \mathrm{ev}_\pi^* \, \mathrm{Aff}(\pi).$$

Thus a function $\varphi : \hat{\mathcal{V}}_0^* \to \mathbf{R}$ is integral affine iff it is of the form $\varphi = \varphi_\pi \circ \mathrm{ev}_\pi$, with $\varphi_\pi \in \mathrm{Aff}(\pi)$. In other words, φ is defined by an exceptional divisor in some blowup.

7.6 Tree Structure on $\mathcal{V}_0$

Next we use Theorem 7.9 to equip $\mathcal{V}_0$ with a tree structure.

7.6.1 Metric Tree Structure

The metric on a dual graph $\Delta(\pi)$ defined in Sect. 7.4.10 turns this space into a finite metric tree in the sense of Sect. 2.2. Further, if $\pi' \geq \pi$, then the embedding $\iota_{\pi'\pi} : \Delta(\pi) \hookrightarrow \Delta(\pi')$ is an isometry. It then follows from the discussion in Sect. 2.2.2 that $\mathcal{V}_0 \simeq \varprojlim \Delta(\pi)$ is a metric tree.

Lemma 7.10 *The ends of $\mathcal{V}_0$ are exactly the valuations that are not quasimonomial.*

Proof The assertion in the lemma amounts to the ends of the tree $\varprojlim \Delta(\pi)$ being exactly the points that do not belong to any single dual graph. It is clear that all points of the latter type are ends. On the other hand, if $t \in \Delta(\pi)$ for some blowup π, then there exists a blowup $\pi' \in \mathfrak{B}_0$ dominating π such that $\iota_{\pi'\pi}(t)$ is not an end of $\Delta(\pi')$. When t is already not an endpoint of $\Delta(\pi)$, this is clear. Otherwise $t = b_i^{-1} e_i$, in which case π' can be chosen as the blowup of a free point on the associated exceptional prime E_i. $\qquad\square$

The hyperbolic space $\mathbf{H} \subseteq \mathcal{V}_0$ induced by the generalized metric on $\mathcal{V}_0$ contains all quasimonomial valuations but also some non-quasimonomial ones, see Sect. 7.7.5.

7.6.2 Rooted Tree Structure

We choose the valuation ord_0 as the root of the tree $\mathcal{V}_0$ and write $\leq$ for the corresponding partial ordering.

The two parametrizations α_π and A_π on the dual graph $\Delta(\pi)$ in Sect. 7.4.11 give rise to parametrizations[23]

$$\alpha : \mathcal{V}_0 \to [-\infty, -1] \quad \text{and} \quad A : \mathcal{V}_0 \to [2, \infty]. \tag{58}$$

The parametrization α gives rise to the generalized metric on $\mathcal{V}_0$ and we have

$$\alpha(v) = -(1 + d(v, \mathrm{ord}_0)). \tag{59}$$

The choice of parametrization will be justified in Sect. 7.8.1. Note that hyperbolic space $\mathbf{H} \subseteq \mathcal{V}_0$ is given by $\mathbf{H} = \{\alpha > -\infty\}$.

There is also a unique, lower semicontinuous *multiplicity* function

$$m : \mathcal{V}_0 \to \mathbf{N} \cup \{\infty\}$$

on $\mathcal{V}_0$ induced by the multiplicity on dual graphs. It has the property that $m(w)$ divides $m(v)$ if $w \leq v$. The two parametrizations α and A are related through the multiplicity by

$$A(v) = 2 + \int_{\mathrm{ord}_0}^{v} m(w) \, d\alpha(w);$$

this follows from (52).

There is also a generalized metric induced by A, but we shall not use it.

---

[23]The increasing parametrization $-\alpha$ is denoted by α and called *skewness* in [FJ04]. The increasing parametrization A is called *thinness* in *loc. cit.* .

7.6.3 Retraction

It will be convenient to regard the dual graph and fan as subsets of the valuation spaces $\mathcal{V}_0$ and $\hat{\mathcal{V}}_0$, respectively. To this end, we introduce

$$|\Delta(\pi)| := \mathrm{emb}_\pi(\Delta(\pi)) \quad \text{and} \quad |\hat{\Delta}^*(\pi)| := \mathrm{emb}_\pi(\hat{\Delta}^*(\pi)).$$

Note that if $\pi' \geq \pi$, then $|\hat{\Delta}^*(\pi)| \subseteq |\hat{\Delta}^*(\pi')|$.

The evaluation maps now give rise to *retractions*

$$r_\pi := \mathrm{emb}_\pi \circ \mathrm{ev}_\pi$$

of $\hat{\mathcal{V}}_0^*$ and $\mathcal{V}_0$ onto $|\hat{\Delta}_0^*|$ and $|\Delta(\pi)|$, respectively. It is not hard to see that $r_{\pi'} \circ r_\pi = r_\pi$ when $\pi' \geq \pi$.

Let us describe the retraction in more detail. Let $\xi = c_\pi(v)$ be the center of v on X_π and let $E_i, i \in J$ be the exceptional primes containing ξ. Write $E_i = (\zeta_i = 0)$ in local algebraic coordinates ζ_i at ξ and set $t_i = v(\zeta_i) > 0$. Then $w := r_\pi(v) \in |\hat{\Delta}^*(\pi)|$ is the monomial valuation such that $w(\zeta_i) = t_i, i \in J$.

It follows from Theorem 7.9 that

$$r_\pi \to \mathrm{id} \quad \text{as } \pi \to \infty.$$

In fact, we have the following more precise result.

Lemma 7.11 *If $v \in \hat{\mathcal{V}}_0^*$ and $\pi \in \mathfrak{B}_0$ is a blowup, then*

$$(r_\pi v)(\mathfrak{a}) \leq v(\mathfrak{a})$$

for every ideal $\mathfrak{a} \subseteq R$, with equality if the strict transform of $\mathfrak{a}$ to X_π does not vanish at the center of v on X_π. In particular, equality holds if $\mathfrak{a}$ is primary and π is a log resolution of $\mathfrak{a}$.

Proof Pick $v \in \hat{\mathcal{V}}_0^*$ and set $w = r_\pi(v)$. Let ξ be the center of v on X_π and $E_i = (\zeta_i = 0), i \in J$, the exceptional primes of π containing ξ. By construction, w is the smallest valuation on $\hat{\mathcal{O}}_{X_\pi, \xi}$ taking the same values as v on the ζ_i. Thus $w \leq v$ on $\hat{\mathcal{O}}_{X_\pi, \xi} \supseteq R$, which implies $w(\mathfrak{a}) \leq v(\mathfrak{a})$ for all ideals $\mathfrak{a} \subseteq R$.

Moreover, if the strict transform of $\mathfrak{a}$ to X_π does not vanish at ξ, then $\mathfrak{a} \cdot \hat{\mathcal{O}}_{X_\pi, \xi}$ is generated by a single monomial in the ζ_i, and then it is clear that $v(\mathfrak{a}) = w(\mathfrak{a})$. $\square$

7.7 Classification of Valuations

Similarly to points in the Berkovich affine line, we can classify semivaluations in the valuative tree into four classes. The classification is discussed in detail in [FJ04] but already appears in a slightly different form in the work of Spivakovsky [Spi90].

One can show that the set of semivaluations of each of the four types below is dense in $\mathcal{V}_0$, see [FJ04, Proposition 5.3].

Recall that any semivaluation $v \in \hat{\mathcal{V}}_0^*$ extends to the fraction field F of R. In particular, it extends to the local ring $\mathcal{O}_0 := \mathcal{O}_{\mathbf{A}^2,0}$. Since $v(\mathfrak{m}_0) > 0$, v also defines a semivaluation on the completion $\hat{\mathcal{O}}_0$.

7.7.1 Curve Semivaluations

The subset $\mathfrak{p} := \{v = \infty\} \subsetneq \hat{\mathcal{O}}_0$ is a prime ideal and v defines a valuation on the quotient ring $\hat{\mathcal{O}}_0/\mathfrak{p}$. If $\mathfrak{p} \neq 0$, then $\hat{\mathcal{O}}_0/\mathfrak{p}$ is principal and we say that v is a *curve semivaluation* as $v(\phi)$ is proportional to the order of vanishing at 0 of the restriction of ϕ to the formal curve defined by $\mathfrak{p}$. A curve semivaluation $v \in \mathcal{V}_0$ is always an endpoint in the valuative tree. One can check that they satisfy $\alpha(v) = -\infty$ and $A(v) = \infty$.

7.7.2 Numerical Invariants

Now suppose v defines a *valuation* on $\hat{\mathcal{O}}_0$, that is, $\mathfrak{p} = (0)$. As in Sect. 6.9 we associate to v two basic numerical invariants: the rational rank and the transcendence degree. It does not make a difference whether we compute these in R, $\mathcal{O}_0$ or $\hat{\mathcal{O}}_0$. The Abhyankar inequality says that

$$\operatorname{tr.deg} v + \operatorname{rat.rk} v \leq 2$$

and equality holds iff v is a quasimonomial valuation.

7.7.3 Divisorial Valuations

A valuation $v \in \hat{\mathcal{V}}_0^*$ is *divisorial* if it has the numerical invariants $\operatorname{tr.deg} v = \operatorname{rat.rk} v = 1$. In this situation there exists a blowup $\pi \in \mathfrak{B}_0$ such that the center of v on X_π is the generic point of an exceptional prime E_i of π. In other words, v belongs to the one-dimensional cone $\hat{\sigma}_i$ of the dual fan $|\hat{\Delta}^*(\pi)|$ and $v = t \operatorname{ord}_{E_i}$ for some $t > 0$. We then set $b(v) := b_i = \operatorname{ord}_{E_i}(\mathfrak{m}_0)$.

More generally, suppose $v \in \hat{\mathcal{V}}_0^*$ is divisorial and $\pi \in \mathfrak{B}_0$ is a blowup such that the center of v on X_π is a closed point ξ. Then there exists a blowup $\pi' \in \mathfrak{B}_0$ dominating π in which the (closure of the) center of v is an exceptional prime of π'. Moreover, by a result of Zariski (cf. [Kol97, Theorem 3.17]), the birational morphism $X_{\pi'} \to X_\pi$ is an isomorphism above $X_\pi \setminus \{\xi\}$ and can be constructed by successively blowing up the center of v.

We will need the following result in Sect. 8.4.

Lemma 7.12 *Let $\pi \in \mathfrak{B}_0$ be a blowup and $v \in \hat{\mathcal{V}}_0^*$ a semivaluation. Set $w :=$ $r_\pi(v)$.*

(i) *if $v \notin |\hat{\Delta}^*(\pi)|$, then w is necessarily divisorial;*

(ii) *if $v \notin |\hat{\Delta}^*(\pi)|$ and v is divisorial, then $b(w)$ divides $b(v)$;*

(iii) *if $v \in |\hat{\Delta}^*(\pi)|$, then v is divisorial iff it is a rational point in the given integral affine structure; in this case, there exists a blowup $\pi' \geq \pi$ such that $|\hat{\Delta}^*(\pi')| = |\hat{\Delta}^*(\pi)|$ as subsets of $\hat{\mathcal{V}}_0^*$ and such that v belongs to a one-dimensional cone of $|\hat{\Delta}^*(\pi')|$;*

(iv) *if $v \in |\hat{\Delta}^*(\pi)|$ is divisorial and lies in the interior of a two-dimensional cone, say $\hat{\sigma}_{12}$ of $|\hat{\Delta}(\pi)|$, then $b(v) \geq b_1 + b_2$.*

Sketch of proof For (i), let ξ be the common center of v and w on X_π. If there is a unique exceptional prime E_1 containing ξ, then it is clear that w is proportional to ord_{E_1} and hence divisorial. Now suppose ξ is the intersection point between two distinct exceptional primes E_1 and E_2. Pick coordinates ζ_1, ζ_2 at ξ such that $E_i = (\zeta_i = 0)$ for $i = 1, 2$. If $v(\zeta_1)$ and $v(\zeta_2)$ are rationally independent, then v gives different values to all monomials $\zeta_1^{\beta_1} \zeta_2^{\beta_2}$, so we must have $v = w$, contradicting $v \notin |\hat{\Delta}^*(\pi)|$. Hence $w(\zeta_1) = v(\zeta_1)$ and $w(\zeta_2) = v(\zeta_2)$ are rationally dependent, so rat. rk $w = 1$. Since w is quasimonomial, it must be divisorial.

For (iii), we may assume that the center of v on X_π is the intersection point between two distinct exceptional primes $E_1 = (\zeta_1 = 0)$ and $E_2 = (\zeta_2 = 0)$ as above. Then v is monomial in coordinates (ζ_1, ζ_2) and it is clear that rat. rk $v = 1$ if $v(\zeta_1)/v(\zeta_2) \in \mathbf{Q}$ and rat. rk $v = 2$ otherwise. This proves the first statement. Now suppose v is divisorial. We can construct π' in (iii) by successively blowing up the center of v using the result of Zariski referred to above. Since v is monomial, the center is always a satellite point and blowing it up does not change the dual fan, viewed as a subset of $\hat{\mathcal{V}}_0^*$.

When proving (ii) we may by (iii) assume that w belongs to a one-dimensional cone $\hat{\sigma}_1$ of $|\hat{\Delta}(\pi)|$. Then $b(w) = b_1$. We now successively blow up the center of v. This leads to a sequence of divisorial valuations $w_0 = w, w_1, \ldots, w_m = v$. Since the first blowup is at a free point, we have $b(w_1) = b_1$ in view of (48). Using (48) and (49) one now shows by induction that b_1 divides $b(w_j)$ for $j \leq m$, concluding the proof of (ii).

Finally, in (iv) we obtain v after a finitely many satellite blowups, so the result follows from (49). $\square$

7.7.4 Irrational Valuations

A valuation $v \in \hat{\mathcal{V}}_0^*$ is *irrational* if tr. deg $v = 0$, rat. rk $v = 2$. In this case v is not divisorial but still quasimonomial; it belongs to a dual fan $|\hat{\Delta}^*(\pi)|$ for some blowup $\pi \in \mathfrak{B}_0$ and for any such π, v belongs to the interior of a two-dimensional cone.

Fig. 15 An infinitely singular valuation. The divisorial valuation v_j is obtained by performing a sequence of $j + 1$ blowups, every other free, and every other a satellite blowup. The picture is not to scale: we have $d(v_{2n}, v_{2n+2}) = d(v_{2n+1}, v_{2n+2}) = 2^{-(2n+1)}$ for $n \geq 0$. Further, $\alpha(v) = -5/3$, $A(v) = -3$ and $d(\mathrm{ord}_0, v) = 2/3$. In particular, v belongs to hyperbolic space **H**

7.7.5 Infinitely Singular Valuations

A valuation $v \in \hat{\mathcal{V}}_0^*$ is *infinitely singular* if it has the numerical invariants rat. rk $v = 1$, tr. deg $v = 0$. Every infinitely singular valuation in the valuative tree $\mathcal{V}_0$ is an end. However, some of these ends still belong to hyperbolic space **H**,

Example 7.13 Consider a sequence $(v_j)_{j=0}^{\infty}$ defined as follows. First, $v_0 = \mathrm{ord}_0 = \mathrm{ord}_{E_0}$. Then $v_j = b_j^{-1} \mathrm{ord}_{E_j}$ is defined inductively as follows: for j odd, E_j is obtained by blowing up a free point on E_{j-1} and for j even, E_j is obtained by blowing up the satellite point $E_{j-1} \cap E_{j-2}$. The sequence $(v_{2j})_{j=0}^{\infty}$ is increasing and converges to an infinitely singular valuation v, see Fig. 15. We have $b_{2n} = b_{2n+1} = 2^{-n}$, $A(v_{2n}) = 3 - 2^{-n}$ and $\alpha(v_{2n}) = -\frac{1}{3}(5 - 2^{1-2n})$. Thus $\alpha(v) = -5/3$ and $A(v) = 3$. In particular, $v \in$ **H**.

For more information on infinitely singular valuations, see [FJ04, Appendix A]. We shall not describe them further here, but they do play a role in dynamics.

7.8 Potential Theory

In Sect. 2.5 we outlined the first elements of a potential theory on a general metric tree and in Sect. 4.9 we applied this to the Berkovich projective line.

However, the general theory applied literally to the valuative tree $\mathcal{V}_0$ does not quite lead to a satisfactory notion. The reason is that one should really view a function on $\mathcal{V}_0$ as the restriction of a homogeneous function on the cone $\hat{\mathcal{V}}_0$. In analogy with the situation over the complex numbers, one expects that for any ideal $\mathfrak{a} \subseteq R$, the function $\log |\mathfrak{a}|$ defined by[24]

$$\log |\mathfrak{a}|(v) := -v(\mathfrak{a})$$

---

[24]The notation reflects the fact that $|\cdot| := e^{-v}$ is a seminorm on R, see (30).

should be plurisubharmonic on $\hat{\mathcal{V}}_0$. Indeed, $\log|a|$ is a maximum of finitely many functions of the form $\log|\phi|$, where $\phi \in R$ is a polynomial. As a special case, the function $\log|m_0|$ should be plurisubharmonic on $\hat{\mathcal{V}}_0$. This function has a pole (with value $-\infty$) at the point triv_0 and so should definitely not be pluriharmonic on $\hat{\mathcal{V}}_0$. However, it is constantly equal to -1 on $\mathcal{V}_0$, and so would be harmonic there with the usual definition of the Laplacian.

7.8.1 Subharmonic Functions and Laplacian on $\mathcal{V}_0$

An *ad hoc* solution to the problem above is to extend the valuative tree $\mathcal{V}_0$ to a slightly larger tree $\tilde{\mathcal{V}}_0$ by connecting the root ord_0 to a "ground" point $G \in \tilde{\mathcal{V}}_0$ using an interval of length one. See Fig. 16.

Denote the Laplacian on $\tilde{\mathcal{V}}_0$ by $\tilde{\Delta}$. We define the class $\mathrm{SH}(\mathcal{V}_0)$ of *subharmonic functions*[25] on $\mathcal{V}_0$ as the set of restrictions to $\mathcal{V}_0$ of functions $\varphi \in \mathrm{QSH}(\tilde{\mathcal{V}}_0)$ with

$$\varphi(G) = 0 \quad \text{and} \quad \tilde{\Delta}\varphi = \rho - a\delta_G,$$

where ρ is a positive measure supported on $\mathcal{V}_0$ and $a = \rho(\mathcal{V}_0) \geq 0$. In particular, φ is affine of slope $\varphi(\mathrm{ord}_0)$ on the segment $[G, \mathrm{ord}_0[= \tilde{\mathcal{V}}_0 \setminus \mathcal{V}_0$. We then define

$$\Delta\varphi := \rho = (\tilde{\Delta}\varphi)|_{\mathcal{V}_0}.$$

For example, if $\varphi \equiv -1$ on $\mathcal{V}_0$, then $\tilde{\Delta}\varphi = \delta_{\mathrm{ord}_0} - \delta_G$ and $\Delta\varphi = \delta_{\mathrm{ord}_0}$.

From this definition and the analysis in Sect. 2.5 one deduces:

Proposition 7.14 *Let $\varphi \in \mathrm{SH}(\mathcal{V}_0)$ and write $\rho = \Delta\varphi$. Then:*

(i) φ *is decreasing in the partial ordering of $\mathcal{V}_0$ rooted in ord_0;*
(ii) $\varphi(\mathrm{ord}_0) = -\rho(\mathcal{V}_0)$;
(iii) $|D_{\vec{v}}\varphi| \leq \rho(\mathcal{V}_0)$ *for all tangent directions $\vec{v}$ in $\mathcal{V}_0$.*

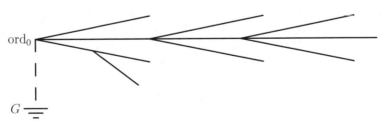

Fig. 16 Connecting the valuative tree $\mathcal{V}_0$ to "ground" gives rise to the auxiliary tree $\tilde{\mathcal{V}}_0$

---

[25]If $\varphi \in \mathrm{SH}(\mathcal{V}_0)$, then $-\varphi$ is a *positive tree potential* in the sense of [FJ04].

As a consequence we have the estimate

$$- \alpha(v)\varphi(\mathrm{ord}_0) \le \varphi(v) \le \varphi(\mathrm{ord}_0) \le 0 \tag{60}$$

for all $v \in \mathcal{V}_0$, where $\alpha : \mathcal{V}_0 \to [-\infty, -1]$ is the parametrization given by (59). The exact sequence in (9) shows that

$$\Delta : \mathrm{SH}(\mathcal{V}_0) \to \mathcal{M}^+(\mathcal{V}_0), \tag{61}$$

is a homeomorphism whose inverse is given by

$$\varphi(v) = \int_{\mathcal{V}_0} \alpha(w \wedge_{\mathrm{ord}_0} v) d\rho(w). \tag{62}$$

In particular, for any $C > 0$, the set $\{\varphi \in \mathrm{SH}(\mathcal{V}_0) \mid \varphi(\mathrm{ord}_0) \ge -C\}$ is compact. Further, if $(\varphi_i)_i$ is a decreasing net in $\mathrm{SH}(\mathcal{V}_0)$, and $\varphi := \lim \varphi_i$, then either $\varphi_i \equiv -\infty$ on $\mathcal{V}_0$ or $\varphi \in \mathrm{SH}(\mathcal{V}_0)$. Moreover, if $(\varphi_i)_i$ is a family in $\mathrm{SH}(\mathcal{V}_0)$ with $\sup_i \varphi(\mathrm{ord}_\infty) < \infty$, then the upper semicontinuous regularization of $\varphi := \sup_i \varphi_i$ belongs to $\mathrm{SH}(\mathcal{V}_0)$.

7.8.2 Subharmonic Functions from Ideals

The definitions above may seem arbitrary, but the next result justifies them. It shows that the Laplacian is intimately connected to intersection numbers and shows that the generalized metric on $\mathcal{V}_0$ is the correct one.

Proposition 7.15 *If $\mathfrak{a} \subseteq R$ is a primary ideal, then the function $\log |\mathfrak{a}|$ on $\mathcal{V}_0$ is subharmonic. Moreover, if $\pi \in \mathfrak{B}_0$ is a log resolution of $\mathfrak{a}$, with exceptional primes $E_i, i \in I$, and if we write $\mathfrak{a} \cdot \mathcal{O}_{X_\pi} = \mathcal{O}_{X_\pi}(Z)$, then*

$$\Delta \log |\mathfrak{a}| = \sum_{i \in I} b_i (Z \cdot E_i)\delta_{v_i},$$

where $b_i = \mathrm{ord}_{E_i}(\mathfrak{m}_0)$ and $v_i = b_i^{-1} \mathrm{ord}_{E_i} \in \mathcal{V}_0$.

Proof Write $\varphi = \log |\mathfrak{a}|$. It follows from Lemma 7.11 that $\varphi = \varphi \circ r_\pi$, so $\Delta\varphi$ is supported on the dual graph $|\Delta(\pi)| \subseteq \mathcal{V}_0$. Moreover, the proof of the same lemma shows that φ is affine on the interior of each 1-dimensional simplex so $\Delta\varphi$ is zero there. Hence it suffices to compute the mass of $\Delta\varphi$ at each v_i.

Note that π dominates π_0, the simple blowup of 0. Let E_0 be the strict transform of the exceptional divisor of π_0. Write $\mathfrak{m}_0 \cdot \mathcal{O}_{X_\pi} = \mathcal{O}_{X_\pi}(Z_0)$, where $Z_0 = -\sum_i b_i E_i$. Since π_0 already is a log resolution of $\mathfrak{m}_0$ we have

$$(Z_0 \cdot E_0) = 1 \quad \text{and} \quad (Z_0 \cdot E_j) = 0, \ j \ne 0. \tag{63}$$

Fix $i \in I$ and let E_j, $j \in J$ be the exceptional primes that intersect E_i properly. First assume $i \neq 0$. Using (63) and $(E_i \cdot E_j) = 1$ for $j \in J$ we get

$$\Delta \varphi\{v_i\} = \sum_{j \in J} \frac{\varphi(v_j) - \varphi(v_i)}{d(v_i, v_j)} = \sum_{j \in J} b_i b_j (\varphi(v_j) - \varphi(v_i)) =$$

$$= \sum_{j \in J} (b_i \operatorname{ord}_{E_j}(Z) - b_j \operatorname{ord}_{E_i}(Z))(E_i \cdot E_j) =$$

$$= b_i (Z \cdot E_i) - \operatorname{ord}_{E_i}(Z)(Z_0 \cdot E_i) = b_i (Z \cdot E_i).$$

If instead $i = 0$, then, by the definition of the Laplacian on $\mathcal{V}_0 \subseteq \tilde{\mathcal{V}}_0$, we get

$$\Delta \varphi\{v_0\} = \sum_{j \in J} \frac{\varphi(v_j) - \varphi(v_0)}{d(v_0, v_j)} + \varphi(\operatorname{ord}_0) = \sum_{j \in J} b_j (\varphi(v_j) - \varphi(v_i)) + \varphi(\operatorname{ord}_0) =$$

$$= \sum_{j \in J} (\operatorname{ord}_{E_j}(Z) - b_j \varphi(\operatorname{ord}_0))(E_j \cdot E_0) + \varphi(\operatorname{ord}_0) =$$

$$= (Z \cdot E_0) - \varphi(\operatorname{ord}_0)(Z_0 \cdot E_0) + \varphi(\operatorname{ord}_0) = (Z \cdot E_0),$$

which completes the proof. (Note that $b_0 = 1$.)   □

Corollary 7.16 *If $v = v_E = b_E^{-1} \operatorname{ord}_E \in \mathcal{V}_0$ is a divisorial valuation, then there exists a primary ideal $\mathfrak{a} \subseteq R$ such that $\Delta \log |\mathfrak{a}| = b_E \delta_{v_E}$.*

Proof Let $\pi \in \mathfrak{B}_0$ be a blowup such that E is among the exceptional primes E_i, $i \in I$. As in Sect. 7.3.5 above, define $\check{E} \in \operatorname{Div}(\pi)$ by $(\check{E} \cdot F) = \delta_{EF}$. Thus $\check{E}$ is relatively nef, so by Proposition 7.4 there exists a primary ideal $\mathfrak{a} \subseteq R$ such that $\mathfrak{a} \cdot \mathcal{O}_{X_\pi} = \mathcal{O}_{X_\pi}(\check{E})$. The result now follows from Proposition 7.15.   □

Remark 7.17 One can show that the function $\log |\mathfrak{a}|$ determines a primary ideal $\mathfrak{a}$ up to integral closure. (This fact is true in any dimension.) Furthermore, the product of two integrally closed ideals is integrally closed. Corollary 7.16 therefore shows that the assignment $\mathfrak{a} \mapsto \Delta \log |\mathfrak{a}|$ is a semigroup isomorphism between integrally closed primary ideals of R and finite atomic measures on $\mathcal{V}_0$ whose mass at a divisorial valuation v_E is an integer divisible by b_E.

Corollary 7.18 *If $\phi \in R \setminus \{0\}$ is a nonzero polynomial, then the function $\log |\phi|$ on $\mathcal{V}_0$ is subharmonic. More generally, the function $\log |\mathfrak{a}|$ is subharmonic for any nonzero ideal $\mathfrak{a} \subseteq R$.*

Proof For $n \geq 1$, the ideal $\mathfrak{a}_n := \mathfrak{a} + \mathfrak{m}_0^n$ is primary. Set $\varphi_n = \log |\mathfrak{a}_n|$. Then φ_n decreases pointwise on $\mathcal{V}_0$ to $\varphi := \log |\mathfrak{a}|$. Since the φ_n are subharmonic, so is φ.   □

Exercise 7.19 If $\phi \in \mathfrak{m}_0$ is a nonzero irreducible polynomial, show that

$$\Delta \log |\phi| = \sum_{j=1}^{n} m_j \delta_{v_j}$$

where v_j, $1 \leq j \leq n$ are the curve valuations associated to the local branches C_j of $\{\phi = 0\}$ at 0 and where m_j is the multiplicity of C_j at 0, that is, $m_j = \mathrm{ord}_0(\phi_j)$, where $\phi_j \in \hat{\mathcal{O}}_0$ is a local equation of C_j. *Hint* Let $\pi \in \mathfrak{B}_0$ be an embedded resolution of singularities of the curve $C = \{\phi = 0\}$.

This exercise confirms that the generalized metric on $\mathcal{V}_0$ is the correct one.

While we shall not use it, we have the following regularization result.

Theorem 7.20 *Any subharmonic function on $\mathcal{V}_0$ is a decreasing limit of a sequence $(\varphi_n)_{n \geq 1}$, where $\varphi_n = c_n \log |\mathfrak{a}_n|$, with c_n a positive rational number and $\mathfrak{a}_n \subseteq R$ a primary ideal.*

Proof By Theorem 2.10 (applied to the tree $\tilde{\mathcal{V}}_0$) any given function $\varphi \in \mathrm{SH}(\mathcal{V}_0)$ is the limit of a decreasing sequence $(\varphi_n)_n$ of functions in $\mathrm{SH}(\mathcal{V}_0)$ such that $\Delta \varphi_n$ is a finite atomic measure supported on quasimonomial valuations. Let $\pi_n \in \mathfrak{B}_0$ be a blowup such that $\Delta \varphi_n$ is supported on the dual graph $|\Delta(\pi_n)|$. Since the divisorial valuations are dense in $|\Delta(\pi_n)|$, we may pick $\psi_n \in \mathrm{SH}(\mathcal{V}_0)$ such that $\Delta \psi_n$ is a finite atomic measure supported on divisorial valuations in $|\Delta(\pi_n)|$, with rational weights, such that $|\psi_n - \varphi_n| \leq 2^{-n}$ on $\mathcal{V}_0$. The sequence $(\psi_n + 3 \cdot 2^{-n})_{n \geq 1}$ is then decreasing and $\Delta(\psi_n + 3 \cdot 2^{-n}) = \Delta \psi_n + 3 \cdot 2^{-n} \delta_{\mathrm{ord}_0}$ is a finite atomic measure supported on divisorial valuations in $|\Delta(\pi_n)|$, with rational weights. The result now follows from Corollary 7.16. $\qquad \square$

Regularization results such as Theorem 7.20 play an important role in higher dimensions, but the above proof, which uses tree arguments together with Lipman's result in Proposition 7.4, does not generalize. Instead, one can construct the ideals $\mathfrak{a}_n$ as *valuative multiplier ideals*. This is done in [FJ05b] in dimension two, and in [BFJ08b] in higher dimensions.

7.9 Intrinsic Description of the Tree Structure on $\mathcal{V}_0$

As explained in Sect. 7.6, the valuative tree inherits a partial ordering and a (generalized) metric from the dual graphs. We now describe these two structures intrinsically, using the definition of elements in $\mathcal{V}_0$ as functions on R. The potential theory in Sect. 7.8 is quite useful for this purpose.

7.9.1 Partial Ordering

The following result gives an intrinsic description of the partial ordering on $\mathcal{V}_0$.

Proposition 7.21 *If $w, v \in \mathcal{V}_0$, then the following are equivalent:*

(i) $v \leq w$ in the partial ordering induced by $\mathcal{V}_0 \simeq \varprojlim \Delta(\pi)$;
(ii) $v(\phi) \leq w(\phi)$ for all polynomials $\phi \in R$;
(iii) $v(\mathfrak{a}) \leq w(\mathfrak{a})$ for all primary ideals $\mathfrak{a} \subseteq R$.

Proof The implication (i) $\Longrightarrow$ (ii) is a consequence of Proposition 7.14 and the fact that $\log |\phi|$ is subharmonic. That (ii) implies (iii) is obvious. It remains to prove that (iii) implies (i). Suppose that $v \not\leq w$ in the sense of (i). After replacing v and w by $r_\pi(v)$ and $r_\pi(w)$, respectively, for a sufficiently large π, we may assume that $v, w \in |\Delta(\pi)|$. Set $v' := v \wedge w$. Then $v' < v$, $v' \leq w$ and $]v', v] \cap [v', w] = \emptyset$. Replacing v by a divisorial valuation in $]v', v]$ we may assume that v is divisorial. By Corollary 7.16 we can find an ideal $\mathfrak{a} \subseteq R$ such that $\Delta \log |\mathfrak{a}|$ is supported at v. Then $w(\mathfrak{a}) = v'(\mathfrak{a}) < v(\mathfrak{a})$, so (iii) does not hold. $\square$

7.9.2 Integral Affine Structure

Next we give an intrinsic description of the integral affine structure.

Proposition 7.22 *If $\pi \in \mathfrak{B}_0$ is a blowup, then a function $\varphi : \hat{\mathcal{V}}_0 \to \mathbf{R}$ belongs to $\mathrm{Aff}(\pi)$ iff it is of the form $\varphi = \log |\mathfrak{a}| - \log |\mathfrak{b}|$, where $\mathfrak{a}$ and $\mathfrak{b}$ are primary ideals of R for which π is a common log resolution.*

Sketch of proof After unwinding definitions this boils down to the fact that any exceptional divisor can be written as the difference of two relatively nef divisors. Indeed, by Proposition 7.4, if Z is relatively nef, then there exists a primary ideal $\mathfrak{a} \subseteq R$ such that $\mathfrak{a} \cdot \mathcal{O}_{X_\pi} = \mathcal{O}_{X_\pi}(Z)$. $\sqcap$

Corollary 7.23 *A function $\varphi : \hat{\mathcal{V}}_0^* \to \mathbf{R}$ is integral affine iff it is of the form $\varphi = \log |\mathfrak{a}| - \log |\mathfrak{b}|$, where $\mathfrak{a}$ and $\mathfrak{b}$ are primary ideals in R.*

7.9.3 Metric

Recall the parametrization α of $\mathcal{V}_0 \simeq \varprojlim \Delta(\pi)$ given by (59).

Proposition 7.24 *For any $v \in \mathcal{V}_0$ we have*

$$\alpha(v) = -\sup \left\{ \frac{v(\phi)}{\mathrm{ord}_0(\phi)} \,\middle|\, p \in \mathfrak{m}_0 \right\} = -\sup \left\{ \frac{v(\mathfrak{a})}{\mathrm{ord}_0(\mathfrak{a})} \,\middle|\, \mathfrak{a} \subseteq R \ \mathfrak{m}_0\text{-primary} \right\}$$

and the suprema are attained when v is quasimonomial.

In fact, one can show that supremum in the second equality is attained *only* if v is quasimonomial. Further, the supremum in the first equality is never attained when v is infinitely singular, but is attained if v is a curve semivaluation (in which case $\alpha(v) = -\infty$), and we allow $\phi \in \mathfrak{m}_0 \cdot \hat{\mathcal{O}}_0$.

Proof Since the functions $\log |\mathfrak{a}|$ and $\log |\phi|$ are subharmonic, (60) shows that $v(\mathfrak{a}) \leq -\alpha(v) \operatorname{ord}_0(\mathfrak{a})$ and $v(\phi) \leq -\alpha(v) \operatorname{ord}_0(\phi)$ for all $\mathfrak{a}$ and all ϕ.

Let us prove that equality can be achieved when v is quasimonomial. Pick a blowup $\pi \in \mathfrak{B}_0$ such that $v \in |\Delta(\pi)|$ and pick $w \in |\Delta(\pi)|$ divisorial with $w \geq v$. By Corollary 7.16 there exists a primary ideal $\mathfrak{a}$ such that $\Delta \log |\mathfrak{a}|$ is supported at w. This implies that the function $\log |\mathfrak{a}|$ is affine with slope $-\operatorname{ord}_0(\mathfrak{a})$ on the segment $[\operatorname{ord}_0, w]$. In particular, $v(\mathfrak{a}) = -\alpha(v) \operatorname{ord}_0(\mathfrak{a})$. By picking ϕ as a general element in $\mathfrak{a}$ we also get $v(\phi) = -\alpha(v) \operatorname{ord}_0(\phi)$.

The case of a general $v \in \mathcal{V}_0$ follows from what precedes, given that $r_\pi v(\phi)$, $r_\pi v(\mathfrak{a})$ and $\alpha(r_\pi(v))$ converge to $v(\phi)$, $v(\mathfrak{a})$ and $\alpha(v)$, respectively, as $\pi \to \infty$. $\qquad\square$

Notice that Proposition 7.24 gives a very precise version of the Izumi-Tougeron inequality (36). Indeed, $\alpha(v) > -\infty$ for all quasimonomial valuations $v \in \mathcal{V}_0$.

7.9.4 Multiplicity

The multiplicity function $m : \mathcal{V}_0 \to \mathbf{N} \cup \{\infty\}$ can also be characterized intrinsically. For this, one first notes that if $v = v_C$ is a curve semivaluation, defined by a formal curve C, then $m(v) = \operatorname{ord}_0(C)$. More generally, one can show that

$$m(v) = \min\{m(C) \mid v \leq v_C\}.$$

In particular, $m(v) = \infty$ iff v cannot be dominated by a curve semivaluation, which in turn is the case iff v is infinitely singular.

7.9.5 Topology

Theorem 7.9 shows that the topology on $\mathcal{V}_0$ induced from $\mathbf{A}^2_{\mathrm{Berk}}$ coincides with the tree topology on $\mathcal{V}_0 \simeq \varprojlim \Delta(\pi)$. It is also possible to give a more geometric description.

For this, consider a blowup $\pi \in \mathfrak{B}_0$ and a *closed* point $\xi \in \pi^{-1}(0)$. Define $U(\xi) \subseteq \mathcal{V}_0$ as the set of semivaluations having center ξ on X_π. This means precisely that $v(\mathfrak{m}_\xi) > 0$, where $\mathfrak{m}_\xi$ is the maximal ideal of the local ring $\mathcal{O}_{X_\pi, \xi}$. Thus $U(\xi)$ is open in $\mathcal{V}_0$. One can in fact show that these sets $U(\xi)$ generate the topology on $\mathcal{V}_0$.

If ξ is a *free* point, belonging to a unique exceptional prime E of X_π, then we have $U(\xi) = U(\vec{v})$ for a tangent direction $\vec{v}$ at v_E in $\mathcal{V}_0$, namely, the tangent direction for which $\operatorname{ord}_\xi \in U(\vec{v})$. As a consequence, the open set $U(\xi)$ is connected and its boundary is a single point: $\partial U(\xi) = \{v_E\}$.

7.10   Relationship to the Berkovich Unit Disc

Let us briefly sketch how to relate the valuative tree with the Berkovich unit disc.
Fix global coordinates (z_1, z_2) on $\mathbf{A}^2$ vanishing at 0 and let $L = K((z_1))$ be the field
of Laurent series in z_1. There is a unique extension of the trivial valuation on K to
a valuation v_L on L for which $v_L(z_1) = 1$. The Berkovich open unit disc over L is
the set of semivaluations $v : L[z_2] \to \mathbf{R}_+$, extending v_L, for which $v(z_2) > 0$. If
v is such a semivaluation, then $v/\min\{1, v(z_2)\}$ is an element in the valuative tree
$\mathcal{V}_0$. Conversely, if $v \in \mathcal{V}_0$ is not equal to the curve semivaluation v_C associated to
the curve $(z_1 = 0)$, then $v/v(z_1)$ defines an element in the Berkovich open unit disc
over L.

Even though L is not algebraically closed, the classification of the points in the
Berkovich affine line into Type 1-4 points still carries over, see Sect. 3.9.1. Curve
valuations become Type 1 points, divisorial valuations become Type 2 points and
irrational valuations become Type 3 points. An infinitely singular valuation $v \in \mathcal{V}_0$
is of Type 4 or Type 1, depending on whether the log discrepancy $A(v)$ is finite or
infinite. The parametrization and partial orderings on $\mathcal{V}_0$ and the Berkovich unit disc
are related, but different. See [FJ04, §3.9, §4.5] for more details.

Note that the identification of the valuative tree with the Berkovich unit disc
depends on a choice of coordinates. In the study of polynomial dynamics in Sect. 8,
it would usually not be natural to fix coordinates. The one exception to this is when
studying the dynamics of a skew product

$$f(z_1, z_2) = (\phi(z_1), \psi(z_1, z_2)),$$

with $\phi(0) = 0$, in a neighborhood of the invariant line $z_1 = 0$. However, it will
be more efficient to study general polynomial mappings in two variables using the
Berkovich affine plane over the trivially valued field K.

As noted in Sect. 6.7, the Berkovich unit disc over the field $K((z_1))$ of Laurent
series is in fact more naturally identified with the space $\mathcal{V}_C$, where $C = \{z_1 = 0\}$.

7.11   Other Ground Fields

Let us briefly comment on the case when the field K is not algebraically closed.

Let K^a denote the algebraic closure and $G = \mathrm{Gal}(K^a/K)$ the Galois group.
Using general theory we have an identification $\mathbf{A}^2_{\mathrm{Berk}}(K) \simeq \mathbf{A}^2_{\mathrm{Berk}}(K^a)/G$.

First suppose that the closed point $0 \in \mathbf{A}^2(K)$ is K-rational, that is, $\mathcal{O}_0/\mathfrak{m}_0 \simeq K$.
Then 0 has a unique preimage $0 \in \mathbf{A}^2(K^a)$. Let $\mathcal{V}_0(K^a) \subseteq \mathbf{A}^2_{\mathrm{Berk}}(K^a)$ denote the
valuative tree at $0 \in \mathbf{A}^2(K^a)$. Every $g \in G$ induces an automorphism of $\mathbf{A}^2_{\mathrm{Berk}}(K^a)$
that leaves $\mathcal{V}_0(K^a)$ invariant. In fact, one checks that g preserves the partial ordering
as well as the parametrizations α and A and the multiplicity m. Therefore, the
quotient $\mathcal{V}_0(K) \simeq \mathcal{V}_0(K^a)$ also is naturally a tree. As in Sect. 3.9.1 we define a

parametrization α of $\mathcal{V}_0(K)$ using the corresponding parametrization of $\mathcal{V}_0(K^a)$ and the degree of the map $\mathcal{V}_0(K^a) \to \mathcal{V}_0(K)$. This parametrization gives rise to the correct generalized metric in the sense that the analogue of Exercise 7.19 holds.

When the closed point 0 is not K-rational, it has finitely many preimages $0_j \in \mathbf{A}^2(K^a)$. At each 0_j we have a valuative tree $\mathcal{V}_{0_j} \subseteq \mathbf{A}^2_{\mathrm{Berk}}(K^a)$ and $\mathcal{V}_0$, which is now the quotient of the disjoint union of the $\mathcal{V}_{0_j}$ by G, still has a natural metric tree structure.

In fact, even when K is not algebraically closed, we can analyze the valuative tree using blowups and dual graphs much as we have done above. One thing to watch out for, however, is that the intersection form on $\mathrm{Div}(\pi)$ is no longer unimodular. Further, when E_i, E_j are exceptional primes intersecting properly, it is no longer true that $(E_i \cdot E_j) = 1$. In order to get the correct metric on the valuative tree, so that Proposition 7.15 holds for instance, we must take into account the degree over K of the residue field whenever we blow up a closed point ξ. The resulting metric is the same as the one obtained above using the Galois action.

7.12 Notes and Further References

The valuative tree was introduced and studied extensively in the monograph [FJ04] by Favre and myself. One of our original motivations was in fact to study superattracting fixed points, but it turned out that while valuations on surfaces had been classified by Spivakovsky, the structure of this valuation space had not been explored.

It was not remarked in [FJ04] that the valuative tree can be viewed as a subset of the Berkovich affine plane over a trivially valued field. The connection that was made was with the Berkovich unit disc over the field of Laurent series.

In [FJ04], several approaches to the valuative tree are pursued. The first approach is algebraic, using *key polynomials* as developed by MacLane [Mac36]. While beautiful, this method is coordinate dependent and involves some quite delicate combinatorics. In addition, even though there is a notion of key polynomials in higher dimensions [Vaq07], these seem hard to use for our purposes.

The geometric approach, using blowups and dual graphs is also considered in [FJ04] but perhaps not emphasized as much as here. As already mentioned, this approach can be partially generalized to higher dimensions, see [BFJ08b], where it is still true that the valuation space $\mathcal{V}_0$ is an inverse limit of dual graphs. The analogue of the Laplace operator on $\mathcal{V}_0$ is then a nonlinear Monge-Ampère operator, but this operator is defined geometrically, using intersection theory, rather than through the simplicial structure of the space. In higher dimensions, the relation between the different positivity notions on exceptional divisors is much more subtle than in dimension two. Specifically, Proposition 7.4 is no longer true.

Granja [Gra07] has generalized the construction of the valuative tree to a general two-dimensional regular local ring.

The valuative tree gives an efficient way to encode singularities in two dimensions. For example, it can be used to study the singularities of planar plurisubharmonic functions, see [FJ05a, FJ05b]. It is also related to many other constructions in singularity theory. We shall not discuss this further here, but refer to the paper [Pop11] by Popescu-Pampu for further references. In this paper, the author, defines an interesting object, the *kite* (cerf-volant), which also encodes the combinatorics of the exceptional primes of a blowup.

In order to keep notes reasonably coherent, and in order to reflect changing trends, I have taken the freedom to change some of the notation and terminology from [FJ04]. Notably, in [FJ04], the valuative tree is simply denoted $\mathcal{V}$ and its elements are called valuations. Here we wanted to be more precise, so we call them semivaluations. What is called subharmonic functions here correspond to positive tree potentials in [FJ04]. The valuation ord_0 is called ν_m in [FJ04].

8 Local Plane Polynomial Dynamics

Next we will see how the valuative tree can be used to study superattracting fixed points for polynomial maps of $\mathbf{A}^2$.

8.1 Setup

Let K be an algebraically closed field, equipped with the trivial valuation. (See Sect. 8.8 for the case of other ground fields.) Further, R and F are the coordinate ring and function field of the affine plane $\mathbf{A}^2$ over K. Recall that the Berkovich affine plane $\mathbf{A}^2_{\mathrm{Berk}}$ is the set of semivaluations on R that restrict to the trivial valuation on K.

8.2 Definitions and Results

We briefly recall the setup from Sect. 1.2 of the introduction. Let K be an algebraically closed field of characteristic zero. Consider a polynomial mapping $f : \mathbf{A}^2 \to \mathbf{A}^2$ over K. We assume that f is *dominant*, since otherwise the image of f is contained in a curve. Consider a (closed) fixed point $0 = f(0) \in \mathbf{A}^2$ and define

$$c(f) := \mathrm{ord}_0(f^*\mathfrak{m}_0),$$

where $\mathfrak{m}_0$ denotes the maximal ideal at 0. We say that f is *superattracting* if $c(f^n) > 1$ for some $n \geq 1$.

Exercise 8.1 Show that if f is superattracting, then in fact $c(f^2) > 1$. On the other hand, find an example of a superattracting f for which $c(f) = 1$.

Exercise 8.2 Show that if f is superattracting and $K = \mathbf{C}$, then there exists a neighborhood $0 \in U \subseteq \mathbf{A}^2$ (in the usual Euclidean topology) such that $f(U) \subseteq U$, and $f^n(z) \to 0$ as $n \to \infty$ for any $z \in U$.

As mentioned in the introduction, the sequence $(c(f^n))_{n \geq 1}$ is supermultiplicative, so the limit

$$c_\infty(f) := \lim_{n \to \infty} c(f^n)^{1/n} = \sup_{n \to \infty} c(f^n)^{1/n}$$

exists.

Exercise 8.3 Verify these statements! Also show that f is superattracting iff $c_\infty(f) > 1$ iff df_0 is nilpotent.

Exercise 8.4 In coordinates (z_2, z_2) on $\mathbf{A}^2$, let f_c be the homogeneous part of f of degree $c = c(f)$. Show that if $f_c^2 \not\equiv 0$, then in fact $f_c^n \not\equiv 0$ for all $n \geq 1$, so that $c(f^n) = c^n$ and $c_\infty = c = c(f)$ is an integer.

Example 8.5 If $f(z_1, z_2) = (z_2, z_1 z_2)$, then $c(f^n)$ is the $(n+2)$th Fibonacci number and $c_\infty = \frac{1}{2}(\sqrt{5} + 1)$ is the golden mean.

For the convenience of the reader, we recall the result that we are aiming for:

Theorem B *The number $c_\infty = c_\infty(f)$ is a quadratic integer: there exists $a, b \in \mathbf{Z}$ such that $c_\infty^2 = a c_\infty + b$. Moreover, there exists a constant $\delta > 0$ such that*

$$\delta c_\infty^n \leq c(f^n) \leq c_\infty^n$$

for all $n \geq 1$.

Here it is the left-hand inequality that is nontrivial.

8.3 Induced Map on the Berkovich Affine Plane

As outlined in Sect. 1.2, we approach Theorem B by studying the induced map

$$f : \mathbf{A}_{\text{Berk}}^2 \to \mathbf{A}_{\text{Berk}}^2$$

on the Berkovich affine plane $\mathbf{A}_{\text{Berk}}^2$. Recall the subspaces

$$\mathcal{V}_0 \subseteq \hat{\mathcal{V}}_0^* \subseteq \hat{\mathcal{V}}_0 \subseteq \mathbf{A}_{\text{Berk}}^2$$

introduced in Sect. 7: $\hat{\mathcal{V}}_0$ is the set of semivaluations whose center on $\mathbf{A}^2$ is the point 0. It has the structure of a cone over the valuative tree $\mathcal{V}_0$, with apex at triv_0. It is clear that

$$f(\hat{\mathcal{V}}_0) \subseteq \hat{\mathcal{V}}_0 \quad \text{and} \quad f(\mathrm{triv}_0) = \mathrm{triv}_0.$$

In general, f does not map the pointed cone $\hat{\mathcal{V}}_0^*$ into itself. Indeed, suppose there exists an algebraic curve $C = \{\phi = 0\} \subseteq \mathbf{A}^2$ passing through 0 and contracted to 0 by f. Then any semivaluation $v \in \hat{\mathcal{V}}_0^*$ such that $v(\phi) = \infty$ satisfies $f(v) = \mathrm{triv}_0$. To rule out this behavior, we introduce

Assumption 8.6 From now on, and until Sect. 8.6 we assume that the germ f is *finite*.

This assumption means that the ideal $f^*\mathfrak{m}_0 \subseteq \mathcal{O}_0$ is primary, that is, $\mathfrak{m}_0^s \subseteq f^*\mathfrak{m}_0$ for some $s \geq 1$, so it exactly rules out the existence of contracted curves. Certain modifications are required to handle the more general case when f is merely dominant. See Sect. 8.6 for some of this.

The finiteness assumption implies that $f^{-1}\{\mathrm{triv}_0\} = \{\mathrm{triv}_0\}$. Thus we obtain a well-defined map

$$f : \hat{\mathcal{V}}_0^* \to \hat{\mathcal{V}}_0^*,$$

which is clearly continuous and homogeneous.

While f preserves $\hat{\mathcal{V}}_0^*$, it does not preserve the "section" $\mathcal{V}_0 \subseteq \hat{\mathcal{V}}_0^*$ given by the condition $v(\mathfrak{m}_0) = 1$. Indeed, if $v(\mathfrak{m}_0) = 1$, there is no reason why $f(v)(\mathfrak{m}_0) = 1$. Rather, we define

$$c(f, v) := v(f^*\mathfrak{m}_0) \quad \text{and} \quad f_\bullet v := \frac{f(v)}{c(f, v)}.$$

The assumption that f is finite at 0 is equivalent to the existence of a constant $C > 0$ such that $1 \leq c(f, v) \leq C$ for all $v \in \mathcal{V}_0$. Indeed, we can pick C as any integer s such that $f^*\mathfrak{m}_0 \supseteq \mathfrak{m}_0^s$. Also note that

$$c(f) = c(f, \mathrm{ord}_0).$$

The normalization factors $c(f, v)$ naturally define a dynamical *cocycle*. Namely, we can look at $c(f^n, v)$ for every $n \geq 0$ and $v \in \mathcal{V}_0$ and we then have

$$c(f^n, v) = \prod_{i=0}^{n-1} c(f, v_i),$$

where $v_i = f_\bullet^i v$ for $0 \leq i < n$.

Apply this equality to $v = \mathrm{ord}_0$. By definition, we have $v_i = f_\bullet^i \mathrm{ord}_0 \geq \mathrm{ord}_0$ for all i. This gives $c(f, v_i) \geq c(f, \mathrm{ord}_0) = c(f)$, and hence $c(f^n) \geq c(f)^n$, as we already knew. More importantly, we shall use the *multiplicative cocycle* $c(f^n, v)$ in order to study the *supermultiplicative sequence* $(c(f^n))_{n \geq 0}$.

8.4 Fixed Points on Dual Graphs

Consider a blowup $\pi \in \mathcal{B}_0$. We have seen that the dual graph of π embeds as a subspace $|\Delta(\pi)| \subseteq \mathcal{V}_0$ of the valuative tree, and that there is a retraction $r_\pi : \mathcal{V}_0 \to |\Delta(\pi)|$. We shall study the selfmap

$$r_\pi f_\bullet : |\Delta(\pi)| \to |\Delta(\pi)|.$$

Notice that this map is continuous since r_π and $f_\bullet$ are. Despite appearances, it does not really define an induced dynamical system on $|\Delta(\pi)|$, as, in general, we may have $(r_\pi f_\bullet)^2 \neq r_\pi f_\bullet^2$. However, the fixed points of $r_\pi f_\bullet$ will play an important role.

It is easy to see that a continuous selfmap of a finite simplicial tree always has a fixed point. (See also Proposition 2.17.) Hence we can find $v_0 \in |\Delta(\pi)|$ such that $r_\pi f_\bullet v_0 = v_0$. There are then three possibilities:

(1) v_0 is divisorial and $f_\bullet v_0 = v_0$;
(2) v_0 is divisorial and $f_\bullet v_0 \neq v_0$;
(3) v_0 is irrational and $f_\bullet v_0 = v_0$.

Indeed, if $v \in \mathcal{V}_0 \setminus |\Delta(\pi)|$ is any valuation, then $r_\pi(v)$ is divisorial, see Lemma 7.12. The same lemma also allows us to assume, in cases (1) and (2), that the center of v_0 on X_π is an exceptional prime $E \subseteq X_\pi$.

In case (2), this means that the center of $f_\bullet v_0$ on X_π is a *free* point $\xi \in E$, that is, a point that does not belong to any other exceptional prime of π.

8.5 Proof of Theorem B

Using the fixed point v_0 that we just constructed, and still assuming f finite, we can now prove Theorem B.

The proof that c_∞ is a quadratic integer relies on a calculation using value groups. Recall that the value group of a valuation v is defined as $\Gamma_v = v(F)$, where F is the fraction field of R.

Lemma 8.7 *In the notation above, we have $c(f, v_0)\Gamma_{v_0} \subseteq \Gamma_{v_0}$. As a consequence, $c(f, v_0)$ is a quadratic integer.*

We shall see that under suitable assumptions on the blowup π we have $c(f, v_0) = c_\infty(f)$. This will show that $c_\infty(f)$ is a quadratic integer.

Proof In general, $\Gamma_{f(v)} \subseteq \Gamma_v$ and $\Gamma_{r_\pi(v)} \subseteq \Gamma_v$ for $v \in \hat{\mathcal{V}}_0^*$. If we write $c_0 = c(f, v_0)$, then this leads to

$$c_0 \Gamma_{v_0} = c_0 \Gamma_{r_\pi f_\bullet v_0} \subseteq c_0 \Gamma_{f_\bullet v_0} = \Gamma_{c_0 f_\bullet v_0} = \Gamma_{f(v_0)} \subseteq \Gamma_{v_0},$$

which proves the first part of the lemma.

Now v_0 is quasimonomial, so the structure of its value group is given by (19). When v_0 is divisorial, $\Gamma_{v_0} \simeq \mathbf{Z}$ and the inclusion $c_0 \Gamma_{v_0} \subseteq \Gamma_{v_0}$ immediately implies that c_0 is an integer. If instead v_0 is irrational, $\Gamma_{v_0} \simeq \mathbf{Z} \oplus \mathbf{Z}$ and c_0 is a quadratic integer. Indeed, if we write $\Gamma_{v_0} = t_1 \mathbf{Z} \oplus t_2 \mathbf{Z}$, then there exist integers a_{ij} such that $c_0 t_i = \sum_{j=1}^2 a_{ij} t_j$ for $i = 1, 2$. But then c_0 is an eigenvalue of the matrix (a_{ij}), hence a quadratic integer. $\square$

It remains to be seen that $c(f, v_0) = c_\infty(f)$ and that the estimates in Theorem B hold. We first consider cases (1) and (3) above, so that $f_\bullet v_0 = v_0$. It follows from (60) that the valuations v_0 and ord_0 are comparable. More precisely, $\mathrm{ord}_0 \le v_0 \le -\alpha_0 \mathrm{ord}_0$, where $\alpha_0 = \alpha(v_0)$. The condition $f_\bullet v_0 = v_0$ means that $f(v_0) = c v_0$, where $c = c(f, v_0)$. This leads to

$$c(f^n) = \mathrm{ord}_0(f^{n*} \mathfrak{m}_0) \le v_0(f^{n*} \mathfrak{m}_0) = (f_*^n v_0)(\mathfrak{m}_0) = c^n v_0(\mathfrak{m}_0) = c^n$$

and, similarly, $c^n \le -\alpha_0 c(f^n)$. In view of the definition of c_∞, this implies that $c_\infty = c$, so that

$$f(v_0) = c_\infty v_0 \quad \text{and} \quad -\alpha_0^{-1} c_\infty^n \le c(f^n) \le c_\infty^n,$$

proving Theorem B in this case.

Case (2) is more delicate and is in some sense the typical case. Indeed, note that we have not made any restriction on the modification π. For instance, π could be a simple blowup of the origin. In this case $|\Delta(\pi)| = \{\mathrm{ord}_0\}$ is a singleton, so $v_0 = \mathrm{ord}_0$ but there is no reason why $f_\bullet \mathrm{ord}_0 = \mathrm{ord}_0$. To avoid this problem, we make

Assumption 8.8 The map $\pi : X_\pi \to \mathbf{A}^2$ defines a log resolution of the ideal $f^*\mathfrak{m}$. In other words, the ideal sheaf $f^*\mathfrak{m} \cdot \mathcal{O}_{X_\pi}$ is locally principal.

Such a π exists by resolution of singularities. Indeed our current assumption that f be a *finite* germ implies that $f^*\mathfrak{m}$ is an $\mathfrak{m}$-primary ideal.

For us, the main consequence of π being a log resolution of $f^*\mathfrak{m}$ is that

$$c(v) = v(f^*\mathfrak{m}_0) = (r_\pi v)(f^*\mathfrak{m}_0) = c(r_\pi v)$$

for all $v \in \mathcal{V}_0$, see Lemma 7.11.

As noted above, we may assume that the center of v_0 on X_π is an exceptional prime E. Similarly, the center of $f_\bullet v_0$ on X_π is a free point $\xi \in E$. Let $U(\xi)$ be the set of all valuations $v \in \mathcal{V}_0$ whose center on X_π is the point ξ. By Sect. 7.9.5, this is a connected open set and its closure is given by $\overline{U(\xi)} = U(\xi) \cup \{v_0\}$. We have $r_\pi U(\xi) = \{v_0\}$, so $c(f, v) = c(f, v_0)$ for all $v \in U(\xi)$ by Lemma 7.11.

We claim that $f_\bullet(\overline{U(\xi)}) \subseteq U(\xi)$. To see this, we could use Sect. 2.6 but let us give a direct argument. Note that $v \geq v_0$, and hence $f(v) \geq f(v_0)$ for all $v \in \overline{U(\xi)}$. Since $c(f, v) = c(f, v_0)$, this implies $f_\bullet v \geq f_\bullet v_0 > v_0$ for all $v \in \overline{U(\xi)}$. In particular, $f_\bullet v \neq v_0$ for all $v \in \overline{U(\xi)}$, so that

$$\overline{U(\xi)} \cap f_\bullet^{-1} U(\xi) = \overline{U(\xi)} \cap f_\bullet^{-1} \overline{U(\xi)}.$$

It follows that $\overline{U(\xi)} \cap f_\bullet^{-1} U(\xi)$ is a subset of $\overline{U(\xi)}$ that is both open and closed. It is also nonempty, as it contains v_0. By connectedness of $\overline{U(\xi)}$, we conclude that $f_\bullet(\overline{U(\xi)}) \subseteq U(\xi)$.

The proof of Theorem B can now be concluded in the same way as in cases (1) and (3). Set $v_n := f_\bullet^n v_0$ for $n \geq 0$. Then we have $v_n \in \overline{U(\xi)}$ and hence $c(f, v_n) = c(f, v_0) =: c$ for all $n \geq 0$. This implies $c(f^n, v_0) = \prod_{i=0}^{n-1} c(f, v_i) = c^n$ for all $n \geq 1$. As before, this implies that $c = c_\infty$ and $-\alpha_0^{-1} c_\infty^n \leq c(f^n) \leq c_\infty^n$, where $\alpha_0 = \alpha(v_0) < \infty$.

8.6 The Case of a Non-Finite Germ

Let us briefly discuss the situation when $f : \mathbf{A}^2 \to \mathbf{A}^2$ is dominant but not finite at a fixed point $0 = f(0)$. In other words, the ideal $f^* \mathfrak{m}_0 \subseteq \mathfrak{m}_0$ is not primary. In this case, the subset $I_f \subseteq \mathcal{V}_0$ given by $c(f, \cdot) = +\infty$ is nonempty but finite. Each element of I_f is a curve valuation associated to an irreducible germ of a curve C at 0 such that $f(C) = 0$. In particular, I_f does not contain any quasimonomial valuations. Write $\hat{I}_f = \mathbf{R}_+^* I_f$, $\hat{D}_f := \hat{\mathcal{V}}_0^* \setminus \hat{I}_f = \{c(f, \cdot) < +\infty\}$ and $D_f := \mathcal{V}_0 \setminus I_f = \hat{D}_f \cap \mathcal{V}_0$. For $v \in \hat{I}_f$ we have $f(v) = \mathrm{triv}_0$. We can view $f : \hat{\mathcal{V}}_0^* \dashrightarrow \hat{\mathcal{V}}_0^*$ as a partially defined map having domain of definition $\hat{D}_f$. On D_f we define $f_\bullet$ as before, namely $f_\bullet v = f(v)/c(f, v)$. One can show that $f_\bullet$ extends continuously through I_f to a map $f_\bullet : \mathcal{V}_0 \to \mathcal{V}_0$. More precisely, any $v \in I_f$ is associated to an analytically irreducible branch of an algebraic curve $D \subseteq \mathbf{A}^2$ for which $f(D) = 0$. The valuation $f(\mathrm{ord}_D)$ is divisorial and has 0 as its center on $\mathbf{A}^2$, hence $f(\mathrm{ord}_D) = r v_E$, where $r \in \mathbf{N}$ and $v_E \in \mathcal{V}_0$ is divisorial. The continuous extension of $f_\bullet$ across v is then given by $f_\bullet v = v_E$. In particular, $f_\bullet I_f \cap I_f = \emptyset$.

Now we can find a log resolution $\pi : X_\pi \to \mathbf{A}^2$ of the ideal $f^* \mathfrak{m}_0$. By this we mean that the ideal sheaf $f^* \mathfrak{m}_0 \cdot \mathcal{O}_{X_\pi}$ on X_π is locally principal and given by a normal crossings divisor in a neighborhood of $\pi^{-1}(0)$. We can embed the dual graph of this divisor as a finite subtree $|\Delta| \subseteq \mathcal{V}_0$. Note that $|\Delta|$ contains all elements of I_f. There is a continuous retraction map $r : \mathcal{V}_0 \to |\Delta|$. Thus we get a continuous selfmap $r f_\bullet : |\Delta| \to |\Delta|$, which admits a fixed point $v \in |\Delta|$. Note that $v \notin I_f$ since $f_\bullet I_f \cap I_f = \emptyset$ and $r^{-1} I_f = I_f$. Therefore v is quasimonomial. The proof now goes through exactly as in the finite case.

8.7   Further Properties

Let us outline some further results from [FJ07] that one can obtain by continuing the analysis.

First, one can construct an *eigenvaluation*, by which we mean a semivaluation $v \in \mathcal{V}_0$ such that $f(v) = c_\infty v$. Indeed, suppose f is finite for simplicity and look at the three cases (1)–(3) in Sect. 8.4. In cases (1) and (3) the valuation v_0 is an eigenvaluation. In case (2) one can show that the sequence $(f_\bullet^n v_0)_{n=0}^\infty$ increases to an eigenvaluation.

Second, we can obtain local normal forms for the dynamics. For example, in Case (2) in Sect. 8.4 we showed that $f_\bullet$ mapped the open set $U(\xi)$ into itself, where $U(\xi)$ is the set of semivaluations whose center of X_π is equal to ξ, the center of $f_\bullet v_0$ on X_π. This is equivalent to the lift $f : X_\pi \dashrightarrow X_\pi$ being regular at ξ and $f(\xi) = \xi$. By choosing X_π judiciously one can even guarantee that $f : (X_\pi, \xi) \to (X_\pi, \xi)$ is a *rigid* germ, a dynamical version of simple normal crossings singularities. Such a rigidification result was proved in [FJ07] for superattracting germs and later extended by Matteo Ruggiero [Rug12] to more general germs.

When f is finite, $f_\bullet : \mathcal{V}_0 \to \mathcal{V}_0$ is a tree map in the sense of Sect. 2.6, so the results in that section apply, but in our approach here we did not need them. In contrast, the approach in [FJ07] consists of first using the tree analysis in Sect. 2.6 to construct an eigenvaluation.

Using numerical invariants one can show that f preserves the type of a valuation in the sense of Sect. 7.7. There is also a rough analogue of the ramification locus for selfmaps of the Berkovich projective line as in Sect. 4.7. At least in the case of a finite map, the ramification locus is a finite subtree given by the convex hull of the preimages of the root ord_0.

While this is not pursued in [FJ07], the induced dynamics on the valuative tree is somewhat similar to the dynamics of a selfmap of the unit disc over $\mathbf{C}$. Indeed, recall from Sect. 7.10 that we can embed the valuative tree inside the Berkovich unit disc over the field of Laurent series (although this does not seem very useful from a dynamical point of view). In particular, the dynamics is (essentially) globally attracting. This is in sharp contrast with selfmaps of the Berkovich projective line that are nonrepelling on hyperbolic space $\mathbf{H}$.

For simplicity we only studied the dynamics of polynomial maps, but the analysis goes through also for formal fixed point germs. In particular, it applies to fixed point germs defined by rational maps of a projective surface and to holomorphic (perhaps transcendental) fixed point germs. In the latter case, one can really interpret $c_\infty(f)$ as a speed at which typical orbits tend to 0, see [FJ07, Theorem B].

8.8 Other Ground Fields

Let us briefly comment on the case when the field K is not algebraically closed. Specifically, let us argue why Theorem B continues to hold in this case.

Let K^a be the algebraic closure of K and $G = \mathrm{Gal}(K^a/K)$ the Galois group. Then $\mathbf{A}^2(K) \simeq \mathbf{A}^2(K^a)/G$ and any polynomial mapping $f : \mathbf{A}^2(K) \to \mathbf{A}^2(K)$ induces a equivariant polynomial mapping $f : \mathbf{A}^2(K^a) \to \mathbf{A}^2(K^a)$.

If the point $0 \in \mathbf{A}^2(K)$ is K-rational, then it has a unique preimage in $0 \in K^a$ and the value of $\mathrm{ord}_0(\phi)$, for $\phi \in R$, is the same when calculated over K or over K^a. The same therefore holds for $c(f^n)$, so since Theorem B holds over K^a, it also holds over K.

In general, $0 \in \mathbf{A}^2$ has finitely many preimages $0_j \in \mathbf{A}^2(K^a)$ but if $\phi \in R$ is a polynomial with coefficients in K, then $\mathrm{ord}_0(\phi) = \mathrm{ord}_{0_j}(\varphi)$ for all j. Again we can deduce Theorem B over K from its counterpart over K^a, although some care needs to be taken to prove that c_∞ is a quadratic integer in this case.

Alternatively, we can consider the action of f directly on $\mathbf{A}^2_{\mathrm{Berk}}(K)$. As noted in Sect. 7.11, the subset of semivaluations centered at 0 is still the cone over a tree and we can consider the induced dynamics. The argument for proving that c_∞ is a quadratic integer, using value groups, carries over to this setting.

8.9 Notes and Further References

In [FJ07] and [FJ11] we used the notation f_*v instead of $f(v)$ as the action of f on the valuative tree is given as a pushforward. However, one usually does not denote induced maps on Berkovich spaces as pushforwards, so I decided to deviate from *loc. cit.* in order to keep the notation uniform across these notes.

In analogy with the degree growth of polynomial maps (see Sect. 10.7) I would expect the sequence $(c(f^n))_{n=0}^\infty$ to satisfy an integral linear recursion relation, but this has not yet been established.[26]

My own path to Berkovich spaces came through joint work with Charles Favre. Theorem B, in a version for holomorphic selfmaps of $\mathbf{P}^2$, has ramifications for problem of equidistribution to the Green current. See [FJ03] and also [DS08, Par11] for higher dimensions.

9 The Valuative Tree at Infinity

In order to study the dynamics at infinity of polynomial maps of $\mathbf{A}^2$ we will use the subspace of the Berkovich affine plane $\mathbf{A}^2_{\mathrm{Berk}}$ consisting of semivaluations centered at infinity. As in the case of semivaluations centered at a point, this is a cone over a

---

[26] The existence of such a relation has been established by W. Gignac and M. Ruggiero in arXiv:1209.3450.

tree that we call the *valuative tree at infinity*.[27] Its structure is superficially similar
to that of the valuative tree at a point, which we will refer to as the *local case*, but,
as we will see, there are some significant differences.

9.1   Setup

Let K be an algebraically closed field of characteristic zero, equipped with the trivial
valuation. (See Sect. 9.8 for the case of other ground fields.) Further, R and F are
the coordinate ring and function field of the affine plane $\mathbf{A}^2$ over K. Recall that the
Berkovich affine plane $\mathbf{A}^2_{\mathrm{Berk}}$ is the set of semivaluations on R that restrict to the
trivial valuation on K.

A *linear system* $|\mathfrak{M}|$ of curves on $\mathbf{A}^2$ is the projective space associated to
a nonzero, finite-dimensional vector space $\mathfrak{M} \subseteq R$. The system is *free* if its
base locus is empty, that is, for every point $\xi \in \mathbf{A}^2$ there exists a polynomial
$\phi \in \mathfrak{M}$ with $\phi(\xi) \neq 0$. For any linear system $|\mathfrak{M}|$ and any $v \in \mathbf{A}^2_{\mathrm{Berk}}$ we write
$v(|\mathfrak{M}|) = \min\{v(\phi) \mid \phi \in \mathfrak{M}\}$.

9.2   Valuations Centered at Infinity

We let $\hat{\mathcal{V}}_\infty \subseteq \mathbf{A}^2_{\mathrm{Berk}}$ denote the set of semivaluations v having center at infinity, that
is, such that $v(\phi) < 0$ for some polynomial $\phi \in R$. Note that $\hat{\mathcal{V}}_\infty$ is naturally a
pointed cone: in contrast to $\hat{\mathcal{V}}_0$ there is no element 'triv$_\infty$'.

The valuative tree at infinity is the base of this cone and we want to realize it as
a "section". In the local case, the valuative tree at a closed point $0 \in \mathbf{A}^2$ was defined
using the maximal ideal $\mathfrak{m}_0$. In order to do something similar at infinity, we fix an
embedding $\mathbf{A}^2 \hookrightarrow \mathbf{P}^2$. This allows us to define the *degree* of a polynomial in R and
in particular defines the free linear system $|\mathcal{L}|$ of *lines*, associated to the subspace
$\mathcal{L} \subseteq R$ of *affine functions* on $\mathbf{A}^2$, that is, polynomials of degree at most one. Note
that $v \in \mathbf{A}^2_{\mathrm{Berk}}$ has center at infinity iff $v(|\mathcal{L}|) < 0$.

We say that two polynomials z_1, z_2 are affine coordinates on $\mathbf{A}^2$ if $\deg z_i = 1$ and
$R = K[z_1, z_2]$. In this case, $F = K(z_1, z_2)$ and $v(|\mathcal{L}|) = \min\{v(z_1), v(z_2)\}$.

Definition 9.1 The *valuative tree at infinity* $\mathcal{V}_\infty$ is the set of semivaluations $v \in \mathbf{A}^2_{\mathrm{Berk}}$ such that $v(|\mathcal{L}|) = -1$.

The role of $\mathrm{ord}_0 \in \mathcal{V}_0$ is played by the valuation $\mathrm{ord}_\infty \in \mathcal{V}_\infty$, defined by

$$\mathrm{ord}_\infty(\phi) = -\deg(\phi). \tag{64}$$

---

[27]The notation in these notes differs from [FJ07,FJ11] where the valuative tree at infinity is denoted
by $\mathcal{V}_0$. In *loc. cit.* the valuation ord_∞ defined in (64) is denoted by $-\deg$.

In particular, $v(\phi) \geq \mathrm{ord}_\infty(\phi)$ for every $\phi \in R$ and every $v \in \mathcal{V}_\infty$. We emphasize that both $\mathcal{V}_\infty$ and ord_∞ depend on a choice of embedding $\mathbf{A}^2 \hookrightarrow \mathbf{P}^2$.

We equip $\mathcal{V}_\infty$ and $\hat{\mathcal{V}}_\infty$ with the subspace topology from $\mathbf{A}^2_{\mathrm{Berk}}$. It follows from Tychonoff's theorem that $\mathcal{V}_\infty$ is a compact Hausdorff space. The space $\hat{\mathcal{V}}_\infty$ is open in $\mathbf{A}^2_{\mathrm{Berk}}$ and its boundary consists of the trivial valuation $\mathrm{triv}_{\mathbf{A}^2}$ and the set of semivaluations centered at a curve in $\mathbf{A}^2$.

As in the local case, we can classify the elements of $\hat{\mathcal{V}}_\infty$ into curve semivaluations, divisorial valuations, irrational valuations and infinitely singular valuations. We do this by considering v as a semivaluation on the ring $\hat{\mathcal{O}}_{\mathbf{P}^2, \xi}$, where ξ is the center of ξ on $\mathbf{P}^2$.

9.3 Admissible Compactifications

The role of a blowup of $\mathbf{A}^2$ above a closed point is played here by a *compactification* of $\mathbf{A}^2$, by which we mean a projective surface containing $\mathbf{A}^2$ as Zariski open subset. To make the analogy even stronger, recall that we have fixed an embedding $\mathbf{A}^2 \hookrightarrow \mathbf{P}^2$. We will use

Definition 9.2 An *admissible compactification* of $\mathbf{A}^2$ is a smooth projective surface X containing $\mathbf{A}^2$ as a Zariski open subset, such that the induced birational map $X \dashrightarrow \mathbf{P}^2$ induced by the identity on $\mathbf{A}^2$, is regular.

By the structure theorem of birational surface maps, this means that the morphism $X \to \mathbf{P}^2$ is a finite composition of point blowups above infinity. The set of admissible compactifications is naturally partially ordered and in fact a directed set: any two admissible compactifications are dominated by a third.

Many of the notions below will in fact not depend on the choice of embedding $\mathbf{A}^2 \hookrightarrow \mathbf{P}^2$ but would be slightly more complicated to state without it.

Remark 9.3 Some common compactifications of $\mathbf{A}^2$, for instance $\mathbf{P}^1 \times \mathbf{P}^1$, are not admissible in our sense. However, the set of admissible compactifications is cofinal among compactifications of $\mathbf{A}^2$: If Y is an irreducible, normal projective surface containing $\mathbf{A}^2$ as a Zariski open subset, then there exists an admissible compactification X of $\mathbf{A}^2$ such that the birational map $X \dashrightarrow Y$ induced by the identity on $\mathbf{A}^2$ is regular. Indeed, X is obtained by resolving the indeterminacy points of the similarly defined birational map $\mathbf{P}^2 \dashrightarrow Y$. See [Mor73, Kis02] for a classification of smooth compactifications of $\mathbf{A}^2$.

9.3.1 Primes and Divisors at Infinity

Let X be an admissible compactification of $\mathbf{A}^2$. A *prime at infinity* of X is an irreducible component of $X \setminus \mathbf{A}^2$. We often identify a prime of X at infinity with its strict transform in any compactification X' dominating X. In this way we can identify a prime at infinity E (of some admissible compactification) with the corresponding divisorial valuation ord_E.

Any admissible compactification contains a special prime L_∞, the strict transform of $\mathbf{P}^2 \setminus \mathbf{A}^2$. The corresponding divisorial valuation is $\mathrm{ord}_{L_\infty} = \mathrm{ord}_\infty$.

We say that a point in $X \setminus \mathbf{A}^2$ is a *free point* if it belongs to a unique prime at infinity; otherwise it is a *satellite point*.

A *divisor at infinity* on X is a divisor supported on $X \setminus \mathbf{A}^2$. We write $\mathrm{Div}_\infty(X)$ for the abelian group of divisors at infinity. If E_i, $i \in I$ are the primes of X at infinity, then $\mathrm{Div}_\infty(X) \simeq \bigoplus_i \mathbf{Z} E_i$.

9.3.2 Intersection Form and Linear Equivalence

We have the following basic facts.

Proposition 9.4 *Let X be an admissible compactification of $\mathbf{A}^2$. Then*

(i) *Every divisor on X is linearly equivalent to a unique divisor at infinity, so $\mathrm{Div}_\infty(X) \simeq \mathrm{Pic}(X)$.*

(ii) *The intersection form on $\mathrm{Div}_\infty(X)$ is nondegenerate and unimodular. It has signature $(1, \rho(X) - 1)$.*

Proof We argue by induction on the number of blowups needed to obtain X from $\mathbf{P}^2$. If $X = \mathbf{P}^2$, then the statement is clear: $\mathrm{Div}_\infty(X) = \mathrm{Pic}(X) = \mathbf{Z} L_\infty$ and $(L_\infty \cdot L_\infty) = 1$. For the inductive step, suppose $\pi' = \pi \circ \mu$, where μ is the simple blowup of a closed point on $X \setminus \mathbf{A}^2$, resulting in an exceptional prime E. Then we have an orthogonal decomposition $\mathrm{Div}_\infty(X') = \mu^* \mathrm{Div}_\infty(X) \oplus \mathbf{Z} E$, $\mathrm{Pic}(X') = \mu^* \mathrm{Pic}(X) \oplus \mathbf{Z} E$ and $(E \cdot E) = -1$.

Statement (ii) about the intersection form is also a consequence of the Hodge Index Theorem and Poincaré Duality. $\qquad\square$

Concretely, the isomorphism $\mathrm{Pic}(X) \simeq \mathrm{Div}_\infty(X)$ can be understood as follows. Any irreducible curve C in X that is not contained in $X \setminus \mathbf{A}^2$ is the closure in X of an affine curve $\{\phi = 0\}$ for some polynomial $\phi \in R$. Then C is linearly equivalent to the element in $\mathrm{Div}_\infty(X)$ defined as the divisor of poles of ϕ, where the latter is viewed as a rational function on X.

Let E_i, $i \in I$ be the primes of X at infinity. It follows from Proposition 9.4 that for each $i \in I$ there exists a divisor $\check{E}_i \in \mathrm{Div}_\infty(X)$ such that $(\check{E}_i \cdot E_i) = 1$ and $(\check{E}_i \cdot E_j) = 0$ for all $j \neq i$.

9.3.3 Invariants of Primes at Infinity

Analogously to the local case (see Sect. 7.3.6) we associate two basic numerical invariants α_E and A_E to any prime E at infinity (or, equivalently, to the associated divisorial valuation $\mathrm{ord}_E \in \hat{\mathcal{V}}_\infty$.

To define α_E, pick an admissible compactification X of $\mathbf{A}^2$ in which E is a prime at infinity. Above we defined the divisor $\check{E} = \check{E}_X \in \mathrm{Div}_\infty(X)$ by duality: $(\check{E}_X \cdot E) = 1$ and $(\check{E}_X \cdot F) = 0$ for all primes $F \neq E$ of X at infinity. Note that if X' is an admissible compactification dominating X, then the divisor $\check{E}_{X'}$ on X' is the pullback of $\check{E}_X$ under the morphism $X' \to X$. In particular, the self-intersection number

$$\alpha_E := \alpha(\mathrm{ord}_E) := (\check{E} \cdot \check{E})$$

is an integer independent of the choice of X.

The second invariant is the *log discrepancy* A_E. Let ω be a nonvanishing regular 2-form on $\mathbf{A}^2$. If X is an admissible compactification of $\mathbf{A}^2$, then ω extends as a rational form on X. For any prime E of X at infinity, with associated divisorial valuation $\mathrm{ord}_E \in \hat{\mathcal{V}}_\infty$, we define

$$A_E := A(\mathrm{ord}_E) := 1 + \mathrm{ord}_E(\omega). \tag{65}$$

This is an integer whose value does not depend on the choice of X or ω. Note that $A_{L_\infty} = -2$ since ω has a pole of order 3 along L_∞. In general, A_E can be positive or negative.

We shall later need the analogues of (40) and (41). Thus let X be an admissible compactification of $\mathbf{A}^2$ and X' the blowup of X at a free point $\xi \in X \setminus \mathbf{A}^2$. Let E' be the "new" prime of X', that is, the inverse image of ξ in X'. Then

$$A_{E'} = A_E + 1, \ b_{E'} = b_E \quad \text{and} \quad \check{E}' = \check{E} - E', \tag{66}$$

where, in the right hand side, we identify the divisor $\check{E} \in \mathrm{Div}_\infty(X)$ with its pullback to X'. As a consequence,

$$\alpha_{E'} := (\check{E}' \cdot \check{E}') = (\check{E} \cdot \check{E}) - 1 = \alpha_E - 1. \tag{67}$$

Generalizing both Sects. 7.3.6 and 9.3.3, the invariants α_E and A_E can in fact be defined for any divisorial valuation ord_E in the Berkovich affine plane.

9.3.4 Positivity

Recall that in the local case, the notion of relative positivity was very well behaved and easy to understand, see Sect. 7.3.5. Here the situation is much more subtle, and this will account for several difficulties.

As usual, we say that a divisor $Z \in \mathrm{Div}(X)$ is *effective* if it is a positive linear combination of prime divisors on X. We also say that $Z \in \mathrm{Div}(X)$ is *nef* if $(Z \cdot W) \geq 0$ for all effective divisors W. These notions make sense also for $\mathbf{Q}$-divisors. It is a general fact that if $Z \in \mathrm{Div}(X)$ is nef, then $(Z \cdot Z) \geq 0$.

Clearly, the semigroup of effective divisors in $\mathrm{Div}_\infty(X)$ is freely generated by the primes E_i, $i \in I$ at infinity. A divisor $Z \in \mathrm{Div}_\infty(X)$ is *nef at infinity* if $(Z \cdot W) \geq 0$ for every effective divisor $W \in \mathrm{Div}_\infty(X)$. This simply means that $(Z \cdot E_i) \geq 0$ for all $i \in I$. It follows easily that the subset of $\mathrm{Div}_\infty(X)$ consisting of divisors that are nef at infinity is a free semigroup generated by the $\check{E}_i$, $i \in I$.

We see that a divisor $Z \in \mathrm{Div}_\infty(X)$ is nef iff it is nef at infinity and, in addition, $(Z \cdot C) \geq 0$ whenever C is the closure in X of an irreducible curve in $\mathbf{A}^2$. In general, a divisor that is nef at infinity may not be nef.

Example 9.5 Consider the surface X obtained by first blowing up any closed point at infinity, creating the prime E_1, then blowing up a free point on E_1, creating the prime E_2. Then the divisor $Z := \check{E}_2 = L_\infty - E_2$ is nef at infinity but Z is not nef since $(Z \cdot Z) = -1 < 0$.

However, a divisor $Z \in \mathrm{Div}_\infty(X)$ that is nef at infinity and effective is always nef: as above it suffices to show that $(Z \cdot C) \geq 0$ whenever C is the closure in X of a curve in $\mathbf{A}^2$. But $(E_i \cdot C) \geq 0$ for all $i \in I$, so since Z has nonnegative coefficients in the basis E_i, $i \in I$, we must have $(Z \cdot C) \geq 0$.

On the other hand, it is possible for a divisor to be nef but not effective. The following example was communicated by Adrien Dubouloz.

Example 9.6 Pick two distinct points ξ_1, ξ_2 on the line at infinity L_∞ in $\mathbf{P}^2$ and let C be a conic passing through ξ_1 and ξ_2. Blow up ξ_1 and let D be the exceptional divisor. Now blow up ξ_2, creating E_1, blow up $C \cap E_1$, creating E_2 and finally blow up $C \cap E_2$ creating F. We claim that the non-effective divisor $Z = 2D + 5L_\infty + 3E_1 + E_2 - F$ on the resulting surface X is nef.

To see this, we successively contract the primes L_∞, E_1 and E_2. A direct computation shows that each of these is a (-1)-curve at the time we contract it, so by Castelnuovo's criterion we obtain a birational morphism $\mu : X \to Y$, with Y a smooth rational surface. Now Y is isomorphic to $\mathbf{P}^1 \times \mathbf{P}^1$. Indeed, one checks that $(F \cdot F) = (C \cdot C) = 0$ and $(F \cdot C) = 1$ on Y and it is easy to see in coordinates that each of F and C is part of a fibration on Y. Now Z is the pullback of the divisor $W = 2D - F$ on Y. Further, $\mathrm{Pic}(Y) \simeq \mathbf{Z}C \oplus \mathbf{Z}F$ and $(W \cdot C) = 1 > 0$ and $(W \cdot F) = 2 > 0$, so W is ample on Y and hence $Z = \mu^* W$ is nef on X.

Finally, in contrast to the local case (see Proposition 7.4) it can happen that a divisor $Z \in \mathrm{Div}_\infty(X)$ is nef but that the line bundle $\mathcal{O}_X(Z)$ has base points, that is, it is not generated by its global sections.

Example 9.7 Consider the surface X obtained from blowing $\mathbf{P}^2$ nine times, as follows. First blow up at three distinct points on L_∞, creating primes E_{1j}, $j = 1, 2, 3$. On each E_{1j} blow up a free point, creating a new prime E_{2j}. Finally blow up a free point on each E_{2j}, creating a new prime E_{3j}. Set $Z = 3L_\infty + \sum_{j=1}^3 (2E_{2j} + E_{1j})$. Then $Z = \sum_{j=1}^3 \check{E}_{3j}$, so Z is nef at infinity. Since Z is also effective, it must be nef.

However, we claim that if the points at which we blow up are generically chosen, then the line bundle $\mathcal{O}_X(Z)$ is not generated by its global sections. To see this, consider a global section of $\mathcal{O}_X(Z)$ that does not vanish identically along L_∞. Such a section is given by a polynomial $\phi \in R$ of degree 3 satisfying $\mathrm{ord}_{E_{ij}}(\phi) = 3-i$, $1 \leq i, j \leq 3$. This gives nine conditions on ϕ. Note that if ϕ is such a section, then so is $\phi - c$ for any constant c, so we may assume that ϕ has zero constant coefficient. Thus ϕ is given by eight coefficients. For a generic choice of points blown up, no such polynomial ϕ will exist. This argument is of course not rigorous, but can be made so by an explicit computation in coordinates that we invite the reader to carry out.

9.4   Valuations and Dual Fans and Graphs

Analogously to Sect. 7.5 we can realize $\hat{\mathcal{V}}_\infty$ and $\mathcal{V}_\infty$ as inverse limits of dual fans and graphs, respectively.

To an admissible compactification X of $\mathbf{A}^2$ we associate a dual fan $\hat{\Delta}(X)$ with integral affine structure $\mathrm{Aff}(X) \simeq \mathrm{Div}_\infty(X)$. This is done exactly as in the local case, replacing exceptional primes with primes at infinity. Inside the dual fan we embed the dual graph $\Delta(X)$ using the integral affine function associated to the divisor $\pi^* L_\infty = \sum_i b_i E_i \in \mathrm{Div}_\infty(X)$. The dual graph is a tree.

The numerical invariants A_E and α_E uniquely to homogeneous functions A and α on the dual fan $\hat{\Delta}(X)$ of degree one and two, respectively and such that these functions are affine on the dual graph. Then A and α give parametrizations of the dual graph rooted in the vertex corresponding to L_∞. We equip the dual graph with the metric associated to the parametrization α: the length of a simplex σ_{ij} is equal to $1/(b_i b_j)$. We could also (but will not) use A to define a metric on the dual graph. This metric is the same as the one induced by the integral affine structure: the length of the simplex σ_{ij} is $m_{ij}/(b_i b_j)$, where $m_{ij} = \gcd\{b_i, b_j\}$ is the multiplicity of the segment.

Using monomial valuations we embed the dual fan as a subset $|\hat{\Delta}(X)|$ of the Berkovich affine plane. The image $|\hat{\Delta}^*(X)|$ of the punctured dual fan lies in $\hat{\mathcal{V}}_\infty$. The preimage of $\mathcal{V}_\infty \subseteq \hat{\mathcal{V}}_\infty$ under the embedding $|\hat{\Delta}^*(X)| \subseteq \hat{\mathcal{V}}_\infty$ is exactly $|\Delta(X)|$. In particular, a vertex σ_E of the dual graph is identified with the corresponding normalized valuation $v_E \in \mathcal{V}_\infty$, defined by

$$v_E = b_E^{-1} \mathrm{ord}_E \quad \text{where } b_E := -\mathrm{ord}_E(|\mathfrak{L}|). \tag{68}$$

Note that $v_{L_\infty} = \mathrm{ord}_{L_\infty} = \mathrm{ord}_\infty$.

We have a retraction $r_X : \hat{\mathcal{V}}_\infty \to |\hat{\Delta}^*(X)|$ that maps $\mathcal{V}_\infty$ onto $|\Delta(X)|$. The induced maps

$$r : \mathcal{V}_\infty \to \varprojlim_X |\Delta(X)| \quad \text{and} \quad r : \hat{\mathcal{V}}_\infty \to \varprojlim_X |\hat{\Delta}^*(X)| \tag{69}$$

are homeomorphisms. The analogue of Lemma 7.12 remains true and we have the following analogue of Lemma 7.11.

Lemma 9.8 *If $v \in \hat{\mathcal{V}}_\infty$ and X is an admissible compactification of $\mathbf{A}^2$, then*

$$(r_X v)(\phi) \leq v(\phi)$$

for every polynomial $\phi \in R$, with equality if the closure in X of the curve $(\phi = 0) \subseteq \mathbf{A}^2$ does not pass through the center of v on X.

The second homeomorphism in (69) equips $\hat{\mathcal{V}}_\infty$ with an integral affine structure: a function φ on $\hat{\mathcal{V}}_\infty$ is integral affine if it is of the form $\varphi = \varphi_X \circ r_X$, where $\varphi_X \in \mathrm{Aff}(X)$.

The first homeomorphism in (69) induces a metric tree structure on $\mathcal{V}_\infty$ as well as two parametrizations[28]

$$\alpha : \mathcal{V}_\infty \to [-\infty, 1] \quad \text{and} \quad A : \mathcal{V}_\infty \to [2, \infty] \tag{70}$$

of $\mathcal{V}_\infty$, viewed as a tree rooted in ord_∞. We extend A and α as homogeneous functions on $\hat{\mathcal{V}}_\infty$ of degrees one and two, respectively.

9.5 Potential Theory

Since $\mathcal{V}_\infty$ is a metric tree, we can do potential theory on it, but just as in the case of the valuative tree at a closed point, we need to tweak the general approach in Sect. 2.5. The reason is again that one should view a function on $\mathcal{V}_\infty$ as the restriction of a homogeneous function on $\hat{\mathcal{V}}_\infty$.

A first guideline is that functions of the form $\log|\mathfrak{M}|$, defined by[29]

$$\log|\mathfrak{M}|(v) = -v(|\mathfrak{M}|) \tag{71}$$

should be subharmonic on $\mathcal{V}_\infty$, for any linear system $|\mathfrak{M}|$ on $\mathbf{A}^2$. In particular, the function $\log|\mathfrak{L}| \equiv 1$ should be subharmonic (but not harmonic). A second guideline is that the Laplacian should be closely related to the intersection product on divisors at infinity.

---

[28]In [FJ04] the parametrization A is called *thinness* whereas $-\alpha$ is called *skewness*.

[29]As in Sect. 7.8 the notation reflects the fact that $|\cdot| := e^{-v}$ is a seminorm on R.

9.5.1 Subharmonic Functions and Laplacian on $\mathcal{V}_\infty$

As in Sect. 7.8.1 we extend the valuative tree $\mathcal{V}_\infty$ to a slightly larger tree $\tilde{\mathcal{V}}_\infty$ by connecting the root ord_∞ to a point G using an interval of length one. Let $\tilde{\Delta}$ denote the Laplacian on $\tilde{\mathcal{V}}_\infty$.

We define the class $\mathrm{SH}(\mathcal{V}_\infty)$ of *subharmonic functions* on $\mathcal{V}_\infty$ as the set of restrictions to $\mathcal{V}_\infty$ of functions $\varphi \in \mathrm{QSH}(\tilde{\mathcal{V}}_\infty)$ such that

$$\varphi(G) = 2\varphi(\mathrm{ord}_\infty) \quad \text{and} \quad \tilde{\Delta}\varphi = \rho - a\delta_G,$$

where ρ is a positive measure supported on $\mathcal{V}_\infty$ and $a = \rho(\mathcal{V}_\infty) \geq 0$. In particular, φ is affine of slope $-\varphi(\mathrm{ord}_\infty)$ on the segment $[G, \mathrm{ord}_\infty[= \tilde{\mathcal{V}}_\infty \setminus \mathcal{V}_\infty$. We then define $\Delta\varphi := \rho = (\tilde{\Delta}\varphi)|_{\mathcal{V}_\infty}$. For example, if $\varphi \equiv 1$ on $\mathcal{V}_\infty$, then $\varphi(G) = 2$, $\tilde{\Delta}\varphi = \delta_{\mathrm{ord}_\infty} - \delta_G$ and $\Delta\varphi = \delta_{\mathrm{ord}_\infty}$.

From this definition and the analysis in Sect. 2.5 one deduces:

Proposition 9.9 *Let $\varphi \in \mathrm{SH}(\mathcal{V}_\infty)$ and write $\rho = \Delta\varphi$. Then:*

(i) φ is decreasing in the partial ordering of $\mathcal{V}_\infty$ rooted in ord_∞;
(ii) $\varphi(\mathrm{ord}_\infty) = \rho(\mathcal{V}_\infty)$;
(iii) $|D_{\vec{v}}\varphi| \leq \rho(\mathcal{V}_\infty)$ for all tangent directions $\vec{v}$ in $\mathcal{V}_\infty$.

As a consequence we have the estimate

$$\alpha(v)\varphi(\mathrm{ord}_\infty) \leq \varphi(v) \leq \varphi(\mathrm{ord}_\infty) \tag{72}$$

for all $v \in \mathcal{V}_\infty$. Here $\alpha : \mathcal{V}_\infty \to [-\infty, +1]$ is the parametrization in (70). It is important to remark that a subharmonic function can take both positive and negative values. In particular, (72) is not so useful when $\alpha(v) < 0$.

The exact sequence in (9) shows that

$$\Delta : \mathrm{SH}(\mathcal{V}_\infty) \to \mathcal{M}^+(\mathcal{V}_\infty), \tag{73}$$

is a homeomorphism whose inverse is given by

$$\varphi(v) = \int_{\mathcal{V}_\infty} \alpha(w \wedge_{\mathrm{ord}_\infty} v)\,d\rho(w). \tag{74}$$

The compactness properties in Sect. 2.5 carry over to the space $\mathrm{SH}(\mathcal{V}_\infty)$. In particular, for any $C > 0$, the set $\{\varphi \in \mathrm{SH}(\mathcal{V}_\infty) \mid \varphi(\mathrm{ord}_\infty) \leq C\}$ is compact. Further, if $(\varphi_i)_i$ is a decreasing net in $\mathrm{SH}(\mathcal{V}_\infty)$, and $\varphi := \lim \varphi_i$, then $\varphi \in \mathrm{SH}(\mathcal{V}_\infty)$. Moreover, if $(\varphi_i)_i$ is a family in $\mathrm{SH}(\mathcal{V}_\infty)$ with $\sup_i \varphi_i(\mathrm{ord}_\infty) < \infty$, then the upper semicontinuous regularization of $\varphi := \sup_i \varphi_i$ belongs to $\mathrm{SH}(\mathcal{V}_\infty)$.

While the function -1 on $\mathcal{V}_\infty$ is not subharmonic, it is true that $\max\{\varphi, r\}$ is subharmonic whenever $\varphi \in \mathrm{SH}(\mathcal{V}_\infty)$ and $r \in \mathbf{R}$.

9.5.2 Laplacian of Integral Affine Functions

Any integral affine function φ on $\hat{\mathcal{V}}_\infty$ is associated to a divisor at infinity $Z \in \mathrm{Div}_\infty(X)$ for some admissible compactification X of $\mathbf{A}^2$: the value of φ at a divisorial valuation ord_{E_i} is the coefficient $\mathrm{ord}_{E_i}(Z)$ of E_i in Z. Using the same computations as in the proof of Proposition 7.15 we show that

$$\Delta\varphi = \sum_{i \in I} b_i (Z \cdot E_i) \delta_{v_i},$$

where $b_i = -\mathrm{ord}_{E_i}(|\mathcal{L}|) \geq 1$ and $v_i = b_i^{-1}\,\mathrm{ord}_{E_i}$. In particular, φ is subharmonic iff Z is nef at infinity.

Recall that we have defined divisors $\check{E}_i \in \mathrm{Div}_\infty(X)$ such that $(\check{E}_i \cdot E_i) = 1$ and $(\check{E}_i \cdot E_j) = 0$ for all $j \neq i$. The integral affine function φ_i on $\mathcal{V}_\infty$ associated to $\check{E}_i$ is subharmonic and satisfies $\Delta\varphi_i = b_i\delta_{v_i}$. In view of (74), this shows that $\min_{\mathcal{V}_\infty} \varphi_i = \varphi_i(v_i) = b_i\alpha(v_i)$. This implies

$$\alpha_{E_i} = (\check{E}_i \cdot \check{E}_i) = \mathrm{ord}_{E_i}(\check{E}_i) = b_i^2\alpha(v_i) = \alpha(\mathrm{ord}_{E_i}). \tag{75}$$

Proposition 9.10 *Let E be a divisor at infinity on some admissible compactification X of $\mathbf{A}^2$. Let $\check{E} \in \mathrm{Div}_\infty(X)$ be the associated element of the dual basis and $v_E = b_E^{-1}\,\mathrm{ord}_E \in \mathcal{V}_\infty$ the associated normalized divisorial valuation. Then $\check{E}$ is nef at infinity and the following statements are equivalent:*

(i) $\check{E}$ is nef;
(ii) $(\check{E} \cdot \check{E}) \geq 0$;
(iii) $\alpha(v_E) \geq 0$.

Proof That $\check{E}$ is nef at infinity is clear from the definition and has already been observed. That (ii) is equivalent to (iii) is an immediate consequence of (75). If $\check{E}$ is nef, then $(\check{E} \cdot \check{E}) \geq 0$, showing that (i) implies (ii). On the other hand, if $\alpha(v_E) \geq 0$, then we have seen above that the minimum on $\mathcal{V}_\infty$ of the integral affine function φ associated to $\check{E}$ is attained at v_E and is nonnegative. Thus $\check{E}$ is effective. Being nef at infinity and effective, $\check{E}$ must be nef, proving that (ii) implies (i). □

9.5.3 Subharmonic Functions from Linear Systems

Let $|\mathfrak{M}|$ be a nonempty linear system of affine curves. We claim that the function $\log|\mathfrak{M}|$, defined by (71) is subharmonic on $\mathcal{V}_\infty$. To see this, note that $\log|\mathfrak{M}| = \max \log|\phi|$, where ϕ ranges over polynomials defining the curves in $|\mathfrak{M}|$. The claim therefore follows from

Exercise 9.11 If $\phi \in R$ is an irreducible polynomial, show that $\log |\phi|$ is subharmonic on V_∞ and that

$$\Delta \log |\phi| = \sum_{j=1}^{n} m_j \delta_{v_j}$$

where $v_j, 1 \le j \le n$ are the curve valuations associated to the all the local branches C_j of $\{\phi = 0\}$ at infinity and where $m_j = (C_j \cdot L_\infty)$ is the local intersection number of C_j with the line at infinity in $\mathbf{P}^2$.

Example 9.12 Fix affine coordinates (z_1, z_2) on $\mathbf{A}^2$ and let $\mathfrak{M} \subseteq R$ be the vector space spanned by z_1 Then $\log |\mathfrak{M}|(v) = \max\{-v(z_1), 0\}$ and $\Delta \log |\mathfrak{M}|$ is a Dirac mass at the monomial valuation with $v(z_1) = 0$, $v(z_2) = -1$.

Example 9.13 Fix affine coordinates (z_1, z_2) on $\mathbf{A}^2$ and let $\mathfrak{M} \subseteq R$ be the vector space spanned by $z_1 z_2$ and the constant function 1. Then $\log |\mathfrak{M}|(v) = \max\{-(v(z_1)+v(z_2)), 0\}$ and $\Delta \log |\mathfrak{M}| = \delta_{v_{-1,1}} + \delta_{v_{1,-1}}$, where v_{t_1,t_2} is the monomial valuation with weights $v_{t_1,t_2}(z_i) = t_i$, $i = 1, 2$.

Proposition 9.14 *Let $|\mathfrak{M}|$ be a linear system of affine curves on $\mathbf{A}^2$. Then the following conditions are equivalent:*

(i) the base locus of $|\mathfrak{M}|$ on $\mathbf{A}^2$ contains no curves;
(ii) the function $\log |\mathfrak{M}|$ is bounded on V_∞;
(iii) the measure $\Delta \log |\mathfrak{M}|$ on V_∞ is supported at divisorial valuations.

Linear systems $|\mathfrak{M}|$ satisfying these equivalent conditions are natural analogs of primary ideals $\mathfrak{a} \subseteq R$ in the local setting.

Sketch of proof That (iii) implies (ii) follows from (74). If the base locus of $|\mathfrak{M}|$ contains an affine curve C, let $v \in V_\infty$ be a curve valuation associated to one of the branches at infinity of C. Then $\log |\mathfrak{M}|(v) = -v(\varphi) = -\infty$ so (ii) implies (i).

Finally, let us prove that (i) implies (iii). Suppose the base locus on $|\mathfrak{M}|$ on $\mathbf{A}^2$ contains no curves. Then we can pick an admissible compactification of $\mathbf{A}^2$ such that the strict transform of $|\mathfrak{M}|$ to X has no base points at infinity. In this case one shows that $\Delta \log |\mathfrak{M}|$ is an atomic measure supported on the divisorial valuations associated to some of the primes of X at infinity. $\square$

In general, it seems very hard to characterize the measures on V_∞ appearing in (iii). Notice that if $\Delta \log |\mathfrak{M}|$ is a Dirac mass at a divisorial valuation v then $\alpha(v) \ge 0$, as follows from (74). There are also sufficient conditions: using the techniques in the proof of Theorem 9.18 one can show that if ρ is an atomic measure with rational coefficients supported on divisorial valuations in the tight tree V'_∞ (see Sect. 9.7) then there exists a linear system $|\mathfrak{M}|$ such that $\log |\mathfrak{M}| \ge 0$ and $\Delta \log |\mathfrak{M}| = n\rho$ for some integer $n \ge 1$.

9.6 Intrinsic Description of Tree Structure on $\mathcal{V}_\infty$

We can try to describe the tree structure on $\mathcal{V}_\infty \simeq \varprojlim |\Delta(X)|$ intrinsically, viewing the elements of $\mathcal{V}_\infty$ purely as semivaluations on the ring R. This is more complicated than in the case of the valuative tree at a closed point (see Sect. 7.9). However, the partial ordering can be characterized essentially as expected:

Proposition 9.15 *If $w, v \in \mathcal{V}_\infty$, then the following are equivalent:*

 (i) *$v \leq w$ in the partial ordering induced by $\mathcal{V}_\infty \simeq \varprojlim |\Delta(X)|$;*
 (ii) *$v(\phi) \leq w(\phi)$ for all polynomials $\phi \in R$;*
 (iii) *$v(|\mathfrak{M}|) \leq w(|\mathfrak{M}|)$ for all free linear systems $|\mathfrak{M}|$ on $\mathbf{A}^2$.*

Proof The implication (i) $\Longrightarrow$ (ii) follows from the subharmonicity of $\log |\phi|$ together with Proposition 9.9 (i). The implication (ii) $\Longrightarrow$ (iii) is obvious. It remains to prove (iii) $\Longrightarrow$ (i).

Suppose $v \not\leq w$ in the partial ordering on $\mathcal{V}_\infty \simeq \varprojlim |\Delta(X)|$. We need to find a free linear system $|\mathfrak{M}|$ on $\mathbf{A}^2$ such that $v(|\mathfrak{M}|) > w(|\mathfrak{M}|)$. First assume that v and w are quasimonomial and pick an admissible compactification X of $\mathbf{A}^2$ such that $v, w \in |\Delta(X)|$. Let $E_i, i \in I$, be the primes of X at infinity. One of these primes is L_∞ and there exists another prime (not necessarily unique) E_i such that $v_i \geq v$. Fix integers r, s with $1 \ll r \ll s$ and define the divisor $Z \in \mathrm{Div}_\infty(X)$ by

$$Z := \sum_{j \in I} \check{E}_j + r \check{E}_i + s \check{L}_\infty.$$

We claim that Z is an *ample* divisor on X. To prove this, it suffices, by the Nakai-Moishezon criterion, to show that $(Z \cdot Z) > 0$, $(Z \cdot E_j) > 0$ for all $j \in I$ and $(Z \cdot C) > 0$ whenever C is the closure in X of a curve $\{\phi = 0\} \subseteq \mathbf{A}^2$.

First, by the definition of $\check{E}_j$ it follows that $(Z \cdot E_j) \geq 1$ for all j. Second, we have $(\check{L}_\infty \cdot C) = \deg \phi$ and $(\check{E}_j \cdot C) = -\mathrm{ord}_{E_i}(\phi) \geq \alpha(v_i) \deg \phi$ for all $j \in I$ in view of (72), so that $(Z \cdot C) > 0$ for $1 \leq r \ll s$. Third, since $(\check{L}_\infty \cdot \check{L}_\infty) = 1$, a similar argument shows that $(Z \cdot Z) > 0$ for $1 \leq r \ll s$.

Since Z is ample, there exists an integer $n \geq 1$ such that the line bundle $\mathcal{O}_X(nZ)$ is base point free. In particular, the corresponding linear system $|\mathfrak{M}| := |\mathcal{O}_X(nZ)|$ is free on $\mathbf{A}^2$. Now, the integral affine function on $|\Delta(X)|$ induced by $\check{L}_\infty$ is the constant function $+1$. Moreover, the integral affine function on $|\Delta(X)|$ induced by $\check{E}_i$ is the function $\varphi_i = b_i \alpha(\cdot \wedge_{\mathrm{ord}_\infty} v_i)$. Since $v_i \geq v$ and $v \not\leq w$, this implies $\varphi_i(v) < \varphi_i(w)$. For $r \gg 1$ this translates into $v(|\mathfrak{M}|) > w(|\mathfrak{M}|)$ as desired.

Finally, if v and w are general semivaluations in $\mathcal{V}_\infty$ with $v \not\leq w$, then we can pick an admissible compactification X of $\mathbf{A}^2$ such that $r_X(v) \not\leq r_X(w)$. By the previous construction there exists a free linear system $|\mathfrak{M}|$ on $\mathbf{A}^2$ such that $r_X(v)(|\mathfrak{M}|) > r_X(w)(|\mathfrak{M}|)$. But since the linear system $|\mathfrak{M}|$ was free also on X, it follows that $v(|\mathfrak{M}|) = r_X(v)(|\mathfrak{M}|)$ and $w(|\mathfrak{M}|) = r_X(w)(|\mathfrak{M}|)$. This concludes the proof. $\quad\square$

The following result is a partial analogue of Corollary 7.23 and characterizes integral affine functions on $\hat{\mathcal{V}}_\infty$.

Proposition 9.16 *For any integral affine function φ on $\hat{\mathcal{V}}_\infty$ there exist free linear systems $|\mathfrak{M}_1|$ and $|\mathfrak{M}_2|$ on $\mathbf{A}^2$ and an integer $n \geq 1$ such that $\varphi = \frac{1}{n}(\log|\mathfrak{M}_1| - \log|\mathfrak{M}_2|)$.*

Proof Pick an admissible compactification X of $\mathbf{A}^2$ such that φ is associated to divisor $Z \in \mathrm{Div}_\infty(X)$. We may write $Z = Z_1 - Z_2$, where $Z_i \in \mathrm{Div}_\infty(X)$ is ample. For a suitable $n \geq 1$, nZ_1 and nZ_2 are very ample, and in particular base point free. We can then take $|\mathfrak{M}_i| = |\mathcal{O}_X(nZ_i)|$, $i = 1, 2$. $\qquad\square$

It seems harder to describe the parametrization α. While (72) implies

$$\alpha(v) \geq \sup_{\phi \in R \setminus 0} \frac{v(\phi)}{\mathrm{ord}_\infty(\phi)}$$

for any v, it is doubtful that equality holds in general.[30] One can show that equality does hold when v is a quasimonomial valuation in the tight tree $\mathcal{V}'_\infty$, to be defined shortly.

9.7 The Tight Tree at Infinity

For the study of polynomial dynamics in Sect. 10, the full valuative tree at infinity is too large. Here we will introduce a very interesting and useful subtree.

Definition 9.17 The *tight tree at infinity* is the subset $\mathcal{V}'_\infty \subseteq \mathcal{V}_\infty$ consisting of semivaluations v for which $A(v) \leq 0 \leq \alpha(v)$.

Since α is decreasing and A is increasing in the partial ordering on $\mathcal{V}_\infty$, it is clear that $\mathcal{V}'_\infty$ is a subtree of $\mathcal{V}_\infty$. Similarly, α (resp. A) is lower semicontinuous (resp. upper semicontinuous) on $\mathcal{V}_\infty$, which implies that $\mathcal{V}'_\infty$ is a closed subset of $\mathcal{V}_\infty$. It is then easy to see that $\mathcal{V}'_\infty$ is a metric tree in the sense of Sect. 2.2.

Similarly, we define $\hat{\mathcal{V}}'_\infty$ as the set of semivaluations $v \in \hat{\mathcal{V}}_\infty$ satisfying $A(v) \leq 0 \leq \alpha(v)$. Thus $\hat{\mathcal{V}}'_\infty = \mathbf{R}^*_+ \mathcal{V}'_\infty$. The subset $\hat{\mathcal{V}}'_\infty \subset \mathbf{A}^2_{\mathrm{Berk}}$ does not depend on the choice of embedding $\mathbf{A}^2 \hookrightarrow \mathbf{P}^2$. In particular, it is invariant under polynomial automorphisms of $\mathbf{A}^2$. Further, $\hat{\mathcal{V}}'_\infty$ is nowhere dense as it contains no curve semivaluations. Its closure is the union of itself and the trivial valuation $\mathrm{triv}_{\mathbf{A}^2}$.

---

[30]P. Mondal has given examples in arXiv:1301.3172 showing that equality does not always hold.

9.7.1 Monomialization

The next, very important result characterizes some of the ends of the tree $\mathcal{V}'_\infty$.

Theorem 9.18 *Let* ord_E *be a divisorial valuation centered at infinity such that* $A(\mathrm{ord}_E) \leq 0 = (\check{E} \cdot \check{E})$. *Then* $A(\mathrm{ord}_E) = -1$ *and there exist coordinates* (z_1, z_2) *on* $\mathbf{A}^2$ *in which* ord_E *is monomial with* $\mathrm{ord}_E(z_1) = -1$ *and* $\mathrm{ord}_E(z_2) = 0$.

This is proved in [FJ07, Theorem A.7]. Here we provide an alternative, more geometric proof. This proof uses the Line Embedding Theorem and is the reason why we work in characteristic zero throughout Sect. 9. (It is quite possible, however, that Theorem 9.18 is true also over an algebraically closed field of positive characteristic).

Proof Let X be an admissible compactification of $\mathbf{A}^2$ on which E is a prime at infinity. The divisor $\check{E} \in \mathrm{Div}_\infty(X)$ is nef at infinity. It is also effective, and hence nef, since $(\check{E} \cdot \check{E}) \geq 0$; see Proposition 9.10.

Let K_X be the canonical class of X. We have $(\check{E} \cdot K_X) = A(\mathrm{ord}_E) - 1 < 0$. By the Hirzebruch-Riemann-Roch Theorem we have

$$\chi(\mathcal{O}_X(\check{E})) = \chi(\mathcal{O}_X) + \frac{1}{2}((\check{E} \cdot \check{E}) - (\check{E} \cdot K_X)) > \chi(\mathcal{O}_X) = 1.$$

Serre duality yields $h^2(\mathcal{O}_X) = h^0(\mathcal{O}_X(K_X - \check{E})) = 0$, so since $h^1(\mathcal{O}_X(\check{E})) \geq 0$ we conclude that $h^0(\mathcal{O}_X(\check{E})) \geq 2$. Thus there exists a nonconstant polynomial $\phi \in R$ that defines a global section of $\mathcal{O}_X(\check{E})$. Since $\check{E}$ is effective, $\phi + t$ is also a global section for any $t \in K$.

Let C_t be the closure in X of the affine curve $(\phi + t = 0) \subset \mathbf{A}^2$. For any t we have $C_t = \check{E}$ in $\mathrm{Pic}(X)$, so $(C_t \cdot E) = 1$ and $(C_t \cdot F) = 0$ for all primes F at infinity different from E. This implies that C_t intersects $X \setminus \mathbf{A}^2$ at a unique point $\xi_t \in E$; this point is furthermore free on E, C_t is smooth at ξ_t, and the intersection is transverse. Since $\mathrm{ord}_E(\phi) = (\check{E} \cdot \check{E}) = 0$, the image of the map $t \mapsto \xi_t$ is Zariski dense in E.

For generic t, the affine curve $C_t \cap \mathbf{A}^2 = (\phi + t = 0)$ is smooth, hence C_t is smooth for these t. By adjunction, C_t is rational. In particular, $C_t \cap \mathbf{A}^2$ is a smooth curve with one place at infinity.

The Line Embedding Theorem by Abhyankar-Moh and Suzuki [AM73, Suz74] now shows that there exist coordinates (z_1, z_2) on $\mathbf{A}^2$ such that $\phi + t = z_2$. We use these coordinates to define a compactification $Y \simeq \mathbf{P}^1 \times \mathbf{P}^1$ of $\mathbf{A}^2$. Let F be the irreducible compactification of $Y \setminus \mathbf{A}^2$ that intersects the strict transform of each curve $z_2 = \mathrm{const}$. Then the birational map $Y \dashrightarrow X$ induced by the identity on $\mathbf{A}^2$ must map F onto E. It follows that $\mathrm{ord}_E = \mathrm{ord}_F$. Now ord_F is monomial in (z_1, z_2) with $\mathrm{ord}_F(z_1) = -1$ and $\mathrm{ord}_F(z_2) = 0$. Furthermore, the 2-form $dz_1 \wedge dz_2$ has a pole of order 2 along F on Y so $A(\mathrm{ord}_F) = -1$. This completes the proof. $\qquad\square$

9.7.2 Tight Compactifications

We say that an admissible compactification X of $\mathbf{A}^2$ is *tight* if $|\Delta(X)| \subseteq \mathcal{V}'_\infty$. Let E_i, $i \in I$ be the primes of X at infinity. Since the parametrization α and the log discrepancy A are both affine on the simplices of $|\Delta(X)|$, X is tight iff $A(v_i) \leq 0 \leq \alpha(v_i)$ for all $i \in I$. In particular, this implies $(\check{E}_i \cdot \check{E}_i) \geq 0$, so the the divisor $\check{E}_i \in \mathrm{Div}_\infty(X)$ is nef for all $i \in I$. Since every divisor in $\mathrm{Div}_\infty(X)$ that is nef at infinity is a positive linear combination of the $\check{E}_i$, we conclude

Proposition 9.19 *If X is a tight compactification of $\mathbf{A}^2$, then the nef cone of X is simplicial.*

See [CPR02, CPR05, GM04, GM05, Mon07] for other cases when the nef cone is known to be simplicial. For a general admissible compactification of $\mathbf{A}^2$ one would, however, expect the nef cone to be rather complicated.

Lemma 9.20 *Let X be a tight compactification of $\mathbf{A}^2$ and ξ a closed point of $X \setminus \mathbf{A}^2$. Let X' be the admissible compactification of $\mathbf{A}^2$ obtained by blowing up ξ. Then X' is tight unless ξ is a free point on a prime E for which $\alpha_E = 0$ or $A_E = 0$.*

Proof If ξ is a satellite point, then X' is tight since $|\Delta(X')| = |\Delta(X)|$.

Now suppose ξ is a free point, belonging to a unique prime on E. Let E' be the prime of X' resulting from blowing up ξ. Then X' is tight iff $\alpha_{E'} := (\check{E}' \cdot \check{E}') \geq 0 \geq A_{E'}$. But it follows from (66) that $A_{E'} = A_E + 1$ and $\alpha_{E'} = \alpha_E - 1$. Hence $\alpha_{E'} \geq 0 \geq A_{E'}$ unless $\alpha_E = 0$ or $A_E = 0$. The proof is complete. □

Corollary 9.21 *If X is a tight compactification of $\mathbf{A}^2$ and $v \in \hat{\mathcal{V}}'_\infty$ is a divisorial valuation, then there exists a tight compactification X' dominating X such that $v \in |\hat{\Delta}^*(X')|$.*

Proof In the proof we shall repeatedly use the analogues at infinity of the results in Sect. 7.7.3, in particular Lemma 7.12.

We may assume $v = \mathrm{ord}_E$ for some prime E at infinity. By Lemma 7.12, the valuation $w := r_X(v)$ is divisorial and $b(w)$ divides $b(v)$. We argue by induction on the integer $b(v)/b(w)$.

By the same lemma we can find an admissible compactification X_0 dominating X such that $|\hat{\Delta}^*(X_0)| = |\hat{\Delta}^*(X)|$, and w is contained in a one-dimensional cone in $|\hat{\Delta}^*(X_0)|$. Then the center of v on X_0 is a free point ξ_0. Let X_1 be the blowup of X_0 in ξ_0. Note that since $v \neq w$ we have $\alpha(w) > \alpha(v) \geq 0 \geq A(v) > A(w)$, so by Lemma 9.20 the compactification X_1 is tight.

If $v \in |\hat{\Delta}^*(X_1)|$ then we are done. Otherwise, set $v_1 = r_{X_1}(v)$. If the center ξ_1 of v on X_1 is a satellite point, then it follows from Lemma 7.12 that $b(v_1) > b(v_0)$. If $b(w) = b(v)$, this is impossible and if $b(w) < b(v)$, we are done by the inductive hypothesis.

The remaining case is when ξ_1 is a free point on E_1, the preimage of ξ_0 under the blowup map. We continue this procedure: assuming that the center of v on X_j is a free point ξ_j, we let X_{j+1} be the blowup of X_j in ξ_j. By (66) we have $A_{E_n} = A_{E_0} + n$. But $A_{E_n} \leq 0$ so the procedure must stop after finitely many steps. When it stops, we either have $v \in |\hat{\Delta}^*(X_n)|$ or the center of v on X_n is a satellite point. In both cases the proof is complete in view of what precedes. □

Corollary 9.22 *If X is a tight compactification of $\mathbf{A}^2$ and $f : \mathbf{A}^2 \to \mathbf{A}^2$ is a polynomial automorphism, then there exists a tight compactification X' such that the birational map $X' \dashrightarrow X$ induced by f is regular.*

Proof Let E_i, $i \in I$ be the primes of X at infinity. Now f^{-1} maps the divisorial valuations $v_i := \mathrm{ord}_{E_i}$ to divisorial valuations $v_i' = \mathrm{ord}_{E_i'}$. We have $v_i' \in \hat{\mathcal{V}}_\infty'$, so after a repeated application of Corollary 9.21 we find an admissible compactification X' of $\mathbf{A}^2$ such that $v_i' \in |\hat{\Delta}^*(X')|$ for all $i \in I$. But then it is easy to check that $f : X' \to X$ is regular. □

Corollary 9.23 *Any two tight compactifications can be dominated by a third, so the set of tight compactifications is a directed set. Furthermore, the retraction maps $r_X : \hat{\mathcal{V}}_\infty \to |\hat{\Delta}^*(X)|$ give rise to homeomorphisms*

$$\hat{\mathcal{V}}_\infty' \xrightarrow{\sim} \varprojlim_X |\hat{\Delta}^*(X)| \quad and \quad \mathcal{V}_\infty' \xrightarrow{\sim} \varprojlim_X |\Delta(X)|,$$

where X ranges over all tight compactifications of $\mathbf{A}^2$.

9.8 Other Ground Fields

Throughout the section we assumed that the ground field was algebraically closed and of characteristic zero. Let us briefly discuss what happens when one or more of these assumptions are not satisfied.

First suppose K is algebraically closed but of characteristic $p > 0$. Everything in Sect. 9 goes through, except for the proof of the monomialization theorem, Theorem 9.18, which relies on the Line Embedding Theorem. On the other hand, it is quite possible that the proof of Theorem 9.18 can be modified to work also in characteristic $p > 0$.

Now suppose K is not algebraically closed. There are two possibilities for studying the set of semivaluations in $\mathbf{A}^2_{\mathrm{Berk}}$ centered at infinity. One way is to pass to the algebraic closure K^a. Let $G = \mathrm{Gal}(K^a/K)$ be the Galois group. Using general theory we have an identification $\mathbf{A}^2_{\mathrm{Berk}}(K) \simeq \mathbf{A}^2_{\mathrm{Berk}}(K^a)/G$ and G preserves the open subset $\hat{\mathcal{V}}_\infty(K^a)$ of semivaluations centered at infinity. Any embedding $\mathbf{A}^2(K) \hookrightarrow \mathbf{P}^2(K)$ induces an embedding $\mathbf{A}^2(K^a) \hookrightarrow \mathbf{P}^2(K^a)$ and allows us to define subsets $\mathcal{V}_\infty(K) \subseteq \hat{\mathcal{V}}_\infty(K)$ and $\mathcal{V}_\infty(K^a) \subseteq \hat{\mathcal{V}}_\infty(K^a)$. Each $g \in G$ maps $\mathcal{V}_\infty(K^a)$ into itself and preserves the partial ordering parametrizations as well as the parametrizations α and A and the multiplicity m. Therefore, the quotient $\mathcal{V}_\infty(K) \simeq \mathcal{V}_\infty(K^a)/G$ also is naturally a tree that we equip with a metric that takes into account the degree of the map $\mathcal{V}_\infty(K^a) \to \mathcal{V}_\infty(K)$.

Alternatively, we can obtain the metric tree structure directly from the dual graphs of the admissible compactifications by keeping track of the residue fields of the closed points being blown up.

9.9   Notes and Further References

The valuative tree at infinity was introduced in [FJ07] for the purposes of studying the dynamics at infinity of polynomial mappings of $\mathbf{C}^2$ (see the next section). It was not explicitly identified as a subset of the Berkovich affine plane over a trivially valued field.

In [FJ07], the tree structure of $\mathcal{V}_\infty$ was deduced by looking at the center on $\mathbf{P}^2$ of a semivaluation in $\mathcal{V}_\infty$. Given a closed point $\xi \in \mathbf{P}^2$, the semivaluations having center at ξ form a tree (essentially the valuative tree at ξ but normalized by $v(L_\infty) = 1$). By gluing these trees together along ord_∞ we see that $\mathcal{V}_\infty$ itself is a tree. The geometric approach here, using admissible compactifications, seems more canonical and amenable to generalization to higher dimensions.

Just as with the valuative tree at a point, I have allowed myself to change the notation from [FJ07]. Specifically, the valuative tree at infinity is (regrettably) denoted $\mathcal{V}_0$ and the tight tree at infinity is denoted $\mathcal{V}_1$. The notation $\mathcal{V}_\infty$ and $\mathcal{V}'_\infty$ seems more natural. Further, the valuation ord_∞ is denoted $-\deg$ in [FJ07].

The tight tree at infinity $\mathcal{V}'_\infty$ was introduced in [FJ07] and tight compactifications in [FJ11]. They are both very interesting notions. The tight tree was studied in [FJ07] using key polynomials, more or less in the spirit of Abhyankar and Moh [AM73]. While key polynomials are interesting, they are notationally cumbersome as they contain a lot of combinatorial information and they depend on a choice of coordinates, something that I have striven to avoid here.

As indicated in the proof of Theorem 9.18, it is possible to study the tight tree at infinity using the basic theory for compact surfaces. In particular, while the proof of the structure result for $\mathcal{V}'_\infty$ in [FJ07] used the Line Embedding Theorem in a crucial way (just as in Theorem 9.18) one can use the framework of tight compactifications together with surface theory to give a proof of the Line Embedding Theorem. (It should be mentioned, however, that by now there are quite a few proofs of the line embedding theorem.)

One can also prove Jung's theorem, on the structure $\mathrm{Aut}(\mathbf{C}^2)$ using the tight tree at infinity. It would be interesting to see if there is a higher-dimensional version of the tight tree at infinity, and if this space could be used to shine some light on the wild automorphisms of $\mathbf{C}^3$, the existence of which was proved by Shestakov and Umirbaev in [SU04].

The log discrepancy used here is a slight variation of the standard notion in algebraic geometry (see [JM12]) but has the advantage of not depending on the choice of compactification. If we fix an embedding $\mathbf{A}^2 \hookrightarrow \mathbf{P}^2$ and $A_{\mathbf{P}^2}$ denotes the usual log discrepancy on $\mathbf{P}^2$, then we have $A(v) = A_{\mathbf{P}^2}(v) - 3v(|\mathfrak{L}|)$.

10 Plane Polynomial Dynamics at Infinity

We now come to the third type of dynamics on Berkovich spaces: the dynamics at infinity of polynomial mappings of $\mathbf{A}^2$. The study will be modeled on the dynamics near a (closed) fixed point as described in Sect. 8. We will refer to the latter situation as the local case.

10.1 Setup

Let K is an algebraically closed field of characteristic zero, equipped with the trivial valuation. (See Sect. 10.8 for the case of other ground fields.) Further, R and F are the coordinate ring and function field of the affine plane $\mathbf{A}^2$ over K. Recall that the Berkovich affine plane $\mathbf{A}^2_{\text{Berk}}$ is the set of semivaluations on R that restrict to the trivial valuation on K.

10.2 Definitions and Results

We keep the notation from Sect. 9 and consider a polynomial mapping $f :$ $\mathbf{A}^2 \to \mathbf{A}^2$, which we assume to be *dominant* to avoid degenerate cases. Given an embedding $\mathbf{A}^2 \hookrightarrow \mathbf{P}^2$, the degree $\deg f$ is defined as the degree of the curve $\deg f^* \ell$ for a general line $\ell \in |\mathfrak{L}|$.

The degree growth sequence $(\deg f^n)_{n \geq 0}$ is submultiplicative,

$$\deg f^{n+m} \leq \deg f^n \cdot \deg f^m,$$

and so the limit

$$d_\infty = \lim_{n \to \infty} (\deg f^n)^{1/n}$$

is well defined. Since f is assumed dominant, $\deg f^n \geq 1$ for all n, hence $d_\infty \geq 1$.

Exercise 10.1 Verify these statements!

Example 10.2 If $f(z_1, z_2) = (z_2, z_1 z_2)$, then $\deg f^n$ is the $(n+1)$th Fibonacci number and $d_\infty = \frac{1}{2}(\sqrt{5} + 1)$ is the golden mean.

Example 10.3 For $f(z_1, z_2) = (z_1^2, z_1 z_2^2)$, $\deg f^n = (n+2)2^{n-1}$ and $d_\infty = 2$.

Exercise 10.4 Compute d_∞ for a skew product $f(z_1, z_2) = (\phi(z_1), \psi(z_1, z_2))$.

Here is the result that we are aiming for.

Theorem C *The number* $d_\infty = d_\infty(f)$ *is a quadratic integer: there exist* $a, b \in \mathbf{Z}$ *such that* $d_\infty^2 = a d_\infty + b$. *Moreover, we are in exactly one of the following two cases:*

(a) there exists $C > 0$ *such that* $d_\infty^n \le \deg f^n \le C d_\infty^n$ *for all* n;
(b) $\deg f^n \sim n d_\infty^n$ *as* $n \to \infty$.

Moreover, case (b) occurs iff f, *after conjugation by a suitable polynomial automorphism of* $\mathbf{A}^2$, *is a skew product of the form*

$$f(z_1, z_2) = (\phi(z_1), \psi(z_1) z_2^{d_\infty} + O_{z_1}(z_2^{d_\infty - 1})),$$

where $\deg \phi = d_\infty$ *and* $\deg \psi > 0$.

The behavior of the degree growth sequence does not depend in an essential way on our choice of embedding $\mathbf{A}^2 \hookrightarrow \mathbf{P}^2$. To see this, fix such an embedding, let $g : \mathbf{A}^2 \to \mathbf{A}^2$ be a polynomial automorphism and set $\tilde{f} := g^{-1} f g$. Then $\tilde{f}^n = g^{-1} f^n g$, $f^n = g \tilde{f}^n g^{-1}$ and so

$$\frac{1}{\deg g \deg g^{-1}} \le \frac{\deg \tilde{f}^n}{\deg f^n} \le \deg g \deg g^{-1}$$

for all $n \ge 1$. As a consequence, when proving Theorem C, we may conjugate by polynomial automorphisms of $\mathbf{A}^2$, if necessary.

10.3 Induced Action

The strategy for proving Theorems C is superficially very similar to the local case explored in Sect. 8. Recall that f extends to a map

$$f : \mathbf{A}_{\mathrm{Berk}}^2 \to \mathbf{A}_{\mathrm{Berk}}^2,$$

given by $f(v)(\phi) := v(f^* \phi)$.

We would like to study the dynamics of f at infinity. For any admissible compactification X of $\mathbf{A}^2$, f extends to a rational map $f : X \dashrightarrow \mathbf{P}^2$. Using resolution of singularities we can find X such that $f : X \to \mathbf{P}^2$ is a morphism. There are then two cases: either $f(E) \subseteq L_\infty$ for every prime E of X at infinity, or there exists a prime E such that $f(E) \cap \mathbf{A}^2 \ne \emptyset$. The first case happens iff f is *proper*.

Recall that $\hat{\mathcal{V}}_\infty$ denotes the set of semivaluations in $\mathbf{A}_{\mathrm{Berk}}^2$ having center at infinity. It easily follows that f is proper iff $f(\hat{\mathcal{V}}_\infty) \subseteq \hat{\mathcal{V}}_\infty$. Properness is the analogue of finiteness in the local case.

10.3.1 The Proper Case

When f is proper, if induces a selfmap

$$f : \hat{\mathcal{V}}_\infty \to \hat{\mathcal{V}}_\infty.$$

Now $\hat{\mathcal{V}}_\infty$ is the pointed cone over the valuative tree at infinity $\mathcal{V}_\infty$, whose elements are normalized by the condition $v(|\mathcal{L}|) = -1$. As in the local case, we can break the action of f on $\hat{\mathcal{V}}_\infty$ into two parts: the induced dynamics

$$f_\bullet : \mathcal{V}_\infty \to \mathcal{V}_\infty,$$

and a multiplier $d(f, \cdot) : \mathcal{V}_\infty \to \mathbf{R}_+$. Here

$$d(f, v) = -v(f^*|\mathcal{L}|),$$

Further, $f_\bullet$ is defined by

$$f_\bullet v = \frac{f(v)}{d(f, v)}.$$

The break-up of the action is compatible with the dynamics in the sense that $(f^n)_\bullet = (f_\bullet)^n$ and

$$d(f^n, v) = \prod_{i=0}^{n-1} d(f, v_i), \quad \text{where } v_i = f_\bullet^i v.$$

Recall that $\mathrm{ord}_\infty \in \mathcal{V}_\infty$ is the valuation given by $\mathrm{ord}_\infty(\phi) = -\deg(\phi)$ for any polynomial $\phi \in R$. We then have

$$\deg f^n = d(f^n, \mathrm{ord}_\infty) = \prod_{i=0}^{n-1} d(f, v_i), \quad \text{where } v_i = f_\bullet^i \, \mathrm{ord}_\infty.$$

Now $v_i \geq \mathrm{ord}_\infty$ on R, so it follows that $\deg f^n \leq (\deg f)^n$ as we already knew. The multiplicative cocycle $d(f, \cdot)$ is the main tool for studying the submultiplicative sequence $(\deg f^n)_{n \geq 0}$.

10.3.2 The Non-Proper Case

When $f : \mathbf{A}^2 \to \mathbf{A}^2$ is dominant but not necessarily proper, there exists at least one divisorial valuation $v \in \hat{\mathcal{V}}_\infty \subseteq \mathbf{A}^2_{\mathrm{Berk}}$ for which $f(v) \in \mathbf{A}^2_{\mathrm{Berk}} \setminus \hat{\mathcal{V}}_\infty$. We can view $f : \hat{\mathcal{V}}_\infty \dashrightarrow \hat{\mathcal{V}}_\infty$ as a partially defined map. Its domain of definition is the open set $\hat{D}_f \subseteq \hat{V}_\infty$ consisting of semivaluations for which there exists an affine function L with $v(f^*L) < 0$. Equivalently, if we as before define $d(f, v) = -v(f^*|\mathcal{L}|)$, then $\hat{D}_f = \{d(f, \cdot) > 0\}$. On $D_f := \hat{D}_f \cap V_\infty$ we define $f_\bullet$ as before, namely $f_\bullet v = f(v)/d(f, v)$.

Notice that $D_{f^n} = \bigcap_{i=0}^{n-1} f_\bullet^{-i} D_f$, so the domain of definition of $f_\bullet^n$ decreases as $n \to \infty$. One may even wonder whether the intersection $\bigcap_n D_{f^n}$ is empty. However, a moment's reflection reveals that ord_∞ belongs to this intersection. More generally, it is not hard to see that the set of valuations $v \in V_\infty$ for which $v(\phi) < 0$ for all nonconstant polynomials ϕ, is a subtree of V_∞ contained in D_f and invariant under f, for any dominant polynomial mapping f.

For reasons that will become apparent later, we will in fact study the dynamics on the even smaller subtree, namely the tight subtree $V_\infty' \subseteq V_\infty$ defined in Sect. 9.7. We shall see shortly that $f_\bullet V_\infty' \subseteq V_\infty'$, so we have a natural induced dynamical system on V_∞' for any dominant polynomial mapping f.

10.4   Invariance of the Tight Tree V_∞'

Theorem B, the local counterpart to Theorem C, follows easily under the additional assumption (not always satisfied) that there exists a quasimonomial valuation $v \in V_0$ such that $f_\bullet v = v$. Indeed, such a valuation satisfies

$$\mathrm{ord}_0 \le v \le \alpha v,$$

where $\alpha = \alpha(v) < \infty$. If $f(v) = cv$, then this gives $c = c_\infty$ and $\alpha^{-1} c_\infty \le c(f^n) \le c_\infty^n$. Moreover, the inclusion $c_\infty \Gamma_v = \Gamma_{f(v)} \subseteq \Gamma_v$ implies that c_∞ is a quadratic integer. See Sect. 8.5.

In the affine case, the situation is more complicated. We cannot just take *any* quasimonomial fixed point v for $f_\bullet$. For a concrete example, consider the product map $f(z_1, z_2) = (z_1^3, z_2^2)$ and let v be the monomial valuation with weights $v(z_1) = 0$, $v(z_2) = -1$. Then $f(v) = 2v$, whereas $d_\infty = 3$. The problem here is that while $v \ge \mathrm{ord}_\infty$, the reverse inequality $v \le C \, \mathrm{ord}_\infty$ does not hold for any constant $C > 0$.

The way around this problem is to use the tight tree V_∞' introduced in Sect. 9.7. Indeed, if $v \in V_\infty'$ is quasimonomial, then either there exists $\alpha = \alpha(v) > 0$ such that $\alpha^{-1} \mathrm{ord}_\infty \le v \le \mathrm{ord}_\infty$ on R, or v is monomial in suitable coordinates on $\mathbf{A}^2$, see Theorem 9.18. As the example above shows, the latter case still has to be treated with some care.

We start by showing that the tight tree is invariant.

Proposition 10.5 *For any dominant polynomial mapping $f : \mathbf{A}^2 \to \mathbf{A}^2$ we have $f(\hat{V}_\infty') \subseteq \hat{V}_\infty'$. In particular, $V_\infty' \subseteq D_f$ and $f_\bullet V_\infty' \subseteq V_\infty'$.*

Sketch of proof It suffices to prove that if $v \in \hat{V}_\infty'$ is divisorial, then $f(v) \in \hat{V}_\infty'$. After rescaling, we may assume $v = \mathrm{ord}_E$. Arguing using numerical invariants as in Sect. 4.4, we show that $f(v)$ is divisorial, of the form $f(v) = r\,\mathrm{ord}_{E'}$ for some prime divisor E' on $\mathbf{A}^2$ (a priori not necessarily at infinity).

We claim that the formula

$$A(f(v)) = A(v) + v(Jf) \tag{76}$$

holds, where Jf denotes the Jacobian determinant of f. Note that the assumption $\alpha(v) \geq 0$ implies $v(Jf) \leq 0$ by (72). Together with the assumption $A(v) \leq 0$, we thus see that $A(f(v)) \leq 0$. In particular, the 2-form ω on $\mathbf{A}^2$ has a pole along E', which implies that E' must be a prime at infinity.

Hence $f(v) \in \hat{\mathcal{V}}_\infty$ and $A(f(v)) \leq 0$. It remains to prove that $\alpha(f(v)) \geq 0$. Let X' be an admissible compactification of $\mathbf{A}^2$ in which E' is a prime at infinity and pick another compactification X of $\mathbf{A}^2$ such that the induced map $f : X \to X'$ is regular. The divisors $\check{E} \in \mathrm{Div}_\infty(X)$ and $\check{E}' \in \mathrm{Div}_\infty(X')$ are both nef at infinity and satisfies $f_*\check{E} = r\check{E}'$. Since $(\check{E} \cdot \check{E}) = \alpha(v) \geq 0$, $\check{E}$ is effective (and hence nef). As a consequence, $\check{E}' = r^{-1} f_*\check{E}$ is effective and hence nef. In particular, $\alpha(f(v)) = r^2(\check{E}' \cdot \check{E}') \geq 0$, which completes the proof.

Finally we prove (76). Write $A_E = A(\mathrm{ord}_E)$ and $A_{E'} = A(\mathrm{ord}_{E'})$. Recall that ω is a nonvanishing 2-form on $\mathbf{A}^2$. Near E' it has a zero of order $A_{E'} - 1$. From the chain rule, and the fact that $f(\mathrm{ord}_E) = r\,\mathrm{ord}_{E'}$, it follows that $f^*\omega$ has a zero of order $r - 1 + r(A_{E'} - 1) = rA_{E'} - 1$ along E. On the other hand we have $f^*\omega = Jf \cdot \omega$ in $\mathbf{A}^2$ and the right hand side vanishes to order $\mathrm{ord}_E(Jf) + A_E - 1$ along E. This concludes the proof. □

10.5 Some Lemmas

Before embarking on the proof of Theorem C, let us record some useful auxiliary results.

Lemma 10.6 *Let $\phi \in R$ be a polynomial, X an admissible compactification of $\mathbf{A}^2$ and E a prime of X at infinity. Let C_X be the closure in X of the curve $\{\phi = 0\}$ in $\mathbf{A}^2$ and assume that C_X intersects E. Then $\deg p \geq b_E$, where $b_E := -\mathrm{ord}_E(|\mathfrak{L}|)$.*

Proof This follows from elementary intersection theory. Let $\pi : X \to \mathbf{P}^2$ be the birational morphism induced by the identity on $\mathbf{A}^2$ and let $C_{\mathbf{P}^2}$ be the closure in $\mathbf{P}^2$ of the curve $\{\phi = 0\} \subseteq \mathbf{A}^2$. Then $\mathrm{ord}_E(\pi^*L_\infty) = b_E$. Assuming that C_X intersects E, we get

$$b_E \leq b_E(C_X \cdot E) \leq (C_X \cdot \pi^*L_\infty) = (C_{\mathbf{P}^2} \cdot L_\infty) = \deg p,$$

where the first equality follows from the projection formula and the second from Bézout's Theorem. □

Applying Lemma 10.6 and Lemma 9.8 to $\phi = f^*L$, for L a general affine function, we obtain

Corollary 10.7 *Let* $f : \mathbf{A}^2 \to \mathbf{A}^2$ *be a dominant polynomial mapping,* X *an admissible compactification of* $\mathbf{A}^2$ *and* E *a prime of* X *at infinity. Assume that* $\deg(f) < b_E$. *Then* $d(f, v) = d(f, v_E)$ *for all* $v \in \mathcal{V}_\infty$ *such that* $r_X(v) = v_E$.

10.6 Proof of Theorem C

If we were to follow the proof in the local case, we would pick a log resolution at infinity of the linear system $f^*|\mathcal{L}|$ on $\mathbf{P}^2$. By this we mean an admissible compactification X of $\mathbf{A}^2$ such that the strict transform of $f^*|\mathcal{L}|$ to X has no base points on $X \setminus \mathbf{A}^2$. Such an admissible compactification exists by resolution of singularities. At least when f is proper, we get a well defined selfmap $r_X f_\bullet :$ $|\Delta(X)| \to |\Delta(X)|$. However, a fixed point v of this map does not have an immediate bearing on Theorem C. Indeed, we have seen in Sect. 10.4 that even when v is actually fixed by $f_\bullet$, so that $f(v) = dv$ for some $d > 0$, it may happen that $d < d_\infty$.

One way around this problem would be to ensure that the compactification X is tight, in the sense of Sect. 9.7.2. Unfortunately, it is not always possible, even for f proper, to find a tight X that defines a log resolution of infinity of $f^*|\mathcal{L}|$.

Instead we use a recursive procedure. The proof below in fact works also when f is merely dominant, and not necessarily proper. Before starting the procedure, let us write down a few cases where we actually obtain a proof of Theorem C.

Lemma 10.8 *Let* X *be a tight compactification of* $\mathbf{A}^2$ *with associated retraction* $r_X : \mathcal{V}_\infty \to |\Delta(X)|$. *Consider a fixed point* $v \in |\Delta(X)|$ *of the induced selfmap* $r_X f_\bullet : |\Delta(X)| \to |\Delta(X)|$. *Assume that we are in one of the following three situations:*

(a) $f_\bullet v = v$ and $\alpha(v) > 0$;
(b) $f_\bullet v \neq v$, $\alpha(v) > 0$, v is divisorial and $b(v) > \deg(f)$;
(c) $\alpha(v) = 0$ and $(r_X f_\bullet)^n w \to v$ as $n \to \infty$ for $w \in |\Delta(X)|$ close to v.

Then Theorem C holds.

Proof Case (a) is treated as in the local situation. Since $\alpha := \alpha(v) > 0$ we have $\alpha^{-1}v \leq \mathrm{ord}_\infty \leq v$ on R. Write $f(v) = dv$, where $d = d(f, v) > 0$. Then

$$\deg f^n = -\mathrm{ord}_\infty(f^{n*}|\mathcal{L}|) \leq -\alpha^{-1}v(f^{n*}|\mathcal{L}|) = -\alpha^{-1}d^n v(|\mathcal{L}|) = \alpha^{-1}d^n.$$

Similarly, $\deg f^n \geq d^n$. This proves statement (a) of Theorem C (and that $d_\infty = d$). The fact that $d = d_\infty$ is a quadratic integers is proved exactly as in the local case, using value groups. Indeed, one obtains $d\Gamma_v \subseteq \Gamma_v$. Since $\Gamma_v \simeq \mathbf{Z}$ or $\Gamma_v \simeq \mathbf{Z} \oplus \mathbf{Z}$, d must be a quadratic integer.

Next we turn to case (b). By the analogue of Lemma 7.12 we may assume that the center of $f_\bullet v$ on X is a free point ξ of E. By Corollary 10.7 we have $d(f, \cdot) \equiv d := d(f, v)$ on $U(\xi)$. As in the local case, this implies that $f_\bullet \overline{U(\xi)} \subseteq U(\xi)$, $d(f^n, v) = d^n$, $d^n \le \deg(f^n) \le \alpha^{-1} d^n$, so that we are in case (a) of Theorem C, with $d_\infty = d$. The fact that $d = d_\infty$ is a quadratic integer follows from $d\,\Gamma_v \subseteq \Gamma_v \simeq \mathbf{Z}$. In fact, $d \in \mathbf{N}$.

Finally we consider case (c). Recall that the statements of Theorem C are invariant under conjugation by polynomial automorphisms. Since X is tight and $\alpha(v) = 0$, we may by Theorem 9.18 choose coordinates (z_1, z_2) on $\mathbf{A}^2$ in which v is monomial with $v(z_1) = 0$, $v(z_2) = -1$. Since v is an end in the $f_\bullet$-invariant tree $\mathcal{V}'_\infty$ and $r_X f_\bullet v = v$, we must have $f_\bullet v = v$. In particular, $f_\bullet v(z_1) = 0$, which implies that f is a skew product of the form

$$ f(z_1, z_2) = (\phi(z_1), \psi(z_1)z_2^d + O_{z_1}(z_2^{d-1})), $$

where $d \ge 1$ and ϕ, ψ are nonzero polynomials. The valuations in $|\Delta(X)|$ close to v must also be monomial valuations, of the form w_t, with $w_t(z_1) = -t$ and $w_t(z_2) = -1$, where $0 \le t \ll 1$. We see that $f(w_t)(z_1) = -t \deg \phi$ and $f(w_t)(z_2) = -(d + t \deg q)$. When t is irrational, $f_\bullet w_t$ must be monomial, of the form $w_{t'}$, where $t' = t\frac{\deg p}{d + t \deg q}$. By continuity, this relationship must hold for all real t, $0 \le t \ll 1$. By our assumptions, $t' < t$ for $0 < t \ll 1$. This implies that either $\deg p < d$ or that $\deg p = d$, $\deg q > 0$. It is then clear that $d_\infty = d$ is an integer, proving the first statement in Theorem C. Finally, from a direct computation, that we leave as an exercise to the reader, it follows that $\deg f^n \sim nd^n$. $\square$

The main case not handled by Lemma 10.8 is the case (b) but without the assumption that $b_E > \deg f$. In this case we need to blow up further.

Lemma 10.9 *Let X be a tight compactification of $\mathbf{A}^2$ with associated retraction $r_X : \mathcal{V}_\infty \to |\Delta(X)|$. Assume that $v = v_E = b_E^{-1} \operatorname{ord}_E \in |\Delta(X)|$ is a divisorial valuation such that $r_X f_\bullet v_E = v_E$ but $f_\bullet v_E \ne v_E$. Then there exists a tight compactification X' of $\mathbf{A}^2$ dominating X and a valuation $v' \in |\Delta(X')| \setminus |\Delta(X)|$ such that $r_{X'} f_\bullet v' = v'$ and such that we are in one of the following cases:*

(a) $f_\bullet v' = v'$ and $\alpha(v') > 0$;
(b) $f_\bullet v' \ne v'$, v' is divisorial, $\alpha(v') > 0$ and $b(v') > b(v)$;
(c) $\alpha(v') = 0$ and $(r_{X'} f_\bullet)^n w \to v'$ as $n \to \infty$ for $w \in |\Delta(X)|$ close to v'.

It is clear that repeated application of Lemma 10.8 and Lemma 10.9 leads to a proof of Theorem C. The only thing remaining is to prove Lemma 10.9.

Proof Write $v_0 = v$. By (the analogue at infinity of) Lemma 7.12 we may find an admissible compactification X_0 dominating X, such that $|\Delta_0| := |\Delta(X_0)| = |\Delta(X)|$, $r_0 := r_{X_0} = r_X$ and such that the center of $v_0 = v$ on X_0 is a prime E_0 of X_0 at infinity. Since $f_\bullet v_0 \ne v_0$, the center of $f_\bullet v_0$ must be a free point $\xi_0 \in E_0$. Let X_1 be the blowup of X_0 at ξ_0, E_1 the exceptional divisor and $v_1 = b_1^{-1} \operatorname{ord}_{E_1}$ the associated divisorial valuation. Note that $b_1 = b_0$ and $\alpha(v_1) = \alpha(v_0) - b_0^{-1}$

by (67). In particular, X_1 is still tight. Write $|\Delta_1| = |\Delta(X_1)|$ and $r_1 := r_{X_1}$. We have $r_1 f_\bullet v_0 \in |\Delta_1| \setminus |\Delta_0| =]v_0, v_1]$. Thus there are two cases:

(1) there exists a fixed point $v' \in]v_0, v_1[$ for $r_1 f_\bullet$;
(2) $(r_1 f_\bullet)^n \to v_1 = r_1 f_\bullet v_1$ as $n \to \infty$;

Let us first look at case (1). Note that $\alpha(v') > \alpha(v_1) \geq 0$. If $f_\bullet v' = v'$, then we are in situation (a) and the proof is complete. Hence we may assume that $f_\bullet v' \neq v'$. Then v' is necessarily divisorial. By Lemma 7.12 we have $b(v') > b_0 = b(v)$. We are therefore in situation (b), so the proof is complete in this case.

It remains to consider case (2). If $\alpha(v_1) = 0$, then we set $X' = X_1$, $v' = v_1$ and we are in situation (c). We can therefore assume that $\alpha(v_1) > 0$. If $f_\bullet v_1 = v_1$, then we set $X' = X_1$, $v' = v_1$ and we are in situation (a). If $f_\bullet v_1 \neq v_1$, so that the center of $f_\bullet v_1$ is a free point $\xi_1 \in E_1$, then we can repeat the procedure above. Let X_2 be the blowup of X_1 at ξ_1, let E_2 be the exceptional divisor and $v_2 = b_2^{-1} \operatorname{ord}_{E_2}$ the associated divisorial valuation. We have $b_2 = b_1 = b$ and $\alpha(v_2) = \alpha(v_1) - b^{-1} = \alpha(v) - 2b^{-1}$ by (67).

Continuing the procedure above must eventually lead us to the situation in (a) or (c). Indeed, all of our compactifications are tight, so in particular all valuations v_n satisfy $\alpha(v_n) \geq 0$. But $\alpha(v_n) = \alpha(v) - nb^{-2}$. This completes the proof. $\qquad\square$

10.7 Further Properties

The presentation above was essentially optimized to give a reasonably short proof of Theorem C. While it is beyond the scope of these notes to present the details, let us briefly summarize some further results from [FJ07, FJ11]. Let $f : \mathbf{A}^2 \to \mathbf{A}^2$ be a polynomial mapping and write f also for its extension $f : \mathbf{A}^2_{\mathrm{Berk}} \to \mathbf{A}^2_{\mathrm{Berk}}$.

To begin, f interacts well with the classification of points: if $v \in \hat{\mathcal{V}}_\infty$ and $f(v) \in \hat{\mathcal{V}}_\infty$ then $f(v)$ is of the same type as v (curve, divisorial, irrational or infinitely singular). This is proved using numerical invariants in the same way as in Sect. 4.4.

At least when f is proper the induced map $f_\bullet : \mathcal{V}_\infty \to \mathcal{V}_\infty$ is continuous, finite and open. This follows from general results on Berkovich spaces, just as in Proposition 4.3. As a consequence, the general results on tree maps in Sect. 2.6 apply.

In [FJ07, FJ11], the existence of an *eigenvaluation* was emphasized. This is a valuation $v \in \mathcal{V}_\infty$ such that $f(v) = d_\infty v$. One can show from general tree arguments that there must exist such a valuation in the tight tree $\mathcal{V}'_\infty$. The proof of Theorem C gives an alternative construction of an eigenvaluation in $\mathcal{V}'_\infty$.

Using a lot more work, the global dynamics on $\mathcal{V}'_\infty$ is described in [FJ11]. Namely, the set $\mathcal{T}_f$ of eigenvaluations in $\mathcal{V}'_\infty$ is either a singleton or a closed interval. (The "typical" case is that of a singleton.) In both cases we have $f_\bullet^n v \to \mathcal{T}_f$ as $n \to \infty$, for all but at most one $v \in \mathcal{V}'_\infty$. This means that the dynamics on the tight tree $\mathcal{V}_\infty$ is globally contracting, as opposed to a rational map on the Berkovich projective line, which is globally expanding.

Using the dynamics on $\mathcal{V}_\infty'$, the cocycle $d(f^n, v)$ can be very well described: for any $v \in \mathcal{V}_\infty'$ the sequence $(d(f^n, v))_{n \geq 0}$ satisfies an integral recursion relation. Applying this to $v = \mathrm{ord}_\infty$ we see that the degree growth sequence $(\deg(f^n))$ satisfies such a recursion relation.

As explained in the introduction, one motivation for the results in this section comes from polynomial mappings of the complex plane $\mathbf{C}^2$, and more precisely understanding the rate at which orbits are attracted to infinity. Let us give one instance of what can be proved. Suppose $f : \mathbf{C}^2 \to \mathbf{C}^2$ is a dominant polynomial mapping and assume that f has "low topological degree" in the sense that the asymptotic degree $d_\infty(f)$ is strictly larger than the topological degree of f, i.e. the number of preimages of a typical point. In this case, we showed in [FJ11] that the functions

$$\frac{1}{d_\infty^n} \log^+ \| f^n \|$$

converge uniformly on compact subsets of $\mathbf{C}^2$ to a plurisubharmonic function G^+ called the *Green function* of f. Here $\|\cdot\|$ is any norm on $\mathbf{C}^2$ and we write $\log^+ \|\cdot\| := \max\{\log |\cdot|, 0\}$. This Green function is important for understanding the ergodic properties of f, as explored by Diller, Dujardin and Guedj [DDG1, DDG2, DDG3].

10.8   Other Ground Fields

Throughout this section we assumed that the ground field was algebraically closed and of characteristic zero. Let us briefly discuss what happens when one or more of these assumptions are not satisfied.

10.9   Other Ground Fields

Throughout the section we assumed that the ground field was algebraically closed and of characteristic zero.

The assumption on the characteristic was used in the proof of formula (76) and hence of Proposition 10.5. The proof of the monomialization result (Theorem 9.18) also used characteristic zero. It would be interesting to have an argument for Theorem C that works in arbitrary characteristic.

On the other hand, assuming that $\mathrm{char}\, K = 0$, the assumption that K be algebraically closed is unimportant for Theorem C, at least for statements (a) and (b). Indeed, if K^a is the algebraic closure of K, then any polynomial mapping $f : \mathbf{A}^2(K) \to \mathbf{A}^2(K)$ induces a polynomial mapping $f : \mathbf{A}^2(K^a) \to \mathbf{A}^2(K^a)$.

Further, an embedding $\mathbf{A}^2(K) \hookrightarrow \mathbf{P}^2(K)$ induces an embedding $\mathbf{A}^2(K^a) \hookrightarrow \mathbf{P}^2(K^a)$ and the degree of f^n is then independent of whether we work over K or K^a. Thus statements (a) and (b) of Theorem C trivially follow from their counterparts over an algebraically closed field of characteristic zero.

10.10 Notes and Further References

The material in this section is adapted from the papers [FJ07, FJ11] joint with Charles Favre, but with a few changes in the presentation. In order to keep these lecture notes reasonably coherent, I have also changed some of the notation from the original papers. I have also emphasized a geometric approach that has some hope of being applicable in higher dimensions and the presentation is streamlined to give a reasonably quick proof of Theorem C.

Instead of working on valuation space, it is possible to consider the induced dynamics on divisors on the Riemann-Zariski space. By this we mean the data of one divisor at infinity for each admissible compactification of $\mathbf{A}^2$ (with suitable compatibility conditions when one compactification dominates another. See [FJ11] for more details and [BFJ08a] for applications of this point of view in a slightly different context.

Acknowledgements I would like to express my gratitude to many people, first and foremost to Charles Favre for a long and fruitful collaboration and without whom these notes would not exist. Likewise, I have benefitted enormously from working with Sébastien Boucksom. I thank Matt Baker for many interesting discussions; the book by Matt Baker and Robert Rumely has also served as an extremely useful reference for dynamics on the Berkovich projective line. I am grateful to Michael Temkin and Antoine Ducros for answering various questions about Berkovich spaces and to Andreas Blass for help with Remark 2.6 and Example 2.7. Conversations with Dale Cutkosky, William Gignac, Olivier Piltant and Matteo Ruggiero have also been very helpful, as have comments by Yûsuke Okuyama. Vladimir Berkovich of course deserves a special acknowledgment as neither these notes nor the summer school itself would have been possible without his work. Finally I am grateful to the organizers and the sponsors of the summer school. My research has been partially funded by grants DMS-0449465 and DMS-1001740 from the NSF.

References

[Aba10] M. Abate, Discrete holomorphic local dynamical systems, in *Holomorphic Dynamical Systems*, Lecture Notes in Mathematics, vol 1998 (Springer, 2010), pp. 1–55

[AM73] S.S. Abhyankar, T.T. Moh, Newton-Puiseux expansion and generalized Tschirnhausen transformation. I, II. J. Reine Angew. Math. **260**, 47–83 (1973); **261**, 29–53 (1973)

[Art66] M. Artin, On isolated rational singularities of surfaces. Am. J. Math. **88**, 129136 (1966)

[BdM09] M. Baker, L. DeMarco, Preperiodic points and unlikely intersections. Duke Math. J. **159**, 1–29 (2011)

[Bak64]   I.N. Baker, Fixpoints of polynomials and rational functions. J. Lond. Math. Soc. **39**, 615–622 (1964)

[Bak06]   M. Baker, A lower bound for average values of dynamical Green's functions. Math. Res. Lett. **13**, 245–257 (2006)

[Bak08]   M. Baker, An introduction to Berkovich analytic spaces and non-Archimedean potential theory on curves. In *p-adic Geometry*, 123–174. Univ. Lecture Ser., vol. 45 (American Mathematical Society, Providence, RI, 2008)

[Bak09]   M. Baker, A finiteness theorem for canonical heights attached to rational maps over function fields. J. Reine Angew. Math. **626**, 205–233 (2009)

[BH05]   M. Baker, L.-C. Hsia, Canonical heights, transfinite diameters, and polynomial dynamics. J. Reine Angew. Math. **585**, 61–92 (2005)

[BR06]   M. Baker, R. Rumely, Equidistribution of small points, rational dynamics, and potential theory. Ann. Inst. Fourier **56**, 625–688 (2006)

[BR10]   M. Baker, R. Rumely, *Potential Theory on the Berkovich Projective Line*. Mathematical Surveys and Monographs, vol. 159 (American Mathematical Society, Providence, RI, 2010)

[Bea91]   A.F. Beardon, *Iteration of Rational Functions*. Graduate Texts in Mathematics, vol. 132 (Springer, New York, 1991)

[BT82]   E. Bedford, B.A. Taylor, A new capacity for plurisubharmonic functions. Acta Math. **149**, 1–40 (1982)

[BT87]   E. Bedford, B.A. Taylor, Fine topology, Shilov boundary and $(dd^c)^n$. J. Funct. Anal. **72**, 225–251 (1987)

[Ben98]   R.L. Benedetto, Fatou components in p-adic dynamics. Ph.D. Thesis. Brown University, 1998. Available at www.cs.amherst.edu/~rlb/papers/

[Ben00]   R.L. Benedetto, p-adic dynamics and Sullivan's no wandering domains theorem. Compositio Math. **122**, 281–298 (2000)

[Ben01a] R.L. Benedetto, Reduction, dynamics, and Julia sets of rational functions. J. Number Theory **86**, 175–195 (2001)

[Ben01b] R.L. Benedetto, Hyperbolic maps in p-adic dynamics. Ergodic Theory Dynam. Syst. **21**, 1–11 (2001)

[Ben02a] R.L. Benedetto, Components and periodic points in non-Archimedean dynamics. Proc. Lond. Math. Soc. **84**, 231–256 (2002)

[Ben02b] R.L. Benedetto, Examples of wandering domains in p-adic polynomial dynamics. C. R. Math. Acad. Sci. Paris **335**, 615–620 (2002)

[Ben05a] R.L. Benedetto, Wandering domains and nontrivial reduction in non-Archimedean dynamics. Ill. J. Math. **49**, 167–193 (2005)

[Ben05b] R.L. Benedetto, Heights and preperiodic points of polynomials over function fields. Int. Math. Res. Not. **62**, 3855–3866 (2005)

[Ben06]   R.L. Benedetto, Wandering domains in non-Archimedean polynomial dynamics. Bull. Lond. Math. Soc. **38**, 937–950 (2006)

[Ben10]   R.L. Benedetto, Non-Archimedean dynamics in dimension one. Lecture notes from the 2010 Arizona Winter School, http://math.arizona.edu/~swc/aws/2010/

[Ber90]   V.G. Berkovich, *Spectral Theory and Analytic Geometry Over Non-Archimedean Fields*. Mathematical Surveys and Monographs, vol. 33 (American Mathematical Society, Providence, RI, 1990)

[Ber93]   R.L. Benedetto, Étale cohomology for non-Archimedean analytic spaces. Publ. Math. Inst. Hautes Études Sci. **78**, 5–161 (1993)

[Ber94]   R.L. Benedetto, Vanishing cycles for formal schemes. Invent. Math. **115**, 539–571 (1994)

[Ber99]   R.L. Benedetto, Smooth p-adic analytic spaces are locally contractible. I. Invent. Math. **137**, 1–84 (1999)

[Ber04]   R.L. Benedetto, Smooth p-adic analytic spaces are locally contractible. II. *Geometric Aspects of Dwork Theory*. (Walter de Gruyter and Co. KG, Berlin, 2004), pp. 293–370

[Ber09]   R.L. Benedetto, A non-Archimedean interpretation of the weight zero subspaces of limit mixed Hodge structures. *Algebra, arithmetic, and geometry: in honor of Yu. I. Manin.* Progr. Math., vol. 269 (Birkhäuser, Boston, MA, 2009), pp. 49–67

[BdFF10]   S. Boucksom, T. de Fernex, C. Favre, The volume of an isolated singularity. Duke Math. J. **161**(8), 1455–1520 (2012)

[BFJ08a]   S. Boucksom, C. Favre, M. Jonsson, Degree growth of meromorphic surface maps. Duke Math. J. **141**, 519–538 (2008)

[BFJ08b]   S. Boucksom, C. Favre, M. Jonsson, Valuations and plurisubharmonic singularities. Publ. Res. Inst. Math. Sci. **44**, 449–494 (2008)

[BFJ12]   S. Boucksom, C. Favre, M. Jonsson, Singular semipositive metrics in non-Archimedean geometry. arXiv:1201.0187. To appear in J. Algebraic Geom.

[BFJ14]   S. Boucksom, C. Favre, M. Jonsson, Solution to a non-Archimedean Monge-Ampère equation. J. Amer. Math. Soc., electronically published on May 22, 2014

[BGR84]   S. Bosch, U. Güntzer, R. Remmert, *Non-Archimedean Analysis* (Springer, Berlin, Heidelberg, 1994)

[BD01]   J.-Y. Briend, J. Duval, Deux caractérisations de la mesure d'équilibre d'un endomorphisme de P^k(C). Publ. Math. Inst. Hautes Études Sci. **93**, 145–159 (2001)

[Bro65]   H. Brolin, Invariant sets under iteration of rational functions. Ark. Mat. **6**, 103–144 (1965)

[CPR02]   A. Campillo, O. Piltant, A. Reguera, Cones of curves and of line bundles on surfaces associated with curves having one place at infinity. Proc. Lond. Math. Soc. **84**, 559–580 (2002)

[CPR05]   A. Campillo, O. Piltant, A. Reguera, Cones of curves and of line bundles at infinity. J. Algebra **293**, 513–542 (2005)

[CG93]   L. Carleson, T. Gamelin, *Complex Dynamics* (Springer, New York, 1993)

[CL06]   A. Chambert-Loir, Mesures et équidistribution sur les espaces de Berkovich. J. Reine Angew. Math. **595**, 215–235 (2006)

[Con08]   B. Conrad, Several approaches to non-archimedean geometry. In *p-adic Geometry (Lectures from the 2007 Arizona Winter School).* AMS University Lecture Series, vol. 45 (Amer. Math. Soc., Providence, RI, 2008)

[CLM07]   T. Coulbois, A. Hilion, M. Lustig, Non-unique ergodicity, observers' topology and the dual algebraic lamination for ℝ-trees. Ill. J. Math. **51**, 897–911 (2007)

[DDG1]   J. Diller, R. Dujardin, V. Guedj, Dynamics of meromorphic maps with small topological degree I: from cohomology to currents. Indiana Univ. Math. J. **59**, 521–562 (2010)

[DDG2]   J. Diller, R. Dujardin, V. Guedj, Dynamics of meromorphic maps with small topological degree II: Energy and invariant measure. Comment. Math. Helv. **86**, 277–316 (2011)

[DDG3]   J. Diller, R. Dujardin, V. Guedj, Dynamics of meromorphic maps with small topological degree III: geometric currents and ergodic theory. Ann. Sci. École Norm. Sup. **43**, 235–278 (2010)

[DS08]   T.-C. Dinh, N. Sibony, Equidistribution towards the Green current for holomorphic maps. Ann. Sci. École Norm. Sup. **41**, 307–336 (2008)

[ELS03]   L. Ein, R. Lazarsfeld, K.E. Smith, Uniform approximation of Abhyankar valuations in smooth function fields. Am. J. Math. **125**, 409–440 (2003)

[Fab09]   X. Faber, Equidistribution of dynamically small subvarieties over the function field of a curve. Acta Arith. **137**, 345–389 (2009)

[Fab13a]   X. Faber, Topology and geometry of the Berkovich ramification locus for rational functions. I. Manuscripta Math. **142**, 439–474 (2013)

[Fab13b]   X. Faber, Topology and geometry of the Berkovich ramification locus for rational functions, II. Math. Ann. **356**, 819–844 (2013)

[Fab14]   X. Faber, Rational functions with a unique critical point. Int. Math. Res. Not. IMRN no. 3, 681–699 (2014)

[Fav05]   C. Favre, Arbres réels et espaces de valuations, Thèse d'habilitation, 2005

[FJ03]   C. Favre, M. Jonsson, Brolin's theorem for curves in two complex dimensions. Ann. Inst. Fourier **53**, 1461–1501 (2003)

[FJ04]   C. Favre, M. Jonsson, *The Valuative Tree*. Lecture Notes in Mathematics, vol. 1853 (Springer, New York, 2004)

[FJ05a] C. Favre, M. Jonsson, Valuative analysis of planar plurisubharmonic functions. Invent. Math. **162**(2), 271–311 (2005)

[FJ05b] C. Favre, M. Jonsson, Valuations and multiplier ideals. J. Am. Math. Soc. **18**, 655–684 (2005)

[FJ07]   C. Favre, M. Jonsson, Eigenvaluations. Ann. Sci. École Norm. Sup. **40**, 309–349 (2007)

[FJ11]   C. Favre, M. Jonsson, Dynamical compactifications of $\mathbf{C}^2$. Ann. Math. **173**, 211–249 (2011)

[FKT11] C. Favre, J. Kiwi, E. Trucco, A non-archimedean Montel's theorem. Compositio **148**, 966–990 (2012)

[FR04]   C. Favre, J. Rivera-Letelier, Théorème d'équidistribution de Brolin en dynamique p-adique. C. R. Math. Acad. Sci. Paris **339**, 271–276 (2004)

[FR06]   C. Favre, J. Rivera-Letelier, Équidistribution quantitative des points de petite hauteur sur la droite projective. Math. Ann. **335**, 311–361 (2006)

[FR10]   C. Favre, J. Rivera-Letelier, Théorie ergodique des fractions rationnelles sur un corps ultramétrique. Proc. Lond. Math. Soc. **100**, 116–154 (2010)

[FLM83] A. Freire, A. Lopez, R. Mañé, An invariant measure for rational maps. Bol. Soc. Bras. Mat. **14**, 45–62 (1983)

[Fol99] G.B. Folland, *Real Analysis: Modern Techniques and Their Applications, 2nd edn.* Pure and Applied Mathematics (New York). A Wiley-Interscience Publication (Wiley, New York, 1999)

[Fre93] D.H. Fremlin, Real-valued measurable cardinals. In *Set Theory of the Reals (Ramat Gan, 1991)*. Israel Math. Conf. Proc., vol. 6 (Bar-Ilan Univ, Ramat Gan, 1993), pp. 151–304. See also www.essex.ac.uk/maths/people/fremlin/papers.htm

[Ful93] W. Fulton, *Introduction to Toric Varieties*. Annals of Mathematics Studies, vol. 131 (Princeton University Press, Princeton, NJ, 1993)

[GM04]   C. Galindo, F. Monserrat, On the cone of curves and of line bundles of a rational surface. Int. J. Math. **15**, 393–407 (2004)

[GM05]   C. Galindo, F. Monserrat, The cone of curves associated to a plane configuration. Comment. Math. Helv. **80**, 75–93 (2005)

[GTZ08] D. Ghioca, T.J. Tucker, M.E. Zieve, Linear relations between polynomial orbits. Duke Math. J. **161**, 1379–1410 (2012)

[GH90]   É. Ghys, P. de la Harpe, *Sur les groupes hyperboliques d'après Mikhael Gromov*. Progress in Mathematics, vol. 83 (Birkhäuser, Boston, 1990)

[Gra07] A. Granja, The valuative tree of a two-dimensional regular local ring. Math. Res. Lett. **14**, 19–34 (2007)

[Gub08] W. Gubler, Equidistribution over function fields. Manuscripta Math. **127**, 485–510 (2008)

[Har77] R. Hartshorne, *Algebraic Geometry*. Graduate Texts in Mathematics, vol. 52 (Springer, New York-Heidelberg, 1977)

[Hsi00] L.-C. Hsia, Closure of periodic points over a non-Archimedean field. J. Lond. Math. Soc. **62**, 685–700 (2000)

[HS01]   R. Hübl, I. Swanson, Discrete valuations centered on local domains. J. Pure Appl. Algebra **161**, 145–166 (2001)

[Izu85] S. Izumi, A measure of integrity for local analytic algebras. Publ. RIMS Kyoto Univ. **21**, 719–735 (1985)

[Jec03] T. Jech, *Set Theory*. The third millenium edition, revised and expanded. Springer Monographs in Mathematics. (Springer, Berlin, 2003)

[JM12]   M. Jonsson, M. Mustaţă, Valuations and asymptotic invariants for sequences of ideals. Ann. Inst. Fourier **62**, 2145–2209 (2012)

[Ked10]   K. Kedlaya, Good formal structures for flat meromorphic connections, I: surfaces. Duke Math. J. **154**, 343–418 (2010)

[Ked11a]   K. Kedlaya, Good formal structures for flat meromorphic connections, II: excellent schemes. J. Am. Math. Soc. **24**, 183–229 (2011)

[Ked11b]   K. Kedlaya, Semistable reduction for overconvergent F-isocrystals, IV: Local semistable reduction at nonmonomial valuations. Compos. Math. **147**, 467–523 (2011)

[KKMS73]   G. Kempf, F.F. Knudsen, D. Mumford, B. Saint-Donat, *Toroidal Embeddings. I.* Lecture Notes in Mathematics, vol. 339 (Springer, Berlin, 1973)

[Kiw06]   J. Kiwi, Puiseux series polynomial dynamics and iteration of complex cubic polynomials. Ann. Inst. Fourier **56**, 1337–1404 (2006)

[Kiw14]   J. Kiwi, Puiseux series dynamics of quadratic rational maps. Israel J. Math. **201**, 631–700 (2014)

[Kis02]   T. Kishimoto, A new proof of a theorem of Ramanujan-Morrow. J. Math. Kyoto **42**, 117–139 (2002)

[Kol97]   J. Kollár, *Singularities of Pairs.* Proc. Symp. Pure Math., vol. 62, Part 1 (AMS, Providence, RI, 1997)

[KM98]   J. Kollár, S. Mori, *Birational Geometry of Algebraic Varieties.* Cambridge Tracts in Mathematics, vol. 134 (Cambridge University Press, Cambridge, 1998)

[Lan02]   S. Lang, *Algebra.* Revised third edition. Graduate Texts in Mathematics, vol. 211 (Springer, New York, 2002)

[Lip69]   J. Lipman, Rational singularities with applications to algebraic surfaces and unique factorization. Publ. Math. Inst. Hautes Études Sci. **36**, 195–279 (1969)

[Lyu83]   M. Lyubich, Entropy properties of rational endomorphisms of the Riemann sphere. Ergodic Theory Dynam. Syst. **3**, 351–385 (1983)

[Mac36]   S. MacLane, A construction for prime ideals as absolute values of an algebraic field. Duke M. J. **2**, 363–395 (1936)

[Mat89]   H. Matsumura, *Commutative Ring Theory.* Cambridge Studies in Advanced Mathematics, vol. 8 (Cambridge University Press, Cambridge, 1989)

[Mil06]   J. Milnor, *Dynamics in One Complex Variable.* Annals of Mathematics Studies, vol. 160 (Princeton University Press, Princeton, NJ, 2006)

[Mon07]   F. Monserrat, Curves having one place at infinity and linear systems on rational surfaces. J. Pure Appl. Algebra **211**, 685–701 (2007)

[Mor73]   J.A. Morrow, Minimal normal compactifications of $\mathbb{C}^2$. In *Complex Analysis, 1972* (Proc. Conf., Rice Univ. Houston, TX, 1972) Rice Univ. Studies **59**, 97–112 (1973)

[MS95]   P. Morton, J.H. Silverman, Periodic points, multiplicities, and dynamical units. J. Reine Angew. Math. **461**, 81–122 (1995)

[Oda88]   T. Oda, *Convex Bodies and Algebraic Geometry. An Introduction to the Theory of Toric Varieties.* Ergebnisse der Mathematik und ihrer Grenzgebiete (3), 15 (Springer, Berlin, 1988)

[Oku11a]   Y. Okuyama, Repelling periodic points and logarithmic equidistribution in non-archimedean dynamics. Acta Arith. **152**(3), 267–277 (2012)

[Oku11b]   Y. Okuyama, Feketeness, equidistribution and critical orbits in non-archimedean dynamics. Math. Z (2012). DOI:10.1007/s00209-012-1032-x

[Par11]   M.R. Parra, The Jacobian cocycle and equidistribution towards the Green current. arXiv:1103.4633

[Pop11]   P. Popescu-Pampu, Le cerf-volant d'une constellation. Enseign. Math. **57**, 303–347 (2011)

[PST09]   C. Petsche, L Szpiro, M. Tepper, Isotriviality is equivalent to potential good reduction for endomorphisms of $\mathbb{P}^N$ over function fields. J. Algebra **322**, 3345–3365 (2009)

[Ree89]   D. Rees, Izumi's theorem. In *Commutative Algebra* (Berkeley, CA, 1987). Math. Sci. Res. Inst. Publ., vol. 15 (Springer, New York 1989), pp. 407–416

[Riv03a]   J. Rivera-Letelier, Dynamique des fonctions rationnelles sur des corps locaux. Astérisque **287**, 147–230 (2003)

[Riv03b] J. Rivera-Letelier, Espace hyperbolique p-adique et dynamique des fonctions rationnelles. Compositio Math. **138**, 199–231 (2003)

[Riv04]   J. Rivera-Letelier, Sur la structure des ensembles de Fatou p-adiques. Available at arXiv:math/0412180

[Riv05]   J. Rivera-Letelier, Points périodiques des fonctions rationnelles dans l'espace hyperbolique p-adique. Comment. Math. Helv. **80**, 593–629 (2005)

[Rob00]   A. Robert, *A Course in p-Adic Analysis*. Graduate Texts in Mathematics, vol. 198 (Springer, New York, 2000)

[Rug12]   M. Ruggiero, Rigidification of holomorphic germs with non-invertible differential. Michigan Math. J. **61**, 161–185 (2012)

[SU04]   I.P. Shestakov, U.U. Umirbaev, The tame and the wild automorphisms of polynomial rings in three variables. J. Am. Math. Soc, **17**, 197–227 (2004)

[Sib99]   N. Sibony, Dynamique des applications rationnelles de $\mathbf{P}^k$. In *Dynamique et géométrie complexes (Lyon, 1997)*. Panor. Synthèses, vol. 8 (Soc. Math. France, Paris, 1999), pp. 97–185

[Sil07]   J.H. Silverman, *The Arithmetic of Dynamical Systems*. Graduate Texts in Mathematics, vol. 241 (Springer, New York, 2007)

[Sil10]   J.H. Silverman, Lecture notes on arithmetic dynamics. Lecture notes from the 2010 Arizona Winter School. math.arizona.edu/~swc/aws/10/

[Spi90]   M. Spivakovsky, Valuations in function fields of surfaces. Am. J. Math. **112**, 107–156 (1990)

[Suz74]   M. Suzuki, Propriétés topologiques des polynômes de deux variables complexes, et automorphismes algébriques de l'espace $\mathbf{C}^2$. J. Math. Soc. Jpn. **26**, 241–257 (1974)

[Tem10]   M. Temkin, Stable modification of relative curves. J. Algebraic Geometry **19**, 603–677 (2010)

[Tem14]   M. Temkin, Introduction to Berkovich Analytic spaces. In this volume

[Thu05]   A. Thuillier, Théorie du potentiel sur les courbes en géométrie analytique non archimédienne. Applications à la théorie d'Arakelov, Ph.D. thesis, University of Rennes, 2005. tel.archives-ouvertes.fr/tel-00010990

[Thu07]   A. Thuillier, Géométrie toroïdale et géométrie analytique non archimédienne. Application au type d'homotopie de certains schémas formels. Manuscripta Math. **123**(4), 381–451 (2007)

[Tou72]   J.-C. Tougeron, *Idéaux de fonctions différentiables*. Ergebnisse der Mathematik und ihrer Grenzgebiete, Band 71 (Springer, Berlin-New York, 1972)

[Tru09]   E. Trucco, Wandering Fatou components and algebraic Julia sets. To appear in Bull. de la Soc. Math. de France

[Vaq00]   M. Vaquié, Valuations. In *Resolution of Singularities (Obergurgl, 1997)*. Progr. Math., vol. 181 (Birkhaüser, Basel, 2000), pp. 539–590

[Vaq07]   M. Vaquié, Extension d'une valuation. Trans. Am. Math. Soc. **359**, 3439–3481 (2007)

[Yua08]   X. Yuan, Big line bundles over arithmetic varieties. Invent. Math. **173**, 603–649 (2008)

[YZ09a]   X. Yuan, S.-W. Zhang, Calabi-Yau theorem and algebraic dynamics. Preprint www.math.columbia.edu/~szhang/papers/Preprints.htm

[YZ09b]   X. Yuan, S.-W. Zhang, Small points and Berkovich metrics. Preprint, 2009, available at www.math.columbia.edu/~szhang/papers/Preprints.htm

[ZS75]   O. Zariski, P. Samuel, *Commutative Algebra*, vol 2. Graduate Texts in Mathematics, vol. 29 (Springer, New York, 1975)

Compactification of Spaces of Representations After Culler, Morgan and Shalen

Jean-Pierre Otal

Contents

1 Introduction

This paper stems from notes of a course given at the "Ultrametric Dynamical Days" in Santiago de Chile in January 2008. The purpose of this course was to explain the compactifications of spaces of representations, as this tool applies for questions in low-dimensional topology and in hyperbolic geometry.

The use of methods from algebraic geometry for studying spaces of representations originates in [13]. Marc Culler and Peter Shalen were motivated by questions from 3-dimensional topology, in particular by the so-called Property P: if $K \subset S^3$ is a nontrivial knot and $K_{p/q}$ denotes the 3-manifold obtained by p/q-Dehn surgery along K, then the fundamental group $\pi_1(K_{p/q})$ is nontrivial when $q \neq 0$. The theory of Culler and Shalen can be very briefly summarized as follows. Let G be the fundamental group of the complement of an hyperbolic knot. Let $\mathcal{R}(G)$ the set of representations of G into SL(2, $\mathbb{C}$) and let $X(G)$ the

J.-P. Otal (✉)
Institut Mathématique de Toulouse, Université Paul Sabatier, 118 route de Narbonne,
Toulouse, France
e-mail: otal@math.univ-toulouse.fr

© Springer International Publishing Switzerland 2015
A. Ducros et al. (eds.), *Berkovich Spaces and Applications*, Lecture Notes
in Mathematics 2119, DOI 10.1007/978-3-319-11029-5_7

quotient of $\mathcal{R}(G)$ by the action of $\mathrm{SL}(2,\mathbb{C})$ by conjugacy: both spaces are affine algebraic sets defined over $\mathbb{Q}$. A theorem of Thurston says that $X(G)$ contains an irreducible component which is an affine curve C. Let R be an irreducible component of $\mathcal{R}(G)$ above C. Let $\mathbb{Q}(C)$ and $\mathbb{Q}(R)$ denote the fields of rational functions defined over $\mathbb{Q}$ on C and on R respectively. Any point at infinity of C can be interpreted as a valuation on $\mathbb{Q}(C)$. There is also a tautological representation of G into $\mathrm{SL}(2,\mathbb{Q}(R))$. A classical construction due to Serre associates to such a valuation v a simplicial tree with an action of $\mathrm{SL}(2,\mathbb{Q}(R))$, and therefore with an action of G. Transversality constructions permit then to deduce an incompressible surface in the knot complement. The surfaces obtained in that way give important topological informations on the knot. For instance one important achievement of this theory is the Cyclic Surgery Theorem of Culler et al. [11]: *if surgery on a nontrival knot produces a manifold with cyclic fundamental group, then the surgery slope is an integer.*

This interpretation of the points at infinity of C was extended by John Morgan and Peter Shalen when G is a general finitely generated discrete subgroup of $\mathrm{SL}(2,\mathbb{C})$ to define a compactification of $X(G)$ [24]. It first led to a new proof of the following Compactness Theorem of Thurston. Let G denote the fundamental group of an acylindrical hyperbolic 3-manifold with incompressible boundary (in terms of the limit set of a convex cocompact model of the manifold, these hypothesis simply mean that this limit set is connected and cannot be disconnected by removing two points). Denote by $\mathcal{DF}(G)$ the space of discrete and faithful representations up to conjugacy of G into $\mathrm{SL}(2,\mathbb{C})$. *Then $\mathcal{DF}(G)$ is compact.*

The approach of Morgan-Shalen of this fundamental result can be sketched as follows. They argue by contradiction and consider an irreducible component C of $X(G)$ such that $C \cap \mathcal{DF}(G)$ is not compact. To a well-chosen subsequence of an unbounded sequence (x_i) in this intersection, they associate a valuation v on $\mathbb{Q}(C)$. Since C is not a curve in general, v is not necessarily discrete: its value group is a totally ordered abelian group Λ. The Serre construction can be adapted to this situation and produces *a Λ-tree $\mathcal{T}_v$*. When Λ is isomorphic to $\mathbb{Z}$, this is an equivalent notion to that of a simplicial tree; when more generally Λ is archimedean, this determines a $\mathbb{R}$-tree. When Λ is not archimedean, it has always a non-trivial archimedean quotient and a non-trivial $\mathbb{R}$-tree $\mathcal{T}$ can always be determined from $\mathcal{T}_v$. The group G acts by isometries on $\mathcal{T}$ minimally and furthermore with the property that any subgroup which stabilizes a non-degenerate segment is virtually abelian. The proof by Morgan-Shalen of Thurston's Compactness Theorem reduces then to the theorem from the domain of $\mathbb{R}$-trees that no action with that property exists when G is the fundamental group of an hyperbolic 3-manifold which is boundary-incompressible and acylindrical. They proved this using tools from 3-dimensional topology and foliations.

This theorem has been now widely generalized by Eliyahu Rips beyond the context of 3-manifolds under the hypothesis that G is not an amalgamated product over a virtually abelian subgroup (the original paper of Rips was not published but his proof is exposed and generalized in [2, 17, 32]). Morgan and Shalen applied in [27] also their theory to the case when G is the fundamental group of a closed surface S. Then some component of the space of real points of $X(S)$ can be identified with the Teichmüller space of S. Their construction provides a compactification where the points at infinity correspond to actions of $\pi_1(S)$ on $\mathbb{R}$-trees with the property that the subgroups which stabilize non-degenerate segments are cyclic. Such a tree is isometric to the dual tree to a measured lamination on S, see also [23, 24, 39]. Thus their theory offered a different perspective on Thurston's compactification of Teichmüller space by measured laminations.

The paper is organized as follows. In Sect. 2 we describe, following Culler-Shalen, the explicit structure of an affine algebraic set on *the space of characters* of a finitely generated group G. Then we describe the general construction of Morgan-Shalen of particular compactifications of affine algebraic sets.

In Sect. 3, we recall classical properties of valuations and we explain the general procedure of Morgan-Shalen for associating to any unbounded sequence of points on an affine algebraic variety X defined over $\mathbb{Q}$ one or more valuations on $\mathbb{Q}(X)$.

Section 4 presents certain basic properties of Λ-trees. These are metric spaces which share many common properties with $\mathbb{R}$-trees but the distance takes values in an abelian totally ordered group Λ. The basic examples are provided by the tree of $SL(2, F)$ when the field F is endowed with a valuation with value group Λ.

In Sect. 5, we explain how Morgan and Shalen interpret the valuation v on the field $\mathbb{Q}(X)$ associated in Sect. 3.5 to an unbounded sequence (x_i) in the variety X which is one component of the space of characters of a group G: the actions of G on $\mathbb{H}^3$ converge (in an appropriate sense) to an action by isometries of G on a Λ-tree (this tree is a subtree of the tree of $SL(2, F)$ with the valuation v). When the sequence (x_i) consists of characters of discrete representations of G, the action has the important property that any subgroup which stabilizes a non-degenerate segment is *virtually abelian*, i.e. contains an abelian subgroup of finite index.

Section 6 exhibits geometric examples of Λ-trees: they arise from codimension one laminations of a closed manifold with a Λ-valued transverse measure. We sketch the proof of the theorem of Skora that any minimal action of a surface group on an $\mathbb{R}$-tree such that the segment stabilizers are virtually abelian is geometric.

The presentation given here follows closely the papers [13] and [24]. This theory is also exposed in [35, 36], and [37] which provides moreover other applications to 3-manifolds. In [21], Morgan showed how actions on trees can be used to compactify the space of representations of a discrete group in $SO(n, 1)$. There are other approaches to attach to sequences of representations of a group an action by isometries on a tree: see Bestvina [1], Paulin [31] for a geometric proof of the convergence to an $\mathbb{R}$-tree in much broader context, and [9] for a proof using non-standard analysis of the convergence to a Λ-tree.

Notation Throughout this paper we follow the conventions of [19, Chapter 1]. We let $\mathbb{A}^n_\mathbb{C}$ be the standard complex affine space of dimension n. Its ring of regular functions is $\mathbb{C}[X_1, \ldots, X_n]$, and the set of its complex points is $\mathbb{C}^n$. An *affine variety* X is an *irreducible* algebraic subspace of some $\mathbb{A}^n_\mathbb{C}$. Its set of complex points $X_\mathbb{C} = X(\mathbb{C})$ is in bijection with the zero locus of a finite family of polynomials $P_1, \ldots, P_k \in \mathbb{C}[X_1, \ldots, X_n]$. When all polynomials have coefficients in a field $k \subset \mathbb{C}$, one says that X is defined over k.

2   The Space of Characters of a Group

We first review the construction of the space of representations of a finitely
generated group G into $SL(2, \mathbb{C})$ as an affine algebraic set. We follow Culler-Shalen
for the presentation of *the space of characters of G*, which is an explicit model for
the algebraic quotient of the space of representations and make then a link between
the two presentations. We then describe the Morgan-Shalen compactification of a
general affine variety.

2.1   The Space of Characters as an Affine Algebraic Set

Let G be a finitely generated group generated by n elements.

Denote by $\mathcal{R}_{\mathbb{C}}(G)$ the set of representations (i.e. of group morphisms) of G into
$SL(2, \mathbb{C})$. Each point in $\mathcal{R}_{\mathbb{C}}(G)$ is determined by its value on the elements of a
generating family of G, that is by a point in $(SL(2, \mathbb{C}))^n$ which is an algebraic
subset of the affine space $\mathbb{A}_{\mathbb{C}}^{4n}$. The points in $\mathcal{R}_{\mathbb{C}}(G)$ hence naturally form an affine
algebraic set $\mathcal{R}(G)$ defined by the vanishing of a family of polynomials with integer
coefficients, each polynomial corresponding to a relation between the generators.
In this way, $\mathcal{R}(G)$ can be identified with an affine algebraic set defined over $\mathbb{Q}$.
This algebraic set does not depend on the particular choice of a generating family:
different choices of generating systems lead to isomorphic algebraic sets.

In the sequel we shall be interested in the quotient of $\mathcal{R}(G)$ under the action
of $SL(2, \mathbb{C})$ by conjugacy, that is, its quotient where one identifies any two
representations ρ_1 and ρ_2 when $\rho_2 = M \circ \rho_1 \circ M^{-1}$ for $M \in SL(2, \mathbb{C})$. This quotient
space has a natural algebraic structure that we now discuss. We first describe the
explicit construction due to Culler and Shalen of *the space of characters*. Then we
will indicate how this construction enters the general theory of algebraic quotients.

For each $g \in G$ the function $\mathcal{R}(G) \to \mathbb{C}$, $\rho \mapsto \mathrm{tr}(\rho(g))$ is a regular function on
$\mathcal{R}(G)$, that is an element of the ring $\mathbb{Q}[\mathcal{R}(G)]$.

Proposition 1   *The ring $\mathbb{Q}[\mathrm{tr}(\rho(g)), g \in G]$ is finitely generated.*

Proof The proposition follows from the classical identity satisfied by all $A, B \in$
$SL(2, \mathbb{C})$: $\mathrm{tr}(AB) + \mathrm{tr}(AB^{-1}) = \mathrm{tr}(A)\,\mathrm{tr}(B)$.   □

Definition 2   Choose a set of generators (X_i) of the ring $\mathbb{Q}[\mathrm{tr}(\rho(g)), g \in G]$, $X_i =$
$\mathrm{tr}(\rho(g_i))$, $i = 1, \ldots, N$. For any element $g \in G$, we denote $\mathcal{T}_g$ a polynomial in
the variables X_i such that $\mathrm{tr}(\rho(g)) = \mathcal{T}_g(X_1, \cdots, X_N)$. Consider the regular map
$t : \mathcal{R}(G) \to \mathbb{C}^N$, $\rho \mapsto t(\rho) = (X_i(\rho))$: its image, denoted by $X(G)$, is the space of
characters of G.

The space of representations $\mathcal{R}(G)$ is clearly an affine algebraic set defined over
$\mathbb{Q}$; the same property holds for the space of characters.

Proposition 3 ([24])   *The space $X(G)$ is an algebraic set defined over $\mathbb{Q}$.*

Before starting the proof we recall a few definitions.

Definition 4 (Reducible representation) A representation $\rho : G \to SL(2, \mathbb{C})$ is *reducible* when there exists a 1-dimensional subspace of $\mathbb{C}^2$ that is invariant by any element of $\rho(G)$. This is equivalent to saying that the representation can be conjugated to take value in the group of upper-triangular matrices.

Recall that the group of isometries of the hyperbolic space $\mathbb{H}^3$ is isomorphic to $PSL(2, \mathbb{C})$. In the model of the upper half-space, the *ideal boundary* of $\mathbb{H}^3$ is identified to $\mathbb{C} \cup \{\infty\} \simeq \mathbb{C}P^1$ with its natural conformal structure. On this boundary, $PSL(2, \mathbb{C})$ acts by Möbius transformations. The group $SL(2, \mathbb{C})$ also acts on $\mathbb{H}^3$ via the quotient map $SL(2, \mathbb{C}) \to PSL(2, \mathbb{C})$. In terms of the action of $SL(2, \mathbb{C})$ on $\mathbb{H}^3$ a representation $\rho : G \to SL(2, \mathbb{C})$ is reducible if and only if $\rho(G)$ has a fixed point in $\mathbb{C}P^1$.

Lemma 5 *A representation $\rho \in \mathcal{R}(G)$ is reducible if and only if for any element g in the commutator subgroup $[G, G]$, one has $tr(\rho(g)) = 2$.*

Proof Since a reducible representation is conjugated to a group of upper-triangular matrices, the trace of any element in $\rho([G, G])$ is equal to 2.

Conversely suppose that each element of $\rho([G, G])$ has trace equal to 2. Suppose also that this group is not reduced to the identity element Id (otherwise $\rho(G)$ would be abelian and would then leave a one-dimensional subspace invariant). Let $h \in \rho([G, G])$, $h \neq id$; since $tr(h) = 2$, h leaves invariant a unique point $p \in \mathbb{C}P^1$. If any element of $\rho([G, G])$ leaves p invariant, then p is also invariant by the entire group $\rho(G)$, since $\rho([G, G])$ is a normal subgroup: in particular, ρ is reducible. Thus we can suppose that some element k of $\rho([G, G])$ does not fix p. Then k and h are two parabolic elements which fix distinct points of $\mathbb{C}P^1$. The *Ping-pong Lemma* (see [33]) produces then an element of $\langle h, k \rangle$ which is hyperbolic, and in particular has a trace $\neq 2$. $\qquad\square$

Lemma 5 says that the set of the characters of reducible representations is the algebraic subset defined by the vanishing of the polynomials $\mathcal{T}_g - 2$ for $g \in [G, G]$.

The proof of Proposition 3 is based essentially on the following result of independent interest.

Proposition 6 *Let (ρ_i) be a sequence of representations in $\mathcal{R}_{\mathbb{C}}(G)$. Suppose that for any $g \in G$, the sequence $(tr(\rho_i(g)))$ is bounded. Then one can conjugate each ρ_i so that the sequence (ρ_i) is bounded, i.e. for all $g \in G$, $(\rho_i(g))$ stays in a compact set of $SL(2, \mathbb{C})$ which depends on g.*

Proof It is sufficient to deal with the following two different cases: either all representations are reducible or they are all irreducible.

Suppose we are in the first case. By conjugating each ρ_i by a suitable matrix, we may also assume that all representations take their values in the subgroup of upper-triangular matrices. By assumption, for each element g of G, the diagonal terms of $\rho_i(g)$ are uniformly bounded. After conjugating ρ_i by a suitable diagonal matrix, one can also get a uniform bound on the upper-right term of $\rho_i(g)$. Applying this to a finite set of generators g_j of G, one concludes that up to conjugacy the family $(\rho_i(g_j))_{i,j}$ is bounded, thus proving Proposition 6 in this case.

Suppose now that each representation ρ_i is irreducible. We proceed by induction on the number of generators of G and use the geometric action of $\mathrm{SL}(2, \mathbb{C})$ on $\mathbb{H}^3$.

Pick $g \in \mathrm{PSL}(2, \mathbb{C})$, $R \geq 0$, and denote by $C_R(g)$ the following subset of $\mathbb{H}^3$:

$$C_R(g) = \{x \in \mathbb{H}^3 \mid d(x, gx) \leq R\}.$$

Each set $C_R(g)$ is closed, convex and invariant under the normalizer of g in $\mathrm{PSL}(2, \mathbb{C})$. Therefore $C_R(g)$ can be described as follows when $g \neq \mathrm{Id}$. If g is hyperbolic, $C_R(g)$ is a neighborhood of constant radius of the axis of g. In the model of the upper half space, and if the axis of g points towards infinity, then $C_R(g)$ is a circular cone based at the finite endpoint of the axis. When g is parabolic, $C_R(g)$ is an horoball centered at the fixed point of g on $\partial \mathbb{H}^3$; when g is elliptic, $C_R(g)$ is a neighborhood of constant radius of the (axis of) fixed points of g. We shall make use of the following observation which follows for instance from hyperbolic trigonometry, see e.g. [33, 43].

Claim 7 For $x \in \partial C_R(g)$, the geodesic segment $x.gx$ is contained in $C_R(g)$ and makes an angle with $\partial C_R(g)$ which tends to $\pi/2$ as $R \to \infty$; furthermore, this convergence is uniform in g, as long as the modulus of $\mathrm{tr}(g)$ is bounded from above.

Suppose that G is generated by n elements $g_1, \ldots, g_n$. Saying that a sequence of representations $\rho_i : G \to \mathrm{SL}(2, \mathbb{C})$ can be conjugated inside $\mathrm{SL}(2, \mathbb{C})$ to become bounded is equivalent to the existence of a constant $R > 0$ such that the intersection of the neighborhoods $C_R(\rho_i(g_j))$ is non-empty: $C_R(\rho_i(g_1), \ldots, \rho_i(g_n)) = \cap_j C_R(\rho_i(g_j)) \neq \emptyset$. To see that, pick for each i a point p_i in this intersection and conjugate ρ_i by an element of $\mathrm{PSL}(2, \mathbb{C})$ mapping p_i to the origin in $\mathbb{H}^3$. Then all isometries $\rho_i(g_j)$ belong to a compact set of $\mathrm{PSL}(2, \mathbb{C})$.

We now observe that each set $C_R(g_1, \cdots, g_k)$ is convex since each $C_R(g_j)$ is. The boundary of $C_R(g_1, \cdots, g_k)$ is also contained in the union of the boundaries of the tubes $C_R(g_j)$. Suppose that Proposition 6 has been proven for all groups generated by $\leq n - 1$ elements, and let G be a group generated by n elements $g_1, \ldots, g_n$. By the induction hypothesis there is an $R > 0$ such that $C_R(\rho_i(g_2), \cdots, \rho_i(g_n))$ and $C_R(\rho_i(g_1))$ are non-empty. Using Claim 7 and the fact that the traces are bounded, we may also choose R sufficiently large so that the angles between $x.\rho_i(g_j)x$ and $\partial C_R(\rho_i(g_j))$ are uniformly close to $\pi/2$ for all $x \in \partial C_R(\rho_i(g_j))$.

For any i, set $d_i = \inf\{d(x, y) \mid x \in C_R(\rho_i(g_1)), y \in C_R(\rho_i(g_2), \cdots, \rho_i(g_n))\}$. Let us first show that for all i this infimum is attained. Pick $x_k \in C_R(\rho_i(g_1))$ and $y_k \in C_R(\rho_i(g_2), \cdots, \rho_i(g_n))$ such that $d(x_k, y_k)$ tends to d_i. Suppose, by contradiction, that d_i is not a minimum. Then x_k and y_k tend to ∞ in $\mathbb{H}^3$. Up to extracting a subsequence, we may assume that they converge in $\mathbb{H}^3 \cup \partial \mathbb{H}^3$ to the same limit x. Since $x_k \in C_R(\rho_i(g_1))$, $x = \lim_k x_k$ is fixed by $\rho_i(g_1)$. In the same way $y_k \in C_R(\rho_i(g_j))$ for $j \geq 2$, hence x is fixed by all $\rho_i(g_j)$'s for $j \geq 2$. We conclude that x is fixed by $\rho_i(G)$, hence the representation ρ_i is reducible, contradicting the hypothesis.

We may thus find points $e_i \in C_R(\rho_i(g_1))$ and $f_i \in C_R(\rho_i(g_2), \cdots, \rho_i(g_n))$ such that $d_i = d(e_i, f_i)$.

If the sequence (d_i) is bounded, say less than A, then $C_{R+A}(g_1, \cdots, g_n)$ is non-empty which proves Proposition 6 in that case.

Suppose that d_i tends to ∞. Denote by k_i the geodesic in $\mathbb{H}^3$ with endpoints e_i and f_i. Since the boundary $\partial C_R(\rho_i(g_1))$ is smooth, k_i is orthogonal to $\partial C_R(\rho_i(g_1))$ at e_i, by the first variation formula. The tangent vector to k_i at e_i is thus equal to the outward-pointing normal to $\partial C_R(\rho_i(g_1))$. Observe that the hyperplane orthogonal to k_i at f_i separates $\mathbb{H}^3$ into two half-spaces, one of which contains k_i, and the other contains $C_R(\rho_i(g_2), \cdots, \rho_i(g_n))$. This implies that the tangent vector to k_i at f_i makes an angle smaller than $\pi/2$ with the inward-pointing normal of $C_R(\rho_i(g_j))$'s for some $j \geq 2$, say of $C_R(\rho_i(g_2))$.

Consider the piecewise geodesic γ which is the concatenation of the four geodesics $\rho_i(g_1^{-1})e_i.e_i$, k_i, $f_i.\rho_i(g_2)f_i$ and $\rho_i(g_2)k_i$. The choice of R implies that for any $x \in \partial C_R(\rho_i(g_j))$, the geodesic $x.\rho_i(g_j)x$ is almost orthogonal to $C_R(\rho_i(g_j))$ at its endpoints. In particular, the angles of two consecutive geodesic segments of γ are almost equal to π. The same remark implies that the union of the segments $\rho_i(g_1g_2)^l(\gamma)$, $l \in \mathbb{Z}$ is a quasi-geodesic [18]. This quasi-geodesic is invariant under $\rho_i(g_1g_2)$ with fundamental domain γ. It is a classical result [33] that in these circumstances, $\rho_i(g_1g_2)$ is an hyperbolic element of $\mathrm{PSL}(2, \mathbb{C})$ with translation distance comparable to the length of γ. Since d_i tends to ∞, this contradicts that the trace of $\rho_i(g_1g_2)$ is bounded.   □

Proof of Proposition 3 The algebraic set $X(G)$ is the union of the images of the irreducible components of $\mathcal{R}(G)$ by the polynomial map t. We show that the image of any irreducible component is an algebraic variety defined over $\mathbb{Q}$. To conclude we use the following classical result from Elimination Theory (see [29]).   □

Lemma 8 *Let $S \subset \mathbb{C}^n$ be an algebraic variety defined over a field $k \subset \mathbb{C}$ and $P : S \to \mathbb{C}^q$ be a polynomial map with coefficients in k. Then the Zariski closure $\overline{P(S)}$ of the image $P(S)$ is an algebraic variety defined over k and there is an algebraic set W strictly contained in $\overline{P(S)}$ and disjoint from $P(S)$ such that $P(S) \cup W = \overline{P(S)}$.*

Proof To show that $X(G)$ is an algebraic set, it suffices to prove that for each irreducible component $\mathcal{R}_0$ of $\mathcal{R}(G)$, the algebraic set $W = \overline{t(\mathcal{R}_0)} - t(\mathcal{R}_0)$ provided by Lemma 8 is empty. Let $x \in W$. Let (x_i) be a sequence in $t(\mathcal{R}_0)$ which converges to x. There is a sequence (ρ_i) in $\mathcal{R}_0$ with $t(\rho_i) = x_i$. By Proposition 6, we can conjugate ρ_i so that some subsequence of (ρ_i) converges to a representation ρ_∞. By continuity $t(\rho_\infty) = \lim t(\rho_i) = x$. Therefore $x \in t(\mathcal{R}_0)$.   □

Remark 9 (Relations with the Geometric Invariant Theory) Another way to describe the algebraic structure of the quotient of $\mathcal{R}(G)$ under the action of $\mathrm{SL}(2, \mathbb{C})$ relies on Geometric Invariant Theory (see [28]). The group $\mathrm{SL}(2, \mathbb{C})$ is reductive and acts rationally in the sense of [28] on $\mathcal{R}(G)$. By a theorem of Hilbert the ring $\mathbb{Q}[\mathcal{R}(G)]^{\mathrm{SL}(2,\mathbb{C})}$ of regular functions on $\mathcal{R}(G)$ which are invariant by $\mathrm{SL}(2, \mathbb{C})$ is thus finitely generated. It is the ring of regular functions of an algebraic variety over $\mathbb{Q}$ which we denote by $\mathcal{R}(G)//\,SL(2,\mathbb{C})$. The canonical map $\mathcal{R}(G) \rightarrow \mathcal{R}(G)//\,SL(2,\mathbb{C})$ which is dual to the inclusion of coordinates rings is onto: this is a particular case of a theorem of D. Mumford [28]. The complex points of $\mathcal{R}(G)//\,SL(2,\mathbb{C})$ are in bijection with the closed orbits of $SL(2,\mathbb{C})$ in $\mathcal{R}_{\mathbb{C}}(G)$. It is important to note that it is possible that two representations in $\mathcal{R}(G)$ are not conjugated one to another, but are mapped however to the same point in $\mathcal{R}(G)//\,SL(2,\mathbb{C})$. This is the case for instance (when $G = \langle g \rangle$) for representations of the form:

$$\rho_1(g) = \begin{pmatrix} \lambda(g) & b(g) \\ 0 & \lambda^{-1}(g) \end{pmatrix} \text{ and } \rho_2(g) = \begin{pmatrix} \lambda(g) & 0 \\ 0 & \lambda^{-1}(g) \end{pmatrix},$$

the representation ρ_2 being in the closure of the $SL(2,\mathbb{C})$-orbit of ρ_1.

In our case, the ring of invariant regular functions is generated by the traces, i.e. $\mathbb{Q}[\mathcal{R}(G)]^{SL(2,\mathbb{C})} = \mathbb{Q}[\mathrm{tr}(\rho(g)), g \in G]$ [24]. Let $X_i = \mathrm{tr}(\rho(g_i))$ for $i \leq N$, be a basis of this ring (see Proposition 1). Let I be the ideal generated by all polynomials P in N variables such that $\rho \mapsto P(\mathrm{tr}(\rho(g_1)), \ldots, \mathrm{tr}(\rho(g_N)))$ is identically zero. Then $\mathcal{R}(G)//\,SL(2)$ is isomorphic to the algebraic subset of $\mathbb{A}_{\mathbb{C}}^N$, $Y(G) = \{(X_i) \,|\, P(X_1, \ldots, X_N) = 0, \forall P \in I\}$. Clearly $X(G) \subset Y(G)$ and the trace map t is equal to the canonical map. It follows from the Mumford's Theorem quoted above that $X(G) = \mathcal{R}(G)//\,SL(2,\mathbb{C})$.

Remark 10 In general $\mathcal{R}(G)$ is reducible and the dimension of its irreducible components may vary. One step of the proof of Thurston's Hyperbolic Dehn Surgery Theorem ([4, 41]) states that if G is the fundamental group of a finite volume hyperbolic 3-manifold with k boundary components, then one irreducible component of $X(G)$ is k-dimensional. This component actually is the one that contains the discrete and faithful representation of G. But one can construct examples of such 3-manifolds where the group G maps onto the free group with two generators $\mathbb{F}^2$. Since any representation of $\mathbb{F}^2$ induces a representation of G, we have an inclusion of $X(\mathbb{F}^2)$ into $X(G)$ and its image is a component disjoint from $\mathcal{DF}(G)$.

2.2   The Tautological Representation of G

This representation will be an important tool in Chapter 4 for the construction of trees.

We identified representations of $G \rightarrow SL(2,\mathbb{C})$ with the complex points in the affine algebraic set $\mathcal{R}(G)$. Fix an element of $g \in G$, and consider the matrix

$$\rho(g) = \begin{pmatrix} a_g(\rho) & b_g(\rho) \\ c_g(\rho) & d_g(\rho) \end{pmatrix}.$$

The coefficients of this matrix are polynomials with integers coefficients in the coordinates of the affine space $(\mathbb{A}^4)^n$ which contains the algebraic set $\mathcal{R}(G)$. Therefore $\rho(g)$ is an element of $\mathrm{SL}(2, \mathbb{Q}[\mathcal{R}(G)])$. The map $g \mapsto \rho(g)$ defines a representation of G into $\mathrm{SL}(2, \mathbb{Q}[\mathcal{R}(G)])$, called *the tautological representation*. If $\mathcal{R}_0$ is an irreducible component of $\mathcal{R}(G)$, we have an obvious restriction map $\mathbb{Q}[\mathcal{R}(G)] \rightarrow \mathbb{Q}[\mathcal{R}_0]$. We can therefore consider, for each irreducible component, the representation of G with values in $\mathrm{SL}(2, \mathbb{Q}[\mathcal{R}_0])$. It will be more convenient in the applications to think of it as a representation into $\mathrm{SL}(2, \mathbb{Q}(\mathcal{R}_0))$.

2.3   A Compactification of Affine Algebraic Sets

Let $X \subset \mathbb{A}_{\mathbb{C}}^N$ be an affine algebraic set defined over a countable field $k \subset \mathbb{C}$. Let $k[X]$ be the ring of the regular functions on X: it is a countable set. Let $\mathcal{F} \subset k[X]$ denote any finite or countable set which generates $k[X]$ as a ring.

Notation Consider the direct product $[0, \infty[^{\mathcal{F}}$ with its product topology. The *projective space* $\mathbb{P}^{\mathcal{F}}$ is defined as the quotient of $[0, \infty[^{\mathcal{F}} \backslash \{0\}$ by the equivalence relation which identifies any sequence (t_f) with the sequence (αt_f) for $\alpha \in \mathbb{R}_+^*$. We denote by π the natural projection $\pi : [0, \infty[^{\mathcal{F}} \backslash \{0\} \longrightarrow \mathbb{P}^{\mathcal{F}}$.

Define a map $\theta_0 : X \rightarrow [0, \infty[^{\mathcal{F}}$ by sending x to $\theta_0(x) = (\log(|f(x)| + 2))_{f \in \mathcal{F}}$, and write $\theta = \pi \circ \theta_0$. The choice of the constant 2 has no particular importance here; any constant > 1 could be used instead.

The projective space $\mathbb{P}^{\mathcal{F}}$ is not compact, however the following holds:

Proposition 11 *The closure of the image $\theta(X)$ in $\mathbb{P}^{\mathcal{F}}$ is compact and metrizable.*

Proof This is a consequence of the

Claim Let $h_1, \cdots, h_m \in \mathcal{F}$ be a finite set of functions which generate $k[X]$. Then for any $f \in \mathcal{F}$, there is a constant c_f such that

$$\log(|f(x)| + 2) \leq c_f \max\{\log(|h_j(x)| + 2)\}.$$

Therefore $\tilde{\theta}(x) = \theta_0(x)/\max\{\log(|h_j(x)| + 2)\}$ is contained in the product $[0, c_f]^{\mathcal{F}}$, which is compact and metrizable. In particular, $\tilde{\theta}(\mathcal{F})$ has compact closure in $[0, \infty[^{\mathcal{F}}$. This closure does not contain the point $\{0\}$: indeed by definition of $\tilde{\theta}$, one of the coordinates of $\tilde{\theta}(x)$ with index $1, 2 \cdots$ or m is equal to 1. Since $\theta(X) = (\pi \circ \tilde{\theta})(X)$, the closure of $\theta(X)$ in $\mathbb{P}^{\mathcal{F}}$ is thus compact and metrizable.   □

Notice that the closure $\overline{\theta(X)}$ might not be a compactification of X since $\theta : X \rightarrow \mathbb{P}^{\mathcal{F}}$ is in general not injective. In order to avoid this difficulty we introduce the one-point-compactification $\hat{X}$ of X and take the fibered product. Concretely let us define $\hat{\theta} : X \rightarrow \hat{X} \times \mathbb{P}^{\mathcal{F}}$ by $\hat{\theta}(x) = (x, \theta(x))$. The map $\hat{\theta}$ is clearly injective. The closure of $\hat{\theta}(X)$ in $X \times \mathbb{P}^{\mathcal{F}}$ is then a metric space which contains a dense open subset homeomorphic to X.

In the definition of θ_0, the constant 2 has no particular role. It can be replaced by a constant > 1 without changing the frontier $\overline{\theta(X)} \setminus \theta(X)$; furthermore the two compactifications are homeomorphic.

Definition 12 The *compactification of X associated to the family* $\mathcal{F}$ is $\overline{X}^{\mathcal{F}} = \overline{\hat{\theta}(X)}$. The *boundary of X* in $\overline{X}^{\mathcal{F}}$ is by definition $B_{\mathcal{F}}(X) = \hat{\theta}(X) \setminus X$ and can be identified with the set of accumulation points of $\theta(X)$ in $\mathbb{P}^{\mathcal{F}}$.

Example 13 Although we shall be more interested in the case where $\mathcal{F}$ is countable, the following simple example sheds some light on the general structure of the compactification described above. Take $X = \mathbb{C}^n$, and let $\mathcal{F} = \{X_1, \ldots, X_n\}$ be a set of coordinates. Then $\mathbb{P}^{\mathcal{F}}$ is naturally homeomorphic to the set of $(s_i) \in [0, 1]^n$ such that $\max s_1 = +1$: it is a piecewise real affine space of dimension $n - 1$ homeomorphic to the closed unit ball in $\mathbb{R}^{n-1}$. The map θ is surjective but not injective. The space $\overline{X}^{\mathcal{F}}$ is homeomorphic to the sphere of real dimension $2n$.

2.4 Thurston's Compactification of Teichmüller Space

Here we explain the link between the previous compactification when G is the fundamental group of surface and *Thurston's compactification of Teichmüller space* [15]. Fix a hyperbolic surface S of finite volume and consider its Teichmüller space $\mathcal{T}(S)$. It is the set of pairs (S', ϕ) where $\phi : S \to S'$ is an orientation-preserving homeomorphism and S' is a hyperbolic surface of finite volume, modulo the equivalence relation $(S_1, \phi_1) \simeq (S_2, \phi_2)$ when $\phi_2 \circ \phi_1^{-1} : S_1 \to S_2$ is isotopic to an isometry.

Now choose a base point in S, and consider the fundamental group $\pi_1(S)$. One has a natural map from $\mathcal{T}(S)$ to the set of real points $X_{\mathbb{R}}(\pi_1(S))$, which is constructed as follows. Pick a point $(S', \phi) \in \mathcal{T}(S)$. Take a universal cover $\mathbb{H} \to S'$ of S' by the upper half-plane. As $\pi_1(S')$ acts by isometries on $\mathbb{H}$, we obtain a representation of $\pi_1(S)$ into $\mathrm{PSL}(2, \mathbb{R})$, identified with the group of isometries of $\mathbb{H}$. Changing the universal cover or the base point amounts to conjugate the representation. It is known also this representation lifts to a representation with values in $SL(2, \mathbb{R})$ ([5] or [24, Proposition III.1.1]), so that we get a well-defined map from $\mathcal{T}(S)$ to $X_{\mathbb{R}}(\pi_1(S))$, the set real points of $R(\pi_1(S))//SL(2)$. This map is injective and it follows from a theorem of André Weil that when S is compact, $\mathcal{T}(S)$ is isomorphic to a connected component of $X_{\mathbb{R}}(\pi_1(S))$ [46] (see also [6,43]). When S has finite volume, the same result of Weil says that $\mathcal{T}(S)$ gets identified with a connected component of an analytic subset of $X_{\mathbb{R}}(\pi_1(S))$, defined by the equality of the traces of $\rho(g)$ to ± 2 for all elements g representing loops around the punctures of S.

Consider now the collection $\mathcal{S}$ of all homotopy classes of simple closed curves on S which are not homotopic to curves around the punctures. There is a natural map $\theta : X_{\mathbb{R}}(\pi_1(S)) \to \mathbb{P}^{\mathcal{S}}$, defined like in Sect. 2.3 whence a map $\theta_{\mathcal{T}} : \mathcal{T}(S) \to \mathbb{P}^{\mathcal{S}}$.

One can construct another map $\mathcal{L} : \mathcal{T}(S) \to \mathbb{P}^S$ as follows. Fix $\sigma = (S', \phi) \in \mathcal{T}(S)$, and let σ be the pull-back by ϕ of the hyperbolic metric on S'. For any $\gamma \in S$, we denote by $\text{length}_\sigma(\gamma)$ the length of the shortest representative on S' of the homotopy class $\phi(\gamma)$. Set $\mathcal{L}(\sigma) = (\text{length}_\sigma(\gamma))_\gamma$. One can prove that this map is injective and that its image is relatively compact in $\mathbb{P}^S$. The set $\mathcal{L}(\mathcal{T}(S))$ is Thurston's compactification of the Teichmüller space of S [15].

Pick a point $\sigma \in \mathcal{T}(S)$ and consider a representation $\rho : \pi_1(S) \to \text{SL}(2, \mathbb{R})$ associated to it. Let $\gamma \in S$. One has:

$$|\operatorname{tr}(\rho(\gamma))| = e^{l_\sigma(\gamma)/2} + e^{-l_\sigma(\gamma)/2}.$$

Therefore there are constants c_1 and c_2 independent of σ and γ such that

$$c_1 \, \text{length}_\sigma(\gamma) \le \log(|\operatorname{tr}(\rho(\gamma))| + 2) \le c_2 \, \text{length}_\sigma(\gamma).$$

This implies that the boundary of $\mathcal{L}(\mathcal{T}(S))$ in $\mathbb{P}^S$ coincides with the boundary of $\theta_T(\mathcal{T}(S))$.

3 Compactification of Affine Algebraic Varieties by Valuations

Morgan and Shalen have shown that points in the boundary $B_{\mathcal{F}}(X)$ of the compactification of the previous chapter can be described in terms of valuations on the field of rational functions $k(X)$. In this section, we prove this result (Proposition 30). We begin by recalling the necessary elements from valuation theory necessary for the proof following [49] and [24].

3.1 Valuations

An *ordered abelian group* is an abelian group with a total order which is compatible with the addition law. An ordered abelian group Λ is *archimedean* when for any positive elements x y in Λ, there is an integer n such that $nx > y$. Any ordered abelian group is ordered isomorphic to a subgroup of $\mathbb{R}$ by an isomorphism which is well-defined up to multiplication by a positive real number.

An important example of ordered abelian group to have in mind for what follows is $\mathbb{R}^p$ with the lexicographic order. In fact, all the ordered abelian groups that will appear in the sequel are ordered isomorphic to subgroups of such a group.

For any field k, write $k^* = k \setminus \{0\}$.

Definition 14 Let F/k be a field extension. A *valuation* on F/k is a surjective group homomorphism $v : F^* \to \Lambda$ where Λ is an ordered abelian group and

(i) for all $f, g \in F^*$, such that $f + g \in F^*$, then $v(f + g) \geq \min(v(f), v(g))$; and
(ii) the restriction $v|k^* = 0$.

The group Λ is called *the value group* of v. Two valuations $v : F^* \to \Lambda$ and $v' : F^* \to \Lambda'$ are *equivalent* if there is an isomorphism of ordered groups $i : \Lambda \to \Lambda'$ such that $v' = i \circ v$. The *trivial valuation* on F/k is the one such that $v|F^* = 0$.

Let v be a valuation on K/k. The *valuation ring of* v is

$$\mathfrak{o}_v = \{f \in F \mid f = 0 \text{ or } v(f) \geq 0\}.$$

This ring contains the field k and possesses a unique maximal ideal, namely

$$\mathfrak{m}_v = \{f \in \mathfrak{o}_v \mid f = 0 \text{ or } v(f) > 0\}.$$

We give examples of valuations in Sect. 3.6 below. For the moment, we present three basic invariants attached to a valuation of F/k.

Definition 15 (The rational rank of a valuation) This is the dimension of the $\mathbb{Q}$-vector space $\mathbb{Q} \otimes \Lambda$: it is denoted by rat. rk(v). The subadditivity property (i) from the definition of a valuation has the following consequence: if $f_1, \cdots, f_n$ are elements of F^* such that the minimum of the valuations $v(f_i)$ is reached by a single function f_j, then $v(f_1 + \cdots + f_n) = v(f_j)$. One deduces from this that if a non-trivial sum $\sum f_j$ is 0, then there exist distinct indices k and l such that $v(f_k) = v(f_l)$. It follows that rat. rk(v) is bounded from above by the transcendence degree of the extension F/k.

Definition 16 (The rank of a valuation) Let Λ be an ordered abelian group. A subgroup Λ' of Λ is said to be *convex* if for all $x > 0$ in Λ' the interval $[0, x] = \{0 \leq y \leq x\}$ is contained in Λ'. When Λ is finitely generated, its convex subgroups form a finite sequence $\Lambda_0 = \{0\} \subset \Lambda_1 \subset \cdots \subset \Lambda_r = \Lambda$. The integer r is called the *rank* of v and is denoted by rk(v). By induction, one proves that r is less than the rational rank of v. Each successive quotient group Λ_{j+1}/Λ_j carries a natural order relation, the *quotient ordering* and is archimedean for that ordering. In particular, any valuation of rank 1 is archimedean and is therefore equivalent to a valuation with value group a subgroup of $\mathbb{R}$.

Definition 17 (The transcendence degree of a valuation) The quotient of the valuation ring $\mathfrak{o}_v$ by its unique maximal ideal $\mathfrak{m}_v$ is a field k_v called *the residue field of* v, which naturally contains k. The *transcendence degree of* v is by definition the transcendence degree of the field extension k_v/k: it is denoted by deg. tr(v). Like the rational rank, the transcendence degree is also bounded from above by the transcendence degree of F/k.

Although we shall not use it in the sequel, we mention that the three invariants above satisfy the following fundamental inequality.

Claim 18 (Abhyankar inequality)

$$\mathrm{rk}(v) + \deg.\mathrm{tr}(v) \leq \mathrm{rat.}\,\mathrm{rk}(v) + \deg.\mathrm{tr}(v) \leq \deg.\mathrm{tr}(F/k)\,.$$

This inequality is actually valid in a much broader context of ring extensions. For fields it is due to Zariski. The valuations for which this inequality is an equality are called *Abhyankar valuations*. We refer to [14, Proposition 2.8] for a geometric description of these valuations which have rank equal to 1.

Let v be a valuation of F/k. Its valuation ring $\mathfrak{o}_v$ has the following property. If $f \notin \mathfrak{o}_v$ then $v(f) < 0$; therefore $1/f \in \mathfrak{m}_v$. This motivates the following definition.

Definition 19 (Places) A *place of F/k* is a pair $(\mathfrak{o}, \mathfrak{m})$ where $\mathfrak{o} \subset F$ is a ring containing k and $\mathfrak{m}$ is a maximal ideal of $\mathfrak{o}$ such that if $f \in F$ and $f \notin \mathfrak{o}$, then $1/f \in \mathfrak{m}$.

Remark Places correspond bijectively to valuations up to equivalence. For any valuation v, $(\mathfrak{o}_v, \mathfrak{m}_v)$ is a place. Conversely, for any place $(\mathfrak{o}, \mathfrak{m})$, there is a valuation v such that $\mathfrak{o}_v = \mathfrak{o}$, and this valuation is unique up to equivalence. One can indeed prove that $\Lambda = F^*/(\mathfrak{o} \setminus \mathfrak{m})$ is an ordered (multiplicative) abelian group [49, p. 35]: positive elements are in bijection with cosets $m(\mathfrak{o} \setminus \mathfrak{m})$, with $m \in \mathfrak{m}^*$. The valuation v is the quotient map $F^* \to F^*/(\mathfrak{o} \setminus \mathfrak{m})$.

However even if both concepts are equivalent, they offer two different perspectives on the same object. A place $\mathfrak{m} \subset \mathfrak{o}$ of F/k determines an homomorphism of $\mathfrak{o}$ to a field, namely to the field $\mathfrak{o}/\mathfrak{m}$. Viewed in this way, a place can be compared with the "evaluation of a function at a point"; it assigns to $f \in F$ an element in $\mathfrak{o}/\mathfrak{m}$ if $f \in \mathfrak{o}$ or ∞ if $f \notin \mathfrak{o}$. While the valuation associated to this place rather measures "the order of vanishing of f at the same point".

3.2 The Riemann-Zariski Space of F/k

Definition 20 Let F/k be a field extension. *The Riemann-Zariski space of F/k is the set $\mathcal{V}(F/k)$ of valuations on F/k up to equivalence. We put on $\mathcal{V}(F/k)$ the topology generated by the open sets $\{v \mid v(f) \geq 0\}$ as f ranges over F^*.*

Theorem 21 *The Riemann-Zariski space of F/k is quasi-compact: from any open cover one can extract a finite cover.*

Proof (see e.g. [44]) Consider the product of discrete spaces $Z = \{-, 0, +\}^{F^*}$ with the product topology of the topologies whose open sets are $\emptyset$, $\{-, 0, +\}$ and $\{0, +\}$. The Riemann-Zariski surface $\mathcal{V}(F/k)$ can be mapped into Z by the map $v \mapsto (s(f))_{F^*}$ where $s(f)$ is the sign of $v(f)$. This map is injective and its image is a closed set. One concludes using Tychonoff's Theorem. $\qquad\square$

Remark The definition of the Riemann-Zariski space and Theorem 21 are due to Zariski [48]. It is an interesting fact that *the Zariski topology* (for algebraic sets) was in fact designed to prove the aforementioned theorem. For the same reason, Zariski was naturally led to enrich the set of points of a fixed algebraic set in order to include all irreducible subvarieties as well. This feature is a characteristic of scheme theory, and was later formalized by Grothendieck in a much broader context.

3.3   The Center of a Valuation on a Projective Model

Let us first review some basic aspects of complex algebraic geometry as developed in [19, Chapter 1].

A projective variety $V \subset \mathbb{C}P^N$ is by definition the set of common zeroes of all polynomials lying in a fixed prime homogeneous ideal I_V of the ring of homogeneous polynomials in $N + 1$ variables with coefficients in the subfield k of $\mathbb{C}$. In other words, it is an irreducible closed subset of $\mathbb{C}P^N$ endowed with its Zariski topology. The *function field of V* is the quotient ring of the ring of rational functions P/Q, where P and Q are homogeneous polynomials of the same degree and coefficients in k and $Q \notin I_V$ by the ideal of the functions P/Q with $P \in I_V$. We denote it $k(V)$.

Let V, M be any two projective varieties contained in the same ambient space $\mathbb{C}P^N$. Then $V \subset M$ if and only if $I_M \subset I_V$. We denote by $\mathfrak{o}_V$ the subring of $k(M)$ which consists of all rational functions P/Q with P and $Q \notin I_V$. This ring is called *the local ring of M at V*. Its maximal ideal $\mathfrak{m}_V$ consists of rational functions P/Q, with $P \in I_V$. The quotient $\mathfrak{o}_V/\mathfrak{m}_V$ is isomorphic to $k(V)$, the function field of V.

Let F/k be any extension of finite transcendence degree. Suppose that k is a subfield of $\mathbb{C}$. A *projective model* of F/k is a pair (M, i) where M is a projective variety $M \subset \mathbb{C}P^N$ defined over k and $i : k(M) \to F$ is an isomorphism between the field of rational functions $k(M)$ and F.

We now explain the definition of Zariski [47, p. 497] of *the center of a valuation* on the function field of a variety.

Lemma 22 *Let M be a projective variety and let v be any non-trivial valuation on $k(M)/k$. Then, there exists a unique proper variety $W \subset M$ whose local ring and maximal ideal satisfy : $\mathfrak{o}_W \subset \mathfrak{o}_v$ and $\mathfrak{m}_W \subset \mathfrak{m}_v$.*

The variety $W = W_{M,v}$ is called *the center of the valuation v in M*.

Proof Let $(X_0, \cdots, X_N)$ be a system of homogeneous coordinates of the projective space which contains M. Consider a coordinate function $X = X_i$ which is minimal in the sense that $v(X_j/X) \geq 0$ for all $j = 0, \cdots, N$.

Since v is a valuation on $k(M)/k$, for each homogeneous polynomial P of degree m, one has $v(P/X^m) \geq 0$. Now define the subset $I \subset k[X_0, \cdots, X_N]$ containing 0 and all polynomials P such that $v(P/X^m) > 0$ with m equal to the degree of P. Then I is an ideal that contains I_M. It is even a prime ideal since if

P and Q are homogeneous polynomials with respective degrees r and s such that $v(PQ/X^{r+s}) > 0$, then either $v(P/X^r) > 0$ or $v(Q/X^s) > 0$. It contains I_M strictly since v is a non-trivial valuation. So there is a proper variety $W \subset M$ such that I is the defining ideal I_W of W.

Let $P/Q \in \mathfrak{o}_W$. Then $v(P/X^d) + v(X^d/Q) \geq 0$ since $Q \notin I_W$. Therefore $\mathfrak{o}_W \subset \mathfrak{o}_v$. The same argument shows that $\mathfrak{m}_W \subset \mathfrak{m}_v$.

For the uniqueness, let $W' \subset M$ be any variety such that $\mathfrak{o}_{W'} \subset \mathfrak{o}_v$ and $\mathfrak{m}_{W'} \subset \mathfrak{m}_v$. Then for any P and any $Q \notin I_{W'}$ both of degree d, one has $v(P/X^d) + v(X^d/Q) \geq 0$, hence $v(X^d/Q) \geq 0$. By the definition of I_W, we infer $Q \notin I_W$, thus $W' \subset W$. Suppose $W \neq W'$. Then one can find a homogeneous polynomial P in $I_{W'}$ but not in I_W. The rational function $P/X^d \in \mathfrak{m}_{W'}$ has positive valuation, but the valuation of its inverse X^d/P is also non-negative, since this element lies in $\mathfrak{o}_W$. This is impossible. Therefore $W = W'$. □

Remark 23 Any rational function $f = P/Q \in F = k(M)$ defines a regular map with values in $\mathbb{C}P^1$ outside the intersection of the zero loci of P and Q. This intersection is called *the indeterminacy locus* of f. When this locus does not contain the center W of v in M, then $f \in \mathfrak{o}_v$ if and only if f takes finite values on an open set of W; and $f \in \mathfrak{m}_v$ if and only if f vanishes on a Zariski open subset of W. Note that when the indeterminacy locus of f contains W one cannot see directly whether or not f belongs to $\mathfrak{o}_v$.

3.4 The Inverse System of the Projective Models of F/k

We now suppose that X is an affine variety defined over k and set $F = k(X)$. Two models (M, i) and (M', i') of F/k are said to be *equivalent* if and only if there is an isomorphism $M \to M'$ which induces the isomorphism $i \circ (i')^{-1}$ between the function fields $k(M)$ and $k(M')$. We denote by $\mathcal{M}$ the set of equivalence classes of models of F/k. Given any two models M and M', there is a birational map which induces $i' \circ (i)^{-1}$ on the function fileds: we denote this map by $j_{M',M} : M' \to M$. The set $\mathcal{M}$ thus carries a natural partial order, defined by $(M, i) \leq (M', i')$ when $j_{M',M}$ is a regular map. When $(M, i) \leq (M', i')$, we say then that M' *dominates* M.

In order to state and prove properly Theorem 24 below it is necessary to work with the scheme-theoretic description of the variety associated to M. Concretely this amounts to replacing M by the set of all its proper subvarieties which are defined over k, and put the Zariski topology on this set (a closed set consists in all proper subvarieties included in a fixed algebraic subset of M). To lighten notation we keep the same letter M for this object. Note that a subvariety is now a point.

If $(M, i) \leq (M', i')$ the regular map $j_{M',M} : M' \to M$ is continuous for this topology. So we can consider the projective limit $\mathbf{M} = \varprojlim_{\mathcal{M}} M$ with the projective limit topology. A point $w \in \mathbf{M}$ is the data for each $M \in \mathcal{M}$ of a subvariety $W_M \subset M$ such that whenever $M \leq M'$, then the regular map $j_{M',M} : M' \to M$ maps $W_{M'}$ to W_M. The set of subvarieties of M with the Zariski topology is quasi-compact.

Therefore the projective subset $\mathbf{M}$, as a closed subset of the direct product of all models of F is also quasi-compact. The following theorem is due to Zariski (see [49]).

Theorem 24 *The projective limit $\mathbf{M}$ is homeomorphic to the Riemann-Zariski space $\mathcal{V}$ of $k(X)/k$.*

Proof One can define a natural map $\mathcal{W} : \mathcal{V} \to \mathbf{M}$ as follows. Pick a valuation $v \in \mathcal{V}$, and for each projective model M, consider the center $W_{M,v}$ of v in M. Suppose that M' dominates M. Then $j_{M',M}$ is regular and maps $W_{M',v}$ to a subvariety $W \subset M$. The local ring o_W is mapped to the local ring of $W_{M',v}$ and its maximal ideal $\mathfrak{m}_W$ is mapped to the maximal ideal of $W_{M',v}$. By the characterization of the center given in Lemma 22, one has : $W = W_{M,v}$. This shows that the collection of subvarieties $(W_{M,v})$ is a point in $\mathbf{M}$. In this way, we get a map $\mathcal{W}$ which is easily seen to be continuous. Since $\mathcal{V}$ is quasi-compact it is enough to prove that this map is bijective to conclude.

Let us first prove that $\mathcal{W}$ is surjective. Let $w = (W_M) \in \mathbf{M}$, and write $\mathcal{R}_w = \cup o_{W_M}$. It is a ring contained in $k(X)$. Let us show that it is a valuation ring. The ideal $\mathfrak{m}_w = \cup \mathfrak{m}_{W_M}$ is a maximal ideal of $\mathcal{R}_w$. We need to show that $(\mathcal{R}_w, \mathfrak{m}_w)$ is a place, that is that, for any $f \in F^*$, either $f \in \mathcal{R}_w$ or $1/f \in \mathfrak{m}_w$. For this, we use the simple fact that for any model M, there exists a (not necessarily smooth) model M' which dominates M and on which f is a regular map $M \to \mathbb{C}P^1$ (this can be seen by taking M' to be the closure of the graph of f in $M \times \mathbb{C}P^1$). Now either f is infinite on $W_{M'}$, in which case $1/f \in o_{W_{M'}} \subset \mathcal{R}_w$; or f is finite on a Zariski dense open set of $W_{M'}$, in which case $f \in \mathfrak{m}_{W_{M'}} \subset \mathfrak{m}_w$. This shows that $\mathcal{R}_w$ is the ring of a valuation v. By the Lemma 22 one has: $\mathcal{W}(v) = w$.

Let us now prove that $\mathcal{W}$ is injective. If $\mathcal{W}(v) = w$ then the local ring $\mathcal{R}_w$ constructed above is necessarily included in R_v. But we saw that $\mathcal{R}_w$ was a valuation ring. Whence $\mathcal{R}_w = R_v$ and v has a unique preimage, so that $\mathcal{W}$ is injective. $\square$

Definition 25 (Valuations centered at ∞) Suppose that F is the function field of an affine variety $V \subset \mathbb{A}_{\mathbb{C}}^N$ defined over k. Then one says that a valuation $v \in \mathcal{V}(K/k)$ is *centered at ∞* when the coordinate ring $k[X]$ is *not* contained in the valuation ring o_v. This is equivalent to saying that the center of v on the model $\overline{X} \subset \mathbb{C}P^N$ is contained in the hyperplane at infinity $\mathbb{C}P^N \setminus \mathbb{A}_{\mathbb{C}}^N$. In particular, the valuations centered at ∞ form a closed subset of $\mathcal{V}(F/k)$.

3.5   Construction of Valuations from Sequences of Points

Let k be a countable field contained in $\mathbb{C}$, and let $X \subset \mathbb{A}_{\mathbb{C}}^N$ be a variety defined over k. Then $k(X)$ is countable.

We now describe a construction due to Morgan-Shalen of valuations of $k(X)/k$ from sequences of points in X. We shall see that all valuations on $k(X)/k$ arise in this way.

We say that a complex point x on a variety X is *k-generic on* X, if it is not contained in any proper subvariety of X defined over k, i.e. if it is dense in the spectrum of $k[X]$ for the Zariski topology. For instance a point in $\mathbb{A}_{\mathbb{C}}^N$ is k-generic if and only if its coordinates generate an extension of $\mathbb{Q}$ of transcendence degree equal to N.

Since X is irreducible, a Baire category argument implies that the set of k-generic points is dense in X for the classical topology. Any element of $k(X)$ can be written as the ratio P/Q of two polynomials, where Q is not in the ideal defining X. If x is a k-generic point, any meromorphic function in $k(X)$ can thus be evaluated at x.

Definition 26 A sequence $(x_i)_{i \in \mathbb{N}}$ in X is a *valuating sequence* if

(i) each x_i is k-generic, and
(ii) for any $f \in k(X)$, $\lim_{i \to \infty} f(x_i)$ exists in $\mathbb{C} \cup \{\infty\}$.

From any sequence of k-generic points which is contained in X, one can extract a valuating sequence using a Cantor diagonal argument.

Definition 27 (The valuation associated to a valuating sequence) Any valuating sequence (x_i) defines a place: the ring is the set $\mathfrak{o}$ of rational functions $f \in k(X)$ for which $\lim_{i \to \infty} f(x_i)$ belongs to $\mathbb{C}$, and its maximal ideal is the kernel of the natural homomorphism $\mathfrak{o} \to \mathbb{C}$, $f \mapsto \lim_{i \to \infty} f(x_i)$. This place corresponds to a valuation called the valuation associated to the valuating sequence (x_i).

In the applications the sequence (x_i) will be unbounded. This implies that the associated valuation is non-trivial since for some coordinate X_j the rational function $1/X_j$ tends to 0 and therefore $X_i \notin \mathfrak{o}_v$. However, for an arbitrary sequence this valuation could be trivial. This happens for instance when (x_i) tends to a k-generic point.

Let us give an application of Theorem 24.

Proposition 28 *Any valuation of $k(X)/k$ is associated to a valuating sequence (x_l) on X. The valuation is centered at infinity if and only if (x_i) tends to infinity in X.*

Proof Let v be a valuation on $k(X)/k$. Let $w = (W_M)$ be the corresponding point in the projective limit $\mathbf{M}$. Since k is countable there is a sequence $M_0 = \overline{X}$, $M_n \le M_{n+1}$ of totally ordered models of $k(X)/k$ such that any model M is dominated by some model M_j. Since the maps $j_{n+1} = j_{M_{n+1}.M_n}$ are regular, in particular continuous, we can choose distances d_n on M_n such that the maps j_n are distance-decreasing. We define by induction a sequence of points $\xi_n \in M_n$ such that ξ_n is k-generic on W_{M_n} and $j_{n+1}(\xi_{n+1}) = \xi_n$. For each n we choose a point $x_n \in M_n$ which is k-generic on M_n and such that $d_n(x_n, \xi_n) \le 1/n$. Then for any $n \ge m$, one has $j_{M_n.M_m}(x_n) \to \xi_m$ as $n \to \infty$. Define $x'_n = j_{M_n.M_0}(x_n)$. The point x'_n is k-generic on $\overline{X}$ hence belongs to X. Now pick $f \in k(X)$ and let $m \in \mathbb{N}$ be an integer such that f defines a regular map $M_m \to \mathbb{C}P^1$. By definition of W_m, $f \in \mathfrak{o}_v$ if and only if $f(\xi_m)$ is finite. This is equivalent to say that the sequence $(f(x_n))$ has a finite limit as $n \to \infty$. In the same way, by definition of $j_{M_n.M_0}$, we have

$f(x_n) = f(x'_n)$. Therefore $f \in \mathfrak{o}_v$ if and only if the sequence $(f(x'_n))$ tends to a finite limit as $n \to \infty$. Hence (x_n) is a valuating sequence and the valuation it defines is v. $\square$

In the next proposition we show that the valuation associated to a valuating sequence (x_i) which tends to infinity in X measures the growth rate of the sequence $f(x_i)$ when $f \in k(X)$. Before doing that, we need to define *the ratio* λ_1/λ_2 of two negative elements λ_1 and λ_2 in the value group Λ of v, when the group is not necessarily archimedean.

Let $\Lambda_0 = \{0\} \subset \Lambda_1 \cdots \subset \Lambda_r = \Lambda$ be the sequence of convex subgroups of Λ. Let $j \geq 1$ be the smallest index for which $\lambda_1 \in \Lambda_j$. Suppose first that j is also the smallest index for λ_2. Then λ_1 and λ_2 map to non-zero elements $\tilde{\lambda}_1, \tilde{\lambda}_2$ in the quotient group Λ_j/Λ_{j-1}. Since this group is archimedean it can be embedded in $\mathbb{R}$ in a unique way up to multiplication by a positive real number, so that the ratio $\tilde{\lambda}_1/\tilde{\lambda}_2$ is a well-defined real number. We set $\lambda_1/\lambda_2 = \tilde{\lambda}_1/\tilde{\lambda}_2 \in \mathbb{R}$. When the smallest index for λ_2 is $< j$, we set $\lambda_1/\lambda_2 = 0$ and when it is $> j$, we set $\lambda_1/\lambda_2 = \infty$. Observe from the definition that if λ_1 and λ_2 are both < 0 and if r and s are positive integers, we have : $\lambda_1/\lambda_2 \leq r/s$ if and only if $r\lambda_2 \leq s\lambda_1$.

Proposition 29 *Let (x_i) be a valuating sequence on X and v be the valuation of $k(X)/k$ associated to it.*

1. *For any $f \in k(X)$, $v(f) \geq 0$ if and only if $\log|f(x_i)|$ is bounded from above.*
2. *Let f and $g \in k(X)$ such that $v(f) \leq 0$ and $v(g) < 0$. Then*

$$\lim_{i \to \infty} \frac{\log|f(x_i)|}{\log|g(x_i)|} = \frac{v(f)}{v(g)}.$$

Proof The first statement is a rephrasing of the definition of v in terms of the valuating sequence. For the second statement, suppose first that $v(f) = 0$; then $\lim_{i \to \infty} |f(x_i)| \in \mathbb{C}^*$. Since $v(g) < 0$, $|g(x_i)| \to \infty$. So both terms in the formula are equal to 0 in this case. Suppose now that $v(f) < 0$. After exchanging the roles of f and g we see that the proof will be complete if we show

$$\overline{\lim_{i \to \infty}} \frac{\log|f(x_i)|}{\log|g(x_i)|} < \frac{v(f)}{v(g)}.$$

Let r and s be positive integers with $v(f)/v(g) < r/s$. Then $rv(g) - sv(f) < 0$ so that $\lim_{i \to \infty} g^r(x_i)/f^s(x_i) = \infty$. In particular $\lim_{i \to \infty}(r|\log g(x_i)| - s|\log f(x_i)|) = \infty$. This implies the required result since this holds for all integers r and s with $r/s > v(f)/v(g)$. $\square$

For any affine variety X defined over a countable field k, we defined in Sect. 2.3 a compactification $\overline{X}^{\mathcal{F}}$ of X which depends on the choice of a family of polynomials $\mathcal{F} \subset k[X]$. We saw that the boundary of this compactification $B_{\mathcal{F}}(X) = \overline{X}^{\mathcal{F}} \setminus X$ can be identified with the set of cluster points of $\theta(X)$ inside $\mathbb{P}^{\mathcal{F}}$. The next proposition states that this boundary can be described in a precise way using valuations on $k(X)/k$.

Recall that the rank of a valuation is equal to 1 when the ordered group Λ is archimedean, in which case one can suppose $\Lambda \subset \mathbb{R}$. Let v be a rank 1 valuation on $k(X)/k$; fix an embedding of its value group into $\mathbb{R}$ (such an embedding is well-defined up to multiplication by a positive real number). Suppose also that v is not supported at ∞. Then the point $U(v) \in \mathbb{P}^{\mathcal{F}}$ with homogeneous coordinates $(-\min(0, v(f)))$ is well-defined. We now extend the definition of $U(v)$ to valuations of higher rank as follows.

Let v be an arbitrary valuation on $k(X)/k$ centered at infinity. Let Λ denote its value group, and $\Lambda_0 = \{0\} \subset \Lambda_1 \subset \cdots \subset \Lambda_r = \Lambda$ be the sequence of its convex subgroups. Let s be the smallest index such that for all polynomials $f \in k[X]$ either $v(f) \geq 0$ or $v(f) \in \Lambda_s$. We define a new valuation $\bar{v}$ by composing v with the quotient map $\Lambda \rightarrow \bar{\Lambda} = \Lambda/\Lambda_{s-1}$. Then $\bar{v}$ is a valuation on $k(X)/k$ which is still centered at infinity. By construction, $\bar{v}$ also enjoys the following property (P): for any polynomial $f \in k[X]$, either $\bar{v}(f) \geq 0$ or $\bar{v}(f)$ belongs to the largest convex archimedean subgroup $\bar{\Lambda}_1$ of $\bar{\Lambda}$.

Denote by $\mathcal{V}_0 \subset \mathcal{V}(k(X)/k)$ the set of the valuations on $k(X)/k$ which are centered at ∞ and satisfy the property (P). One can then copy the definition that we gave for valuations of rank 1 and define a natural map $U : \mathcal{V}_0 \rightarrow \mathbb{P}^{\mathcal{F}}$ by setting: $U(v) = (-\min(0, v(f)))$. Note that $U(v)$ does not depend on the choice of an embedding of $\bar{\Lambda}_1$ in $\mathbb{R}$.

Proposition 30 *The map U maps continuously and surjectively $\mathcal{V}_0$ onto $B_{\mathcal{F}}(X)$.*

Proof It is clear that U is continuous. We claim that if (x_i) is a valuating sequence for a valuation v and $x_i \rightarrow \infty$, then $\lim_{i\rightarrow\infty} \theta(x_i) = U(\bar{v})$, where $\bar{v}$ is the quotient valuation defined by composing v with an homomorphism $\Lambda \rightarrow \Lambda/\Lambda_{s-1}$ as defined above.

Suppose that this claim is proved. Then pick $v \in \mathcal{V}_0$ and take a valuating sequence (x_i) for v. As v is centered at infinity the sequence (x_i) tends to infinity, and thus $U(v) = \lim_{i\rightarrow\infty} \theta(x_i)$ is a cluster point of $\theta(X)$. We conclude that $U(v) \in B_{\mathcal{F}}(X)$ which shows $U(\mathcal{V}_0) \subset B_{\mathcal{F}}(X)$. Conversely pick a point $\xi \in B_{\mathcal{F}}(X)$ and a sequence (x_i) tending to ∞ such that $\theta(x_i) \rightarrow \xi$. Using a Cantor diagonal argument we can approximate the sequence (x_i) by a valuating sequence (x_i') contained in X such that $\theta(x_i') \rightarrow \xi$. By what precedes we conclude that $U(\bar{v}) = \xi$.

To prove the claim we proceed as follows. Choose a function $g \in \mathcal{F}$ such that $\bar{v}(g) < 0$. Then $|g(x_i)| \rightarrow \infty$ as $i \rightarrow \infty$. The point ξ is the limit of the sequence with homogeneous coordinates $(\log(|f(x_i)| + 2)/\log(|g(x_i)| + 2))$.

For a function $f \in \mathcal{F}$ such that $\lim_{i\rightarrow\infty} f(x_i) \in \mathbb{C}$, then

$$\lim_{i\rightarrow\infty} \frac{\log(|f(x_i)| + 2)}{\log(|g(x_i)| + 2)} = 0$$

which is also $-\min(0, \bar{v}(f))/\bar{v}(g)$. When $\lim_{i\to\infty} f(x_i) = \infty$, then $v(f) < 0$. By
Proposition 29,

$$\lim_{i\to\infty} \frac{\log(|f(x_i)| + 2)}{\log(|g(x_i)| + 2)} = \frac{v(f)}{v(g)}.$$

From the definition of the ratio of two elements in an ordered group, this limit is
also $\bar{v}(f)/\bar{v}(g)$. $\square$

3.6 Examples of Valuations

3.6.1 Discrete Valuations of Rank 1

A valuation v on a field K is said *discrete of rank* 1 if its value group is equal to $\mathbb{Z}$. In
a geometric context the main examples arise as follows. Suppose X is an algebraic
variety over $\mathbb{C}$ and D is an irreducible divisor on X. Then the function attaching to
any rational function $f \in \mathbb{C}(X)$ its order of vanishing $\mathrm{ord}_D(f)$ along D defines a
discrete valuation of rank 1. More generally if $\pi : Y \to X$ is a birational morphism
and D an irreducible divisor in Y, then the function $\mathrm{ord}_D(f \circ \pi)$ is a discrete
valuation of rank 1 on $\mathbb{C}(X)$. Such valuations are called *divisorial* valuations.

In fact any discrete valuation v of rank 1 on the field of rational functions of an
algebraic variety X such that $\deg.\mathrm{tr}(v) = \dim X - 1$ is divisorial (see [49]).

3.6.2 (Quasi)-Monomial Valuations

Let $X \subset \mathbb{A}^N_{\mathbb{C}}$ be an affine variety of dimension n defined over $\mathbb{C}$. Take local
coordinates $x_1, \ldots, x_n$ at a smooth point $p \in X$ and fix non-negative real numbers
$s_1, \ldots, s_n \geq 0$. Pick any function f that is regular at p, and expand it locally in
power series $f(x) = \sum_I a_I x^I$, with $a_I \in \mathbb{C}$, and $x^I = \prod x_j^{i_j}$ if $I = (i_1, \ldots, i_n)$.
Denote $s.I = \sum_1^n s_k i_k$. Then the function $v_{s,x}(f) = \min\{s \cdot I, a_I \neq 0\}$ defines a
valuation $v_{s,x}$ of rank 1 whose value group is equal to $\sum_{i=1}^n \mathbb{Z}s_i$. It is not difficult
to check that any such valuation satisfies $\mathrm{rk}(v) + \deg.\mathrm{tr}(v) = \dim X$. Conversely
suppose v is a valuation of rank 1 on $\mathbb{C}(X)$ for which $\mathrm{rat}.\mathrm{rk}(v) + \deg.\mathrm{tr}(v) =
\dim(X)$. Then, one can show the existence of a birational morphism $\pi : Y \to X$,
a smooth point $p \in Y$, and local coordinates near p, such that the valuation
$f \mapsto v_{s,x}(f \circ \pi)$ is equivalent to v (see [14]).

3.6.3 More Complicated Examples

In general, the structure of a valuation can be quite complicated, even on the field of
rational functions of a smooth algebraic variety X. If X is a curve then any valuation

is discrete of rank 1. When X is a surface, a complete classification can be obtained (see [44, §3.2]). In higher dimension though, the picture is less clear for the moment.

Let us mention that given any sequence of integers $m_i \geq 1$ one can construct a valuation of rank 1 on the ring $\mathbb{C}[x_1, x_2]$ with value group $\sum_i \dfrac{1}{m_1 \ldots m_i} \mathbb{Z}$ (see [49, §15]). In particular choosing $m_i = p$ for all i, one may obtain a valuation with value group $\mathbb{Z}[\frac{1}{p}]$.

4 Λ-Trees

In the preceding two chapters we have constructed a compactification of the space of representations of a group G into $\mathrm{SL}(2, \mathbb{C})$ and we interpreted the boundary points in terms of valuations. In the next chapter, we will describe a construction introduced by Morgan-Shalen of a geometric object associated to a valuation v on the character variety [24]. This construction extended one by Bruhat-Tits, and also by Serre of the simplicial tree associated to a discrete valuation of rank 1. Here the geometric object is a Λ-*tree*, an object similar to a tree with the difference that the distance takes its values in a general abelian ordered group Λ. We review now the basic elements of this theory (see [8, 22]).

4.1 Λ-Trees

We begin with the

Definition 31 (Λ-metric space) Let Λ be an abelian ordered group; denote by Λ^+ the set of its non-negative elements. A Λ-metric space is a set Z with a map $d :$ $Z \times Z \to \Lambda^+$ which satisfies the axioms of a distance: $d(x, y) = 0$ if and only if $x = y$; $d(x, y) = d(y, x)$; and the triangular inequality $d(x, z) \leq d(x, y) + d(y, z)$.

The simplest example is given by the group Λ itself, which is a Λ-metric space with the distance $d(\lambda_1, \lambda_2) = |\lambda_2 - \lambda_1| = \max\{\lambda_2 - \lambda_1, \lambda_1 - \lambda_2\}$. The isometry group of Λ with this distance is generated by the translations $\lambda \to \lambda + \delta$ (of length $\delta \in \Lambda$) together with the involution $\lambda \to -\lambda$.

A subset $I \subset \Lambda$ is called *an interval* when it is *convex*, that is if for all $\lambda_1 < \lambda_2 \in I$ the subset $[\lambda_1, \lambda_2] = \{\lambda \in \Lambda | \lambda_1 \leq \lambda \leq \lambda_2\}$ is contained in I. A *closed interval* is an interval of the form $[\lambda_1, \lambda_2] = \{\lambda | \lambda_1 \leq \lambda \leq \lambda_2\}$ that is when it contains both its upper and lower bounds. One defines the analogous notions of open, left-, right-open interval and denote them in a natural way by $]\lambda_1, \lambda_2]$, $]\lambda_1, \lambda_2]$, etc. Note that in a non-archimedean group Λ, a bounded interval is not always of this type. Take for instance $\{p\} \times \mathbb{Z}$ in $\mathbb{Z} \times \mathbb{Z}$ endowed with the lexicographic order.

Let Z be a Λ-metric space. A *segment* in Z is a subset isometric to an interval I contained in Λ; a segment is closed (resp. open) when the interval I is closed (resp.

open). When the upper or lower bounds of I are contained in I, the corresponding points in Z are called *endpoints*. A segment is *non-degenerate* when it is not reduced to a single point.

Definition 32 A Λ-*tree* is a Λ-metric space T which satisfies the following three axioms:

(i) T is "uniquely connected by segments", i.e. any two points x, y are the endpoints of a closed segment; such a closed segment is unique and will be denoted $x \cdot y$;

(ii) when two segments have a common endpoint, then their intersection is a segment;

(iii) when two segments $x \cdot y$ and $y \cdot z$ have only the point y in common, then the union $x \cdot y \cup y \cdot z$ is a segment.

Remark When $\Lambda = \mathbb{Z}$, we get back the notion of simplicial tree: any $\mathbb{Z}$-tree is isometric to the set of vertices of a simplicial tree. The notion of Λ-tree in the case $\Lambda = \mathbb{R}$ coincides with the standard notion of $\mathbb{R}$-tree as defined in [20]. In the sequel, we shall be interested in Λ-trees when Λ is the value group of a valuation which is not necessarily of rank 1.

A *broken segment* is a map $\mu : [\lambda, \lambda'] \to T$ such that there exists a subdivision of $[\lambda, \lambda']$, $\lambda_0 = \lambda < \lambda_1 < \cdots < \lambda_n = \lambda'$ for which the restriction μ to the interval $[\lambda_i, \lambda_{i+1}]$ is an isometry to a closed segment of T.

Proposition 33 *Let T be a Λ-tree and let $\mu : [\lambda, \lambda'] \to X$ be a broken segment. Then*

(i) the segment between $\mu(\lambda)$ and $\mu(\lambda')$ is contained in the image of μ;

(ii) if $d(\mu(\lambda), \mu(\lambda')) = \lambda' - \lambda$, then μ is an isometry to its image.

Proof We argue by induction on the integer n appearing in the definition of a broken segment. The case $n = 1$ is trivial. We consider the case $n = 2$. By assumption, $\mu([\lambda, \lambda_1])$ and $\mu([\lambda_1, \lambda'])$ are segments. By the property (ii) of Λ-tree, their intersection is a segment which can be equally parameterized by the restrictions $\mu|[\alpha, \lambda_1]$, for some $\alpha \in [\lambda, \lambda_1]$ or by $\mu|[\lambda_1, \beta]$ with $\beta \in [\lambda_1, \lambda']$. Then the images $\mu([\lambda, \alpha])$ and $\mu([\beta, \lambda'])$ are segments which intersect only at $\mu(\alpha) = \mu(\beta)$. By the property (iii) of a Λ-tree, the union of these segments is a segment, so it is equal to the segment $\mu(\lambda).\mu(\lambda')$. This proves Proposition 33 (i). Since the length of the segment $\mu(\lambda).\mu(\lambda')$ is equal to $\lambda' - \lambda - 2(\lambda_1 - \alpha)$, Proposition 33 (ii) follows also. The case $n > 2$ reduces to the case $n = 2$, using the induction hypothesis. $\square$

4.2 Classification of Isometries of a Λ-Tree

Let T be a Λ-tree. An isometry of T is a bijection $g : T \to T$ such that $d(gx, gy) = d(x, y)$ for all x, y in T. Isometries of T fall into three categories: *elliptic isometries*, *phantom inversions*, and *hyperbolic isometries*. In this section we explain this trichotomy and describe in detail the structure of hyperbolic isometries.

Let g be an isometry of T. Pick a point $x \in T$ and consider the segment $x.gx$. The image of this segment by g is the segment $gx.g^2x$. From the axioms of a Λ-tree we deduce that the intersection $x.gx \cap x.g^2x$ is a closed segment: it is equal to $gy.gx$ for some $y \in x.gx$. The classification of g will be done according to the value of $d(y, x)$ compared to that of $d(x, gx)$.

Case 1. Suppose $2d(x, y) = d(x, gx)$. Then y is the midpoint of the segment $x.gx$ and it is fixed by g. One says that g is *an elliptic isometry*. Observe that since g is an isometry, the set $Fix(g)$ of its fixed points is a convex subset of T: for any two points x and y in $Fix(g)$, the segment $x \cdot y$ is contained in $Fix(g)$.

Case 2. Suppose $2d(y, x) > d(x, gx)$. Then y and gy are exchanged by g. Therefore the segment $y.gy$ is mapped to itself by g but g reverses the order on that segment. If the distance $d(x, gx)$ is divisible by 2 in Λ then g fixes this midpoint. In that case g is an elliptic isometry. If this distance is not divisible by 2 then g has no fixed points. It is called a *phantom inversion*.

Case 3. Suppose $2d(y, x) < d(x, gx)$. Therefore g has no fixed points and is not a phantom inversion. Then y and gy are distinct and $y \in [x, gy]$, $gy \in [y, gx]$. The two closed segments $y.gy$ and $gy.g^2y$ thus intersect only at gy. Observe that the smallest convex subgroup Λ_y containing $d(y, gy)$ is the reunion $\bigcup_{n\in\mathbb{Z}}[nd(y, gy), (n+1)d(y, gy)]$. By induction on n we can define a map from Λ_y to T whose image is the reunion of segments $A_y = \bigcup_{\mathbb{Z}} g^n y.g^{n+1}y$ and which conjugates the action of g to the translation of length $d(y, gy)$ on Λ_y. In this case g is called a *hyperbolic isometry*.

Let us introduce the following notion. A *partial axis* for a hyperbolic isometry g is a segment $A \subset T$ which is invariant by g and such that there exists an isometry from A onto an interval $I \subset \Lambda$ which conjugates the restriction $g|A$ to a translation on I. An example of a partial axis is the segment A_y defined above. *An axis for g is* a partial axis A which is *maximal* for the inclusion: any partial axis which contains A is equal to A.

Proposition 34 *If an isometry of T has partial axes, then it has a unique axis, and this axis contains all partial axes.*

Proof We first prove the following lemma.

Lemma 35 *Suppose that A_1 and A_2 are two partial axes for an isometry g. Then $A_1 \cup A_2$ is contained in a partial axis.*

Proof Denote by δ_i the translation distance of g on A_i. Two cases appear depending on wether the intersection $A_1 \cap A_2$ is empty or not.

First Case: suppose that $A_1 \cap A_2 = \emptyset$. This situation occurs for instance when T is the group $\Lambda = \mathbb{Z} \times \mathbb{Z}$ endowed with the lexicographic order, g is the translation by $(0, 1)$, and A_1, A_2 are two distinct vertical segments $a_1 \times \mathbb{Z}$, $a_2 \times \mathbb{Z}$.

Choose arbitrary points $a_1 \in A_1$, $a_2 \in A_2$. Suppose first that some closed segment $k_1.a_2$ contained in $a_1.a_2$ has the property that its intersection with A_1 is reduced to k_1. In that case we shall reach a contradiction. By Proposition 33, the segment $a_2.ga_2$ is equal to the union of $a_2.k_1$, $k_1.gk_1$ and $gk_1.gk_2$; it follows that g moves the point a_2 by a length $d(a_2, ga_2) = 2d(a_2, k_1) + \delta_1$. Suppose now that the segment $a_2.k_1$ contains a point $a'_2 \in A_2$ distinct from a_2 : in particular $d(a'_2, k_1) < d(a_2, k_1)$. Since $d(a'_2, ga'_2) = 2d(a'_2, k_1) + \delta_1 < d(a_2, ga_2)$, the restriction of g to A_2 cannot be conjugated to a translation. This contradiction implies that $a_2.k_1 \cap A_2 = \{a_2\}$. But then for any point $a'_2 \neq a_2$ on A_2, the segment $k_1.a'_2$ is the union of $k_1.a_2$ and the segment $a_2.a'_2$. In particular g moves the point a'_2 at a distance $d(a'_2, ga'_2) = 2d(k_1, a'_2) + \delta_1$ which is strictly greater than $d(a_2, ga_2)$: this is again a contradiction. We deduce that the intersections $a_1.a_2 \cap A_1$ and $a_1.a_2 \cap A_2$ are *ends* of A_1 and A_2 respectively. This means the following.

For $j = 1, 2$, pick an isometry i_j from an interval $I_j \subset \Lambda$ to A_j which conjugates $g|A_j$ to a translation on I_j. Write $\alpha_j = i_j^{-1}(a_j) \in I_j$. Then after possibly exchanging the roles of A_1 and A_2, one has: $i_1^{-1}(a_1.a_2 \cap A_1) = \{\lambda \in I_1 | \lambda \geq \alpha_1\}$ and $i_2^{-1}(a_1.a_2 \cap A_2) = \{\lambda \in I_2 | \lambda \leq \alpha_2\}$. In particular $A'_j = A_j - (A_j \cap a_1.a_2) \cup \{a_j\}$, $j = 1, 2$ is a segment with endpoint a_j. The set $A_1 \cup A_2$ is thus the union of three segments A'_1, $a_1.a_2$ and A'_2 each intersecting the next at its endpoints: it is therefore a segment. This segment is invariant by g and g acts on it as a translation (since it does so on A_1 and on A_2). Therefore $A_1 \cup A_2 \cup a_1.a_2$ is a partial axis for g.

Second case: suppose that $A_1 \cap A_2 \neq \emptyset$. Denote as before $i_1 : I_1 \to A_1$ an isometry between an interval $I_1 \subset \Lambda$ and A_1. Then $i_1^{-1}(A_1 \cap A_2)$ is an interval J contained in I_1. The axioms of a Λ-tree imply that one of the three following possibilities does occur. (1): J has one endpoint, (2): J has two endpoints, or (3): each end of J is mapped by i_1 to an end of A_1 or A_2 (an *upper end* in a segment is a subsegment of the form $\{a | a \geq a_0\}$ for some a_0; one defines in an analogous way the notion of *lower end*.)

Since $A_1 \cap A_2$ is invariant under g, the interval J is invariant by a non-trivial translation of Λ: this rules out (1) and (2). When both ends of J are mapped to ends of A_1 (resp. A_2), then $J = A_1$ and therefore $A_1 \subset A_2$ (resp. $A_2 \subset A_1$). When one end of J is mapped to an end of A_1, and the other to an end of A_2, then $A_1 \cup A_2$ is a segment. In that case $A_1 \cap A_2$ is a partial axis for g. □

We also notice the following consequence of the proof. Suppose that A_1 and A_2 are two partial axes for g with $A_1 \cap A_2 \neq \emptyset$. Then any isometry $i_1 : I_1 \to A_1$ which conjugates g to a translation on I_1 extends in a unique way to an isometry from an interval which contains I_1 to $A_1 \cup A_2$.

We continue the proof of Proposition 34. Let $A_0 \subset T$ be a partial axis for g. Let $i_0 : I_0 \to A_0$ be the isometry between an interval $I_0 \subset \Lambda$ and A_0 which conjugates g and the translation of length δ. For any partial axis A_j which contains A_0 there is a unique interval $I_j \subset \Lambda$ and a unique isometry $i_j : I_j \to A_j$ such that $I_0 \subset I_j$ and $i_j|_{I_0} = i_0$. Let $\mathcal{J}$ be the set of all partial axes A_j which contain A_0. Denote by I the union of all intervals I_j as $j \in \mathcal{J}$; and by A the union of all partial axes

A_j, for $j \in \mathcal{J}$. Then I is an interval of Λ since the interval which parameterizes $A_1 \cup A_2$ contains $I_1 \cup I_2$ thanks to the previous observation. Also the uniqueness of the map i_j implies that the restrictions of the isometries i_1 and i_2 to $I_1 \cap I_2$, are equal. Therefore the maps i_j, $j \in \mathcal{J}$ can be glued together to define an isometry $i : I \to T$, the image of which is equal to A. It is clear that A is invariant by g and that the isometry i conjugates the translation by δ on I to the restriction of g to A. This shows that A is a partial axis. Since any partial axis for g is contained in a partial axis element of $\mathcal{J}$, we conclude that A is an axis. $\qquad\square$

Definition 36 (Length of an isometry) Let g be an isometry of a Λ-tree. The *length of g* is by definition the infimum of $d(x, gx)$ over all $x \in T$. We denote it by l_g. When g is an elliptic isometry the infimum is attained and $l_g = 0$ by definition. When g is a phantom inversion l_g may or may not be equal to 0 and the infimum may or may not be attained.

Suppose g is an hyperbolic isometry which translates by a distance δ along its axis A_g; for all $x \in A_g$, $d(x, gx) = \delta$. When $x \notin A_g$, then the segment $x.gx$ intersects a partial axis A_x for g along a closed segment (see the description of the hyperbolic isometries given above). Therefore there is some closed segment with endpoints x and a point $k \in A_x$ such that the intersection of $x.k$ and A_x is reduced to k. It follows that $d(x, g.x) = \delta_x + 2d(x, k)$, where δ_x is the length of g along the partial axis A_x. Since each partial axis for g is contained in A_g, we infer $\delta_x = \delta$. In particular for any $x \notin A_g$, $d(x, gx) > \delta$. We summarize what we proved in the following proposition.

Proposition 37 *Let g be a hyperbolic isometry of T. Then $l_g = \inf\{d(x, gx), x \in T\}$ is attained and $l_g > 0$. The axis A_g of g coincides with the set of points x such that $d(x, gx) = l_g$. Furthermore, for any point $x \notin A_g$ there exists $k \in A_g$ such that $x.k \cap A_g = \{k\}$ and $d(x, gx) = l_g + 2d(x, k)$.*

4.3 Groups Acting on Λ-Trees

Let us introduce the following

Definition 38 (Length function) Let G be a group acting by isometries on a Λ-tree T. The length function of this action is the function $G \to \Lambda^+$, $g \mapsto l_g$. Since l_g only depends on the conjugacy class of g we can view this function as a function $\mathcal{C} \to \Lambda^+$, where $\mathcal{C}$ is the set of conjugacy classes of G.

Proposition 39 *Let G be a finitely generated group which acts by isometries on a Λ-tree and without phantom inversion. Then G has a fixed point if and only each of its elements has a fixed point, i.e. $l_g = 0$ for all $g \in G$.*

Proof We argue by induction on the number of generators of G. Suppose that G is generated by n elements $g_1, \dots, g_n$ and that $l_g = 0$ for all g. Then by assumption, the set of the fixed points of the group generated by $\langle g_1, \dots, g_{n-1} \rangle$ is non-empty: it is a non-empty convex set C. To prove the proposition, we show that the fixed point set of g_n intersects C. We begin with the following observation.

Lemma 40 *Let g, h be isometries of T and let x, $y \in T$ be fixed points of g and h respectively. Then either the segment $x \cdot y$ contains a common fixed point of g and h, or hg is an hyperbolic isometry.*

Proof Let $x, y \in T$ with $gx = x$ and $hy = y$. The intersection $x \cdot y \cap x \cdot gy$ is a closed segment $x \cdot x'$. Since g is an isometry, any point in $x \cdot x'$ is fixed by g and the intersection $x' \cdot y \cap x' \cdot gy$ is equal to x'. The same property is satisfied by the intersection $x \cdot y \cap hx \cdot y$. In this way, we find a segment $x' \cdot y'$ contained in $x \cdot y$ where

(i) $gx' = x'$ and $hy' = y'$;
(ii) $x' \cdot y' \cap x' \cdot gy' = \{x'\}$; $x' \cdot y' \cap hx' \cdot y' = \{y'\}$.

If the segment $x'.y'$ is degenerate then $x' = y'$ is fixed by g and h. If it is non-degenerate, consider the union $x' \cdot y' \cup y' \cdot hx'$: it is the segment $x' \cdot hx'$. One deduces from (ii) that $x' \cdot y'$ intersects its image by hg only at hx'. This implies that the non-degenerate segment $x' \cdot hx'$ is a fundamental domain for hg; in particular hg is an hyperbolic isometry.   □

Let $x \in T$ fixed by $\langle g_1, \ldots, g_{n-1} \rangle$; let $y \in T$ fixed by g_n. After applying the Lemma to these x and y, we obtain either a point fixed by the entire group G or an hyperbolic element of the form $g_n g_i$ for some $i \leq n - 1$.   □

Definition 41 (Minimal action) Let $G \times T \to T$ be an isometric action without phantom inversion of a group G on a Λ-tree T. The action is said to be minimal if any Λ-tree $T' \subset T$ which is invariant by G is equal to T.

Observe that there always exists an invariant subtree contained in T on which the action of G is minimal.

Recall that a group is virtually abelian when it contains an abelian subgroup with finite index.

Definition 42 (Small actions) An action $G \times T \to T$ is small when for any non-degenerate closed segment $x \cdot y \subset T$, the subgroup of G which fixes $x \cdot y$ pointwise is virtually abelian.

Note that any free action is small.

Remark We will be mainly interested in groups which can be embedded into $SL(2, \mathbb{C})$ as discrete subgroups. Such a group is either virtually abelian or contains a subgroup isomorphic to a free group with two generators. In particular, if such a group is not virtually abelian its commutator subgroup is not virtually abelian either. This motivates the assumption made in the next result.

Proposition 43 *Let $G \times T \to T$ be an action of a group G on a Λ-tree T which is small and minimal. Suppose that g and h are hyperbolic elements of G such that*

$\langle g, h \rangle$ *and its commutator subgroup both are not virtually abelian. Then,*

1. *if the axis A_g and A_h intersect, then their intersection $A_g \cap A_h$ is a closed segment, and*
2. *if the axes A_g and A_h are disjoint, there is a closed non-degenerate segment $k.l \subset T$ with $k \cdot l \cap A_g = \{k\}$, $k \cdot l \cap A_h = \{l\}$.*

Proof Suppose that $A_g \cap A_h \neq \emptyset$. Since A_g is an axis, there exists an interval $I \subset \Lambda$ and an isometry $\iota : I \to A_g$ which conjugates g to the translation by l_g on I. Then $\iota^{-1}(A_g \cap A_h)$ is a convex subset of I, and is therefore equal to a sub-interval $J \subset I$. Denote by $\{0\} = \Lambda_0 \subset \Lambda_1 \ldots \subset \Lambda_r = \Lambda$ the sequence of convex subgroups of Λ. Up to translation, one can assume that J is contained in Λ_s but not in Λ_{s+1}. Pick $\xi \in J$, and consider the interval $[\xi, \infty[\cap J \subset \Lambda_s$ with $[\xi, \infty[= \{\lambda \in \Lambda_s | \xi \leq \lambda \leq \infty\}$. We will show that this interval is closed. The same reasoning applied to the interval $]\infty, \xi] \cap J$ will imply (1).

Suppose by contradiction that the interval $[\xi, \infty[\cap J$ is not a closed interval of Λ_s. We may then assume that the segment $\iota([\xi, \infty[\cap J)$ is an end of A_g and that it is mapped into itself by all positive powers of g.

— Suppose $\iota([\xi, \infty[\cap J)$ is also an end of A_h, stabilized by all positive powers of h. Since g and h act by translation on $\iota([\xi, \infty[\cap J)$, any element in the commutator group of $\langle g, h \rangle$ acts as the identity on an end of $\iota([\xi, \infty[\cap J)$. Since the action of G on T is small, the commutator subgroup of $\langle g, h \rangle$ is virtually abelian. This contradicts our assumption.

— Suppose now that $\iota([\xi, \infty[\cap J)$ is not an end of A_h. One can find a point $y \in A_h$ such that the segment $x \cdot y$ contains $\iota([\xi, \infty[\cap J)$. Then $x \cdot y$ contains the upper end of A_g. By invariance, $gx \cdot gy$ also contains the upper end of A_g. Since $x \cdot gy = x \cdot gx \cup gx \cdot gy$, the intersection $x \cdot y \cap x \cdot gy$ contains the upper end of A_g. As T is a Λ-tree, the intersection $x \cdot y \cap x \cdot gy$ is a closed segment $x \cdot z$. If $z \in A_g$, then $gz \in A_g$, and so $x \cdot gz$ is contained in $x \cdot y \cap x \cdot gy$. But z is between x and gz, hence $x \cdot gz$ is strictly contained in $x \cdot y \cap x \cdot gy$. We conclude that $z \notin A_g$.

On the other hand, one has $z \in x \cdot y$ hence $gz \in x \cdot gy$. Since g is an isometry $d(gz, gy) = d(z, y)$, $d(x, y) = d(gx, gy) = d(x, y) - d(x, gx)$, therefore, $d(z, y) = d(z, gy) - d(x, gx)$. We infer $gz \in z \cdot gy$, and it follows that $d(z, gz)$ is equal to $d(x, gx)$ the length of g, hence $z \in A_g$. This gives a contradiction and proves that $\iota([\xi, \infty[\cap J)$ is a closed segment. This concludes the proof of (1).

To prove (2), choose points a and b in T with $a \in A_g$ and $b \in A_h$. Let $k \in A_h$ be the point provided by Proposition 37 such that the intersection of the segment $k \cdot b$ and A_g is equal to $\{k\}$. In a similar way, let $l \in A_h$ be the point such that the intersection of $a \cdot l$ and A_h is reduced to $\{l\}$. Then the segment $k \cdot l$ satisfies (2). $\square$

The segment $k \cdot l$ defined in Proposition 43 (2) is the unique shortest one among all segments $a.b$ connecting a point $a \in A_g$ and a point $b \in A_h$. Indeed, the axioms of a Λ-tree easily imply that $a.b$ is the union of the three segments $a \cdot k$, $k \cdot l$ and

$l \cdot b$, with $a \cdot k \cap k \cdot l = \{k\}$ and $k \cdot l \cap l \cdot b = \{l\}$. We will call the length of $k.l$ the *distance* between the two axes A_g and A_h and we will denote it by $d(A_g, A_h)$.

Lemma 44 *For any hyperbolic isometries a, b on a Λ-tree, the distance between the two axes A_a and A_b is determined by the lengths of the elements a, b and ab. More precisely, one has*

$$d(A_a, A_b) = \frac{1}{2}\max(0, l_{ab} - l_a - l_b). \tag{1}$$

Proof The proof is the same as for $\mathbb{R}$-trees. We are reduced to the case when the axes are disjoint. Then let $k \cdot l$ be the shortest segment with $k \in A_a$ and $b \in A_b$ provided by Proposition 43 (2). A fundamental domain for ab on its axis is then the union of the segments $a^{-1}k \cdot k$, $k \cdot l$ $l \cdot bl$ and $bl \cdot bk$. Therefore the length of ab is $l_a + l_b + 2d(k, l)$, proving (1).   □

4.4   Conjugated Actions

We will now generalize to arbitrary Λ-trees a fundamental result first proved by Culler and Morgan in the case of $\mathbb{R}$-trees (see [12, Theorem 3.7]). Our proof in the general case closely follows their one, and relies in an essential way on Proposition 43.

Theorem 45 ([12]) *Let $\mathcal{T}_1$ and $\mathcal{T}_2$ be Λ-trees and $\rho_1 : G \times \mathcal{T}_1 \to \mathcal{T}_1$ and $\rho_2 : G \times \mathcal{T}_2 \to \mathcal{T}_2$ be actions of a group G on $\mathcal{T}_1$ and on $\mathcal{T}_2$ which are small and minimal. Then there is an isometry between $\mathcal{T}_1$ and $\mathcal{T}_2$ which conjugates ρ_1 and ρ_2 if and only if the length-functions of these actions are equal.*

Proof We follow the main steps of the proof in [12]. One ingredient in their proof is the classification (up to isometry) of groups acting on trees in terms of their *displacement functions*. This classification is due to Chiswell [7, 8], see also [30, p. 22]. In what follows, we do not rely on this classification although our arguments are very close in essence.

— One can find two elements $g, h \in G$ which form *a good pair of isometries of* $\mathcal{T}_1$. This means that they satisfy the following three properties:

1. the axes A_g^1 and A_h^1 have non-empty intersection ;
2. the intersection of $A_g^1 \cap A_h^1$ is a closed segment and g and h move in the same direction along this segment;
3. the length of $A_g^1 \cap A_h^1$ is shorter than l_g and l_h.

To see this, pick two hyperbolic isometries a and b in G, and denote by A_a^1 and A_b^1 their respective axes in $\mathcal{T}_1$. If $A_a^1 \cap A_b^1 = \emptyset$, then one checks using Proposition 43 (2) that the axes of the isometries a and ab have non-empty intersection. The intersection of these axes in then a closed segment by

Proposition 43 (1). Up to replacing a by a^{-1} if necessary, both isometries a and ab translate in the same direction. After taking sufficiently large positive powers of these isometries, (3) is satisfied.

— Since the length-functions of ρ_1 and ρ_2 are equal, g and h form also a good pair of isometries of $\mathcal{T}_2$.

— Let $p_i \in \mathcal{T}_i$ be the unique point common to the three axes A_g^i, A_h^i and $A_{gh^{-1}}^i$ (it is the "upper" endpoint of the closed segment $A_g^i \cap A_h^i$ if the isometries move in the positive direction). Then there is a formula for the displacement distance of p_i under an element $k \in G : d(p_i, kp_i)$ is the maximum of the distances between one of the axes A_g^i, A_h^i, $A_{gh^{-1}}^i$ and the image by k of one of these axes. To see this, consider the segment $p_i.kp_i$. Then for at least one of the axis $C \in \{A_g^i, A_h^i, A_{gh^{-1}}^i\}$, the intersection $C \cap p_i.kp_i$ is equal to $\{p_i\}$. A similar result holds for the intersection with the other three axes.

— It follows that the *displacement function* (or *Chiswell length-function*) of p_i in $\mathcal{T}_i$, that is the function from G to Λ, defined by $k \mapsto d(p_i, kp_i)$ is determined by the length-function of the action $\rho_i : G \times \mathcal{T}_i \to \mathcal{T}_i$. In particular the displacement function of p_1 in $\mathcal{T}_1$ is equal to that of p_2 in $\mathcal{T}_2$.

— Let us now briefly explain the argument of Chiswell for constructing at this point an explicit isometry from $\mathcal{T}_1$ to $\mathcal{T}_2$. Pick any $x_1 \in \mathcal{T}_1$. By minimality, there exists $g \in G$ such that $x_1 \in p_1 \cdot gp_1$. Since $d(p_1, gp_1) = d(p_2, gp_2)$, there is a unique point denoted $x_2 \in \mathcal{T}_2$ such that x_2 lies in the segment $p_2 \cdot gp_2$ and satisfies $d(p_2, x_2) = d(p_1, x_1)$. Let us check that this definition does not depend on the choice of g. Let g' be any other element of G such that $x_1 \in p_1 \cdot g'p_1$. Since the displacement functions of G in $\mathcal{T}_1$ and in $\mathcal{T}_2$ are the same, one has $d(p_1, gp_1) = d(p_2, gp_2)$, $d(p_1, g'p_1) = d(p_2, g'p_2)$ and $d(gp_1, g'p_1) = d(gp_2, g'p_2)$. Therefore the length of the common part of the segments $p_1 \cdot gp_1$ and $p_1 \cdot g'p_1$ is equal to that of the common part of $p_2 \cdot gp_2$ and $p_2 \cdot g'p_2$. This implies that the definition of the point x_2 is the same whether one uses g' or g. Set $\Psi(x_1) = x_2$. One easily checks that Ψ is an isometry. By definition $\Phi(gp_1) = gp_2$, thus $\Psi : Gp_1 \to Gp_2$ conjugates the restrictions of the two actions ρ_1 and ρ_2. Since $\Psi : \mathcal{T}_1 \to \mathcal{T}_2$ is an isometry, $\Psi : \mathcal{T}_1 \to \mathcal{T}_2$ conjugates the two actions ρ_1 and ρ_2. This proves Proposition 45.   □

5 The Tree of SL(2, F) Associated to a Valuation of F/k

In this section, we explain how Morgan and Shalen defined a Λ-tree from a valuation on a F/k with value group Λ [24]. Their construction generalized to arbitrary valuations, one by Bruhat and Tits, and Serre [34].

5.1   The Λ-Tree of a Lattice in a 2-Dimensional Vector Space

Let F/k be a field extension. Let v be a valuation of F/k, $\mathfrak{o}_v$ its valuation ring, and Λ its value group. Denote by V the n-dimensional F-vector space F^n. A *lattice* or $\mathfrak{o}_v$-*lattice* in V is a $\mathfrak{o}_v$-module $L \subset V$ of the form $L = \mathfrak{o}_v e_1 \oplus \cdots \oplus \mathfrak{o}_v e_n$ for some basis $e_1, \cdots, e_n$ of V. When we take as basis the canonical basis of $V = F^n$, we obtain the *standard lattice*. By definition, any lattice is a free $\mathfrak{o}_v$-lattice of rank n.

We say that two lattices L and L' in V are *equivalent* when they differ by an homothety of V: for some $\alpha \in F$, $L' = \alpha L$. Denote by $[L]$ the equivalence class of L.

Our aim is now to show that when $n = 2$, the set $\mathcal{T}_v$ of equivalence classes of lattices in V has the structure of a Λ-tree. We define first a Λ-distance on $\mathcal{T}_v$. Let L, L' be two lattices in V. Up to replacing L' by an homothetic lattice we can suppose that $L' \subset L$ and that L' has a basis

$$(e_1', e_2') = \begin{pmatrix} a & b \\ c & d \end{pmatrix} (e_1, e_2)$$

where (e_1, e_2) is a basis of L and where the coefficients a, b, c and d are in $\mathfrak{o}_v$. Up to permuting the basis vectors if necessary, we may also impose the condition $v(a) = \min(v(a), v(b), v(c), v(d))$ (with the convention $v(0) = +\infty$). Then a divides in the ring $\mathfrak{o}_v$ all others coefficients of the matrix so that after operations on the rows and columns, we obtain a new basis for L'

$$(e_1', e_2') = \begin{pmatrix} a & 0 \\ 0 & d - \frac{b}{a}c \end{pmatrix} (e_1, e_2). \tag{2}$$

In that case, the representative $L_0' = \dfrac{1}{a} L'$ of the equivalence class $[L']$ satisfies $L/L_0' \simeq \mathfrak{o}_v/\beta\mathfrak{o}_v$ with $\beta = \dfrac{da - bc}{a^2}$.

We introduce the following terminology: a lattice M' contained in a lattice M such that $M/M' \simeq \mathfrak{o}_v/\gamma\mathfrak{o}_v$ is said *cocyclic* in M. We claim that there exists a unique lattice in the equivalence class $[L']$ which is cocyclic in L. We already proved the existence of such a cocyclic lattice L_0'. Suppose now that L_1' is another lattice cocyclic in L. Then $L_1' = \alpha L_0'$ for some $\alpha \in F^*$. Assume by contradiction that $v(\alpha) \neq 0$. We can assume that $v(\alpha) > 0$, so that $L_1' \subset L_0' \subset L$ and $L_1' = \alpha\mathfrak{o}_v e_1 \oplus \alpha\beta\mathfrak{o}_v e_2$. It follows that $L/L_1' \simeq \mathfrak{o}_v/\alpha\mathfrak{o}_v \oplus \mathfrak{o}_v/\alpha\beta\mathfrak{o}_v$. This contradicts the fact that L_1' is cocyclic in L since $v(\alpha) > 0$ and $v(\beta) \geq 0$. Whence $v(\alpha) = 0$, which means that α is a unit in $\mathfrak{o}_v$, and we conclude $L_1' = L_0'$ as required.

Note that different choices of the basis of L might possibly affect β, but only by multiplying it with a unit, therefore $v(\beta)$ is well-defined. For any two equivalence classes $[L], [L'] \in \mathcal{T}_v$ given by two basis satisfying relation (2), we set:

$$d([L], [L']) = v(\beta) = v(ad - bc) - 2\min(v(a), v(b), v(c), v(d)). \tag{3}$$

The preceding discussion shows that this expression does not depend on the particular representatives of the classes $[L]$ and $[L']$.

Proposition 46 *The function* $d : \mathcal{T}_v \times \mathcal{T}_v \to \Lambda$ *is a* Λ-*distance for which* $\mathcal{T}_v$ *is a* Λ-*tree. This tree is called the tree of* SL$(2, F)$ *associated to the valuation v.*

Proof We first prove that d is a Λ-distance. We keep the same notation as above. Suppose $d([L], [L']) = 0$ for two lattices L, L', with L' cocyclic in L. Then $L/L' \simeq \mathfrak{o}_v/\beta\mathfrak{o}_v$ with $v(\beta) = 0$, hence $L = L'$. The symmetry axiom is easy to check. To check the triangular inequality, consider three equivalence classes x, x' and x'' in $\mathcal{T}_v$. Let L' be a representative of x' and (e, f) be a basis of L', such that a representative of x is the lattice L generated by e and αf, for some $\alpha \in \mathfrak{o}_v$. Similarly take (u, v) a basis of L' and a representative of x'' generated by the two vectors u and βv, for some $\beta \in \mathfrak{o}_v$. We can write $(u, v) = \begin{pmatrix} a & b \\ c & d \end{pmatrix}(e, f)$ for some invertible matrix with coefficients in $\mathfrak{o}_v$, hence

$$(u, \beta v) = \begin{pmatrix} a & \frac{b}{\alpha} \\ \beta c & \frac{\beta d}{\alpha} \end{pmatrix}(e, \alpha f).$$

A particular representative of $[L'']$ contained in L has basis $(\alpha u, \alpha\beta v) = M(e, \alpha f)$ with $M = \begin{pmatrix} \alpha a & b \\ \alpha\beta c & \beta d \end{pmatrix}$. Since all coefficients of M lie in $\mathfrak{o}_v$ we can apply (3), and we infer $d([L], [L'']) \le v(\det(M)) = v(ad - bc) + v(\alpha) + v(\beta) = v(\alpha) + v(\beta)$. This proves the triangular inequality, and shows that $\mathcal{T}_v$ is a Λ-metric space.

Suppose that equality holds in the triangular inequality for three distinct equivalence classes: $d(x, x'') = d(x, x') + d(x', x'')$. Keeping the same notation as above, the three lattices x, x' and x'' can be represented respectively by $L = \mathfrak{o}_v e \oplus \mathfrak{o}_v \alpha f$, $L' = \mathfrak{o}_v \alpha u \oplus \mathfrak{o}_v \alpha v$ and $L'' = \mathfrak{o}_v \alpha u \oplus \mathfrak{o}_v \alpha\beta v$ with $v(\alpha), v(\beta) > 0$. Now if equality holds in the triangular inequality then the minimum of the valuations of the coefficients of the matrix M must be 0. This implies $v(b) = 0$ hence (u, v) can be chosen to be $u = f, v = e$.

We conclude that $d(x, x'') = d(x, x') + d(x', x'')$ if and only if there is a basis (e_1, e_2) of L such that L', L'' are generated respectively by $(e_1, \alpha e_2)$ and $(e_1, \alpha\beta e_2)$ with $d(x, x') = v(\alpha)$ and $d(x', x'') = v(\beta)$.

We now check that $\mathcal{T}_v$ satisfies the three axioms of a Λ-tree.

— We first construct a segment connecting any two points $x, x' \in \mathcal{T}_v$. Let L', L be any representatives of x and x' such that $L' \subset L$ and cocyclic in L. There is a basis (e, f) of L such that L' is the lattice $\mathfrak{o}_v e \oplus \mathfrak{o}_v \beta f$, with $\beta \in \mathfrak{o}_v$ and $d(x, x') = v(\beta)$. For any z in the interval $[0, v(\beta)] \subset \Lambda_v$ choose a $\gamma_z \in \mathfrak{o}_v$ with $v(\gamma_z) = z$. Consider the lattice $L_z = \mathfrak{o}_v e \oplus \mathfrak{o}_v \gamma_z f$. By construction one has: $d([L_z], [L_{z'}]) = |v(\gamma'_z) - v(\gamma_z)|$. So the map $z \mapsto [L_z]$ from $[0, v(\beta)]$ to $\mathcal{T}_v$ is an isometry onto a segment which connects x and x'. The fact that this segment is unique follows from the characterization of the equality case in the triangular inequality. Observe that the segment $x \cdot x'$ coincides with the set of equivalence classes of lattices L'' such that $L' \subset L'' \subset L$.

— Let $x_1 \cdot z$ and $x_2 \cdot z$ be any two closed segments having common endpoint z. Consider the representatives L_1, L_2 and L_0 of x_1, x_2 and z such that $L_1 \subset L_0$ and $L_2 \subset L_0$ and both lattices L_1, L_2 are cocyclic in L_0. Then $x_i \cdot z = \{[L] \mid L_i \subset L \subset L_0\}$ for $i = 1, 2$ therefore $x_1.z \cap x_2.z = \{[L] \mid L_1 + L_2 \subset L \subset L_0\}$. The sum $L_1 + L_2$ is an $\mathfrak{o}_v$-module contained in L_0 which is torsion free and finitely generated: therefore $L_1 + L_2$ is a lattice. So the intersection of $x_1 \cdot z$ and $x_2 \cdot z$ is the segment with endpoints $[L_0]$ and $[L_1 + L_2]$.

— Suppose that $x_1 \cdot z$ and $x_2 \cdot z$ are any two closed segments intersecting only at z. We need to show that the union of $x_1 \cdot z$ and $z \cdot x_2$ is again a segment. Keeping previous notation, we know that $L_1 + L_2$ represents z, i.e. is equivalent to L_0. But since $L_1 + L_2$ contains L_1 and is contained in L_0, it must be cocyclic in L_0, hence $L_1 + L_2 = L_0$. We can write $L_1 = \mathfrak{o}_v e_1 \oplus \mathfrak{o}_v \beta_1 f_1$ and $L_2 = \mathfrak{o}_v e_2 \oplus \mathfrak{o}_v \beta_2 f_2$ where (e_1, f_1) and (e_2, f_2) are basis of L_0. We have $v(\beta_i) > 0$ since the segments are non-degenerate. Let $\mathfrak{m}_v = v^{-1}\{> 0\}$ be the maximal ideal of $\mathfrak{o}_v$. Since $\mathfrak{o}_v e_1 + \mathfrak{o}_v e_2 + \mathfrak{m} L_0 = L_0$, we must have $\mathfrak{o}_v e_1 + \mathfrak{o}_v e_2 = L_0$. Hence the vectors e_1 and e_2 form a basis of the lattice L_0. This allows to write L_1 and L_2 as $L_1 = \mathfrak{o}_v e_1 \oplus \mathfrak{o}_v \beta_1 e_2$ and $L_2 = \mathfrak{o}_v e_2 \oplus \mathfrak{o}_v \beta_2 e_1$. From the previous description of the segments in $\mathcal{T}_v$ we deduce that $[L_0]$ belongs to the segment $[L_1] \cdot [L_2]$. Therefore the union of $x_1 \cdot z$ and $z \cdot x_2$ is the segment $[L_1 \cdot L_2]$. $\square$

5.2   The Action of SL$(2, F)$

The group GL(2,F) naturally acts on the set of lattices preserving equivalence classes. It acts therefore on the Λ-tree $\mathcal{T}_v$ and this action is isometric. The action of GL(2,F) is transitive on lattices and therefore also on $\mathcal{T}_v$: any point in $\mathcal{T}_v$ is equivalent to the class x_0 of the standard $\mathfrak{o}_v$-lattice in F^2. Pick any element $g = \begin{pmatrix} a & b \\ c & d \end{pmatrix}$ of SL$(2, F)$. Since the determinant of g has valuation 0, relation (3) gives:

$$d(x_0, gx_0) = -2 \min(v(a), v(b), v(c), v(d)).$$ It follows that $d(x, gx)$ is divisible by 2 in Λ for any point $x \in \mathcal{T}_v$. In particular g is not a phantom inversion. Whence SL$(2, F)$ acts on $\mathcal{T}_v$ by hyperbolic or elliptic isometries.

5.2.1   Elliptic Isometries of $\mathcal{T}_v$

Let $g \in$ SL$(2, F)$ be an elliptic isometry. Let $x = [L] \in \mathcal{T}_v$ be a point fixed by g, so that $gL = \alpha L$ for some $\alpha \in k^*$. Suppose that $v(\alpha) \neq 0$. After possibly replacing g by g^{-1} we can assume that $v(\alpha) \geq 0$; then $gL \subset L$. Since g has determinant 1, we have $L/gL = \{0\}$. If $v(\alpha) > 0$, then $L/\alpha L$ is not zero, since its dimension as a vector space over the field $\mathfrak{o}_v/\mathfrak{m}_v$ is equal to 2. Therefore $v(\alpha) = 0$, and this means that $gL = L$: the expression of g on any basis of L has coefficients in o_v. Note that L is the image of the standard lattice by some element of $GL(2, F)$.

We have thus proved that the elements of $SL(2, F)$ acting by elliptic transformations on $\mathcal{T}_v$ are those which are conjugated in $GL(2, F)$ to elements of $SL(2, o_v)$. In particular their trace lies in o_v. Conversely any element of $SL(2, F)$ whose trace belongs to o_v is conjugated in $GL(2, F)$ to an element in $SL(2, o_v)$ (consider the matrix of this element on the basis (e, ge) for a vector e which is not an eigenvector of g).

The following is now a direct consequence of Proposition 39.

Proposition 47 *Let $G \subset SL(2, F)$ be a finitely generated subgroup such that the trace of any $g \in G$ belongs to o_v (i.e. $v(\mathrm{tr}(g)) \geq 0$). Then G is conjugated in $GL(2, F)$ to a subgroup of $SL(2, o_v)$.*

5.2.2 Hyperbolic Isometries of $\mathcal{T}_v$

Take $g \in SL(2, F)$. For any $x \in \mathcal{T}_v$ let $g = \begin{pmatrix} a & b \\ c & d \end{pmatrix}$ be the matrix representation of g in a basis of a representative of x. By (3), we have $d(x, gx) = -2\min(v(a), v(b), v(c), v(d))$. In particular $d(x, gx) \geq -2\min(v(a), v(d)) \geq -2v(a+d) = -2v(\mathrm{tr}(g))$. Pick a vector e which is not an eigenvector of g. The o_v-lattice generated by e and $g \cdot e$ satisfies: $d([L], [gL]) = -2v(\mathrm{tr}(g))$. If $v(\mathrm{tr}(g)) < 0$, the minimum displacement of g is thus positive: thus g is an hyperbolic isometry with translation distance $-2v(\mathrm{tr}(g))$ and axis consisting of all the equivalence classes of lattices of the form $o_v e \oplus o_v ge$.

5.2.3 Segment-Stabilizers

Consider the subgroup $G_{x \cdot x'}$ of $SL(2, F)$ which fixes a non-degenerate segment $s = x \cdot x'$. Let $[L]$ and $[L']$ be respective representatives of x and x' with L' cocyclic in L and pick $(e, f) \in L$ such that $L' = o_v e \oplus o_v \cdot \beta f$. The matrix expression on the basis (e, f) of the stabilizer of x is the group $SL(2, o_v)$ and $G_{x \cdot x'}$ is isomorphic to the subgroup G_s of $SL(2, o_v)$ which fixes L'. Therefore G_s consists of all matrices $\begin{pmatrix} a & b \\ c & d \end{pmatrix} \in SL(2, o_v)$ with $v(c) \geq v(\beta)$. The group G_s is thus identified with the preimage of the triangular group $\begin{pmatrix} \alpha & \beta \\ 0 & \delta \end{pmatrix}$ under the canonical homomorphism $SL(2, o_v) \to SL(2, o_v/\beta o_v)$.

We have thus proved.

Proposition 48 *Let $G_{x \cdot x'} \subset SL(2, F)$ be the stabilizer of a non-degenerate segment $x \cdot x'$ of length $v(\beta)$, with $\beta \in F$. Then $G_{x \cdot x'}$ is conjugated in $GL(2, F)$ to a subgroup G_s of $SL(2, o_v)$ such that the commutator subgroup $[G_s, G_s]$ maps to the identity under the homomorphism $SL(2, o_v) \to SL(2, o_v/\beta o_v)$.*

5.3 Applications to the Character Variety

Let X_0 be an irreducible component of the character variety $X(G)$. Recall that to any valuating sequence of points (x_i) in X_0 tending to infinity is associated a valuation on $\mathbb{Q}(X_0)/\mathbb{Q}$ (see Sect. 3.5). Our aim is to attach to any such valuation an action of G on a Λ-tree.

It is technically easier to work with the space of representations than with the character variety. In doing so, we shall use the following classical result, see [49, Chapter VI.4], or [3, Chapitre 6].

Proposition 49 *Let K/k be a field extension and v be a valuation of K/k. Let K' be a finite type extension of K. Then there is a valuation v' on K'/k extending v. Furthermore one can choose v' such that its value group Λ' is a finite extension of the value group Λ of v: the index of Λ in Λ' is finite.*

Proof The extension K'/K is the composition of a purely transcendental extension and a finite extension. By induction, it suffices to consider the case of an extension of transcendental degree 1. The case of a purely transcendental extension $K' = K(x)$ can be solved in an explicit way. Define for a non-zero polynomial $P(x) = \sum a_j x^j$, $v'(P(x)) = \min_j \big(v(a_j)\big)$. Clearly this defines a valuation on $K[x]$ that extends to $K(x)$ and has the same value group as v.

Suppose now that K'/K is a finite extension. Consider the set $\mathcal{R}$ of all rings $R \subset K'$ that contain $\mathfrak{o}_v = \{v \geq 0\} \subset K$ and such that $R.\mathfrak{m}_v \neq R$ with $\mathfrak{m}_v = \{v > 0\}$. The set $\mathcal{R}$ is partially ordered by the inclusion and one easily checks that any totally ordered subset (R_α) has an upper bound. Hence by Zorn's Lemma, $\mathcal{R}$ admits a maximal element R' which satisfies $R' \cdot \mathfrak{m}_v \neq R'$.

Claim 50 The ring R' has the following property: if $x \in K'$ is $\neq 0$, then either x or $\dfrac{1}{x}$ is in R'.

Granting this claim we finish the proof. Denote by $\mathfrak{m}'$ the subset of R' consisting of 0 and all inverses of elements of $K' \setminus R'$. Since R' satisfies the claim $\mathfrak{m}'$ is a maximal ideal (see for instance [49, Theorem 1, Ch. VI]). Therefore R' is a place of K', whence the valuation ring of a valuation v'. Let us prove that $K \cap R' = \mathfrak{o}_v$. Suppose by contradiction that $x \in K \setminus \mathfrak{o}_v$ lies in R'. Then we have $x^{-1} \in \mathfrak{m}_v$, so $1 = x \cdot x^{-1}$ lies in $R'.\mathfrak{m}_v$, which is absurd. In the same way we get $K \cap \mathfrak{m}' = \mathfrak{m}_v$ and v' is an extension of v. Notice that the extension v' is not unique (see [45] for a description of the possible extensions in the degree 1 case). Suppose that Λ'/Λ contains m elements represented by $v(x_1), \cdots, v(x_m)$. Then the x_j's must be linearly independent over K; in particular, m is less than the degree of the algebraic extension K'/K. Therefore the index of Λ in Λ' is less than the degree of K'/K. $\square$

Proof of the claim By the maximality property of R' it suffices to show that either $R'[x] \cdot \mathfrak{m}_v \neq R'[x]$ or $R'[x^{-1}] \cdot \mathfrak{m}_v \neq R'[x^{-1}]$. Assume by contradiction that none is satisfied. Then there are polynomial relations $1 = a_0 + \cdots + a_n x^n$, $1 = b_0 + \cdots + b_m (x^{-1})^m$, with coefficients a_i and b_j in the ideal $\mathfrak{m}_v$. We may assume that

$m \leq n$ and that the degrees m and n are the smallest possible. The first equation gives: $1 - b_0 = a_0(1 - b_0) + \cdots + a_n(1 - b_0)x^n$ and the second $(1 - b_0)a_n x^n = b_1 a_n x^{n-1} + \cdots + b_m a_n x^{n-m}$. After adding these equations and simplifying we get a polynomial equation of degree $\leq n - 1$ satisfied by x. This is a contradiction.   □

We are now ready for the main result of this section.

Theorem 51 *Let $\rho_i : G \to \mathrm{SL}(2, \mathbb{C})$ be a sequence of representations which does not contain a sequence which converges up to conjugacy. Then there is a Λ-tree T and a minimal action by isometries $G \times T \to T$ such that a subsequence of (ρ_i) converges to it in the following sense: for any conjugacy classes g, h in G with $\delta_T(h) > 0$, one has:*

$$\lim_{i \to \infty} \frac{\log |(\mathrm{tr}(\rho_i(g))| + 2)}{\log(|\, \mathrm{tr}(\rho_i(h))| + 2)} = \frac{\delta_T(g)}{\delta_T(h)}. \tag{4}$$

Proof Denote by $x_i = t(\rho_i) \in X(G)$ the character of ρ_i. We may assume that (x_i) is contained in the same irreducible component X_0 of $X(G)$. By Proposition 6 and since (ρ_i) does not contain subsequences which converge up to conjugacy, the sequence of characters (x_i) is unbounded. We may assume that (x_i) is a valuating sequence and defines a valuation v on $\mathbb{Q}(X_0)/\mathbb{Q}$. Recall that we defined $\mathcal{I}_g(\rho) = \mathrm{tr}(\rho(g))$. By Proposition 29, if $v(\mathcal{I}_h) < 0$ then

$$\lim_{i \to \infty} \frac{\log(|\mathcal{I}_g(x_i)| + 2)}{\log |(\mathcal{I}_h(x_i)| + 2)} = \frac{v(\mathcal{I}_g)}{v(\mathcal{I}_h)}. \tag{5}$$

Let $\mathcal{R}_0$ be a component of $\mathcal{R}(G)$ mapped by t onto a Zariski open subset of X_0. By Proposition 49, we may pick a lift v' of v to a valuation on $\mathbb{Q}(\mathcal{R}_0)/\mathbb{Q}$. Denote by $\mathcal{T}$ the tree of $\mathrm{SL}(2, \mathbb{Q}(\mathcal{R}_0))$ associated to v'. We may identify G with a subgroup of $\mathrm{SL}(2, \mathbb{Q}(\mathcal{R}_0))$ via the tautological representation. Therefore G acts on the Λ-tree $\mathcal{T}$. Since (x_i) is unbounded this action has no global fixed point. Up to replacing $\mathcal{T}$ by a subtree we may assume that this action is minimal.

Now note that for any $g \in G$, we have $-2\min(0, v(\mathrm{tr}(g))) = -2\min(0, v'(\mathrm{tr}(g))) = \delta_T(g) > 0$ (see Sect. 5.2). We conclude by comparing (4) and (5) using $\mathcal{I}_g(t(\rho)) = \mathrm{tr}(\rho(g))$.   □

To the sequence (ρ_i), we have thus associated an action of G on a Λ-tree. This action—through the length-function—gives a precise information on the growing rate of traces of group elements. If one wants to describe only the top order terms, one possibility is to look at the embedding of the space of characters into $\mathbb{P}^{\mathcal{F}}$ as in Chapter 1.

To simplify notation, we will now set $X(G) = X$. Consider as set of functions $\mathcal{F} \subset \mathbb{Q}[X]$ the set $\{\mathcal{I}_g\}$ where g ranges over all conjugacy classes of G. Let $\bar{X}^{\mathcal{F}}$ be the Morgan-Shalen compactification of X as explained in Sect. 2.3, and denote by

$B(X) = \bar{X}^{\mathcal{F}} \setminus X$ the frontier. Then we may interpret the points of $B(X)$ as length-functions of an action of G on a Λ-tree where the ordered group Λ is archimedean.

The following proposition is [24, Th.II.4.3].

Proposition 52 *Any point in the frontier $B(X)$ is the projectivized length-function of an action of G on a Λ-tree with Λ archimedean; this action may be taken minimal and without phantom inversion.*

Proof Note first that the frontier $B(X)$ is the union of the frontiers of the (finitely many) irreducible components of $X(G)$. Let X_0 be any such irreducible component and let ξ be a point in the frontier of X_0. By Proposition 29, ξ has homogeneous coordinates $(-\min(0, v(\operatorname{tr}(g)))) \in \mathbb{P}^{\mathcal{F}}$, where v is a valuation on $\mathbb{Q}(X_0)/\mathbb{Q}$ such that for any $g \in G$, $v(\mathcal{I}_g)$ is either positive or in the smallest isolated subgroup Λ_1 of the value group Λ.

As in the proof of the preceding Theorem and using the tautological representation, we may identify G with a subgroup of $\operatorname{SL}(2, \mathbb{Q}(\mathcal{R}_0))$ where $\mathcal{R}_0$ is a component of $\mathcal{R}(G)$ which dominates X_0. Pick any valuation v' on $\mathbb{Q}(\mathcal{R}_0)/\mathbb{Q}$ extending v and let $\bar{v}$ be the quotient valuation obtained after post-composing v' with the quotient homomorphism $\Lambda \to \Lambda/\Lambda_1$. Denote by $\mathcal{T}_{\bar{v}}$ the tree of $\operatorname{SL}(2, \mathbb{Q}(\mathcal{R}_0))$ associated to $\bar{v}$. Since for any element $g \in G$ $\bar{v}(\operatorname{tr}(g)) \geq 0$, g has a fixed point on $\mathcal{T}_{\bar{v}}$. By Proposition 47 the entire group G has a fixed point, which is to say that the tautological representation of G is conjugated to a representation $\rho : G \to \operatorname{SL}(2, \mathfrak{o}_{\bar{v}})$. In particular, the matrix of $\rho(g)$ in the standard basis has coefficients $a_g, b_g, c_g, d_g \in \mathfrak{o}_{\bar{v}}$, and $v'(a_g), v'(b_g), v'(c_g), v'(d_g)$ are either positive or lie in Λ_1.

Let now $\mathcal{T}_v$ denote the tree of $\operatorname{SL}(2, \mathbb{Q}(\mathcal{R}_0))$ associated this time to the valuation v'. Denote by x_0 the equivalence class of the standard lattice. Since $d(x_0, \rho(g)(x_0)) = -2\min(v'(a_g), v'(b_g), v'(c_g), v'(d_g))$, the distance between any two points of the orbit Gx_0 is in Λ_1. Observe now that the union of all segments $x_0.\rho(g)(x_0)$ is a Λ_1-tree $\mathcal{T}$ with Λ_1 archimedean. The group G acts on $\mathcal{T}$ without phantom inversion and its length-function is projectively the point ξ. □

5.4 Limit Trees of Discrete and Faithful Representations

A representation $\rho : G \to \operatorname{SL}(2, \mathbb{C})$ is *discrete* if its image is a discrete subgroup, and *faithful* if it is injective. Denote $\mathcal{DF}(G) \subset \mathcal{R}(G)$ the set of all the discrete and faithful representations.

Proposition 53 ([10]) *Let G be a finitely generated group which is not virtually abelian. Then $\mathcal{DF}(G)$ is a closed subset of $\mathcal{R}(G)$.*

Theorem 51 admits the following refinement when (ρ_i) is a sequence of discrete and faithful representations.

Proposition 54 *Let (ρ_i) be a sequence of representations in $\mathcal{DF}(G)$ which does not contain any subsequence which converges up to conjugacy. Let T be any Λ-tree provided by Theorem 51. Then any subgroup of G which stabilizes a non-trivial segment is virtually abelian.*

Proof We keep the notation of the proof of Theorem 51. Recall that v is the valuation associated to a valuating sequence (x_i') which approximates sufficiently closely the sequence of characters $t(\rho_i)$ so that by Proposition 6, $v(f) \geq 0$ if and only if the sequence $\mathrm{tr}(\rho_i(f))$ is bounded. Let $x \cdot y$ be a non-trivial segment of T and denote $G_{x \cdot y}$ the subgroup of G which fixes $x \cdot y$. Since $G_{x \cdot y}$ fixes x, it is conjugated to a subgroup G_s of $\mathrm{SL}(2, \mathfrak{o}_v)$, hence the traces $\mathrm{tr}(\rho_i(g))$ are bounded for any $g \in G_s$. By Proposition 6 the sequence of representations $\rho_i|G_s$ contains a sequence which converges to a representation $\rho_\infty : G_s \to \mathrm{SL}(2, \mathbb{C})$. Since G_s stabilizes the segment $x \cdot y$, its commutator subgroup $[G_s, G_s]$ is mapped to the identity by the homomorphism $\mathrm{SL}(2, \mathfrak{o}_v) \to \mathrm{SL}(2, \mathfrak{o}_v/\beta\mathfrak{o}_v)$ (Proposition 48). Therefore for any $g \in [G_s, G_s]$, $\mathrm{tr}(\rho_\infty(g)) = 2$. Suppose by contradiction that G_s is not virtually abelian. By Proposition 53, $\rho_\infty|G_s$ is then a discrete and faithful representation. Since $\mathrm{tr}(\rho_\infty(g)) = 2$, $\rho_\infty([G_s, G_s])$ must be a parabolic subgroup and so it fixes a unique point in $\mathbb{C}P^1$. This point must also be fixed by $\rho_\infty(G_s)$. But any discrete subgroup of $\mathrm{SL}(2, \mathbb{C})$ which acts on $\mathbb{C}P^1$ with a fixed point is virtually abelian. A contradiction. □

6   Geometric Actions of Groups on Λ-Trees

The study of unbounded sequences of points in $\mathcal{R}(G)$ led to isometric group actions on trees (Theorem 51). In general, this tree is a Λ-tree where Λ is a non-archimedean group. However the tree associated to the quotient rank 1 valuation already carries a lot of informations (Proposition 52). In this section, we address the following two questions: which groups G do have a non-trivial action on a Λ-tree (for example on a $\mathbb{R}$-tree) such that all segment stabilizers are virtually abelian? When such an action exists, can one describe it in a geometric way? We begin with the case when G is the fundamental group of a n-dimensional manifold. Then we will focus on *surface groups*, which are the groups isomorphic to the fundamental group of a closed surface. In that case, the dual trees to Λ-measured laminations on the surface give models for the actions on Λ-trees with cyclic edge stabilizers and which satisfy also an extra geometric condition (see Theorem 62). When Λ is archimedean, this geometric condition can be removed: it is theorem due to Skora (see Theorem 63).

6.1   Λ-*Measured Laminations*

6.1.1   Basics

Let M be a manifold of dimension n. A (codimension 1) *lamination* of M is a closed subset $L \subset M$ such that there is a cover of M by open sets V_i which satisfy:

1. for each i, $(V_i, V_i \cap L)$ has a product structure, i.e. there is a compact set $F_i \subset]0, 1[$ and an homeomorphism ϕ_i between the pairs $(V_i, V_i \cap L)$ and $(U_i \times]0, 1[, U_i \times F_i)$;
2. if two open sets U_j and U_k have non-empty intersection, then over their intersection, the homeomorphism $\phi_k \circ \phi_j^{-1} : \phi_j(L \cap U_i \cap U_j) \to \phi_k(L \cap U_i \cap U_j)$ preserves the product structure.

The open sets V_j in this definition are called *flow-boxes for* L. Consider a flow-box V_j for L. The set Y_j of the connected components $]0, 1[\backslash F_j$ is ordered by the order of the interval $]0, 1[$. Let Λ be a countable totally ordered abelian group. A Λ-*measure* on $]0, 1[$ supported on F_j is a monotonic bijection v_j from Y_j to an interval of Λ. This allows to assign to any interval $[a, b] \subset]0, 1[$ with endpoints disjoint from F_j, a number in Λ^+, namely the absolute value $|v_j(a) - v_j(b)|$. This defines a finitely additive measure. Also the measure of an interval $[a, b]$ is 0 if and only if $[a, b]$ is disjoint from F_j.

A Λ-*measure transverse to* L is the data, for each flow-box $V_j = U_j \times]0, 1[$, of a Λ-measure μ_j on $]0, 1[$ such that the obvious compatibility relations are satisfied for the flow-boxes which intersect. A Λ-*measured lamination* is the data (L, μ) of a lamination L and a Λ-measure μ transverse to L .

A continuous path $c : [0, 1] \to M$ in M is *transverse to* L if $c(0)$ and $c(1)$ are disjoint from L and if c can be decomposed as a product $c_1.c_2. \cdots .c_n$ of paths c_j such that for each j, c_j is contained in a flow-box U_i and the projection $c_j \cap L \to F_j$ is strictly monotone. A Λ-measure transverse to L induces a finitely additive measure on c with *total mass* denoted $\mu(c)$. If c and c' are two paths homotopic by an homotopy c_t such that c_t remains transverse to L, the total mass $\mu(c_t)$ remains constant.

6.1.2   The Dual Tree to a Λ-Measured Lamination

Denote by $\tilde{M}$ the universal cover of M. The preimage of a Λ-measured lamination in $\tilde{M}$ is a Λ-measured lamination.

Proposition 55 *Let* $\mathcal{L} = (L, \mu)$ *be a* Λ-*measured lamination. Let* $\tilde{L}$ *denote the preimage of* L *in* $\tilde{M}$. *Suppose that* $\tilde{L}$ *satisfies the following axioms:*

1. *the leaves of* $\tilde{L}$ *are closed subsets of* $\tilde{M}$;
2. *given any two connected components* x *and* y *of* $\tilde{M} - \tilde{L}$, *there exists a path* $c : [0, 1] \to \tilde{M}$ *with endpoints contained in* x *and* y *respectively, and which is transverse to* $\tilde{L}$ *and which intersects each leaf of* $\tilde{L}$ *at most once.*

Then the set $\mathcal{T}_{\mathcal{L}}$ of the connected components of $\tilde{M} \setminus \tilde{L}$ has a natural structure of Λ-tree.

Proof First, the set $\mathcal{T}_{\mathcal{L}}$ is a Λ-metric space. For complementary components x and y of $\tilde{M} \setminus \tilde{L}$, define $d(x, y)$ to be the mass of a path provided by (2). This does not depend on the choice of the path, by (1) and the invariance property of a transverse Λ-measure. To check the triangular inequality choose points x, y and z in components of $\tilde{M} \setminus \mathcal{L}$ and paths xy, yz, zx connecting two-by-two these points which intersect exactly once the leaves of $\tilde{L}$. We can suppose that these arcs are smooth and the closed curve $xy \cup yz \cup zx$ bounds a smooth disc Δ in $\tilde{M}$. Since $\tilde{L}$ has empty interior, we can suppose that this disc is transverse to $\tilde{L}$. Then any leaf of the induced lamination of Δ with one endpoint on the path xz has its other endpoint either on the path xy or on yz. The triangular inequality is a consequence of this.

We now define segments in $\mathcal{T}_{\mathcal{L}}$. Let c be a path in $\tilde{M}$ joining two connected components x and y of $\mathcal{T}_{\mathcal{L}}$ which satisfies (2). Then $w \mapsto d(x, w)$ is an isometry from the set of components of $\mathcal{T}_{\mathcal{L}}$ that c intersects to the interval $[0, d(x, y)]$. When c is contained in a flow-box, this follows from the definition of a Λ-measured lamination and the general case follows from this one, since c can be written as a composition $c_1 \cdots c_k$ where each path c_j is contained in a flow-box. Therefore any two distinct points can be joined by a segment. We call *special* those segments obtained on that way. We show now that any segment in $\mathcal{T}_{\mathcal{L}}$ is special. Suppose that there is another segment S in $\mathcal{T}_{\mathcal{L}}$ with endpoints x and y. Let w be a point on S; since S is a segment, $d(x, w) + d(w, y) = d(x, y)$. Consider the two paths provided by (2) which connect respectively x to w and w to y. We can suppose that the endpoints of these paths in the component w are the same: their composition is a path $c' \subset \tilde{M}$. Let Δ be a smooth disk which realizes an homotopy fixing endpoints between c and c'. Since $\tilde{L}$ has empty interior, Δ can be chosen to be transverse to $\tilde{L}$; then $\tilde{L} \cap \Delta$ is a lamination of Δ. Each leaf of this lamination which has one endpoint in c has its other endpoint on the special segment, because c satisfies (2). Since $\mu(c') = d(x, y)$, any leaf starting from c' ends on c. In particular, w is on the special segment.

We check now that $\mathcal{T}_{\mathcal{L}}$ satisfies the first axiom of a Λ-tree: the intersection of two segments $x \cdot y$ and $x \cdot z$ is a segment issued from x. Let x, y and $z \in \mathcal{T}_{\mathcal{L}}$. Let Δ be a triangle in $\tilde{M}$ whose sides are paths c, c' and c'' in $\tilde{M}$, whose endpoints are successively x, y, z and which intersect each leaf of L almost once. This disc can be made transverse to $\tilde{L}$. The induced lamination of $\tilde{\Delta}$ has the property that each leaf with an endpoint on one side has its other endpoint on another side. The set of the leaves from the side c to the side c'' is ordered by the inclusion of the disc containing x and bounded by the leaf on Δ. Some leaf is maximal for this order among the leaves from c to c'. The component of $\tilde{M} \setminus \tilde{L}$ bounded by this leaf determines a vertex of $\mathcal{T}_{\mathcal{L}}$ such that $x \cdot y \cap x \cdot z = x \cdot z$.

The second axiom can be proven in the same way. $\square$

Remark When $\Lambda \subset \mathbb{R}$, a Λ-measure with support a closed subset $F \subset]0, 1[$ extends in a unique way to a σ-additive measure on $]0, 1[$ with support F. In that situation

$\mathcal{L} = (L, \mu)$ is a Λ-measured lamination; it has *a transverse measure*, i.e. each transversal I carries a Radon measure, supported on $L \cap I$ and these measures are invariant under the holonomy pseudo-group (see [25]). When a lamination is the support of a transverse measure, it has the following structure by [25, Theorem 3.2]: it is the union of finitely many components and each component is either a family of parallel compact leaves or an exceptional minimal set. When the ordered group Λ is not contained in $\mathbb{R}$, there are no similar descriptions. In particular, when M is a surface, a Λ-measured lamination can contain a leaf which spirals towards a closed curve. Such an example is given in e.g. [23, p. 179], for $\Lambda = \mathbb{Z} + \mathbb{Z}$ with the lexicographic order. This example can be modified to exhibit a $\mathbb{Z} + \mathbb{Z}$-measured lamination containing a "Reeb component", and in particular a leaf whose both ends spiral towards different isotopic closed curves in the same direction. This prevents axiom (2) in Proposition 55 to be satisfied and also the existence of a dual tree.

6.2   Construction of Laminations by Transversality

The geometric model that we will look for describing a given action of a fundamental group $\pi_1(M)$ on an abstract Λ-tree is the action on the dual tree of a Λ-measured lamination on M. We will explain in the next section how this model can be found when M is a closed surface. The first step in the construction of the Λ-measured lamination, valid in any dimension, is "by transversality" following a method that was first introduced by John Stallings in the context of simplicial trees [40]. When $\Lambda \subset \mathbb{R}$ and when M has dimension equal to 3, the construction of the $\mathbb{R}$-measured lamination is done in [26, Proposition 3.1]

Definition 56 Let $\mathcal{L} = (L, \mu)$ be a Λ-measured lamination contained in M and let $\mathcal{T}$ be a Λ-tree. A *transverse map* $\phi : \tilde{M} \setminus \tilde{L} \to \mathcal{T}$ is a locally constant map such that any point $x \in L$ is contained in a flow-box $V \simeq U \times]0, 1[$ for $\tilde{L}$ on which the restriction $\phi|(V \setminus \tilde{L})$ is the composition of the projection $(U \times]0, 1[) \setminus \tilde{L} \to]0, 1[\setminus F$ and a map $]0, 1[\setminus F \to \mathcal{T}$ which induces a monotone bijection between the components of $]0, 1[\setminus F$ and a segment in $\mathcal{T}$.

Proposition 57 *Let M be a closed n-manifold. Let $\pi_1(M) \times \mathcal{T} \to \mathcal{T}$ be an action of $\pi_1(M)$ on a Λ-tree $\mathcal{T}$. Then there exists a Λ-measured lamination $\mathcal{L} = (L, \mu)$ contained in M and a transverse map from the set of connected components of $\tilde{M} \setminus \tilde{L}$ to $\mathcal{T}$ which is $\pi_1(M)$-equivariant.*

Proof We select a triangulation τ of M and we denote by $\tilde{\tau}$ its lift to a triangulation to $\tilde{M}$. We construct the support of the measured lamination L by defining its intersection with the i-skeleton of τ inductively.

Over the 0-skeleton we choose any equivariant map ϕ_0 from $\tilde{\tau}_0$ to $\mathcal{T}$. In order to extend this map to a map ϕ_1 defined over the 1-skeleton, we use the following result.

Lemma 58 *Let Λ be a countable ordered group and let $[a, b] \subset \Lambda$ be a closed interval. There is a Λ-measure μ on $[0, 1]$ with the following properties:*

1. *the support of μ is a closed subset $F \subset]0, 1[$ with empty interior;*

2. *the function $t \mapsto a + \int_0^t d\mu$ is a monotone bijection from the connected components of $]0, 1[\setminus F$ to $[a, b]$.*

Furthermore if μ_1 and μ_2 are two Λ-measures on $]0, 1[$ satisfying properties (1) and (2), then there is an orientation-preserving homeomorphism $h : [0, 1] \to [0, 1]$ that carries μ_1 to μ_2.

Let τ_1 be an edge of the 1-skeleton of τ; let $\tilde{\tau}_1$ be one of its lifts to the universal cover, and $\tilde{a}, \tilde{b}$ the endpoints of $\tilde{\tau}_1$. Set $a = \phi_0(\tilde{a})$, $b = \phi_0(\tilde{b})$. The edge $\tilde{\tau}_1$ can be identified with $[0, 1]$. An application of Lemma 58 gives a Λ-measure with support contained in $]0, 1[$ and a bijection from the set of connected components of the complement of the support of this measure to $[a, b]$. This support will be the intersection of $\tilde{L}$ with $\tilde{\tau}_1$ and we define ϕ_1 to be the bijection. The measures and the map ϕ_1 can clearly be made equivariant since the covering group $\pi_1(M)$ acts freely on the preimage of the 1-skeleton of τ.

For extending ϕ_1 over the 2-skeleton, we consider a 2-simplex τ_2; let $\tilde{\tau}_2$ be one of its lifts to $\tilde{M}$. The boundary of $\tilde{\tau}_2$ is the union of 3 edges e_1, e_2 and e_3 of $\tilde{\tau}_1$. Each edge e_j contains a closed subset F_j and ϕ_1 identifies the set of connected components of $e_j \setminus F_j$ with a segment $I_j \subset T$. By the axioms of a Λ-tree, there exists a unique point $x \in T$ which is common to all segments $\phi_1(e_j)$. Let $c_j \subset I_j$ be the connected component of $e_j \setminus F_j$ which is mapped by ϕ_1 to v. The uniqueness part of Lemma 58 provides, for each j homeomorphisms between the connected components of $e_j \setminus v_j$ which are mapped to the same segment of T and which identify the traces of $\cup F_j$ on these intervals. Using these homeomorphisms one can construct a lamination L of $\tilde{\tau}_2$ which intersects $\partial \tau_2$ along $\cup F_j$ and a map ϕ_2 from the components of $\tilde{\tau}_2 \setminus \tilde{L}$ which extends ϕ_1. These laminations and maps can be chosen equivariant. When M is a surface, this construction ends the proof of Lemma 57. For the higher dimensional case, the proof goes on by induction; we refer to [23, Theorem II.3] for the details. $\qquad \square$

Remark Suppose that the Λ-measured lamination $\mathcal{L} = (L, \mu)$ constructed in the proof above satisfies the axioms of Proposition 55. Then $\mathcal{L}$ has a dual tree and ϕ induces a map $T_{\mathcal{L}} \to T$ which is *a morphism of trees* [23, p. 175]. We do not recall the definition of morphism of trees here; we just indicate that in our case, this property of ϕ is equivalent to the fact that $\phi : \tilde{M} \setminus \tilde{L} \to T$ is a transverse map. It follows from the construction of $\mathcal{L}$ that $T_{\mathcal{L}}$ is a union of segments on which ϕ is an isometry.

6.3 Actions of Surface Groups

We consider in this section a closed surface S of negative Euler characteristic. For simplicity we assume that S is closed.

We denote by $\mathcal{DF}(\pi_1(S))$ the set of discrete and faithful representations of $\pi_1(S)$ into $\mathrm{SL}(2,\mathbb{C})$ and by $\mathcal{DF}(\pi_1(S))_\mathbb{R}$ those with image contained in $\mathrm{SL}(2,\mathbb{R})$. In [23], the Λ-trees which are limits of a diverging sequence of discrete and faithful representations (ρ_i) belonging to $\mathcal{DF}(\pi_1(S))_\mathbb{R}$ are characterized. Let us explain this.

Definition 59 Let $\pi_1(S) \times \mathcal{T} \to \mathcal{T}$ be an action of $\pi_1(S)$ on a Λ-tree $\mathcal{T}$. This action is called *geometric* if there is a Λ-measured lamination $\mathcal{L} = (L, \mu)$ on S which has a dual tree $\mathcal{T}_\mathcal{L}$ and a $\pi_1(S)$-equivariant isometry between $\mathcal{T}$ and $\mathcal{T}_\mathcal{L}$.

Theorem 60 ([23]) *Let (ρ_i) be a sequence in $\mathcal{DF}(\pi_1(S))_\mathbb{R}$ which converges to a minimal action of $\pi_1(S)$ on a Λ-tree $\mathcal{T}$ in the sense of Theorem 51. Then $\pi_1(S) \times \mathcal{T} \to \mathcal{T}$ is geometric.*

We sketch the proof of this theorem and begin with the following complement to Proposition 57.

Proposition 61 *Let $\pi_1(S) \times \mathcal{T} \to \mathcal{T}$ be a minimal action of $\pi_1(S)$ by isometries on a Λ-tree. Then there is a Λ-measured lamination $\mathcal{L}$ on S which has a dual tree $\mathcal{T}_\mathcal{L}$ and a morphism $\mathcal{T}_\mathcal{L} \to \mathcal{T}$ which is $\pi_1(S)$-equivariant.*

Proof In the case of surfaces the axioms for a lamination that guarantee the existence of a dual tree can be formulated differently. Let us identify the universal cover $\tilde{S}$ with the unit disk after choosing an arbitrary metric of constant curvature -1. Denote by $(S^1 \times S^1 \setminus \mathrm{diagonal})/\mathbb{Z}_2$ the set of pairs of distinct points of S^1. Let L be a codimension 1 lamination of S and let $\tilde{L}$ its preimage in $\tilde{S}$. Suppose that L satisfies the following set of axioms.

1. there is a finite cover of S by flow-boxes for L such that each leaf of $\tilde{L}$ intersects the lift of a flow-box in a connected set;
2. each leaf of $\tilde{L}$ is proper;
3. each leaf has two distincts ends defining a point in $(S^1 \times S^1 \setminus \mathrm{diagonal})/\mathbb{Z}_2$;
4. the map $\tilde{L} \to (S^1 \times S^1 \setminus \mathrm{diagonal})/\mathbb{Z}_2$ which assigns to a point of $\tilde{L}$ the ends of the leaf of $\tilde{L}$ which passes through it is continuous.

Then it is not difficult to show that L satisfies also the axioms of Proposition 55 (see [23, Theorem I.4.2]). Therefore, if L is the support of a Λ-measured lamination $\mathcal{L}$, then $\mathcal{L}$ has a dual Λ-tree.

In order to construct the lamination $\mathcal{L} = (L, \mu)$, consider the lamination $\mathcal{L}' = (L', \mu')$ provided by Proposition 57. This lamination might contain closed leaves that are homotopic to 0 on S. One proves that unions of those leaves consist of finitely many sublaminations of L' which are the union of parallel leaves [23, Lemma III.1.4]. By removing those sublaminations and modifying accordingly the transverse map, one obtains a Λ-measured lamination $\mathcal{L} = (L, \mu)$ and a transverse map $\tilde{S} \setminus \tilde{L} \to T$. At this stage, one can prove that each half-leaf of $\tilde{L}$ has one endpoint in S^1; however this does not suffice for verifying (2) and in particular the lamination might contain *Reeb components*. A new modification of $\mathcal{L}$ is required which simplifies the annuli bounded by leaves of $\mathcal{L}$ to guarantee that the lamination has a dual tree (see [23]). By Remark 6.2, there is an equivariant morphism $T_{\mathcal{L}} \to T$. □

Let us go back to the sketch of the proof of Theorem 60. By Proposition 54, a first property of T is that any subgroup of $\pi_1(S)$ which stabilizes a non-trivial segment of T is virtually abelian. It is not difficult to check that in $\pi_1(S)$, any virtually abelian subgroup is cyclic; therefore non-trivial segment stabilizers are cyclic. A second one is the following. Say first that $g \in \pi_1(S)$ is *hyperbolic* when for some $\rho \in \mathcal{DF}(\pi_1(S))_{\mathbb{R}}$, $\rho(g)$ acts on $\mathbb{H}$ leaving invariant a geodesic, called its *axis*. This property is independent on the representation $\rho \in \mathcal{DF}(\pi_1(S))_{\mathbb{R}}$. One denotes by $l_\rho(g)$ the translation distance of the hyperbolic element $\rho(g)$. Let g and h be hyperbolic elements of $\pi_1(S)$ whose axis intersect in $\mathbb{H}$. This property also is independent on ρ. By the triangular inequality, the translation distance of the elements g, h and gh satisfy for any $\rho \in DF(\pi_1(S))$: $l_\rho(gh) \leq l_\rho(g) + l_\rho(h)$. Therefore, since T is the limit of the sequence (ρ_i), the translation distances in T satisfy $\delta_T(gh) \leq \delta_T(g) + \delta_T(h)$. A geometric interpretation of this inequality is that any partial axis (or the fixed points sets) of g and h in T have non-empty intersection.

Therefore Theorem 60 follows from the following

Theorem 62 ([23, 38]) *Let $\pi_1(S) \times T \to T$ be an action of $\pi_1(S)$ on a Λ-tree which satisfies*

1. *the action is minimal and without phantom inversions,*
2. *the edge stabilizers are cyclic and*
3. *if g and h are hyperbolic elements whose axes intersect in $\mathbb{H}$, then any partial axes of g and h in T intersect.*

Then the action $\pi_1(S) \times T \to T$ is geometric.

Sketch of proof Following [23] we start with the lamination $\mathcal{L}$ obtained in Proposition 61 and with the $\pi_1(S)$-equivariant morphism $\iota : T_{\mathcal{L}} \to T$. If ι is not an embedding, there is a vertex v and two distinct segments issued from v such that $\iota(e_1) \cap \iota(e_2)$ is a segment which is not reduced to v. The vertex v of $T_{\mathcal{L}}$ corresponds to a connected component C of $\tilde{S} \setminus \tilde{L}$; for each segment e_i there is a leaf $\tilde{l}_i$ of $\tilde{L}$ in the boundary $\tilde{C}$. Since $e_1 \neq e_2$, $\tilde{l}_1 \neq \tilde{l}_2$.

One distinguishes two cases according to the behaviour of the projections l_i of $\tilde{l}_i$ on S: first when both projections are compact and second when both are non-compact. Using the equivariance of the map $\tilde{\iota}$ one sees that these two cases cover all possible situations. These two cases contradict respectively assumptions 1. and 2., see [23, Chapter IV]. □

In [23] the question whether the conclusion of Theorem 62 was still true only assuming condition 1. was left open. It was solved when Λ is archimedean by Richard Skora who proved:

Theorem 63 ([39]) *Let $\pi_1(S) \times T \to T$ be a minimal action of $\pi_1(S)$ by isometries on an $\mathbb{R}$-tree without phantom inversion and such that the edge stabilizers are cyclic. Then this action is geometric.*

Furthermore there is an action on a Λ-tree which satisfies (1) but which is not geometric.

The proof by Skora also uses the $\mathbb{R}$-measured lamination $\mathcal{L}$ and the morphism $\iota : \mathcal{T}_{\mathcal{L}} \to \mathcal{T}$. Assuming by contradiction again that ι is not an isometry, then two germs of segments e_1 and e_2 in $\mathcal{T}_{\mathcal{L}}$ issued from some vertex v are identified; e_1 (resp. e_2) corresponds to a leaf $\tilde{l}_1$ (resp. $\tilde{l}_2$) in the boundary of the component v of $\tilde{S} \setminus \tilde{L}$. Since $\mathcal{L}$ is $\mathbb{R}$-measured, the structure theorem [25, Theorem 3.2] says that L is the union of finitely many disjoint sublaminations which either are formed by parallel closed leaves or are exceptional minimal sublaminations [25, Theorem 3.2]. The case when each leaf l_i is closed contradicts immediately that the edge stabilizers are cyclic, as in the previous proof. The case when each leaf l_i is contained in a minimal exceptional is handled differently: Skora uses interval exchanges to produce a subinterval of $\iota(e_1) \cap \iota(e_2)$ which has a non-cyclic edge stabilizer [39] (see also [30]).

6.4   Actions of 3-Manifolds Groups

Let M be a compact 3-manifold. One says that M is *boundary-incompressible* when any closed curve embedded in ∂M which is homotopic to 0 in M bounds a 2-disc properly embedded in M. One says that M is *acylindrical* when any properly embedded annulus or any embedded 2-torus can be homotoped relatively to its boundary into ∂M.

Theorem 64 *Let M be a compact boundary-incompressible and acylindrical 3-manifold. Then the space of discrete and faithful representations $\mathcal{DF}(\pi_1(M))$ is compact.*

This theorem was first proven by Thurston in [42]: it is one of the main steps in his proof of the Hyperbolization Theorem for non-Haken 3-manifolds. Morgan and Shalen gave an entirely different proof in [26] using trees. Their proof can be outlined as follows.

Sketch of proof Set $G = \pi_1(M)$. Suppose by contradiction that $\mathcal{DF}(G)$ is not compact. Then by Proposition 54, there exists an $\mathbb{R}$-tree $\mathcal{T}$ and a non-trivial action of G on $\mathcal{T}$ by isometries which is small and minimal. The proof reduces then to show that such an action does not exist when M satisfies the hypothesis of the theorem.

By Proposition 57, there is a measured lamination $\mathcal{L} = (L, \mu)$ in M and a transverse map $\tilde{M} \setminus \tilde{L} \to \mathcal{T}$. The construction gave that $\mathcal{L}$ is *carried with positive weights* by a *branched surface* [16]. A *branched surface* is a 2-complex B embedded in M according to the following local model. In a chart homeomorphic to $(x, y, z) \in \mathbb{R}^3$, B is the union of the horizontal plane $z = 0$ and the graphs of two functions: the graph of a positive function over $x \geq 0$, tangent to the horizontal plane along $x = z = 0$ and the graph of a negative function over $y \geq 0$ tangent to the horizontal plane along $y = z = 0$. Particular neighborhoods of this 2-complex in M have a natural decomposition in intervals. A measured lamination $\mathcal{L} = (L, \mu)$ is *carried by* B if it is contained in such a neighborhood of this type so that L is transverse to the intervals. A difficult part of the proof is to show that one can choose the lamination to be carried by a branched surface which is *incompressible* (see [16]). This property implies in particular that the fundamental group of each leaf l of L maps injectively into $\pi_1(M) = G$. By equivariance of the transverse map, $\pi_1(l)$ stabilizes a non-degenerate segment of $\mathcal{T}$. The property of the segment stabilizers implies that $\pi_1(l)$ is virtually abelian. Since l is a surface, it follows that $\pi_1(l)$ is either cyclic or trivial. Therefore the fundamental group of any leaf of L is cyclic or trivial.

On the other hand, the incompressibility of B implies that no closed surface carried by B is homeomorphic to a disk or a 2-sphere. Under these circumstances, Theorem 5.1 of [25] implies that any surface carried by B has zero Euler characteristic. By approximating $\mathcal{L}$, one constructs compact surfaces carried by B. Such a surface is a union of annuli or tori. This contradicts the acylindricity of M.   □

The hypothesis on M in Theorem 64 are of a topological nature, but they can also be formulated in a group-theoretic way. The property that M is boundary-incompressible and irreducible is equivalent to the fact that G is not isomorphic to a free product. The acylindricity of M is equivalent to the non-existence of a non-trivial splitting of G as an amalgamated product over a subgroup isomorphic to $\mathbb{Z}$ or to $\mathbb{Z}^2$. After giving their proof based on 3-dimensional arguments, Morgan and Shalen conjectured the following vast generalization which is now a theorem of Rips.

Theorem 65 *Let G be a finitely presented group which can not be written as an amalgamated free product over a virtually abelian group. Then there are no non-trivial actions of G on an $\mathbb{R}$-tree such that the stabilizers of segments are virtually abelian.*

We refer to [2] for a generalization and to [32] for a survey of the proof.

Acknowledgements I would like to thank Charles Favre, Jan Kiwi, and Juan Rivera-Letelier for the invitation to give a course during the "Ultrametric Dynamics Days" in Santiago de Chile in January 2008. This paper owes a lot to Charles who encouraged me to develop my set of notes to an actual paper and made several important suggestions. The referee, through his constant criticism, also helped to bring the paper to its actual form.

References

1. M. Bestvina, Degenerations of the hyperbolic space. Duke Math. J. **56**(1), 143–161 (1988)
2. M. Bestvina, M. Feighn, Stable actions of groups on real trees. Invent. Math. **121**(2), 287–321 (1995)
3. N. Bourbaki, *Algèbre Commutative*, Chapitres 5–7 (Masson, 1985)
4. R. Benedetti, C. Petronio, *Lectures on Hyperbolic Geometry*, Universitext (Springer, Berlin, 1991)
5. M. Culler, Lifting representations to covering groups. Adv. Math. **59**, 54–70 (1986)
6. R. Canary, D. Epstein, P. Green, Notes on notes of Thurston, in *Analytical and Geometrical Aspects of Hyperbolic Spaces* (Cambridge University Press, Cambridge, 1987), pp. 3–92
7. I. Chiswell, Abstract length functions in groups. Proc. Cambridge Phil. Soc. **80**, 451–463 (1976)
8. I. Chiswell, *Introduction to Λ-Trees* (World Scientific, Singapore, 2001)
9. I. Chiswell, Non standard analysis and the Morgan-Shalen compactification. Quart. J. Math. Oxford Ser. (2) **42**(167), 257–270 (1991)
10. V. Chuckrow, On Schottky groups with applications to Kleinian groups. Ann. Math. **88**, 47–61 (1968)
11. M. Culler, C. Gordon, J. Luecke, P. Shalen, Dehn surgery on knots. Ann. Math. (2) **125**, 237–300 (1987)
12. M. Culler, J. Morgan, Group actions on ℝ-trees. Proc. Lond. Math. Soc. **55**, 571–604 (1987)
13. M. Culler, J. Morgan, Varieties of group representations and splittings of 3-manifolds. Ann. Math. **117**, 109–146 (1983)
14. L. Ein, R. Lazarsfeld, K. Smith, Uniform approximation of Abhyankar valuation ideals in smooth function fields. Am. J. Math. **125**(2), 409–440 (2003)
15. A. Fathi, F. Laudenbach, V. Poenaru, *Travaux de Thurston sur les surfaces*, Astérisque, pp. 66–67 (1979), SMF Paris
16. W. Floyd, U. Oertel, Incompressible surfaces via branched surfaces. Topology **23**, 117–125 (1984)
17. D. Gaboriau, G. Levitt, F. Paulin, Pseudogroups of isometries of ℝ and Rips' theorem on free actions on ℝ-trees. Isr. J. Math. **87**, 403–428 (1994)
18. M. Gromov, Hyperbolic groups. in *Essays in Group Theory*, ed. by S.M. Gersten (ed.), MSRI Publ., vol. 8 (Springer, Berlin, Heidelberg, New York, 1987), pp. 75–263
19. R. Hartshorne, *Algebraic Geometry*. Graduate Texts in Mathematics, vol. 52 (Springer, New York, 1977)
20. R. Lyndon, Length functions in groups. Math. Scand. **12**, 209–234 (1963)
21. J. Morgan, Group actions on trees and the compactification of the space of classes of SO(n, 1)-representations. Topology **25**, 1–33 (1986)
22. J. Morgan, Λ-trees and their applications. Bull. Am. Math. Soc. **26**, 87–112 (1992)
23. J. Morgan, J.-P. Otal, Relative growth rates of closed geodesics on closed surfaces. Comment. Math. Helvitici **68**, 171–208 (1993)
24. J. Morgan, P. Shalen, Degenerations of hyperbolic structures, I: Valuations, trees and surfaces. Ann. Math. **120**, 401–476 (1984)
25. J. Morgan, P. Shalen, Degenerations of hyperbolic structures II : Measured laminations in 3-manifolds. Ann. Math. **127**, 403–456 (1988)
26. J. Morgan, P. Shalen, Degenerations of hyperbolic structures III: Actions of 3-manifold groups on trees and Thurston's compactness theorem. Ann. Math. **127**, 457–519 (1988)
27. J. Morgan, P. Shalen, Free actions of surface groups on ℝ-trees. Topology **30**(2), 143–154 (1991)
28. D. Mumford, *Geometric Invariant Theory*. Ergebnisse der Mathematik und ihrer Grenzgebiete, vol. 34 (Springer, New York, 1965)
29. D. Mumford, *Algebraic Geometry I: Complex Projective Varieties*. Grundlehren der Math. Wiss., vol. 221 (Springer, New York, 1976)

30. J.-P. Otal, *Le théorème d'hyperbolisation pour les variétés fibrées de dimension 3*. Astérisque no. 235 (Soc. Math. France, Paris, 1996)

31. F. Paulin, Topologie de Gromov équivariante, structures hyperboliques et arbres réels. Inv. Math. **94**, 53–80 (1988)

32. F. Paulin, Actions de groupes sur les arbres. Séminaire Bourbaki 1995–96, exp. 808, Astérisque no. 241 (Soc. Math. France, Paris, 1997)

33. J.G. Ratcliffe, *Foundations of Hyperbolic Manifolds*. Graduate Texts in Math., vol. 149 (Springer, New York, 1994)

34. J.-P. Serre, *Arbres, Amalgames, SL(2)*. Astérisque no. 46 (Soc. Math. France, Paris, 1977)

35. P. Shalen, Dendrology of groups: An introduction. in *Essays in Group Theory*, ed. by S.M. Gersten. Mathematical Sciences Research Institute Publications, vol. 8 (Springer, New York, 1987)

36. P. Shalen, Dendrology and its applications, in *Group Theory from a Geometrical Viewpoint* (Trieste, 1990) (World Scientific, River Edge, NJ, 1991), pp. 543–616

37. P. Shalen, Representations of 3-manifolds groups, in *Handbook of 3-Manifold Groups* (North-Holland, Amsterdam, 2002), pp. 955–1044

38. R. Skora, Geometric actions of surface groups on Λ-trees. Comment. Math. Helvetici **65**, 519–533 (1990)

39. R. Skora, Splittings of surfaces. J. Am. Math. Soc. **9**(2), 605–616 (1996)

40. J. Stallings, A topological proof of Grushko's theorem on free products. Math. Zeit. **90**, 1–8 (1965)

41. W. Thurston, *The Geometry and Topology of 3-Manifolds*. Princeton Lecture Notes (1978–1981), http://library.msri.org/books/gt3m/

42. W. Thurston, Hyperbolic structures on 3-manifolds I: Deformations of acylindrical manifolds. Ann. Math. **124**, 203–246 (1986)

43. W. Thurston, *Three-Dimensional Geometry and Topology* (Princeton University Press, Princeton, NJ, 1995)

44. M. Vaquié, Valuations and local uniformization. Singularity theory and its applications, pp. 477–527. Adv. Stud. Pure Math., vol. 43 (Math. Soc. Japan, Tokyo, 2006)

45. M. Vaquié, Extension d'une valuation. Trans. Am. Math. Soc. **359**(7), 3439–3481 (2007)

46. A. Weil, On discrete subgroups of Lie groups. Ann. Math. (2) **72**, 369–384 (1960)

47. O. Zariski, Foundations of a general theory of birational correspondences. Trans. Am. Math. Soc. **53**, 490–542 (1943)

48. O. Zariski, The compactness of the Riemann manifold of an abstract field of algebraic functions. Bull. Am. Math. Soc. **50**, 683–691 (1944)

49. O. Zariski, P. Samuel, *Commutative Algebra*, vol. 2. Graduate Texts in Mathematics no 29 (Springer, New York-Heidelberg-Berlin, 1975)

LECTURE NOTES IN MATHEMATICS

Edited by J.-M. Morel, B. Teissier; P.K. Maini

Editorial Policy (for Multi-Author Publications: Summer Schools / Intensive Courses)

1. Lecture Notes aim to report new developments in all areas of mathematics and their applications - quickly, informally and at a high level. Mathematical texts analysing new developments in modelling and numerical simulation are welcome. Manuscripts should be reasonably selfcontained and rounded off. Thus they may, and often will, present not only results of the author but also related work by other people. They should provide sufficient motivation, examples and applications. There should also be an introduction making the text comprehensible to a wider audience. This clearly distinguishes Lecture Notes from journal articles or technical reports which normally are very concise. Articles intended for a journal but too long to be accepted by most journals, usually do not have this "lecture notes" character.

2. In general SUMMER SCHOOLS and other similar INTENSIVE COURSES are held to present mathematical topics that are close to the frontiers of recent research to an audience at the beginning or intermediate graduate level, who may want to continue with this area of work, for a thesis or later. This makes demands on the didactic aspects of the presentation. Because the subjects of such schools are advanced, there often exists no textbook, and so ideally, the publication resulting from such a school could be a first approximation to such a textbook. Usually several authors are involved in the writing, so it is not always simple to obtain a unified approach to the presentation.

   For prospective publication in LNM, the resulting manuscript should not be just a collection of course notes, each of which has been developed by an individual author with little or no coordination with the others, and with little or no common concept. The subject matter should dictate the structure of the book, and the authorship of each part or chapter should take secondary importance. Of course the choice of authors is crucial to the quality of the material at the school and in the book, and the intention here is not to belittle their impact, but simply to say that the book should be planned to be written by these authors jointly, and not just assembled as a result of what these authors happen to submit.

   This represents considerable preparatory work (as it is imperative to ensure that the authors know these criteria before they invest work on a manuscript), and also considerable editing work afterwards, to get the book into final shape. Still it is the form that holds the most promise of a successful book that will be used by its intended audience, rather than yet another volume of proceedings for the library shelf.

3. Manuscripts should be submitted either online at www.editorialmanager.com/lnm/ to Springer's mathematics editorial, or to one of the series editors. Volume editors are expected to arrange for the refereeing, to the usual scientific standards, of the individual contributions. If the resulting reports can be forwarded to us (series editors or Springer) this is very helpful. If no reports are forwarded or if other questions remain unclear in respect of homogeneity etc, the series editors may wish to consult external referees for an overall evaluation of the volume. A final decision to publish can be made only on the basis of the complete manuscript; however a preliminary decision can be based on a pre-final or incomplete manuscript. The strict minimum amount of material that will be considered should include a detailed outline describing the planned contents of each chapter.

   Volume editors and authors should be aware that incomplete or insufficiently close to final manuscripts almost always result in longer evaluation times. They should also be aware that parallel submission of their manuscript to another publisher while under consideration for LNM will in general lead to immediate rejection.

4. Manuscripts should in general be submitted in English. Final manuscripts should contain at least 100 pages of mathematical text and should always include

   – a general table of contents;
   – an informative introduction, with adequate motivation and perhaps some historical remarks: it should be accessible to a reader not intimately familiar with the topic treated;
   – a global subject index: as a rule this is genuinely helpful for the reader.

   Lecture Notes volumes are, as a rule, printed digitally from the authors' files. We strongly recommend that all contributions in a volume be written in the same LaTeX version, preferably LaTeX2e. To ensure best results, authors are asked to use the LaTeX2e style files available from Springer's web-server at
   ftp://ftp.springer.de/pub/tex/latex/svmonot1/ (for monographs) and
   ftp://ftp.springer.de/pub/tex/latex/svmultt1/ (for summer schools/tutorials).
   Additional technical instructions, if necessary, are available on request from:
   lnm@springer.com.

5. Careful preparation of the manuscripts will help keep production time short besides ensuring satisfactory appearance of the finished book in print and online. After acceptance of the manuscript authors will be asked to prepare the final LaTeX source files and also the corresponding dvi-, pdf- or zipped ps-file. The LaTeX source files are essential for producing the full-text online version of the book. For the existing online volumes of LNM see:
   http://www.springerlink.com/openurl.asp?genre=journal&issn=0075-8434.
   The actual production of a Lecture Notes volume takes approximately 12 weeks.

6. Volume editors receive a total of 50 free copies of their volume to be shared with the authors, but no royalties. They and the authors are entitled to a discount of 33.3 % on the price of Springer books purchased for their personal use, if ordering directly from Springer.

7. Commitment to publish is made by letter of intent rather than by signing a formal contract. Springer-Verlag secures the copyright for each volume. Authors are free to reuse material contained in their LNM volumes in later publications: a brief written (or e-mail) request for formal permission is sufficient.

Addresses:
Professor J.-M. Morel, CMLA,
École Normale Supérieure de Cachan,
61 Avenue du Président Wilson, 94235 Cachan Cedex, France
E-mail: morel@cmla.ens-cachan.fr

Professor B. Teissier, Institut Mathématique de Jussieu,
UMR 7586 du CNRS, Équipe "Géométrie et Dynamique",
175 rue du Chevaleret,
75013 Paris, France
E-mail: teissier@math.jussieu.fr

For the "Mathematical Biosciences Subseries" of LNM:

Professor P. K. Maini, Center for Mathematical Biology,
Mathematical Institute, 24-29 St Giles,
Oxford OX1 3LP, UK
E-mail: maini@maths.ox.ac.uk

Springer, Mathematics Editorial I,
Tiergartenstr. 17,
69121 Heidelberg, Germany,
Tel.: +49 (6221) 4876-8259
Fax: +49 (6221) 4876-8259
E-mail: lnm@springer.com